Quantum Mechanics for Applied Nanotechnology

Quantum Mechanics for Applied Nanotechnology

Editor:

Olga Moreira, Ph.D.

ArclerPress

www.arclerpress.com

Quantum Mechanics for Applied Nanotechnology

Editor: Olga Moreira, Ph.D.

708 3rd Avenue, 6th Floor
New York
NY 10017
United States of America
www.arclerpress.com

ISBN: 978-1-68094-353-5

Library of Congress Control Number: 2016950941

About the Editor

Olga Moreira, Ph.D.

Olga Moreira obtained her Ph.D. in Astrophysics from the University of Liege (Belgium) in 2010, her BSc. in Physics and Applied Mathematics from the University of Porto (Portugal). Her post-graduate travels and international collaborations with the European Space Agency (ESA) and European Southern Observatory (ESO) led to great personal and professional growth as a scientist. Currently, she is working as an independent researcher, technical writer, and editor in the fields of Mathematics, Physics, Astronomy and Astrophysics.

Contents

List of Contributors

Debasis Bera
Department of Materials Science and Engineering, University of Florida, Gainesville, FL, USA

Lei Qian
Department of Materials Science and Engineering, University of Florida, Gainesville, FL, USA

Teng-Kuan Tseng
Department of Materials Science and Engineering, University of Florida, Gainesville, FL, USA

Paul H. Holloway
Department of Materials Science and Engineering, University of Florida, Gainesville, FL, USA

Ákos **Nemcsics**
Institute for Microelectronics and Technology, Óbuda University Budapest, Hungary
CER, Institute for Technical Physics and Materials Science, Hungarian Academy of Sciences, Budapest, Hungary

Khalil Ebrahim Jasim
Department of Physics, College of Science, University of Bahrain, Kingdom of Bahrain

Lijia Shao
Jiangsu Affiliated Cancer Hospital with Nanjing Medical University, Jiangsu Institute of Cancer Prevention and Cure, Nanjing 210009, China

Yanfang Gao
Jiangsu Affiliated Cancer Hospital with Nanjing Medical University, Jiangsu Institute of Cancer Prevention and Cure, Nanjing 210009, China

Feng Yan
Jiangsu Affiliated Cancer Hospital with Nanjing Medical University, Jiangsu Institute of Cancer Prevention and Cure, Nanjing 210009, China

Alireza Valizadeh
Department of Medical Nanotechnology, Faculty of Advanced Medical

Haleh Mikaeili
Tuberculosis and Lung Disease Research Center of Tabriz University of Medical Sciences, Tabriz 51656-65811, Iran

Mohammad Samiei
Faculty of Dentistry, Tabriz University of Medical Sciences, Tabriz 51664, Iran.

Samad Mussa Farkhani
Department of Medical Nanotechnology, Faculty of Advanced Medical Science, Tabriz University of Medical Sciences, Tabriz 51664, Iran.

Nosratalah Zarghami
Department of Medical Nanotechnology, Faculty of Advanced Medical Science, Tabriz University of Medical Sciences, Tabriz 51664, Iran.

Mohammad kouhi
Department of Physics, Tabriz Branch, Islamic Azad University, Tabriz, Iran.

Abolfazl Akbarzadeh
Department of Medical Nanotechnology, Faculty of Advanced Medical Science, Tabriz University of Medical Sciences, Tabriz 51664, Iran

Soodabeh Davaran
Department of Medical Nanotechnology, Faculty of Advanced Medical Science, Tabriz University of Medical Sciences, Tabriz 51664, Iran.

Biswajit Mukherjee
Department of Pharmaceutical Technology, Jadavpur University, Kolkata, India

Niladri Shekhar Dey
Department of Pharmaceutical Technology, Jadavpur University, Kolkata, India

Ruma Maji
Department of Pharmaceutical Technology, Jadavpur University, Kolkata, India

Priyanka Bhowmik
Department of Pharmaceutical Technology, Jadavpur University, Kolkata, India

Pranab Jyoti Das
Department of Pharmaceutical Technology, Jadavpur University, Kolkata, India

Paramita Paul
Department of Pharmaceutical Technology, Jadavpur University, Kolkata, India

Denis Uhland
3rd Physics Institute, University of Stuttgart and Stuttgart Research Center of Photonic Engineering (SCoPE) and IQST, Pfaffenwaldring 57, Stuttgart, 70569, Germany

Torsten Rendler
3rd Physics Institute, University of Stuttgart and Stuttgart Research Center of Photonic Engineering (SCoPE) and IQST, Pfaffenwaldring 57, Stuttgart, 70569, Germany

Matthias Widmann
3rd Physics Institute, University of Stuttgart and Stuttgart Research Center of Photonic Engineering (SCoPE) and IQST, Pfaffenwaldring 57, Stuttgart, 70569, Germany

Sang-Yun Lee
3rd Physics Institute, University of Stuttgart and Stuttgart Research Center of Photonic Engineering (SCoPE) and IQST, Pfaffenwaldring 57, Stuttgart, 70569, Germany

Jörg Wrachtrup
3rd Physics Institute, University of Stuttgart and Stuttgart Research Center of Photonic Engineering (SCoPE) and IQST, Pfaffenwaldring 57, Stuttgart, 70569, Germany
Max Planck Institute for Solid State Research, Heisenbergstraße 1, Stuttgart, 70569, Germany

Ilja Gerhardt
3rd Physics Institute, University of Stuttgart and Stuttgart Research Center of Photonic Engineering (SCoPE) and IQST, Pfaffenwaldring 57, Stuttgart, 70569, Germany
Max Planck Institute for Solid State Research, Heisenbergstraße 1, Stuttgart, 70569, Germany

Arnav R. Mistry
Department of Endodontology, Temple University Kornberg School of Dentistry, 3223 North Broad Street, Philadelphia, PA 19140, USA

Daniel Uzbelger Feldman
Department of Endodontology, Temple University Kornberg School of Dentistry, 3223 North Broad Street, Philadelphia, PA 19140, USA

Jie Yang
Oral and Maxillofacial Radiology Department, Temple University Kornberg School of Dentistry, 3223 North Broad Street, Philadelphia, PA 19140, USA

Eric Ryterski
E3 Medical, Inc., 941 Garfield Avenue, Louisville, CO 80027, USA

Giuseppe de Risi
Dipartimento Interareneo di Fisica, Università degli Studi e Politecnico di Bari, via Amendola 173, I-70126 Bari, Italy

Lorenzo Luigi Columbo
Dipartimento Interareneo di Fisica, Università degli Studi e Politecnico di Bari, via Amendola 173, I-70126 Bari, Italy
CNR—Istituto di Fotonica e Nanotecnologie UOS Bari, via Amendola 173, I-70126 Bari, Italy

Massimo Brambilla
Dipartimento Interareneo di Fisica, Università degli Studi e Politecnico di Bari, via Amendola 173, I-70126 Bari, Italy
CNR—Istituto di Fotonica e Nanotecnologie UOS Bari, via Amendola 173, I-70126 Bari, Italy

Dmitry A. Kalashnikov
Data Storage Institute, Agency for Science Technology and Research (A-STAR), 5 Engineering Drive I, 117608, Singapore

Zhenying Pan
Data Storage Institute, Agency for Science Technology and Research (A-STAR), 5 Engineering Drive I, 117608, Singapore

Arseniy I. Kuznetsov
Data Storage Institute, Agency for Science Technology and Research (A-STAR), 5 Engineering Drive I, 117608, Singapore

Leonid A. Krivitsky
Data Storage Institute, Agency for Science Technology and Research (A-STAR), 5 Engineering Drive I, 117608, Singapore

Chandler Downs
Renewable Energy and Applied Photonics Laboratories, Electrical and Computer Engineering Department, Tufts University, 161 College Ave., Medford, MA 02115, USA

Thomas E. Vandervelde
Renewable Energy and Applied Photonics Laboratories, Electrical and Computer Engineering Department, Tufts University, 161 College Ave., Medford, MA 02115, USA

Vladimir Mitin
University at Buffalo, SUNY, 332 Bonner Hall, Buffalo, NY 14260-1920, USA

Andrei Antipov
University at Buffalo, SUNY, 332 Bonner Hall, Buffalo, NY 14260-1920, USA

Andrei Sergeev
University at Buffalo, SUNY, 332 Bonner Hall, Buffalo, NY 14260-1920, USA

Nizami Vagidov
University at Buffalo, SUNY, 332 Bonner Hall, Buffalo, NY 14260-1920, USA

David Eason
University at Buffalo, SUNY, 332 Bonner Hall, Buffalo, NY 14260-1920, USA

Gottfried Strasser
University at Buffalo, SUNY, 332 Bonner Hall, Buffalo, NY 14260-1920, USA

Bruno Romeira
Centro de Electrónica, Optoelectrónica e Telecomunicaçõe s (CEOT), Departamento de Física, Universidade do Algarve, Campus de Gambelas, 8005-139 Faro, Portugal

Luis M. Pessoa
INESC TEC, Faculdade de Engenharia, Universidade do Porto, 4200-465 Porto, Portugal

Henrique M. Salgado
INESC TEC, Faculdade de Engenharia, Universidade do Porto, 4200-465 Porto, Portugal

Charles N. Ironside
School of Engineering, University of Glasgow, Glasgow G12 8LT, UK

José M. L. Figueiredo
Centro de Electrónica, Optoelectrónica e Telecomunicaçõe s (CEOT), Departamento de Física, Universidade do Algarve, Campus de Gambelas, 8005-139 Faro, Portugal

Mengxing Wang
Spintronics Interdisciplinary Centre & School of Electronic and Information Engineering, Beihang University, Beijing 10191, China

Yue Zhang
Spintronics Interdisciplinary Centre & School of Electronic and Information Engineering, Beihang University, Beijing 10191, China

Xiaoxuan Zhao
Spintronics Interdisciplinary Centre & School of Electronic and Information Engineering, Beihang University, Beijing 10191, China

Weisheng Zhao
Spintronics Interdisciplinary Centre & School of Electronic and Information Engineering, Beihang University, Beijing 10191, China
Institute of Fundamental Electronics (IEF), University of Paris-Sud, Orsay 91405, France

Andrew Lyle
Department of Electrical and Computer Engineering, University of Minnesota, 200 Union St SE, Minneapolis, Minnesota 55455, USA

Jonathan Harms
Department of Electrical and Computer Engineering, University of Minnesota, 200 Union St SE, Minneapolis, Minnesota 55455, USA

Todd Klein
Department of Electrical and Computer Engineering, University of Minnesota, 200 Union St SE, Minneapolis, Minnesota 55455, USA

August Lentsch
Department of Electrical and Computer Engineering, University of Minnesota, 200 Union St SE, Minneapolis, Minnesota 55455, USA

Angeline Klemm
Department of Electrical and Computer Engineering, University of Minnesota, 200 Union St SE, Minneapolis, Minnesota 55455, USA

Daniel Martens
Department of Electrical and Computer Engineering, University of Minnesota, 200 Union St SE, Minneapolis, Minnesota 55455, USA

Jian-Ping Wang
Department of Electrical and Computer Engineering, University of Minnesota, 200 Union St SE, Minneapolis, Minnesota 55455, USA

W. J. Zhang
State Key Laboratory of Functional Materials for Informatics, Shanghai Institute of Microsystem and Information Technology (SIMIT), Chinese Academy of Sciences (CAS), 865 Changning Rd., Shanghai 200050, China

H. Li
State Key Laboratory of Functional Materials for Informatics, Shanghai Institute of Microsystem and Information Technology (SIMIT), Chinese Academy of Sciences (CAS), 865 Changning Rd., Shanghai 200050, China

L. X. You
State Key Laboratory of Functional Materials for Informatics, Shanghai Institute of Microsystem and Information Technology (SIMIT), Chinese Academy of Sciences (CAS), 865 Changning Rd., Shanghai 200050, China

Y. H. He
State Key Laboratory of Functional Materials for Informatics, Shanghai Institute of Microsystem and Information Technology (SIMIT), Chinese Academy of Sciences (CAS), 865 Changning Rd., Shanghai 200050, China

L. Zhang
State Key Laboratory of Functional Materials for Informatics, Shanghai Institute of Microsystem and Information Technology (SIMIT), Chinese Academy of Sciences (CAS), 865 Changning Rd., Shanghai 200050, China

X.Y. Liu
State Key Laboratory of Functional Materials for Informatics, Shanghai Institute of Microsystem and Information Technology (SIMIT), Chinese Academy of Sciences (CAS), 865 Changning Rd., Shanghai 200050, China

X. Y. Yang
State Key Laboratory of Functional Materials for Informatics, Shanghai Institute of Microsystem and Information Technology (SIMIT), Chinese Academy of Sciences (CAS), 865 Changning Rd., Shanghai 200050, China

J. J. Wu
State Key Laboratory of Functional Materials for Informatics, Shanghai Institute of Microsystem and Information Technology (SIMIT), Chinese Academy of Sciences (CAS), 865 Changning Rd., Shanghai 200050, China

Q. Guo
State Key Laboratory of Functional Materials for Informatics, Shanghai Institute of Microsystem and Information Technology (SIMIT), Chinese Academy of Sciences (CAS), 865 Changning Rd., Shanghai 200050, China

S. J. Chen
State Key Laboratory of Functional Materials for Informatics, Shanghai Institute of Microsystem and Information Technology (SIMIT), Chinese Academy of Sciences (CAS), 865 Changning Rd., Shanghai 200050, China

Z. Wang
State Key Laboratory of Functional Materials for Informatics, Shanghai Institute of Microsystem and Information Technology (SIMIT), Chinese Academy of Sciences (CAS), 865 Changning Rd., Shanghai 200050, China

X. M. Xie
State Key Laboratory of Functional Materials for Informatics, Shanghai Institute of Microsystem and Information Technology (SIMIT), Chinese Academy of Sciences (CAS), 865 Changning Rd., Shanghai 200050, China

Bo-Bo Wei
Department of Physics, The Chinese University of Hong Kong, Hong Kong, China
Centre for Quantum Coherence, The Chinese University of Hong Kong, Hong Kong, China

Christian Burk
3rd Institute of Physics and Research Center SCOPE, University Stuttgart, Stuttgart, 70569, Germany

Jörg Wrachtrup
3rd Institute of Physics and Research Center SCOPE, University Stuttgart, Stuttgart, 70569, Germany

Ren-Bao Liu
Department of Physics, The Chinese University of Hong Kong, Hong Kong, China
Centre for Quantum Coherence, The Chinese University of Hong Kong, Hong Kong, China
Institute of Theoretical Physics, The Chinese University of Hong Kong, Hong Kong, China
Shenzhen Research Insitute, The Chinese University of Hong Kong, Shenzhen, Guangdong 518057, China

Hai-Ou Li
Key Laboratory of Quantum Information, University of Science and Technology of China, Chinese Academy of Sciences, Hefei, P.R. China

Tao Tu
Key Laboratory of Quantum Information, University of Science and Technology of China, Chinese Academy of Sciences, Hefei, P.R. China

Gang Cao
Key Laboratory of Quantum Information, University of Science and Technology of China, Chinese Academy of Sciences, Hefei, P.R. China

Lin-Jun Wang
Key Laboratory of Quantum Information, University of Science and Technology of China, Chinese Academy of Sciences, Hefei, P.R. China

Guang-Can Guo
Key Laboratory of Quantum Information, University of Science and Technology of China, Chinese Academy of Sciences, Hefei, P.R. China

Guo-Ping Guo
Key Laboratory of Quantum Information, University of Science and Technology of China, Chinese Academy of Sciences, Hefei, P.R. China

Stefania Castelletto
School of Aerospace, Mechanical and Manufacturing Engineering RMIT University, Melbourne, Australia

Lorenzo Rosa
Swinburne University of Technology, Centre for Micro-Photonics (H74), Hawthorn, Australia
University of Parma, Department of Information Engineering, Parma, Italy

Brett C. Johnson
School of Physics, The University of Melbourne, Parkville, Australia

Preface

Quantum dots (QDs) are one of the first nanotechnologies to be integrated into the designing of pharmaceutical and biomedical instruments. This book starts by reviewing their structure, quantum confinement effects, band-gap characteristics and applications to the various branches in nanotechnology (e.g. bioimaging, solar cells, light emitting diodes [1]). Fabrication methods of QDs are also included in this review, in particular, the droplet epitaxial method [2]. In the case of bulk materials, the energy band-gap is fixed by the choice of material. However, QDs are a special class of semiconductors, they can be regarded as an artificial molecule with a band-gap dependent on its size. The usage of QDs as an attempt to replace bulk materials (e.g. silicon and copper indium gallium selenide) in solar cells is discussed in some detail [3]. QDs size-dependent bandgap allows the fabrication of nanostructured or multi-junction solar cells and, therefore, a variety of materials can be used to improve efficiency by harvesting multiple portions of the solar spectrum. Experiments have demonstrated that, due to unique optical traits within luminescence characteristics and electronic properties, QDs possess multiple potential applications, such as labeling and tracking single biomolecules and cells in real time, in vitro as well as in vivo; assessing cell growth in damaged tissue; pH probes for the study of enzyme reaction kinetics; imaging and sensing of infectious diseases; as well as the detection of cancer biomarkers [4, 5]. A discussion about the toxicity of QDs applications to nanosize drug delivery systems is also discussed [6]. Next, this book overviews of the recent advances in quantum sensors. A technique of filtering a single molecule with an atomic vapor notch results in an improvement of about 15% in a confocal and in a wide-field configuration. This can potentially enhance DNA detection efficiency, better sensing and localization accuracies in material science and applications in microbiology. [7]. The combination of a low milliamperes X-ray source and a microlens-based sensor may potentially reduce dose exposure without compromising the resolution of dental digital images [8]. Ultra-stable quantum cascade lasers can be integrated into a real-time, nanometric displacement sensor with a resolution on the order of $\lambda/1000$ [9]. A spectroscopy method combining the use of entangled photon pairs and plasmonic nanoparticle arrays with Fano resonance have been shown to be robust, even in excessively noisy conditions, when compared with conventional broadband transmission spectroscopy [10]. The second part of this book, overviews physics and the technological advances in infrared [11, 12] optical [13] photodetectors based on the quantum tunneling as well as the wide-range of applications these detectors possess: e.g. navigation, night vision, weapons detection, communications, aerospace, medical imaging, atmospheric sounding, pollution control, meteorology, environmental monitoring, astronomy, etc. Last but not the least, the book focuses on the quantum information processing. Magnetic tunnel junction (MJC) have the potential of becoming the basic component of random access memory, logic circuits in quantum computing [15]. MTJ based nanomagnet arrays can offer a highly tunable system with an electrical I/O for logic and sensing applications, including two or three-dimensional field sensing [16]. The efficient detection of single photons are also crucial for quantum information processing. Superconducting nanowire single-photon detectors at a wavelength of 940 nm showed high detection efficiencies (~60%) and fast recovery times (~12 ns) promising a great future for quantum computing [17]. Nitrogen-vacancy centers in diamond are revealed to be a good candidate for solid-state quantum information processing due to their long coherence time and high feasibility in initialization, control, and readout of their spin states. To realize quantum computation, the effects of interactions between qubits and their environment must be minimized, the weak spin-orbit coupling and largely eliminated hyperfine interaction in graphene provides the solution to building highly controllable systems [18]. Finally, the use of Silicon carbide as an alternative material to host deep optically active defects is also suitable for optical and spin quantum bits [19].

Editor
Olga Moreira, Ph.D.

INTRODUCTION

The term quantum mechanics (QM) generally refers to a mathematical formulation of quantum theory. It often describes and predicts the movement and behavior of particles at the atomic and subatomic levels. These include particles such as atoms, electrons, protons, and photons. According to QM, the behavior and movement of particles at these microscopic levels are counter-intuitive, radically differing from anything observed in everyday life. This requires a theory like quantum mechanics in order to better explain this different world.

In the macroscopic world, there usually are two types of phenomenon that are classified by scientists: particles and waves. Particles can be thought of as being localized, transporting mass and energy in their movements. Waves are a type of phenomenon that never is localized and has no mass, but nevertheless carries energy. The micro-world particles that are examined in quantum mechanics are different from those of the macro-world, because in certain situations, particles may act like waves, while conversely, waves may act like particles. According to QM, this would mean that on the quantum level an apple could behave like a wave, while a wave could behave like an apple.

Quantum mechanics uses the quantum unit to roughly assign and measure quantities of the physical properties of these particles. Particles generally are measured by considering the wave function recorded at the time of the measurement. Yet because particles may behave like waves, the simultaneous location or velocity of a particle can never completely be determined.

This apparent conundrum is known as Heisenburg's uncertainty principle, which in other words states that position and momentum cannot be exactly measured simultaneously. This is not entirely because there are any limitations in measuring equipment or ability, but because this uncertainty is believed to be inherent to the particles themselves. Quantum mechanics therefore is a mathematical theory that tries to calculate probabilities according to the wave function.

Many scientists do not agree with the postulates of QM. Albert Einstein was a scientist who sometimes was critical of quantum theory, as it conflicted with his own theory. Thus, one of the biggest problems in contemporary physics is to create a theory that would unify the ideas of quantum mechanics with those of Einstein's general theory of relativity. Physicists such as Julian Barbour have proposed some innovative solutions to this problem, basically stating that if humans consider that time itself does not exist, there is no obstacle to unifying quantum mechanics and the theory of relativity.

Quantum mechanics, science dealing with the behavior of matter and light on the atomic and subatomic scale. It attempts to describe and account for the properties of molecules and atoms and their constituents electrons, protons, neutrons, and other more esoteric particles such as quarks and gluons. These properties include the interactions of the particles with one another and with electromagnetic radiation (i.e., light, X-rays, and gamma rays).

The behavior of matter and radiation on the atomic scale often seems peculiar, and the consequences of quantum theory are accordingly difficult to understand and to believe. Its concepts frequently conflict with common-sense notions derived from observations of the everyday world. There is no reason, however, why the behavior of the atomic world should conform to that of the familiar, large-scale world. It is important to realize that quantum mechanics is a branch of physics and that the business of physics is to describe and account for the way the world on both the large and the small scale actually is and not how one imagines it or would like it to be.

The study of quantum mechanics is rewarding for several reasons. First, it illustrates the essential methodology of physics. Second, it has been enormously successful in giving correct results in practically every situation to which it has been applied.

There is, however, an intriguing paradox. In spite of the overwhelming practical success of quantum mechanics, the foundations of the subject contain unresolved problems in particular, problems concerning the nature of measurement. An essential feature of quantum mechanics is that it is generally impossible, even in principle, to measure a system without disturbing it; the detailed nature of this disturbance and the exact point at which it occurs are obscure and controversial. Thus, quantum mechanics attracted some of the ablest scientists of the 20th century, and they erected what is perhaps the finest intellectual edifice of the period.

Energy Quantization

In classical mechanics, a mechanical system can be in a state of every possible energy, with the proviso that the energy is bounded from below by the minimum of the potential. Not so in quantum mechanics; only specific bound state energies exist. Let us take the hydrogen atom as an example. The spectrum of the light emitted by an excited hydrogen atom is shown in Fig. 1.1. 2 As we shall see shortly, light can be described as being made up of particles called photons, and light of frequency v is made up of photons with energy E = hv, where h is a dimensional constant called the Planck constant with units J s in SI (International System of Units), h = 6.62606878 × 10^{-34} Js. The energy of a photon emitted in the decay of a hydrogenic state of energy E_i to a state of lower energy E_f is equal to the energy difference $E_i - E_f$,

$$h\nu = E_i - E_f. \quad (1)$$

Figure 1.1 shows a discrete spectrum, i.e., it is composed of well-defined frequencies. Hence, energies of the hydrogen atom are discrete. This discrete nature of the energies of an electron around a proton is not understandable from a classical mechanics perspective, wherein states of the hydrogen atom should be able to take on all possible energy values. The "quantization" of the observed energies as determined from the emission spectrum just doesn't make sense from a classical mechanics point of view. This situation of discrete energies exists not only for hydrogen atoms, but for all atoms and molecules, and in fact for all bound states of quantum systems. This said, we further note that a continuum of energies is possible for unbound states (in the case of the hydrogen atom, these correspond to states where the electron is not bound to the proton – they are scattering states with positive energy, as opposed to the bound states that have negative energy relative to a proton and an electron at rest and infinitely separated in distance.

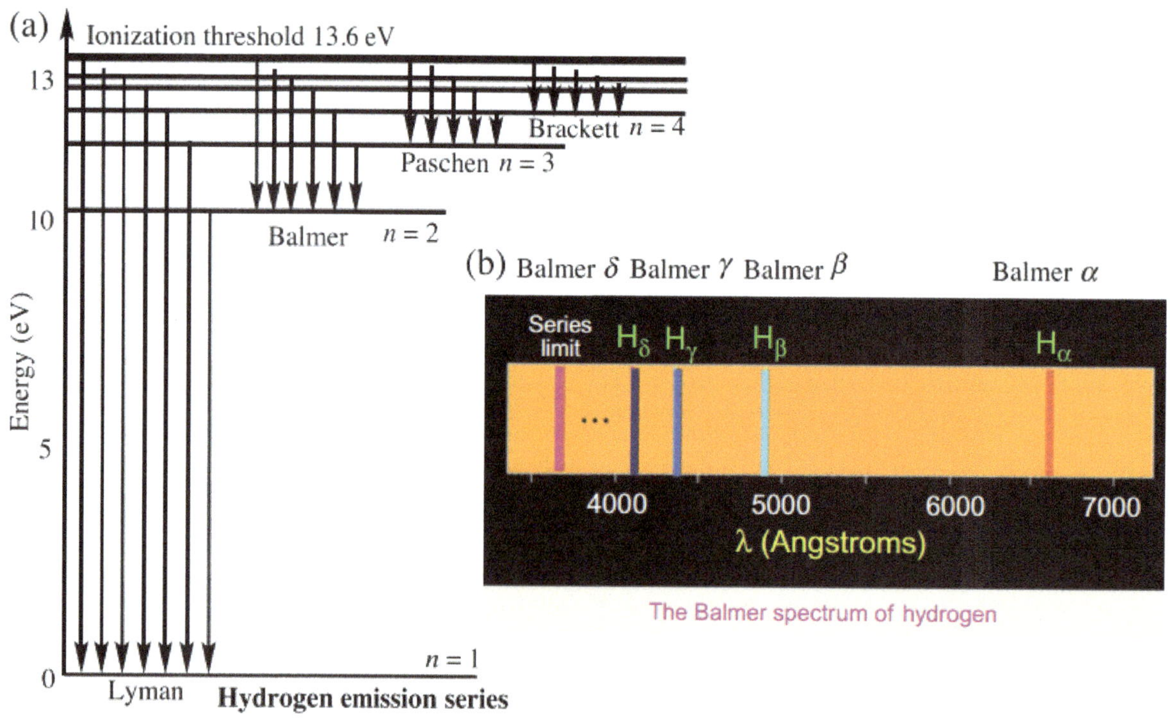

Figure 1.1: The emission spectrum of hydrogen. (a) Lyman, Balmer, Paschen, and Brackett transitions in hydrogen. (b) Two views of the Balmer spectrum in hydrogen versus wavelength. The $n' \rightarrow$ n transitions, where n, n' are principal quantum numbers, are called "Balmer" when n = 2. Balmer lines with n' = 3 are called α, β for n' = 4, γ for n' = 5, etc. The H in H_α, H_β, etc., stands for hydrogen.

Nanotechnology and Information Technology

Nanoscience and nanotechnology can be defined as the science and technology of manipulating materials on an atomic scale. Nanotechnology is used to develop molecular devices such as molecular gears and other molecular machines, nanometer-scale electronic systems, nanocomputers, quantum computers based on qubits (quantum bits), microscopic size atomic clocks, and protein-based molecular devices. For the latter, it might be possible to exploit a host of examples of natural self-replicating machines (e.g., bacteria, viruses) to make such devices efficiently and cheaply.

Nanotechnology makes use of existing micromanipulation techniques, such as the scanning tunneling microscopy (STM) and atomic force microscopy (AFM), deep ultraviolet lithography, electron beam lithography, focused ion beam machining, nanoimprint lithography, atomic layer deposition, molecular vapor deposition, molecular beam epitaxy, and molecular self-assembly techniques such as those employing diblock copolymers. Although some of these techniques were developed before the onset of the nano-era, they are all an integral part of the field. They will undoubtedly evolve and improve, even as new techniques are developed. In recent years, the quest for new devices capable of manipulating electron spins via magnetic fields, in a similar manner to electron charge controlled by electric fields, are beginning to be developed; this new area is called spintronics. Conventional electronic devices rely on the transport of electrical charge carriers – electrons or holes – in semiconductors (e.g., silicon). Exploiting the spin of the electron rather than (or in addition to) its charge is expected to lead to a remarkable new generation of spintronic devices that will be smaller, more versatile, and more robust than those currently used in silicon chips and circuit elements. Moreover, an electron spin is a natural candidate for a qubit, and hence, spintronics might be the ultimate scientific basis for quantum information devices. The main idea is that information can be stored in electron spins in a particular spin orientation (up or down), and can then be carried along a wire, and eventually read at a terminal. Spin orientation of conduction electrons survives for a relatively long time (nanoseconds, compared to femtoseconds for electron momentum decay). This makes spintronic devices particularly attractive for memory storage and magnetic sensors applications, and, potentially for quantum computing and quantum information, where an electron spin can serve as a qubit.

CHAPTER 1 QUANTUM DOTS AND THEIR MULTIMODAL APPLICATIONS

Debasis Bera, Lei Qian, Teng-Kuan Tseng and Paul H. Holloway

Keywords: Quantum dots, semiconducting nanomaterials, electroluminescence, photoluminescence, solar cells, biological imaging

INTRODUCTION

Nanostructured materials [1,2,3,4] are of interest because they can bridge the gap between the bulk and molecular levels and leads to entirely new avenues for applications, especially in electronics, optoelectronics and biology. When a solid exhibits a distinct variation of optical and electronic properties with a variation of particle size <100 nm, it can be called a nanostructure, and is categorized as (1) two dimensional, e.g., thin-films or quantum wells, (2) one dimensional, e.g., quantum wires, or (3) zero dimensional or dots. During the last two decades, a great deal of attention has been focused on the optoelectronic properties of nanostructured semiconductors or quantum dots (Qdots) as many fundamental properties are size dependent in the nanometer range. A Qdot is zero dimensional relative to the bulk, and the limited number of electrons results in discrete quantized energies in the density of states (DOS) for nonaggregated zero dimensional structures [5,6]. (Although it is zero dimensional to bulk, it is regarded as a box in quantum mechanics; size of the box is important and discussed later). Sometimes, the presence of one electronic charge in the Qdots repels the addition of another charge and leads to a staircase-like I-V curve and DOS. The step size of the staircase is proportional to the reciprocal of the radius of the Qdots. The boundaries, as to when a material has the properties of bulk, Qdot or atoms, are dependent upon the composition and crystal structure of the compound or elemental solid. An enormous range of fundamental properties can be realized by changing the size at a constant composition and some of these are discusses. Qdots can be broadly categorized into either elemental or compound systems. In this review, we emphasize compound semiconductor-based nanostructured materials and their multimodal applications based on optoelectronic and optical properties.

A process for synthesizing PbS Qdots was developed more than 2000 years ago using low-cost natural materials like PbO, $Ca(OH)_2$ and water [7]. The *Romans* and *Greeks* used these materials as cosmetics to dye their hair. In more recent history, control of the size of Qdots in silicate glasses is one of the oldest and most frequently used techniques to control the color of glass. In the early 20thcentury, CdS and CdSe were incorporated into silicate glasses to get red-yellow colors. In 1932,*Rocksby* [8] used x-ray diffraction (XRD) to determine that precipitates of CdS and CdSe caused the colors. Earlier, semiconductor particles doped glasses were also used in optics as filters. A blue shift of the optical spectrum for nanometer sized CuCl in silicate glass was reported in 1981 by*Ekinov* and *Onushchenko* [9]. In 1982, *Efros* and *Efros* [10] advanced the postulate that quantum size effects (the change of optical and optoelectronic properties with size) could be used to control the color of glass by either changing the size or stoichiometry of CdS_xSe_{1-x}. In 1991, the change in color of colloidal solutions of semiconductor was discussed by *Rosetti et al.* [11]. Several different synthesis methods were developed during this period [12,13,14,15]. Over the last two decades, experimental and theoretical research on these nanoparticles has increased significantly [1,16,17,18] in order to explore many basic properties [19,20] of Qdots and attracted by commercialization efforts [21,22]. In this review, we discuss briefly the structure, properties, processing and performance of the Qdots in multimodal applications.

Citation: Debasis Bera, Lei Qian, Teng-Kuan Tseng and Paul H. Holloway, Quantum Dots and Their Multimodal Applications, doi:10.3390/ma3042260.

STRUCTURE OF QUANTUM DOTS

Core Structure

As mentioned before, Qdots have dimensions and numbers of atoms between the atomic-molecular level and bulk material with a band-gap that depends in a complicated fashion upon a number of factors, including the bond type and strength with the nearest neighbors. For isolated atoms, sharp and narrow luminescent emission peaks are observed. However, a nanoparticle, composed of approximately 100–10000 atoms, exhibits distinct narrow optical line spectra. This is why, Qdots are often described as artificial atoms (δ-function-like DOS) [18]. A significant amount of current research is aimed at using the unique optical properties of Qdots in devices, such as light emitting devices (LED), solar cells and biological markers. The most fascinating change of Qdots with particle size <~30 nm is the drastic differences in the optical absorption, exciton energies and electron-hole pair recombination. Use of these Qdot properties requires sufficient control during their synthesis, because their intrinsic properties are determined by different factors, such as size, shape, defect, impurities and crystallinity. The dependence on size arises from (1) changes of the surface-to-volume ratio with size, and from (2) quantum confinement effects (discussed later). Nevertheless, Qdots exhibit different color of emission with change in size. Figure 1 shows change of photoluminescence (PL) emission color with size for CdSe Qdots.

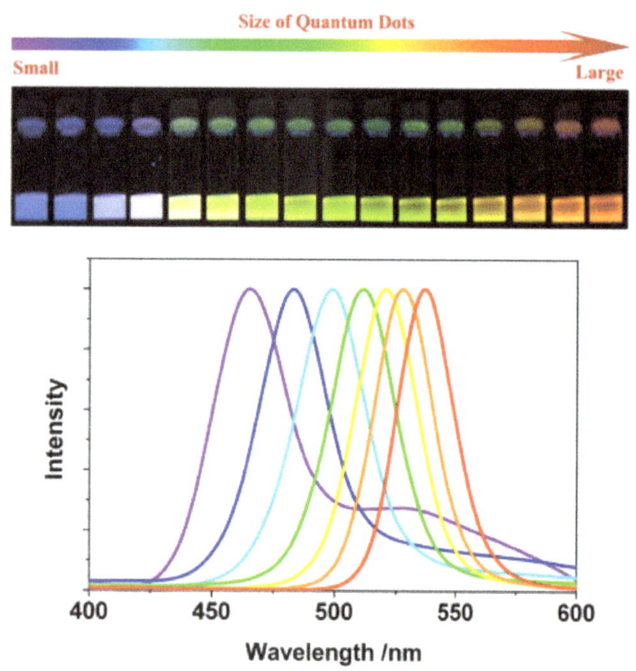

Figure 1. Top: Sixteen emission colors from small (blue) to large (red) CdSe Qdots excited by a near-ultraviolet lamp; size of Qdots can be from ~1 nm to ~10 nm (depends on several parameters, see text for details). Bottom: Photoluminescence spectra of some of the CdSe Qdots [23].

Size versus Density of States

A most unique property of the Qdots is quantum confinement, which modifies the DOS near the band-edges. Schematic diagrams of the DOS as a function of energy in Figure 2 show that Qdots lie between the discrete atomic and continuous bulk materials. Quantum confinement effects are observed when the size is sufficiently small that the energy level spacing of a nanocrystal exceeds kT (where k is *Boltzmann's* constant and T is temperature). Energy differences > kT restrict the electron and holes mobility in the crystal. Among many properties that exhibit a dependence upon size in Qdots, two are of particular importance. The first is a blue shift (increase) of band-gap energy when the nanoparticle diameters are below a particular value that depends on the type of semiconductor. This is called a quantum confinement effect [24,25] and is discussed below in detail. This effect allows tuning of the energy gap with changes in the Qdot size. The band-gap energy also depends on the composition of the semiconductors as well as the size. The second important property is the observation of discrete, well separated energy states due to the small number of atoms in Qdots compared to the bulk. This leads to the electronic states of each energy level exhibiting wave functions that are more atomic-like. Since the Qdots solutions for *Schrödinger* wave equation are very similar to those for electrons bound to a nucleus, Qdots are called artificial atom, and atomic-like sharp emission peaks are possible. Typical intraband energy level spacings for Qdots are in the range of 10–100 meV. Band-gap can also be tuned by alloying [26,27] the core of the Qdots (discussed below).

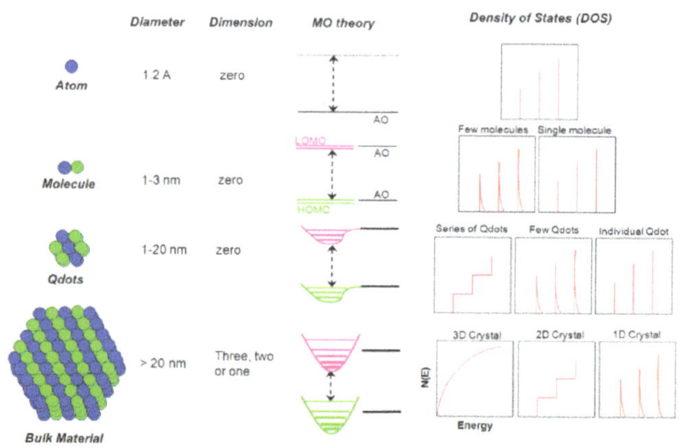

Figure 2. Schematic illustration of the changes of the density of quantum states (DOS) with changes in the number of atoms in materials (MO: molecular orbital; HOMO: highest occupied MO; LUMO: lowest unoccupied MO; AO: atomic orbital) [6,23].

Phases and Phase Transitions

II-VI compound semiconductors include the cations of zinc, cadmium and/or mercury combined with anionic oxygen, sulfur, selenium and/or tellurium. These semiconductors generally crystallize in both face-centered cubic (zinc blende) and hexagonal (wurtzite) crystal structures. For example, the equilibrium crystal structures of both ZnO and ZnS are hexagonal, although ZnS often exhibits a metastable cubic or a mixed hexagonal/cubic structures. The II-VI compound semiconductors may exhibit very good luminescence because they have a direct band-gap. In addition, many of the II-VI semiconductors are often used as a host for luminescent activators (discussed later), e.g., ZnS doped with Mn^{2+} which emits yellow light [28]. Near band-edge emission from excitons can be observed from II-VI semiconductors, especially at low temperatures from those materials with a low exciton binding energy.

Qdots exhibits solid-solid phase transition like bulk semiconductors, and these transitions have a substantial influence on the optical properties of Qdots. Phase transitions in bulk materials can be induced by varying pressure, temperature and composition [29,30]. Bulk CdSe may exhibit either a hexagonal wurtzite or a rock salt cubic structure with a direct or indirect band-gap, respectively. Above a pressure of ~ 3 GPa, the CdSe bulk semiconductor can be converted reversibly from low pressure wurtzite to the high pressure rock salt structures [31]. The low intensity optical emission from the rock salt form of CdSe is in the near infrared (NIR) spectral region at 0.67 eV (1.8 μm). Using high pressure XRD and optical absorption, *Tolbert* and *Alivisatos* showed that the wurzite to rock salt structural transformation also occurred in CdSe Qdots [29,30]. The ratio of oscillator strength between direct and indirect structures was unchanged with size of Qdot.

Doping in Quantum Dots

Doping in Qdots is an important aspect when Qdots are used for various technological applications [32,33,34], especially, optoelectronic, magnetic, biological and spintronic applications. These impurities, called activators, perturb the band structures by creating local quantum states that lies within the band-gaps. In the Qdots, the dopants are found to be auto-ionized without thermal activation due to quantum confinement. When quantum confinement energy (increase of band-gap energy with decreasing size) exceeds *Coulombic* interaction between carrier (hole or electron) and impurity (*n*-type or *p*-type), auto-ionization occurs. Several transition elements such as, Cr [35,36], Mn [35,37,37,38,39], Fe, Co [40], Cu [35,41,42] and Ag [43], and other elements, such as, P [44], B [44], Na [45] and Li [45] were doped in Qdots, for different applications. Optical properties of Qdots can be varied by changing the amounts [46] and the positions (see Figure 3) [47] of dopants in the Qdots. In optoelectronic application of Qdots, doping can play an important role. Conduction in doped Qdot films depends on the size uniformity of Qdots and proximity of neighboring Qdots, so that the orbital overlap among Qdots is maximized. Optical properties of Qdots can be enhanced by doping in Qdots. We discuss more on the optical properties of the doped Qdots in 'properties of Qdots' section. Application of doped Qdots can be found mostly in Section 5.5.

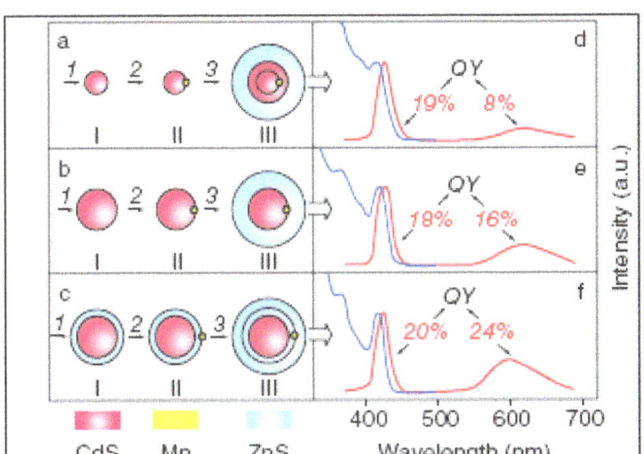

Figure 3. Differences in optical properties of Mn-doped CdS/ZnS core/shell Qdots with different Mn positions; (a) doped in CdS core, (b) at interface of the core/shell structure, and (c) in ZnS shell structure. Core diameter was 3.8 nm and ZnS shell thickness was 1.5 nm. Photoluminescence and photoluminescence excitation spectra are shown in (d), (e) and (f) for IIIa, IIIb and IIIc structure, respectively; (QY: quantum yield). (Reprinted with permission from [47]. Copyright 2006 American Chemical Society).

Alloying of Quantum Dots

Keeping the size of a Qdot constant, the band-gap can be engineered by alloying of the core [48,49,50,51,52,53]. Composition of the core materials and/or ratio of alloying materials can change the optoelectronic properties of a Qdot. Such a process has been extensively investigated for the last few years. The process is particularly interesting, because: (1) these nanostructures of semiconductors provide different and nonlinear optoelectronic properties [54] and, (2) Qdot alloyed with multiple semiconductors exhibits a mixed or intermediate optoelectronic properties. NIR emission (600 nm–1350 nm) was achieved by synthesizing alloyed CdHgTe Qdots [55] and varying the stoichiometric ratio of two binary semiconductors. In addition, (3) PL emission efficiencies can be improved by minimizing bulk and surface defects [56], and (4) narrow full-width-half-maximum (FWHM) of PL can be achieved. For example, weak quantum confinement regime (explained in 'structure of Qdots' section) $Zn_xCd_{1-x}S$ Qdots were synthesized [57] using a wide band-gap and small *Bohr* radii ZnS and CdS semiconductors. Since these Qdots are in the weak quantum confinement regime (discussed in Section 3.1), inhomogeneous broadening of PL due to size fluctuation is greatly reduced. These Qdots are also being used in many applications including biological imaging [51,58,59].

Table 1. Change of band-gaps with composition of $Zn_xCd_{1-x}S$ alloy (Adapted from Reference [54]).

Semiconductor/ Qdot	Value of 'x' in Qdot composition	Calculated Particle size Diameter /nm	Band-gap (eV)
ZnS		bulk	3.7
		2.7	4.1
CdS		bulk	2.45
		2.6	2.9
HgS		bulk	~0.0
$Zn_xCd_{1-x}S$	0.14	2.6	3.0
	0.14	3.2	3.0
	0.14	3.5	3.0
	0.14	4.3	2.95
	0.15	2.7	3.0
	0.25	4.0	3.05
	0.34	3.7	3.15
	0.44	4.7	3.5
	0.61	3.9	4.0
$Hg_xCd_{1-x}S$	0.0025	4.0	4.5
	0.005	4.0	4.45
	0.05	4.0	4.4
	0.01	4.0	4.35
	0.2	4.0	3.8
	0.5	4.0	3.25
	0.75	4.0	3.15

Surface Structure

Due to the high surface-to-volume ratio of Qdots, electronic quantum states associated with the surface (called surface states) have significant effects on the optical properties of Qdots. For example, roughly 15% of the atoms in a 5 nm CdS Qdot are at the surface [60]. Such a high surface-to-volume ratio may allow an enhanced or reduced transfer rate of photogenerated charge carriers due to the high density of surface sites. The surface states of the Qdots may influence the optical absorption (photoluminescence excitation – PLE), quantum efficiency, luminescent intensity and spectrum and aging effects [61]. In general, surface states arise from unsatisfied bonds at the reconstructed surface, and may be affected by nonstoichiometry and voids. The energies of these surface states generally lie in the band-gap of the Qdots [62]. Therefore, they can trap charge carriers (electron or hole) and behave as reducing (electron) or oxidizing (hole) agents. These electrochemical reactions or behavior at the surface significantly can affect the overall conductivity and optical properties of Qdots. As a result, surface states have significant effects on the optical and optoelectronic properties of the Qdots. Surface passivation of Qdots can confine the carrier inside the core and improves the optical properties of Qdots. But these passivation layer acts as either insulator or barrier for the conduction of charge.

Surface Passivation

As discussed above, surface defects in Qdots act as temporary 'traps' for the electron, hole or excitons, quenching radiative recombination and reducing the quantum yields (QYs). Therefore, capping or passivation of the surface is crucial for development of photostable Qdots. In principle, a perfectly passivated surface of a Qdot has all dangling bonds saturated and, therefore, exhibits no surface state, and all near band-edge states are quantum-confined internally. For a compound semiconductor, if the anion dangling bonds at the surface are not passivated, a band of surface states is expected in the gap just above the valence band-edge. However, passivation of anions with surface cations would also leave dangling bonds that would lead to a broad band of surface states just below the conduction band-edge. Therefore, surface modification of Qdots is very demanding and is generally carried out by depositing an organic or inorganic capping layer on the Qdots.

Organically Capped Quantum Dots

Generally, monodispersed Qdots are developed by introducing organic molecules that adsorb on the Qdot surface and act as capping agents [63,64,65]. Some advantages of organic capping layers include simultaneous achievements of colloidal suspension and the ability to bio-conjugate the Qdots. However, the selection of organic ligands that bond with surface atoms of the Qdots is a very delicate issue. In general, phosphenes, (e.g., tri-n-octyl phosphene oxide–TOPO) or mercaptans (-SH) are the most widely used ligands. Most of the organic capping molecules are distorted in shape and larger than a surface site. As a result, coverage of surface atoms with the organic capping molecules may be sterically hindered. Another crucial issue is the simultaneous passivations of both anionic and cationic surface sites using such capping agents, which is achievable but still complex. Some dangling bonds on the surface are always present when the surface is passivated by organic agents. Finally, the organic capped Qdots are photo-unstable. The bonding at the interface between the capping molecules and surface atoms is generally weak leading to the failure of passivation and creation of new surface states under ultraviolet (UV) irradiation. The surface states of Qdots are known to be sites of preferential photodegradation and luminescence quenching. Figure 4(a) illustrates a Qdot passivated with organic molecules. Chemically reduced bovine serum albumin (BSA) has been used simultaneously to passivate and functionalize the surface of CdTe Qdots to make them water soluble [66]. Denatured BSA (dBSA) conjugated to the CdTe Qdots surface improved the chemical stability and the PL-QY [66]. This study showed that over a pH range of 6 to 9, the solution of dBSA-coated CdTe Qdots were stable and bright, but higher and lower pH values led to dramatic decreases in PL intensity and chemical stability. Similarly, concentrations of dBSA that were too high or too low in the Qdots solution resulted in a decreased PL-QY. Recently, DNA passivated CdS Qdots were reported to be stable in and nontoxic to biological systems [67].

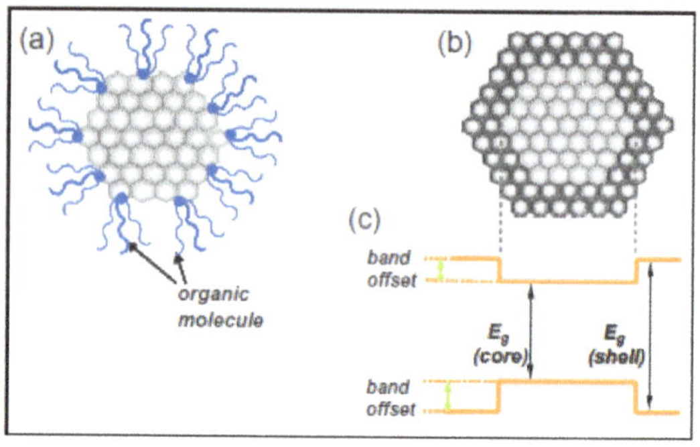

Figure 4. Schematic illustration of (a) an organically capped Qdot and (b) an inorganically passivated Qdot (core/shell structure of Qdot). (c) An energy diagram shows the band-gap difference of core and shell of inorganically passivated Qdots; (E_g: band-gap) [6]

Inorganically Passivated Quantum Dots

A second approach to passivation of the Qdots surface is the use of inorganic layers, particularly a material with a larger band-gap. The passivating shell is grown either epitaxially (as depicted inFigure 4(b)) or as a non-epitaxial crystalline or amorphous layer on the core. The QY of Qdots is increased by a defect-free, uniform shell coating. When the shell material adapts the lattice parameters of the core during epitaxial growth, coherency strains result and can play an important role in the properties of these core/shell systems. For example, strain may cause the absorption and emission spectra of core/shell Qdots to be red-shifted [68]. The maximum PL efficiency or QY of core/shell Qdots is also dependent upon the thickness of the shell layer. The thickness is less than two monolayers for optimum properties of CdSe/CdS core/shell nanoparticles. Thicker capping layers lead to formation of misfit dislocations which are also nonradiative recombination sites that decrease the PL-QY. More information on shell thickness and PL-QY can be found in section 2.2.2.

Epitaxial Growth: Generally, a wider band-gap shell material is desired to create a potential barrier around the Qdot core to confine the exciton (see Figure 4(c)). Confinement of the charge carriers into the core region by the band offset potentials result in efficient and photostable luminescence from Qdots. Additional factors to consider when selecting the Qdot inorganic shell material include whether it is hydrophobic or hydrophilic. Most inorganic core/shell Qdots are not compatible with dispersion in water due to the hydrophobic surface property of the shell. For biological application of Qdots, however, an appropriate water-compatible coating, such as an amorphous silica layers, is necessary [69]. For better passivation, the shell material should have a lattice parameter within 12% of the core to encourage epitaxy and minimize strain, and a thickness below the critical value that results in misfit dislocations [70]. The lattice mismatch between CdSe and ZnS (10.6%) is larger than that between CdSe and ZnSe (6.3%) and CdSe and CdS (3.9%), but the band-gap is also larger leading to better exciton confinement. ZnSe is also a good shell material for CdSe, because it has a wider band-gap (2.72 eV) than that of CdSe (1.76 eV). ZnSe also has the same anion (Se) which leads to a larger offset in the conduction bands and therefore to better confinements of the excitons. Luminescence QYs for CdSe/ZnSe core/shell Qdots have been reported in the range of 60–85% [71]. Finally, the smaller lattice mismatch between a CdSe core and a CdS shell facilitates epitaxial growth of a CdS shell. CdSe/CdS core/shell Qdots typically display higher PL-QY with longer PL lifetimes. Our recent investigation shows multi-fold enhancement of CdSe luminescence when amorphous silica layer is introduced onto the CdSe Qdots [72]. Nanocrystalline ZnSe particles have been encapsulated by graphite [73], and enhanced blue emission was observed as compared to unencapsulated ZnSe nanoparticles. Mid-gap defect orange emission was quenched by carbon passivation.

Although significant improvement was observed by introducing an inorganic shell layer, several evidences on incomplete passivation by the shell are reported in literature. Such an incomplete passivation causes different results like (a) less QY than that from complete passivation [74], (b) a significant amount of permanently dark Qdots [74], (c) photooxidation of core and/or shell [75,76], (d) fluctuation of intensity from core/shell Qdots due to trapping of carriers by the surface states [77].

Inverted core/shell Qdots, e.g., ZnSe/CdSe (with a larger band-gap for the core) Qdots show very interesting optoelectronic properties. They exhibits either type I or type II interfacial band offsets depending on the core radius and the shell thickness [78]. Type I offset is an opposite offset for both the valence and conduction bands. This is the case for bulk ZnSe/CdSe interfaces, where the ZnSe valence band-edge is lower than that in CdSe (energy offset ~0.14eV), while the conduction band-edge is higher (energy offset ~0.86eV). Such an energy alignment results in confinement of both electrons and holes inside the CdSe core which reduces their interactions with surface trap states and improves their QYs. However, the situation can change in the case of nanostructures in which the alignment of quantized energy states is determined not only by bulk energy offsets, but also by the confinement energies determined by the heterostructure dimensions. Core/shell Qdots with type II offsets (valence and conduction band offsets in the same direction) can also provide "spatially indirect" states, in which electrons are spatially confined to the core (or shell) and holes confined to the shell (or core). The emission energy from type II core/shell nanostructures is smaller than the band-gap of either the core or the shell material due to the interfacial energy offsets. Because of the reduced electron-hole wave function overlap, these structures show extended exciton lifetimes and are useful in photovoltaic and photocatalysis applications [79]. With a large red-shift in emission from type-II core/shell Qdots, NIR emission may be possible for *in vivo*bioanalytical and biomedical applications.

Non-epitaxial Growth: As mentioned earlier, Qdots are often synthesized in nonpolar, nonaqueous solvents leaving them hydrophobic. In addition, except for some oxide based Qdots, which are assumed to be less toxic, most of the Qdots contain toxic ions (e.g., cadmium (Cd), selenium (Se) and tellurium (Te). Therefore, an oxide-based coating is important to reduce the toxicity in biological applications. Furthermore, proper functionalization of Qdots is very important for biological applications, as mentioned earlier. To address these issues, a silica shell is grown on the Qdots. A high resolution transmission electron microscopic (HRTEM) image of silica coated CdSe Qdots is shown in Figure 5(a). Aqueous-based synthesis methods generally are used to produce silica-capped Qdots [80]. Recently, it is shown that a silica shell can prevent the leakage of toxic Cd^{2+} from infrared (IR)-emitting CdTe Qdots. Cytotoxicity and the potential interference of Qdots with cellular processes are the subject of intensive studies [81,82]. The silica shell also allows easy functionalization with biomolecules such as proteins [83,84] and results in greater photostability. The luminescent properties of silica-coated Qdots depend on the charge trapped on the surface as well as the local electric field. The field dependent emission from Qdots is called quantum-confined *Stark* effect [85]. External electric field or internal local field results in shifts of both emission wavelength and intensity (shown in Figure 5(b)). By neutralizing a surface positive charge, we recently found that the emission from CdSe Qdots was blue shifted, and the QY increased dramatically. Although there is no attempt found in the literature, the electric field induced change of emission from Qdots can be potentially useful for biological imaging and sensing.

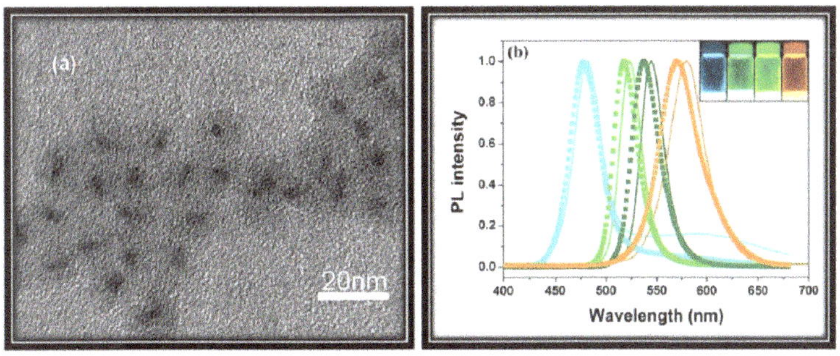

Figure 5. (a) High resolution transmission electron micrograph of silica-coated CdSe Qdots; (b) Photoluminescence (PL) spectra from CdSe (solid) and silica coated CdSe (dotted) Qdots with variation of core size; inset photograph shows emitting color under long wavelength UV-light [23].

CdS Qdots have also been coated with metal shells resulting in large and fast third-order optical nonlinearity due to surface plasmon resonance (SPR). The collective charge oscillation causes a large resonant enhancement of the local field inside and near the particle which may be used in surface-enhanced *Raman* scattering and in nonlinear optical devices. Qdots coated with a noble metal have been shown to exhibit coupling between the plasmon resonance from the metal and the quantum size effect of the Qdots that give rise to new properties. *Jeang et al.* reported a red-shift of the exciton absorption peak for the Ag/CdS nanocomposite [86]. *Je et al.* investigated the local field-induced optical properties of CdS/Ag core/shell nanocomposites [87]. They confirmed by theoretical calculations that the strong local field created confined *Wannier-Stark* states (*i.e.,* energy spectrum of a crystalline solid in an electric field) that explains the red shift of the exciton peak in the nanocomposite.

Multi-Shell Structure

Double shell Qdots are being studied for improved optical properties. As discussed above, the lattice mismatch and differences in band-gap are important to the properties of core/shell Qdots. The band-gaps and band offsets of the core and shell materials are also critical to suppression of tunneling of charge carriers from the core to the surface states of the shell. In the case of CdSe/CdS, the lattice mismatch is small but so are the band offsets. For CdSe/ZnS Qdots, the reverse is true with the lattice mismatch being large along with the band offsets. The advantages of both shell materials are combined in core/shell/shell CdSe/CdS/ZnS Qdots [88,89]. In these double shell nanostructures, the lattice strain at the interface is reduced while large band offsets are maintained.

Qdots may exhibit significantly low QY due to *Auger* recombination. This process is strongly affected by the confinement. Suppression of non-radiative decay due to *Auger* recombination can be achieved by minimizing wave function overlap of charge carriers [90]. Recently quantum well quantum dots (QWQD) has been introduced to circumvent this issue [90]. In QWQD, a hollow spherical quantum well (QW) surrounds a large band-gap center core Qdot and an outer large band-gap shell passivates the surface, minimizing wave function overlap in the QW [91]. Therefore, higher PL-QY can be achieved from the QWQD compared to the bare Qdots or core/shell Qdots. It is, therefore, of utmost importance to control the structure of QWQD very precisely in order to maximize the QY from Qdots. Multi-shell structured Qdots are also used in biological imaging and investigation in order to achieve bioconjugation [23,92,93,94,95].

Characterization of Shell Structures

Recently, chemical distribution of shell materials on the CdSe/ZnS core/shell Qdots was studied by *Yu et al.* [96] by using scanning transmission electron microscopy (STEM) coupled with electron energy loss spectroscopy (EELS). According to their analyses on electron EELS spectra and simultaneous annular dark field (ADF) signal, the ZnS shell was well-defined around the CdSe core. However, the distribution of shell material was highly anisotropic possibly due to differences in chemical activity of the crystal faces of the core CdSe [97]. Accuracy of the measurement is questionable as collection of the localized EELS spectra was carried out from a subnanometer area on a single Qdot which can be easily overwhelmed by either near-by areas of same Qdots or neighbor Qdots due to movement of Qdots under a high-energy electron beam. Z-contrast STEM analysis possesses several advantages over conventional transmission electron microscopy (TEM) [98]. Figure 6 shows a Z-contrast STEM *versus* a high resolution TEM or HRTEM micrograph of a CdSe Qdot. In Z-contrast STEM, incoherently scattered electrons are collected by a high-angle annular dark field (HAADF) detector. According to the *Rutherford* scattering equation, the intensity of scattered electrons is proportional to square of the atomic number for sufficiently high angular range. Therefore, mass-contrast can be observed directly from images. Although *McBride et al.*[99,100] demonstrated the imaging of both bare CdSe Qdots and core/shell Qdots by using an aberration-corrected Z-contrast STEM, the real challenge is to interpret bonding information from TEM analysis. In addition, Z-contrast STEM images of Qdots with an amorphous shell, e.g., silica, is near-impossible.

X-ray photoelectron spectroscopy (XPS) is a critical tool to analyze the surface of Qdots [101,102]. One of the earliest and most extensive XPS studies on the nature of the CdSe Qdots was carried out by *Katari et al.* [101]. They showed that the XPS core level positions for Cd and Se from covalently bound monodispersed CdSe to Au on Si substrate were in agreement with those of bulk CdSe. However, both Cd and Se were oxidized in the case of completely washed and air-exposed CdSe Qdots [101]. XPS was also used to verify the shelling in core/shell Qdots [76,103,104]. All abovementioned studies demonstrated that composition of surface ligand and shell coverage, and surface oxidation of Qdots can be analyzed by XPS. However, XPS cannot be used to analyze specifically surface, shell structure and core-shell interface bonding of Qdots, because the escape length in conventional XPS is comparable to the size of the Qdots. In addition, the smallest spot size in XPS analysis is much larger (~1 μm) than the Qdots.

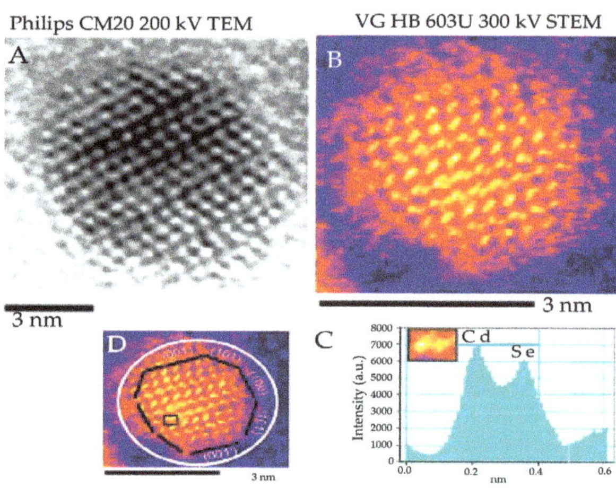

Figure 6. (A) High resolution transmission electron microscopic micrograph *vs.* (B) Z-contrast scanning transmission electron microscopic (STEM) micrograph of a CdSe Qdots. (C) STEM image shows mass distribution to determine a crystallite facet; (D) CdSe crystal facets are shown in STEM micrograph [reprinted from [98] with permission from Elsevier].

As discussed in previous sections, thickness of the shell has profound effects on luminescence properties of Qdots. For example, a single monolayer of surface passivating inorganic shell on a Qdot increases the QY by a factor of 3 [70]. However thick-shell on Qdots can reduce [70] the QY of the Qdot significantly by formation of misfit dislocations which are also nonradiative recombination sites that decrease the QY. An optimum thickness is important in order to maximize the PL-QY. Actual thickness of the shell of a core/shell Qdot is difficult to measure as (1) thickness is very small compared to core to be observed in TEM, and (2) it may be epitaxially grown so that could not be resolved in XRD studies. Traditional characterization techniques, such as, TEM, XRD, and XPS confirms the addition of a few monolayers onto Qdots. Currently, thickness of the shell of a core/shell system is approximated either by comparing TEM micrographs of bare and core/shell Qdots or by calculating differences from average particle sizes of core and core/shell Qdots using XRD [97]. However, these characterization techniques do not provide any information about interface bonding of core and shell. This warrants the need of new characterization techniques that can provide more complete information on the shell structure in the core/shell Qdots and QWQD.

PROPERTIES

Quantum Confinement Effects and Band-Gap

Quantum confinement generally results in a widening of the band-gap with a decrease in the size of the Qdots. The band-gap in a material is the energy required to create an electron and a hole at rest (*i.e.,* with zero kinetic energy) at a distance far enough apart that their *Coulombic attraction* is negligible. If one carrier approaches the other, they may form a bound electron-hole pair, *i.e.,* an exciton, whose energy is a few meV lower than the band-gap. This exciton behaves like a hydrogen atom, except that a hole, not a proton, forms the nucleus. Obviously, the mass of a hole is much smaller than that of a proton, which affects the solutions to the *Schrödinger* wave equation. The distance between the electron and hole is called the exciton *Bohr* radius (r_B). If m_e and m_h are the effective masses of electrons and holes, respectively, the exciton *Bohr* radius for bulk semiconductor can be expressed by Equation 1, where ε, ħ, and e are the optical dielectric constant, reduced*Planck's* constant and the charge of an electron, respectively.

$$r_B = \frac{\hbar^2 \varepsilon}{e^2}\left(\frac{1}{m_e} + \frac{1}{m_h}\right) \qquad (1)$$

If the radius (R) of a Qdot approaches r_B, *i.e.,* $R \approx r_{B,}$ or $R < r_{B,}$ the motion of the electrons and holes are confined spatially to dimension of the Qdot which causes an increase of the excitonic transition energy and the observed blue shift in the Qdot band-gap and luminescence. The exciton *Bohr* radius is a threshold value, and the confinement effect becomes important when the Qdot radius is smaller. For small Qdots, the exciton binding energy and biexciton binding energy (exciton-exciton interaction energy) is much larger that for bulk materials [105]. Note that for a material with a relatively higher ε or smaller m_e and m_h, the r_B is larger. Two detailed theoretical approaches are used to better predict the exciton properties, specifically the effective mass approximation (EMA) model and linear combination of atomic orbital (LCAO) theory. Below, we discuss these two theories in brief.

Effective Mass Approximation Model

This approach, based on the 'Particle-in-Box Model', is the most widely used model to predict quantum confinement. It was first proposed by *Efros* and *Efros* [10] in 1982 and later modified by*Brus* [106]. It assumes a particle in a potential well with an infinite potential barrier at the particle boundary. For a particle free to assume any position in the box the relationship between its energy (E) and wave vector (k) is given by Equation 2.

$$E = \frac{\hbar^2 k^2}{2m^*} \qquad (2)$$

In the EMA model, this relationship (Equation 2) is assumed to hold for an electron or hole in the semiconductor, therefore the energy band is parabolic near the band-edge. The shift of band-gap energy (ΔE_g) due to confinement of the exciton in a Qdot with a diameter R can be expressed as follows (Equation 3), where, μ is the reduced mass of an electron-hole pair and E^*_{Ry} is *Rydberg* energy.

$$\Delta E_g = \frac{\hbar^2\pi^2}{2\mu R^2} - \frac{1.8e^2}{\varepsilon R} = \frac{\hbar^2\pi^2}{2R^2}\left(\frac{1}{m_e} + \frac{1}{m_h}\right) - \frac{1.78e^2}{\varepsilon R} - 0.248E^*_{Ry} \qquad (3)$$

The first term of the Equation 3 represents a relation between 'particle-in-a-box' quantum localization energy or confinement energy and the radius of the Qdot (R), whereas the second term shows the *Columbic* interaction energy with a R^{-1} dependence. The *Rydberg* energy term is size independent and is usually negligible, except for semiconductors with small dielectric constant [60]. Based on Equation 3, the first excitonic transition (*i.e.,* the band-gap) increases as the Qdot radius (R) decreases (quantum localization term shifts to higher energy with lower R value (R^{-2}) and *Columbic* terms shifts excited electronic state to lower value (R^{-1})). However, the EMA model breaks down in the small Qdot regime [14,60] because the *E-k* relationship can no longer be approximated as parabolic. Figure 7 shows such a deviation of theoretically predicted band-gaps for CdS Qdots from the experimental values.

Linear Combination of Atomic Orbital Theory–Molecular Orbital Theory

A model based on a linear combination of atomic orbitals–molecular orbitals (LCAO-MO) provides a more detailed basis for predicting the evolution of the electronic structure of clusters from atoms and/or molecules to Qdots to bulk materials, and predicting the dependence of band-gap on size of the crystals. Figure 8 shows the results of this approach pictorially. In a diatomic Si molecule, the atomic orbitals (AO) of two individual atoms are combined, producing bonding and anti-bonding molecular orbitals. In this approach, nanosized Qdots are considered as large molecules. As the number of atoms increase, the discrete energy band structure change from large steps to small energy steps, *i.e.,* to a more continuous energy band. The occupied (bonding) molecular orbital quantum states (equivalent to the valence band) are called the highest occupied molecular orbital (HOMO) levels. The unoccupied antibonding orbitals (equivalent to the conduction band) are called the lowest unoccupied molecular orbital (LUMO) levels. The energy difference between the top of the HOMO and bottom of the LUMO (equal to the band-gap) increases and the bands split into discrete energy levels reduced mixing of AOs for a small number of atoms. Therefore, the small size of the Qdots results in quantized electronic band structures intermediate between the atomic/molecular and bulk crystalline MOs.

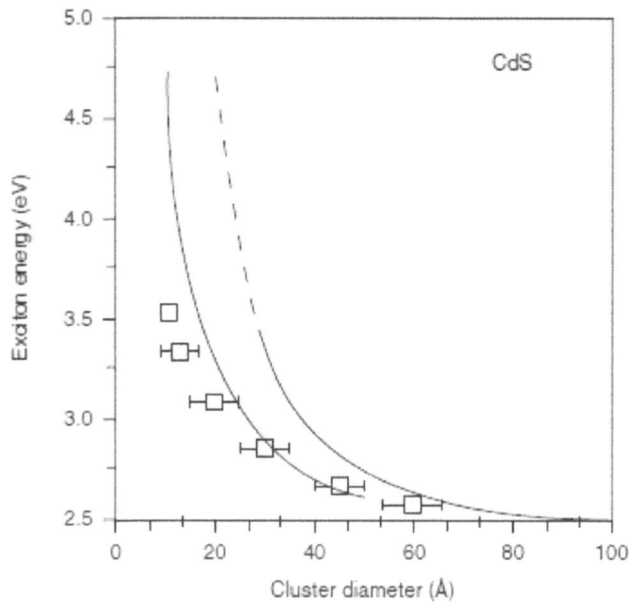

Figure 7. Experimentally and theoretically determined band-gap as a function size of CdS Qdots. Broken line: calculated parameters based on effective mass approximation, solid-line: tight-bonding calculation; squares: experimental data (Reprinted with permission from [60]. Copyright 1991 American Chemical Society).

Compared to the effective mass approximation, the LCAO-MO model provides a methodology to calculate the electronic structure of much smaller Qdots. In contrast, this method cannot be used to calculate the energy levels of large Qdots due to mathematical complexity and limitations of the computing systems. Nevertheless, the degree of quantum confinement is determined by the ratio of the radius of a Qdot (R) to bulk excitonic *Bohr* radius (r_B). At crystal sizes greater than the excitonic*Bohr* diameter ($2r_B$), semiconductor crystals exhibit translational motion confinement of the fully coupled exciton due to a strong *Coulombic* interaction between the electron and holes, *i.e.,* exhibits single-particle confinement behavior (sometimes called the strong confinement regime). In the intermediate size range ($R \leq r_B$), the transition energies of photoexcited carriers in the crystal are determined by the relative strengths of the kinetic energy of confinement and the electron-hole interaction.

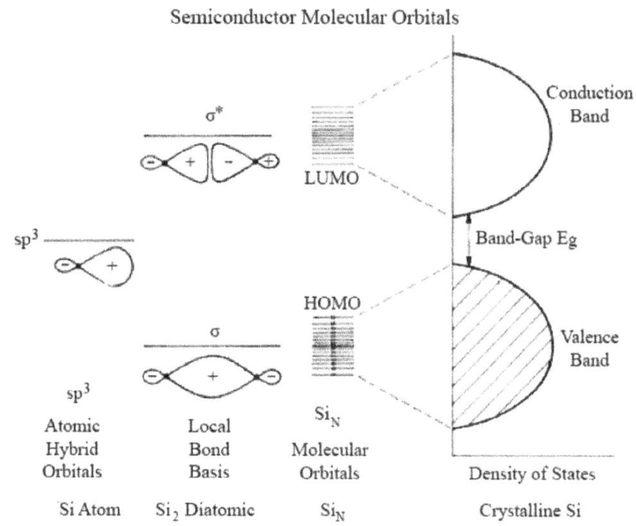

Figure 8. Combination of atomic orbital to molecular orbital and to band-gap of silicon molecule (Reprinted with permission from [107]. Copyright 1990 American Chemical Society).

Band-gap of Qdots can be determined by electrochemical measurement using Qdots films. Cyclic voltametry (CV) are often employed [108,109,110,111,112,113] to determine the oxidation and reduction potential of the film of Qdots to be measured using a standard three-electrode cell. CV is a dynamic electrochemical method in which current-potential curves are traced at a pre-defined scan rates. Qdots-coated gold plate, platinum wire or indium tin oxide film on glass substrates are often used as working electrodes and platinum electrode acts as a counter electrode. The cell potential is generally normalized to reference electrode using Fc/Fc^+ couple. Band-gaps of CdS [109], CdSe [110,113], CdTe [111,112], and Qdots are determined using CV. Figure 9 shows calculated ionization potentials of different sized CdSe Qdots.

Luminescence Properties

After excited by an external energy, e.g., photon for photoluminescence, electric field for electroluminescence, primary electron for cathodoluminescence *etc.*, electron and hole possess high energies due to transitions of electron from ground state to an excited state. The energies associated with such optical absorptions are directly determined by the electronic structure of the material. The excited electron and hole may form an exciton, as discussed above. The electron may recombine with the hole and relax to a lower energy state, ultimately reaching the ground state. The excess energy resulting from recombination and relaxation may be either radiative (emits photon) or nonradiative (emits phonons or *Auger* electrons). Some radiative events from band-edge, defects and nonradiative processes are discussed in brief.

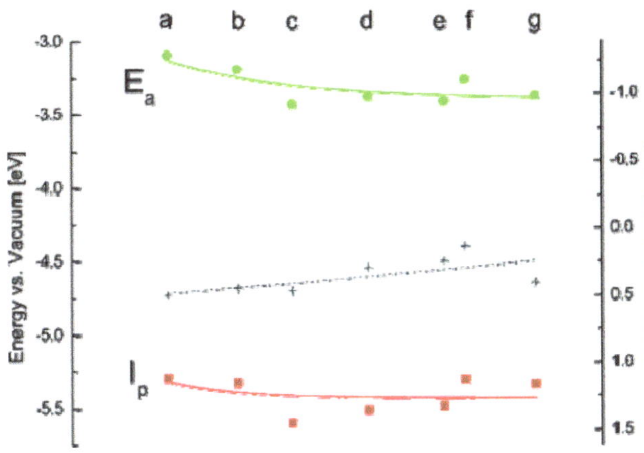

Figure 9. Experimental values of ionization potential (I_p) and electron affinity (E_a) of different sized TOPO coated CdSe Qdots (a: 2.3 nm, b: 3.1 nm, c: 3.2 nm, d: 3.5 nm, e: 3.7 nm, f: 3.8 nm, g: 4.0 nm) (Reprinted with permission from [62]. Copyright 2005 American Chemical Society).

Radiative Relaxation

Radiative relaxation results in spontaneous luminescence from Qdots. Such luminescence may result from band-edge or near band-edge transitions or from defect and/or activator quantum states. We discuss such emissions in the following sections.

Band-Edge Emission

The most common radiative relaxation processes in intrinsic semiconductors and insulators are band-edge and near band-edge (exciton) emission. The recombination of an excited electron in the conduction band with a hole in the valence band is called band-edge emission. As noted above, an electron and hole may be bound by a few meV to form an exciton. Therefore, radiative recombination of an exciton leads to near band-edge emission at energies slightly lower than the band-gap. The lowest energy states in Qdots are referred as $1s_e$-$1s_h$ (also called exciton state). The full width at half maximum (FWHM) of a room-temperature band-edge emission peak from Qdots varies from 15 to 30 nm depending on the average size of particles. For ZnSe Qdots, however, the luminescence can be tuned by size over the spectral range 390–440 nm with FWHM as narrow as 12.7–16.9 nm [114,115]. The optical absorption spectrum reflects the band structure of the materials. While PL from bulk semiconductors is fairly simple and well-understood, and can be explained by parabolic band theory, the PL from Qdots raises several questions. For example, radiative lifetime of 3.2 nm sized CdSe Qdots can be 1 µs at 10K compared to bulk (~1 ns) [116,117]. This was explained by the fact that there were surface states that involved in emission [116]. Band structures of semiconductors are often determined from either absorption spectra or PLE spectra. The study [116] also showed that these two spectra exhibited different characteristics when these spectra were acquired at 15 K. The PLE spectrum was associated with couple of additional peaks along with $1s_e$-$1s_h$. *Bawendi et al.* assigned these peaks as formally forbidden $1s_e$-$1p_h$ and $1s_e$-$2s_h$. It was also observed experimentally that the *Stokes* shift was size dependent. For a large size CdSe Qdots (5.6 nm), the *Stokes* shift was found to be 2 meV whereas for a same Qdots of size 1.7nm, the value could be 20 meV. Such a discrepancy was explained, theoretically and experimentally, in terms of increase of distance between optically active state and optically forbidden ground exciton state with decreasing the size of Qdots [117,118].

As mentioned in previous sections, Qdots have a number of advantages over organic dyes in bio-applications, e.g., better photostability, wide absorption edges, and narrow, tunable emission. However, they may exhibit a random, intermittent luminescence which is called ‘blinking’. In blinking, a Qdot emits lights for a time followed by a dark period (shown in Figure 10). In 1996,*Nirmal et al.* [77] observed this switching between an emitting and a non-emitting state from a single CdSe Qdot at room temperature. The postulated mechanism of blinking was a photoinduced ionization process [119] which leads to a charged Qdots that results in a separation between electrons and holes. Based on this model, the Qdots would be dark for the lifetime of the ionized state. Nonradiative *Auger* recombination process would be expected to dominate the quenching of ionized Qdots [119]. However, the experimental results don’t completely support this model. For example, a photo-induced *Auger* process should exhibit a quadratic dependence of the average blinking time on excitation intensity, whereas the experimental result showed a linear behavior. In addition, the bright and dark periods followed an inverse power law [120] given by Equation 4, where, P(t) is a probability of the blinking period, m is an exponent between 1 and 2 and A is a constant.

$$P(t)=A.t^{-m} \quad (4)$$

Several additional mechanisms have been proposed to explain the blinking [121,122,123,124], including thermally activated ionization, electron tunneling through fluctuating barriers or into a uniform distribution of traps, or resonant electron tunneling between the excited states of Qdots and dark-trap states that wander randomly in energy. Despite tremendous efforts, the blinking effect is still not properly explained.

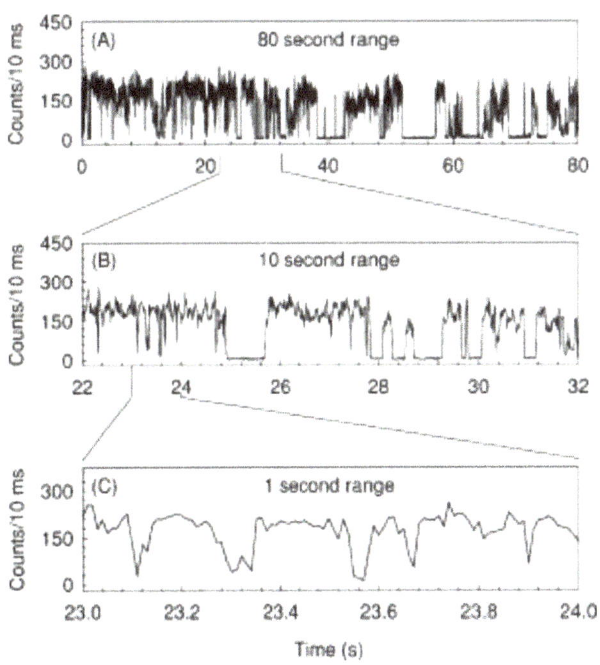

Figure 10. Blinking effect during luminescence from a single 2.9 nm sized CdSe Qdot (reprinted with permission from [121]. Copyright 2000, American Institute of Physics).

Defect Emission

Radiative emission from Qdots also comes from localized impurity and/or activator quantum states in the band-gap. Defect states lie inside the bands themselves [124]. Depending on the type of defect or impurity, the state can act as a donor (has excess electrons) or an acceptor (has a deficit of electrons). Electrons or holes are attracted to these sites of deficient or excess local charge due to*Coulombic* attraction. Similar to the case of excitons, trapped charge on defect/impurity sites can be modeled as a hydrogenic system, where binding energy is reduced by the dielectric constant of the material [125]. These defects states can be categorized into either shallow or deep levels, where shallow level defect states have energies near the conduction band or valence band-edge. In most cases, shallow defect exhibits radiative relaxation at temperatures sufficiently low so that thermal energies (kT) do not excite the carriers out of the defects or traps states. Deep levels, on the other hand, are so long-lived that they typically experience nonradiative recombination.

Luminescence from these defect levels can be used to identify their energy and their concentration is proportional to the intensity. Both PL spectral distribution and intensity change with changes of the excitation energy due to contributions from different defect energy levels and the band structure of the host. The excitation energy also determines the initial photoexcited states in the sample, but this state is short-lived because of thermalization of the photoexcited carriers *via*phonon emission, as discussed above. Relaxation to within kT of the lowest vibrational level of the excited states is usually orders of magnitude faster than the recombination event [125].

Defect states are expected at the surface of a Qdot despite the use of various passivation methods, because of the large surface-to-volume ratio, discussed above. The concentration of surface states on the Qdots is a function of the synthesis and passivation processes. These surface states act as traps for charge carriers and excitons, which generally degrade the optical and electrical properties by increasing the rate of nonradiative recombination. However, in some cases, the surface states can also lead to radiative transitions, such as in the case of ZnO nanostructures (Figure 11). Powders of ZnO have a green emission from defects along with a band-edge near UV emission (the band-gap of ZnO is 3.37 eV or 386 nm) at room temperature [126,127,128,129,130]. It is also reported that the green emission suppressed the band-edge emission. Theoretical and experimental studies [131,132] showed that the defect states in a ZnO Qdot can be of several types including neutral, singly or doubly charged Zn vacancies (V_{Zn}), neutral or singly charged oxygen vacancies (V_O), singly charged or neutral interstitial Zn (Zn_i), interstitial O (O_i), a complex of $V_O$and Zn_i (V_OZn_i), a complex of V_{Zn} and Zn_i ($V_{Zn}Zn_i$), and substitution O at Zn position (O_{Zn}). According to *Aleksandra et al.* [132], the singly charged oxygen vacancy (V_{O+}) is located at 2.28 eV below the conduction band in the ZnO band-gap and results in an emission at ~ 540 nm. The most widely, but not universally, accepted mechanism for green luminescence from ZnO is the electron-hole recombination on singly ionized oxygen vacancies. In solution-based synthesis, the oxygen vacancies appear to be intrinsic and may result from heterogeneous nucleation and growth, enhanced by the large surface area. If the radiative center is associated in part with the surface, their concentration would be expected to decrease with aggregation of Qdots as observed [133].

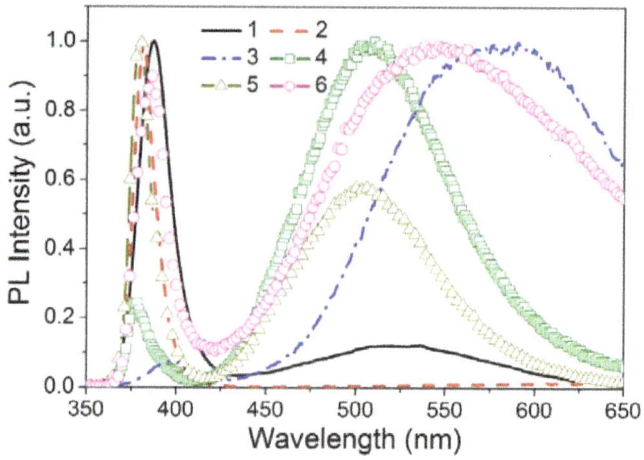

Figure 11. Room temperature PL spectra of various ZnO nanostructures: (1) tetrapods, (2) needles, (3) nanorods, (4) shells, (5) highly faceted rods, and (6) ribbons/combs, [reproduced with permission from [128]. Copyright Wiley-VCH Verlag GmbH & Co. KGaA].

Activator Emission

Luminescence from intentionally incorporated impurities is called extrinsic luminescence. The predominant radiative mechanism in extrinsic luminescence is electron-hole recombination which can occur *via* transitions from conduction band to acceptor state, donor state to valance band or donor state to acceptor state. In some cases, this mechanism is localized on the activator atom center. In some many cases, the selection rule is relaxed due to mixing of orbitals, such as *d-p*mixing in a crystal or ligand field where the orbitals are split into hyperfine structures. Therefore,*d-d* transition is allowed in some cases for transition elements. For Mn^{2+}, the lifetime of the luminescence [76] is in the order of millisecond due to the forbidden *d-d* transition. Similarly, *f-f*transition are also often observed for rare earth elements (e.g., Tm^{3+}, Er^{3+}, Tb^{3+},and Eu^{3+}), although the *f* levels are largely unaffected by the crystal field of the host due to shielding by the outer *s*- and*p*-orbitals [134]. Due this shielding effect, *f-f* transitions typically have atomic-like sharp peaks in the emission spectra.

The optical properties of doped ZnO Qdots have also been widely investigated [135,136,137]. Doping with Er or Mn has been reported to result in preferential orientation of nanorods perpendicular to the substrate [138]. ZnO Qdots have also been doped with rare-earth elements, such as Tb [139], Ce [140], Eu [141] and Dy [142]. In the case of Tb-doped ZnO Qdots, emissions from both Tb and defect states were observed. The emission from Tb was found to increase with increasing Tb concentrations, while that from defect states decreased. Eu-related emission was observed from ZnO:Eu nanorods for a suitable excitation wavelength. However, Dy-doped ZnO nanowires exhibited a relatively strong UV emission with a very weak emission from Dy. The effects of doping Mn in ZnO nanoparticles depend strongly on the synthesis conditions [143]. The Mn was found to quench green emission [143], while others reported either a reduction in both UV and defect emissions [144] or a blue shift and increase in UV peak intensity [145]. Very similar spectra from ZnO and Mn-doped ZnO were observed after annealing at 800 °C [146]. Other dopants, such as sulfur and copper, have been studied in ZnO Qdots. Increased intensity and changes in spectral distribution of the broad green defect emission with S doping has been reported [147,148].

Doped ZnS Qdots [149,150] are very important semiconductor nanomaterials, with Mn^{2+}-doped ZnS Qdots being one of those most studied as a phosphor [151]. In 1994, *Bhagrava et al.* [152] reported very high PL-QY (~18%) from ZnS:Mn Qdots. Coincident with the intensity enhancement, they reported shorter luminescent lifetimes for the Mn^{2+} emission (decrease from hundreds of microseconds for the bulk to nanoseconds in nanocrystals) [153,154]. The increased intensity was attributed to an efficient energy transfer from the ZnS host to Mn^{2+} ions facilitated by mixed electronic states. Hybridization of atomic orbitals of ZnS and *d*-orbitals of Mn^{2+} in the nanoparticles was suggested to also be responsible for the relaxation of selection rules for the spin-forbidden $^4T_1$ $^6A_1$ transition of Mn^{2+}, leading to the short emission lifetimes. Subsequent research demonstrated that while the QY of passivated ZnS:Mn Qdots could be high, the luminescent lifetimes were not significantly smaller from those of the bulk material. The luminescence properties were, however, found to be dependent upon the S^{2-} and Mn^{2+} concentrations, as well as, the structural properties of the Qdots. The $^4T_1$ $^6A_1$ Mn^{2+} emission intensity generally increases with increasing doping Mn^{2+} concentration [155] and a quenching of Mn^{2+} emission was observed at high Mn^{2+} concentrations (>0.12 at %). The local environment around the Mn^{2+} in the Qdots has been studied using X-ray absorption fine structure (XAFS) and electron spin resonance (ESR). XAFS data showed that the Mn^{2+} substituted on the tetrahedral Zn^{2+} site in the lattice. ESR data were consistent with this conclusion, showing a spectrum for Mn^{2+} spins typical of a tetrahedral crystal field [156]. In some cases for ZnS:Mn Qdots, the ESR spectra show a Mn^{2+} signal with octahedral symmetry (See Figure 12), but the location of this defect site is not fully understood. It has also been suggested [156] that this signal resulted from Mn^{2+} on the surface of the Qdots *versus* in the interior, but this assignment has been disputed and attributed to Mn-Mn clustering at high concentrations.

Quantum Yield of Quantum Dots

The accurate quantum yield or QY measurements are crucial for Qdots. It is commonly noted that the values of the QYs do not agree among several reports. This may be due to one of the following reasons or a combination thereof: (1) different approaches to measuring QYs, (2) inappropriate concentrations or optical density (OD) of sample or standards, where *Beer's* law does not follow or concentration quenching occurs, (3) change of slits between samples and standards during measurements; (4) use of different excitation wavelength for PLE or first absorption peak, (5) no overlap between emission wavelengths of samples and standards, and/or (6) instrumental error/s, such as, wavelength shift, instability of source light *etc.* It is also known that inorganic semiconducting Qdots do not behave like fluorescent molecules, such as dyes. In previous, we have used the equation provided by IUPAC [157] for determining QY of Qdots [69,94,97,102,158,159]. The procedure to determine the QY is by comparing the integrated emission intensity from the Qdots to that from standards. The optical densities of the Qdots and the standard/s are determined. The absorbance value for both samples and standards should be kept below 0.08 at the excitation wavelength. In previous, we used the following equation to measure QY [102,159].

$$QY = QY_{St} \frac{1-10^{-A_{St}}}{1-10^{-A}} \times \frac{\eta^2}{\eta_{St}^2} \times \frac{I}{I_{St}} \quad (5)$$

In Equation 5, QY and QY_{St} are quantum yields (St: Standards), A and A_{St} are absorbance values at the excitation wavelength, η and η_{St} are refractive indices of the solvents, and I and I_{St} are integrated emission areas for the Qdots samples and the standards, respectively. Same excitation wavelength should be used for both the sample and standards.

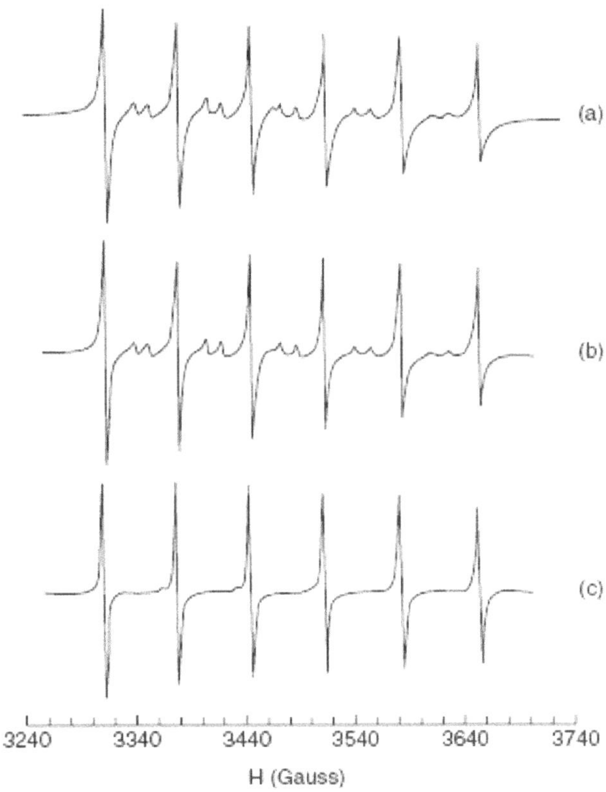

Figure 12. EPR spectra for Mn^{2+}in ZnS:Mn sample measured at room temperature; (a), 0.003% Mn^{2+}(experimental), (b) 0.008% Mn^{2+}(experimental), and (c) 0.008% Mn^{2+} (simulated) (reprinted with Permission from [156]. Copyright 2004 American Chemical Society).

Reported Quantum Yield

In this section, we review some of the QY values that are reported in literature. In the following table (Table 2), we summarized some of the research reports from literatures.

Table 2. Some of the literature reported Qdot Quantum Yield data for selected systems.

Quantum Dot	Size (nm)	Emission	Quantum Yield	Standard (QY)	Specification	Ref
CdSe/ZnS	2.7–3	Excitation 470 nm Emission range: 480–850 nm	50%	Rhodamine 560 in ethanol	Shell thickness: ~0.6nm	[103]
CdSe	4.2		20%		Bare	[104]
CdSe/ZnS	4.2		50%		1.5 monolayer of ZnS	[104]
CdSe/CdS	2.3 3.0		59% 84%		2.1 MLr CdS 1.8 MLr CdS	[70]
ZnSe	4.3–6	360–420 nm	20–50%			[160]
CdSe/ZnS	2.0 2.6 4.6 5.6	O.D. 0.1	36% 49% 30% 27%	Rhodamine 590, 610, 640		[161]
CdSe/ZnS	3.7		66%		1.6 MLr ZnS	[162]
ZnSe:Mn	2.7–6		22% at RT 75% at 50K			[163]
CdSe	7.5	O.D. 0.1	85%	Coumarin 540 (62% @ 458nm), Rhodamin 6G (95% @ 528nm), 3B (50% @ 550nm), 640 (100% @ 570 nm), LD 690 (63% @616 nm)	As synthesized	[164]
CdSe	7.5	O.D. 0.1	85%	Coumarin 540 (62% @ 458nm), Rhodamin 6G (95% @ 528nm), 3B (50% @ 550nm), 640 (100% @ 570 nm), LD 690 (63% @616 nm)	As synthesized	[164]
PbSe	4–5		6–20%			[165]
CdSe	2–8	excitation: 400 nm; OD: ≤ 0.1)	50–80%	Rhodamin B in ethanol (90% @ 400 nm)		[166]
CdSe CdSe/CdS CdSe/CdS/ZnCdS CdSe/CdS/ZnCdS/ZnS	3.8 5.2 7.6 8.9		30% 60% 65% 80%		Bare CdS: 2ML ZnCdS: 2ML ZnS: 2.5 ML	[89]
CdSe/ZnS ZnSe/CdSe/ZnS			90% (550–650 nm) 70% (510–560 nm)			[167]
CdSe CdSe/CdS CdSe/CdS/ZnS	4.0 5.5 6.8	Same OD with standard	16% 38% 75%	Rhodamine 6G (95%)	Bare CdS: 2ML ZnS: 2ML	[168]
CdSe CdSe/SiO_2		OD: <0.06 and >0.01 Excitation: 350 nm (B) 450 nm (G) 500 (R)	22% (bare 523 & 581 nm) 82% (542 nm)	9,10-diphenyl-anthracene in cyclohexane (90% @ 350nm); Fluorescein in 0.1 M NaOH (95% @450 nm); Rhodamin 6G in methanol(95% @ 500 nm)	Shell thickness: ~6nm	[72]

OD: optical density; ML: monolayer; 1 CdS ML [168]: 0.35 nm; 1 ZnS ML [168]: 0.31 nm

First, we start our discussion with QY data from CdSe Qdots. The QY of CdSe/ZnS core/shell Qdots was measured using rhodamine (RD) 560 in ethanol at excitation 560 nm (emission 480–850 nm) [103]. Integrated emission intensities of RD 590 or RD 640 and Qdots with same OD at excitation wavelength were used to determine QYs of Qdots [104]. QY of CdSe/CdS Qdots was found to be as high as 84% [70]. *Peng et al.* determined QY by comparing integrated intensities of Qdot PL to two standards: RD 6G and RD 640 in methanol at optical densities of all solutions less than 0.3 at the excitation wavelengths [70]. For Blue emitting ZnSe, *Hines et al.* [160] used stilbene 420 in methanol found as high as 50% QY. *Norris et al.* [163], on the other hand, found ~22% QY in room temperature from ZnSe:Mn Qdots. *Peng et al.* [164] described QY measurement process where the optical density at the excitation wavelength was kept same value with Qdots. They also kept ODs at the first exciton absorption peak of Qdots and peak absorption peaks of dyes at below 0.1.Figure 13 is adapted from *Dr. X. Peng*'s highly cited JACS paper [164], where it has been shown that the composition of Qdots and stoichiometric ratio of precursors are important parameters to improve the QY values of Qdots. *Meijerink's* group [166] from the *Netherlands* reported the same.*Dijken et al.* studied the QY of ZnO Qdots [169]. QYs of the visible emission from ZnO nanoparticles prepared by colloidal solution were found to decrease from 20% to 12% as the particle radius increased from 0.7 to 1 nm [169].

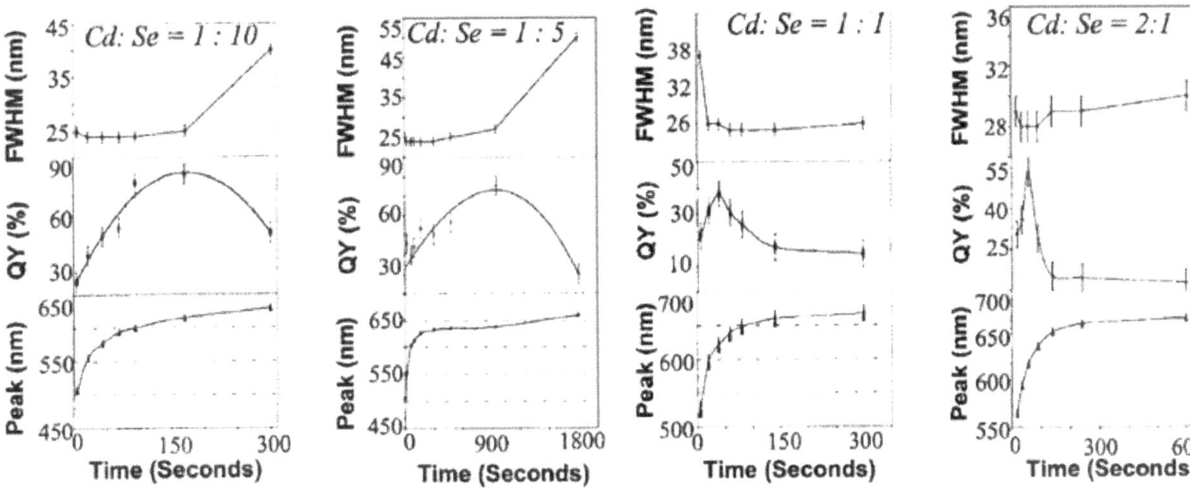

Figure 13. Change of quantum yield with reaction time and growth of particles, and stoichiometric ratio of anion and cation precursors (reprinted with permission from [164]. Copyright 2002 American Chemical Society).

The PL-QYs of ZnSe Qdots can be increased by coating with a wider band-gap semiconductor, such as ZnS which has an ~5% lattice mismatch with ZnSe [170]. After passivation with 1.8 monolayers of ZnS shell, the QYs of ZnSe/ZnS core/shell Qdots increased 450% to a value of ~32%. We recently reported QYs of ~5% and 13% for 3–4 nm ZnO Qdots [102] and ZnO/MgO core/shell structure [159], and the QY decreased with aging time in air due to a reduced concentration of radiative traps [159]. *Bol* and *Meijerink* [171] compared the QY for Qdots coated with poly(vinybutryral) (PVB), poly(vinylalcohol) (PVA), methacrylic acid (MA), and sodium polyphosphate (PP), and the QYs of 0.3% to 1% for unpassivated Qdots was increased to ~4% for ZnS:Mn capped with PP. The Mn^{2+} emission at 580nm from ZnS:Mn/ZnS core/shell Qdots was found to be seven times more intense than from unpassivated ZnS:Mn. The enhanced intensity was believed to result from suppression of nonradiative transitions by the undoped ZnS-shell. Qdots of ZnS:Mn coated by a SiO_2 shell showed an enhanced PL intensity as compared to bare Qdots [172]. We recently reported [72] the QY of bare CdSe Qdots to be between 9–21% whereas for silica coated Qdot exhibited QY more than 80%.

Change of Quantum Yield under Ultraviolet Irradiation

A significant increase in the luminescence QY of ZnS Qdots was observed due to UV irradiation.*Becker* and *Bard* attributed this phenomenon to irradiation-induced oxygen absorption that blocked nonradiative recombination at surface states [173]. *Henglein et al.* proposed [174] that photoanodic dissolution of the ZnS Qdots was induced by irradiation in the presence of oxygen, which led to the improved efficiencies. *Dunstan et al.* explained the increased efficiencies in terms of a photocorrosion process that created new recombination centers [175]. In our group, *Yang, et al.* [97] used XPS data to show that 400 nm irradiation in air converted ZnS shells to $ZnSO_4$ which increased the QY of CdS:Mn/ZnS core/shell structures. UV irradiation in argon did not result in the formation of $ZnSO_4$ nor did it change the QY. UV irradiation of organically passivated ZnS Qdots can either increase or decrease the QY. *Bhargava et al.* [37] observed an increase upon UV irradiation and speculated that increased cross-linking and polymerization of the passivating organic molecules was the mechanism. Recently, *Bol* and *Meijrink* [176] postulated that enhanced emission from organically passivated surfaces resulted from either UV curing of samples coated with PVB, PVA, or MA, or from a photochemical reaction at Qdot surfaces coated with PP or on unpassivated samples. After UV curing, QYs of ~10% were obtained. However, prolonged UV irradiation (hours to days) in the presence of water and oxygen led to a decreased QY. As in the case of *Yang, et al.*[97], photochemical reactions could produce $ZnSO_4$ or $Zn(OH)_2$, which presumably served as a passivating layer around Qdots (Figure 14), but too thick a layer can lead to lower QYs. In the case of ZnS:Mn/ZnS core/shell structures, luminescence intensity was not changed significantly as a result of UV irradiation [177]. For ZnS:Mn/SiO_2 core/shell nanoparticles, UV irradiation increased the PL intensity [172]. The luminescent intensity from ZnS:Mn^{2+} colloid solutions decreased after the colloids were kept in room temperature air [178]. This was presumably due to the deterioration of surface structure, which led to an increase of nonradiative relaxation paths. *Eychmüller et al.*reported a PL peak at ~390 nm from thiol passivated, water soluble, sulfur alloyed ZnSeS colloidal Qdots (~ 2–3nm in size) [179]. UV irradiation after synthesis improved their QY to 25–30% for the band-gap UV emission. Irradiation resulted in incorporation of sulfur into the ZnSeS Qdots.

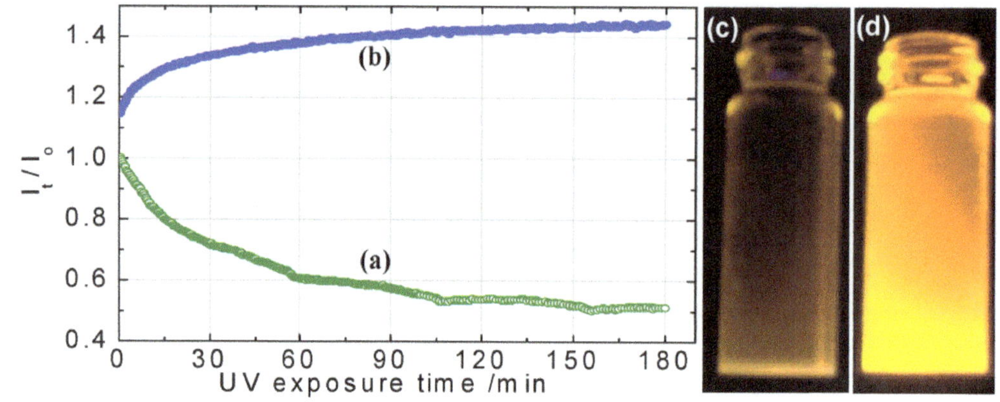

Figure 14. Variation of PL intensity in room temperature air from (a) n-dodecanethiol and (b) ZnS-passivated CdS:Mn Qdots *vs.* time of exposure to 400 nm UV light. The monitored wavelengths are 580 nm and 585 nm for n-dodecanthiol and ZnS passivated CdS:Mn Qdots, respectively. Comparison of relative brightness from (c) n-dodecanthiol passivation and (d) ZnS-passivated CdS:Mn Qdots under 366 nm UV-illumination [180] (reproduced by permission of The Electrochemical Society).

Non-radiative Process in Quantum Dots

Absorption of energy by a luminescent material may not result in emission of light. Electrons and holes in excited states may return to lower energy and ground states by radiative and/or nonradiative relaxations. Deep level traps have a tendency to undergo nonradiative recombination by emitting phonons. Experimental data show that the time required for nonradiative recombination is short (e.g., tens of picoseconds [119]). Non-radiative relaxation may be categorized as internal conversion, external conversion or *Auger* recombination [181]. Nonradiative recombination though crystalline and/or molecular vibrations is a common phenomenon in internal conversion. The difference between the energy absorbed by Qdots, hv, and the band-gap, E_g, is generally converted into heat by electron-phonon scattering processes. Even for a radiative process in an indirect semiconductor, a phonon is generated due to the change of k-value. Furthermore, strain in a lattice can create a local potential well that can also trap electrons and holes and result in a nonradiative transition. Nonradiative recombination also occurs at surface states. As reported above, 15–30% of atoms in a Qdot are at the surface and represent defects due to unsaturated dangling bonds. These defects are dominant channels for nonradiative decay of carriers. The electronic surface states are filled below the *Fermi* level with electrons from the core of the Qdots. Accumulation of charge at the surface creates an electric field or a depletion region that leads to bending of the valence and conduction band-edges. Electron and hole carriers generated in this region are swept in opposite direction by the electric field, prohibiting radiative recombination. This leads to the concept of a 'dead layer' [125]. Capping of these defects with organic ligands or inorganic shells lead to an improvement in luminescent efficiency as discussed below and insection 2.2.1.

Strong carrier-to-carrier interaction can lead to an *Auger* nonradiative process (rather than releasing the energy of recombination as a photon or phonon, the excess energy is transferred to another electron). The *Auger* electron loses its surplus energy by creation of phonons. The *Auger* recombination process involves two electrons and a hole in the conduction and valence bands, respectively. (Sometimes *Auger* process involves two holes and one electron). *Auger* recombination can also create a hole deep in the valence band or can be observed for electrons and holes on localized activator levels [181]. In an *Auger* transition, the momentum and energy must be conserved. Therefore, indirect semiconductors show much higher *Auger* recombination rates as compared to direct band-gap materials. This is due to the fact that a momentum change is necessary for *Auger* recombination, and is also required for the transition in indirect band-gap semiconductors. *Auger* process differs greatly between nano-systems and bulk systems of same composition as the efficiency of *Auger* process depends on *Coulomb* electron-electron interaction. In atomic or nano-systems, electron-electron coupling is stronger than the electron-photon coupling. Therefore, the rate of *Auger* transition is higher compared to radiative transition [182].

SYNTHESIS PROCESSES

Several routes have been used to synthesize Qdots. Generally, the techniques for synthesis of Qdots are categorized either as a top-down or bottom-up approach. Below, we discuss both approaches in brief.

Top-Down Synthesis Processes

In the top-down approaches, a bulk semiconductor is thinned to form the Qdots. Electron beam lithography, reactive-ion etching and/or wet chemical etching are commonly used to achieve Qdots of diameter ~30 nm. Controlled shapes and sizes with the desired packing geometries are achievable for systematic experiments on quantum confinement effect. Alternatively, focused ion or laser beams have also been used to fabricate arrays of zero-dimension dots. Major drawbacks with these processes include incorporation of impurities into the Qdots and structural imperfections by patterning. Etching, known for more than 20 years, plays a very important role in these nanofabrication processes. In dry etching, a reactive gas species is inserted into an etching chamber and a radio frequency voltage is applied to create a plasma which breaks down the gas molecules to more reactive fragments. These high kinetic energy species strike the surface and form a volatile reaction product to etch a patterned sample. When the energetic species are ions, this etching process is called reactive ion etching (RIE). With a masking pattern, selective etching of the substrate is achieved. Fabrication of GaAs/AlGaAs quantum structures as small as 40 nm has been reported using RIE with a mixture of boron trichloride and argon [183]. This RIE process has been used to produce close-packed arrays for testing of lasing in Qdot semiconductors. Close packed arrays of ZnTe Qdots with interdot distance of 180 nm to 360 nm were produced by RIE using $CH_4$and H_2 [184].

Focused ion beam (FIB) techniques also offer the possibility of fabricating Qdots with extremely high lateral precision. Highly focused beams from a molten metal source (e.g., Ga, Au/Si, Au/Si/Be, or Pd/As/B) may be used directly to sputter the surface of the semiconductor substrate. The shape, size and inter-particle distance of the Qdots depend on the size of the ion beam but a minimum beam diameter of 8–20 nm has been reported for both lab and commercial systems, allowing etching of Qdots to dimensions of <100 nm [185]. The FIB technique can also be used to selectively deposit material from a precursor gas with a resolution of ~100 nm. Scanning ion beam images (analogous to scanning electron microscope images) can be developed by ion beam nanofabrication at the desired, predetermined locations with high resolution [185]. However this is a slow, low throughput process employing expensive equipment that leaves residual surface damage. Another method to achieve patterns with Qdots dimensions is the use of electron beam lithography followed by etching or lift-off processes. This approach offers a high degree of flexibility in the design of nanostructured systems. Any shape of Qdots, wires, or rings with precise separation and periodicity may be realized with this technique. This method was successfully employed for the synthesis of III-V and II-VI Qdots with particle sizes as small as 30 nm.

Bottom-up Approach

A number of different self-assembly techniques have been used to synthesize the Qdots, and they may be broadly subdivided into wet-chemical and vapor-phase methods. Microemulsion, sol-gel, competitive reaction chemistry, hot-solution decomposition, and electrochemistry are generally placed in the category of (1) wet-chemical methods. Self-assembly of nanostructures in material grown by molecular beam epitaxy (MBE), sputtering, liquid metal ion sources, or aggregation of gaseous monomers are generally categorized under (2) vapor-phase methods.

Wet-Chemical Methods

Wet-chemical methods mainly follow the conventional precipitation methods with careful control of parameters for a single solution or mixture of solutions. The precipitation process invariably involves both nucleation and limited growth of nanoparticles. Nucleation may be categorized as homogeneous, heterogeneous or secondary nucleation [186]. Homogeneous nucleation occurs when solute atoms or molecules combine and reach a critical size without the assistance of a pre-existing solid interface. By varying factors, such as temperature, electrostatic double layer thickness, stabilizers or micelle formation, concentrations of precursors, ratios of anionic to cationic species and solvent, Qdots of the desired size, shape and composition can be achieved. Some of the common synthesis processes are briefly discussed below.

Sol-Gel Process

Sol-gel techniques have been used for many years to synthesize nanoparticles including Qdots [61,102,133,159]. In a typical technique, a sol (nanoparticles dispersed in a solvent by *Brownian*motion) is prepared using a metal precursor (generally alkoxides, acetates or nitrates) in an acidic or basic medium. The three main steps in this process are hydrolysis, condensation (sol formation) and growth (gel formation). In brief, the metal precursor hydrolyzes in the medium and condenses to form a sol, followed by polymerization to form a network (gel). This method has been used to synthesize II-VI & IV-VI Qdots, such as CdS [12], ZnO [61,102,133,159], PbS [187]. As an example, ZnO Qdots have been prepared by mixing solutions of Zn-acetate in alcohol and sodium hydroxide, followed by control aging in air [61]. The process is simple, cost-effective and suitable for scale-up. The main disadvantages of the sol-gel process include a broad size distribution and a high concentration of defects [187]. Therefore, this synthesis technique is used sparingly.

Microemulsion Process

Microemulsion processes are popular methods for synthesizing Qdots at room temperature. The processes can be categorized as either normal microemulsions, *i.e.,* oil-in-water, or as reverse microemulsions, *i.e.,* water-in-oil. In some cases, other polar solvents, e.g., alcohol, may be used instead of water. The reverse micelle process is popular for synthesizing Qdots, where two immiscible liquids (polar water and nonpolar long-chain alkane) are mixed and stirred to form emulsion. Nanometer water droplets dispersed in *n*-alkane solutions can be achieved using surfactants, like aerosol OT (AOT), cetyl trimethyl-ammonium bromide (CTAB), sodium dodecyl sulphate (SDS) or triton-X. Since the surfactants are terminated by hydrophilic and hydrophobic groups on opposite ends, numerous tiny droplets called micelles are formed in the continuous oil medium. These micelles are thermodynamically stable and can act as 'nanoreactors'. Mixing of vigorous stirred micellar solutions leads to a continuous exchange of reactants due to dynamic collisions. Growth of the resultant Qdots is limited by the size of the micelle which is controlled by the molar ratio of water and surfactant (W). The relation between W and the radius (r) of the micelle has been reported as [188]

$$\left(\frac{r+15}{r}\right)^3 - 1 = \frac{27.5}{W} \quad (6)$$

The reverse micelle technique has been used to prepare II-VI core and core/shell Qdots, such as CdS [63], CdS:Mn/ZnS [76,97,189,190], ZnS/CdSe [13], CdSe/ZnSe [188], ZnSe [115] and IV-VI Qdots [108]. Some advantages of this process are easy control of the Qdot size by changing the molar ratio of water to surfactant, a narrow distribution of size as compared to the sol gel process, and ease of dispersion of the Qdots. Some disadvantages include low yield and incorporation of impurities and defects.

Hot-Solution Decomposition Process

High temperature (~300 °C) pyrolysis of organometallic compound is a well-established route for the production of Qdots and was first discussed in detail in 1993 by *Bawendi* and co-workers [14]. Precursors, such as alkyl [14], acetate [191], carbonate [191] and oxides [164,191] of Group II elements, are mixed with Group VI phosphene or bis(trimethyl-silyl) precursors. A typical procedure involves first degassing and drying of trioctyl-phosphine oxide (TOPO, a coordinating solvent) at 200–350 °C under vacuum (1 Torr or 7.5 × 10^{-6} Pa) in a three-neck round flask in a dry box. A mixture of Cd-precursor and tri-*n*-octyl-phosphine (TOP) selenide is prepared in a dry box and injected with vigorous stirring into the flask at a temperature of ~300 °C. The simultaneous injection of precursors into the flask along with TOPO results in homogeneous nucleation to form Qdots, with the subsequent growth of Qdots through '*Ostwald ripening*' being relatively slow. In*Ostwald ripening*, the higher free energy of smaller Qdots makes them lose mass to large size Qdots, eventually disappearing. The net result is a slow increase of the size of Qdots at the reaction temperature of ~230–250 °C (depending on precursor, coordinating agents and solvents). The coordinating TOPO solvent stabilizes the Qdot dispersion, improves the passivation of the surface, and provides an adsorption barrier to slow the growth of the Qdots. The final size of the Qdots is mainly controlled by the reaction time and temperature. Aliquots may be removed from the flask at regular intervals during the first few hours and the optical absorption edge used to achieve a desired particle size. This method has been extensively used to synthesize II-VI [56,111,160,192,193,194,195,196], IV-VI [197] and III-V Qdots [198]. The size, shape, and control of the overall reaction depends not only on process parameters and precursors, solvents and coordinating agents, but also on the purity of the coordinating solvent, such as TOPO. It has been reported that technical grade TOPO (90% pure) were better for synthesizing uniform Qdots than the pure TOPO [162,191].

An advantage of this synthesis route is that it provides sufficient thermal energy to anneal defects and results in monodispersed Qdots (typically standard deviation about the average size of 5%). Since growth of the particles in this process is relatively slow and can be controlled by modulating the temperature, a series of Qdot sizes can be prepared from the same precursor bath. Using this process large quantities of Qdots [199] and alloying process [52] have been demonstrated. Some of the disadvantages of this method include higher costs due to the use of high temperature, toxicity of some of the organometallic precursors, and generally poor dispersions in water. Table 3 shows a chronological summary of this synthesis technique with different growth parameters and precursors to produce different Qdots.

Table 3. Synthesis of different sized Qdots using hot-solution decomposition reaction.

Year	Qdots	Precursor	Process parameters	Particle Size (nm)	Ref
1990	GaAs	$GaCl_3$, $(TMS)_3As$ in Quinoline	240 °C for 3days; flame anneal at 450 °C	2.4	[200]
1990	ZnS, ZnSe, CdS, CdSe, CdTe, HgTe	$M(ER)_2$; R: n-butyl phenyl; E: S, Se, Te; M: Cd, Zn, Hg and/or phosphine complexes; Co.Sol.: DEPE	DEPE and $M(ER)_2$ reacted, (Temp. range: 250–300 °C)	2.5–5 nm	[201]
1993 -1998	CdS, CdSe, CdTe	Me_2Cd, silylchalconides, Phosphine chalconides; Co.sol: TOPO & TOP/TBP	300–350 °C at 1 atm at Ar (TOPO degassing); 230–260 °C (growth temp.)	1.2–11.5	[14,101,202]
1994	GaAs	$GaCl_3$/ GaI_3, diglyme, As, toluene, Na-K alloy	As, Na-K alloy mixture refluxed to 100 °C in Ar for 2 days; $GaCl_3$/GaI_3 diglyme mixture added, heated from 0 °C to RT to 111 °C. for 2 days	6–10	[203]
1995	InP, GaP, $GaInP_2$	Mixture of chloro-indium/gallium oxalate ($GaCl_3$ for GaP) and $(TMS)_3P$ in CH_3CN ; Co.sol: TOPO & TOP	270–360 °C at airless condition for 3 days; Qdots dispersed in methanol	2.6–4.6 (InP), 3 (GaP), 6.5 ($GaInP_2$)	[204]
1996	InP, InAs	$InCl_3$, TOPO, $(TMS)_3P$/$(TMS)_3As$	$InCl_3$ & TOPO heat at 100 °C for 12 h, $(TMS)_3P$ added, after 3hr heated to 265 °C for 6 days	2–6	[205,206]
1996	CdSe/ZnS	Me_2Cd,Me_2 Zn, Se, (TMS)2S, Co.sol: TOPO, TOP	Single step synthesis Core: 350 °C at 1 atm at Ar, growth: 310 °C Shell: 300 °C	2.7–4	[103]

1997	CdSe/ZnS	Me_2Cd, Me_2 Zn, Se, (TMS)2S, Co.sol: TOPO, TOP	Two step synthesis (airless) Core growth: 290–300 °C Shell growth: 140 °C for 2.3 nm & 220°C for 5.5 nm	2.3–5.5	[104]
1997	CdSe/CdS	Me_2Cd, Se, (TMS)2S, Co.sol: TOPO, TBP	Two step process: Core: 300 °C; Shell: 100 °C	2.5–4	[70]
1998	ZnSe	Me_2Zn, Se, HDA, TOP	HDA dried & degassed at 150 °C for hrs in vacuum and heated to 310 °C at 1 atm in Ar; Core growth with Zn & Se precursor at 270 °C.	4.3–6 nm	[160]
1996–1999	InAs/InP InAs/CdSe	$(TMS)_3As$, Indium (III) chloride, TOP (TMS)3P, Me_2Cd; TBP-Se	Two-step Process (airless) Core growth: 260 °C; Shell: dropwise addition; 260 °C	2.5–6 nm (InAs); 1.7 (core/shell)	Core [202,206] Core/shell [207]
2000	CdSe	Me_2Cd, Se, TBP, TOPO, HPA	TOPO (+HPA 1.5–3 wt%) degassed at 360 °C (or 310 °C, 280 °C); Core growth: 300 °C (or 280 °C or 250 °C)	~6nm	[208,209]
2001	ZnSe:Mn	Me_2Mn, Et_2Zn, TOP, Se, HDA	Dimethyl Mn, TOP, Se, Diethyl Zn mixture added to HDA at310 °C in N_2. Growth: 240–300 °C	2.7–6.3	[163]
2001-2003	CdSe/ZnS	Me_2Cd, Se, TOP, TOPO, HDA, (TMS)2S, Me_2Zn	Two Step: Core: reaction & growth: 270–310 °C; Shell: slow addition of Zn & S precursor at 180–220 °C	4.5–5 nm	[162,210]
2001	CdSe	Scheme 1: $Cd(Ac)_2$,, SA/ TOPO; 2: $Cd(Ac)_2$, SA; 3: $CdCO_3$, SA/TOPO; 4: $CdCO_3$, LA/TOPO; 5: CdO, SA/TOPO; 6: $Cd(Ac)_2$, tech TOPO; 7: CdO, TDPA/TOPO	Solvent & Cd-precursor heated to 250–360 °C at Ar; TOP-Se or TBP-Se injected; Growth temp: 200–320 °C (if DDA involve, temp: ~220 °C	2–25nm	[191]
2001	CdS, CdSe, CdTe	CdO, TOPO, HPA/TDPA, S, Se, Te & TOP	One pot: CdO, HPA/TDPA heated 300 °C; Core with chalconide precursor: reaction: 270 °C, & growth 250 °C	2–8 nm	[211]
2002	CdSe	CdO, Se, TOPO, TBP, HDA, ODA, SA	CdO & SA, heated to 150 °C in Ar; after CdO dissolution, cool to RT; TOPO & HAD added & heated to 320 °C in Ar; TBP-Se added, Growth 290 °C		[164]
2003	PbS	PbO, OA, $(TMS)_2S$, TOP	PbO dissolved in oleic acid at 150 °C in Ar; $(TMS)_2S$ & TOP injected	5nm	[197]
2003	CdSeS	CdO, OA, TOA, Se, S, TOP	CdO+ OA+TOA heated at 300 °C in N_2, TOP-S, TOP-Se injected	~5nm	[26]
2005	PbSe	Pb-acetate trihydrate, OA, Se, TOP	Single Step: Pb acetate + Co.sol degassed at 100–120 °C at 300–500 mTorr for 2h; reaction and growth: 140 °C	5 nm	[212]
2006	CdSe	CdO, OA, TOA, C_8SH or $C_{18}SH$,	CdO + OA + TOA heated at 300 °C, TOA + C_8SH or $C_{18}SH$ injected	3, 4, 6 nm	[213]

Ac: acetate; Co.sol: coordinating solvent; DDA: dodecylamine; DEPE: 1,2-bis(diethyl-phosphino)-ethane; DMPA: 2,2 –dimethoxy-2-phenylacetophenone; EGDMA: ethylene glycol domethacrylate; Et: ethyl; HDA: hexadecylamine; HPA: hexyl-phosphonic acid; LA: lauric acid; Me: methyl; MMA: methylmethacrylate; MPA: marcaptopropionic acid; OA: oleic acid; ODA: octadecylamine; ODE: 1-octadecene; SA: stearic acid; TBP: tri-n-butyl phosphine; TDPA: tetradecylphosphonic acid; TMS: trimethyl-silyl; TOA: trioctyl amine; TOP: tri-n-octyl-phosphine; TOPO: tri-n-octyl-phophine oxide (tech TOPO: Technical grade TOPO) 1 Torr = 7.5 × 10^{-6} Pa

Other Synthesis Processes

Sonic waves or microwaves [214,215] have been passed through a mixture of precursors in water to grow Qdots. These waves provide energy to dissociate the precursor and water molecules which results in the growth of Qdots. Ultrasound waves have been reportedly used to synthesize Qdots in the size range of 1–5 nm by formation, growth and implosive collapse of bubbles in a liquid [214]. Such acoustic cavitation generates a localized hotspot through adiabatic compression within the gas inside the collapsing bubble, enabling the reactions that form Qdots. In one approach, acetate precursors of metal ions were dispersed in a solution and seleno-urea was added and sonicated for an hour under an argon atmosphere

[214]. The temperature of the solution rose to 80 °C during the time required to produce Qdots.

Hydrothermal synthesis methods [216,217] or similar synthesis process [218] have been used to produce Qdots. These are crystallization of inorganic salts from aqueous solution, controlled by pressure and temperature. The solubility of inorganic compounds typically decreases as the temperature and/or pressure is lowered, leading to crystalline precipitates. By changing pressure, temperature, reaction and aging time and reactants, different shapes and sizes of the Qdots can be achieved. Several other wet-chemical processes for synthesizing Qdots [28,219,220] were also reported in literatures. Passing H_2S gas through precursors has also been used to prepare sulfide-based Qdots [221,222].

Vapor-Phase Methods

Initially, *in situ* self-assembled nanostructured materials were produced by hetero-epitaxial growth of highly strained materials. These vapor-phase methods for producing Qdots begin with processes in which layers are grown in an atom-by-atom process. Consequently, self-assembly of Qdots occurs on a substrate without any patterning [223,224,225,226,227]. In general, the layered materials grow as a uniform, often epitaxial layer (*Frank-van der Merwe* mode–FvdM) [228], initially as a smooth layer sometimes followed by nucleation and growth of small islands (*Volmer-Weber*mode–VW) [229], or as small (Qdots) islands directly on the substrate (*Stranski-Krastonow*mode–SK) [230]. Depending on the interfacial/surface energies and lattice mismatch (*i.e.,* lattice strain), one of these growth modes is observed. For example, Qdots may be formed by SK growth with an overlayer material that has a good lattice match with the substrate, but the substrate surface energy (σ_1) is less than the sum of the interfacial energy between the substrate and overlayer (γ_{12}) and the overlayer surface energy (σ_2), *i.e.,* when $_1 < \, _2 + \, _{12}$ [231]. In other cases, formation of Qdots was due to relaxation of strain required to maintain epitaxy. In the case of substrates with an overlayer with a large lattice mismatch and appropriately small surface and interface energies, initial growth of the overlayer occurs through by a layer-by-layer FvdM growth. However, when the film is sufficiently thick (a few monolayers) to induce a large strain energy, the system lowers its total free energy by breaking the film into isolated islands or Qdots (*i.e.,* the VW mode). *Kim et al.* [232] synthesized ZnSe/ZnS Qdots with the SK growth mode using a metal-organic chemical vapor deposition (CVD) technique in an atomic layer epitaxy (ALE) mode. The mean dot height was 1–1.9 nm. An apparent temperature dependent, anomalous behavior of confined carriers in the ZnSe Qdots was observed and attributed to thermalized carrier hopping between Qdots. The carrier hopping resulted in a substantial decrease of the PL peak energy and line width when the sample was cooled from room temperature to 140 K.

Molecular beam epitaxy (MBE) has been used to deposit the overlayers and grow elemental, compound or alloy semiconductor nanostructured materials on a heated substrate under ultra high vacuum (~ 10^{-10} Torr or 7.5 × 10^{-16} Pa) conditions [233,234]. The basic principle of the MBE process is evaporation from an apertured source (*Knudsen* effusion cell) to form a beam of atoms or molecules. The beams in the MBE process can be formed from solids (e.g., elemental Ga and As are used to produce GaAs Qdots) or a combination of solid plus gases (e.g., AsH_3, PH_3, or metal-organics such as tri-methyl gallium or tri-ethyl gallium). The metal-organic sources may leave high concentrations of carbon in the Qdots. MBE has been mainly used to self-assemble of Qdots from III-V semiconductors [223] and II-VI semiconductors [224,235] using the large lattice mismatch e.g., InAs on GaAs has a 7% mismatch and leads to SK growth, as discussed above.

Layer growth by physical vapor deposition (PVD) results from condensation of a solid from vapors produced by thermal evaporation or by sputtering [186]. Different techniques have been used to cause evaporation, such as electron beam heating, resistive or *Joule* heating, arc-discharge and pulsed laser ablation. In any case, the factors discussed above (strain and surface energies) control the formation of Qdots from the deposited thin films. CVD is another method to form thin films from which Qdots can be self-assembled. In CVD, precursors are introduced in a chamber at a particular pressure and temperature and they diffuse to the heated substrate, react to form a film, followed by gas-phase byproducts desorbing from the substrate and being removed from the chamber. InGaAs and AlInAs Qdots have been synthesized using either surface energy or strained-induced SK growth processes [236]. Although, self-assembling of Qdots using vapor-phase methods is effective in producing Qdots arrays without template, fluctuation in size of Qdots often results in inhomogeneous optoelectronic properties.

APPLICATION

Quantum Dots for Electroluminescence Device Fabrication

Qdots-based light emitting diodes (QLEDs) have attracted intense research and commercialization efforts over the last decade [22]. Infact, QLEDs have several advantages compared to organic LEDs (OLEDs). These are as follows: (i) FWHM of the emission peak from Qdots is only 20–30 nm, compared with >50 nm for their organic counterpart, which is necessary for a high quality image; (ii) Inorganic materials usually show better thermal stability than organic materials. Under operating at high brightness as well and/or high current, *Joule* heat is one of the predominant problems for device degradation. With better thermal stability, inorganic materials based devices are expected to exihibit longer lifetimes. (iii) The display color of OLEDs generally changes with time due to the different lifetime of the red, green and blue pixels [237]. However, one can obtain all of three premium colors from Qdots with the same composition changing the particle size (due to the quantum confinement effect, as discussed above). The same chemical composition should exhibit similar degradation with time. (iv) The QLED device can produce IR emission while the organic molecules in OLEDs usually exhibit wavelengths shorter than 1 μm. (v) The spin statistics are not restrictive for Qdots, *i.e.,* external quantum efficiency (EQE) of 100% can be achieved (in principle). The EQE of QLEDs can be expressed [238] as:ηExt=ηr·ηINT·η·ηOUT, where, η_r is the probability of holes and electron forming exciton, η_{INT} is the internal PL-QY and η & η_{OUT} are the probability of radiative decay and out-coupling efficiency, respectively. The value of η_r for fluorescent organics theoretically is limited to 25%

as the ratio of singlet to triplet states is 1:3 and only singlet state recombinations result in luminescence. However, for phosphorescent organics it is found > 25% due to spin-orbit coupling [239]. Note that phosphorescent organics lead to rapid degradation of the host material. With respect to high η_{Ext}, the value of η_{OUT} for planar devices is typically found to be ~20% [240]. The η_{OUT} efficiency can be enhanced by incorporating a microcavity structure [241]. For QLEDs, the value of η_{INT} (QY) can approach 100% and for a device with the appropriate electron and hole energies, the value of η_r can also be ~100%. It has been observed that a QLED can emit light under both forward and reverse bias [220]. Reason for this behavior is uncertain and many explanation are plausible, such as, different rates of electron and hole injections, different carrier mobility in the electron and hole transport layers, energy level offsets of the different layers, the emitting layer composition, surface and, uniformity and thickness of the layers. The valence (HOMO) and conduction (LUMO) band energies of some of the polymers that have been used in inorganic/organic QLEDs and solar cell are tabulated in Table 4. Table 5 shows the valence band and conduction band-energies reported in literature for different sized and structured Qdots, and Table 6 shows the work functions of some of the commonly used materials in QLED solar cell and OLEDs.

Table 4. Valence band (HOMO) and conduction band (LUMO) energies for some of the commonly used organics for QLEDs, solar cells and OLEDs.

Organics	Conduction Band (eV)	Valence Band (eV)	Reference
Alq3	3.1	5.8	[242,243]
CBP	2.9	6.0	[244]
PBD	2.6	6.1	[245]
PCBM	4.0	6.5	[246]
PPV	2.5	5.1	[247,248]
PVK	2.2	5.3	[245]
TAZ	3.0	6.5	[249]
TFB	2.2	5.4	[250]
TPBI	2.7	6.2	[244,251]
TPD	2.1	5.4	[242]
Poly TPD	2.3/2.5	5.2/5.4	[244,252]

Alq3: tris-(8-hydroxyquinoline) aluminum; CBP: 4,4',N,N'-diphenylcarbazole; t-Bu-PBD: 2-(4-biphenylyl)- 5-(4-tert-butylphenyl)-1,3,4 oxadiazole; PCBM: [6,6]-phenyl C61 butyric acid methyl ester; PPV: poly(phenylene vinylene); PVK: poly(vinyl-carbazole); TAZ: 3-(4-Biphenylyl)- 4-phenyl-5-tert-butylphenyl- 1, 2, 4-triazole; TFB: Poly[(9,9-dioctylfluorenyl-2,70diyl)-co-(4-4'-(N-(4-sec-butylphenyl)) diphenylamine)]; TPBI: 1,3,5-tris(N-phenylbenzimidazole-2-yl)-benzene; TPD: N, N'-diphenyl-N, N'-bis(3-methylphenyl)-(1, 1'-biphenyl)-4, 4'-diamine;

Table 5. Valence band and conduction band energies for different sized and structured CdSe Qdots.

Qdots	Conduction Band (eV)	Valence Band (eV)	Particle Size (nm)	Emission	Ref.
CdSe	4.4	6.5	5		[248]
CdSe/CdS	4.4	6.5	4.6		[251]
CdSe/CdS	4.7	6.8	4		[253]
CdSe/CdS/ZnS	4.8	6.8	6.8	600	[250]
CdSe/ZnS	4.4	6.5			[247]
CdSe/ZnS	4.3	6.5		550 nm	[242]
CdSe/ZnS	4.8	6.5			[245]
CdSe/ZnS	4.6 (CdSe)	6.8 (CdSe)	5.8		[249]
CdSe/ZnS	4.7	6.7			[254]
CdSe/ZnS/CdS	3.9	6.0	3–8.3	G, Y, O, R	[252]

Table 6. Work function of some of the commonly used materials as anode or cathode in QLED, solar cell and OLEDs.

Materials	Work function (eV)
Al	4.1
Ag	4.6
Ca	2.9
ITO	4.7
LiF/Al	2.8
Mg	3.7
PEDOT:PSS	5

ITO: indium tin oxide; PEDOT:PSS: poly(3,4-ethylene-dioxy-thiophene) poly(styrene-sulfonate).

Early Qdots-based EL devices were fabricated from self-assembled Qdots, grown using vapor-phase processes. These self-assembled structures were used to study the device physics [255].*Nakamura et al.* [235] fabricated an EL device with a structure of In/n-GaAs/n-ZnSe:Cl (1μm)/self-assembled CdS/pZnSe:N (1 μm)/Au. The blue-green emission peak wavelength varied with position in the wafer, possibly due to inhomogeneity of the CdS Qdot dimension. Recently, QLEDs have been developed for display and lighting sources [22,242,250,252, 256,257,258,259,260,261] and device structure and reported efficiency are tabulated in Table 7. In 1994, *Colvin et al.* [256] demonstrated a Qdot-polymer based hybrid EL device. They used spin-coated poly(phenylene vinylene) (PPV) polymer as the hole transport layer with indium tin oxide (ITO) and Mg as anode and cathode, respectively. The turn-on voltage was 4.0 V and EQE was 0.001–0.01. PPV was also used by many others for the fabrication of QLED devices [247,248,253,262,263], where the Qdots layers were used as an emitting (EML) as well as a charge transport layer, but the EQE was low (<1%). *Coe et. al* [242] created Qdot monolayer by spin-coating TOPO capped CdSe/ZnS core/shell Qdot along with a N,N'-diphenyl-N,N' bis(3-methylphenyl)-(1,1'-biphenyl-4,4'-diamine (TPD) layer followed by a 40 nm thermally evaporated layer of tris-(8-hydorxyquinoline) aluminum (Alq3) as electron transport layer (ETL). They reported 2000 cd/m^2 brightness at 125 mA/cm^2.*Tessler et al.* [264] reported NIR (1300–1400 nm) emitting QLEDs from InAs/ZnSe core/shell Qdots in a multilayer hybrid device structure: ITO//poly(3,4-ethylene-dioxy-thiophene (PEDOT)//(mixture of InAs/ZnSe Qdots and poly[2-methoxy-5-(2-ethylhexyloxy)-1,4-phenylenevinylene] (MEHPPV) or poly[(9,9-dihexylfluorenyl-2,7-diyl)-co-(1,4-{benzo-[2,1',3]thiadiazole})] (F6BT))//Mg:Al. Although such devices exhibited EQE values of approximately 0.5%, the turn-on voltage was high (15 V). In 2005, NIR emission from an EL device was achieved using monodispersed 5 nm sized PbSe Qdot [165]. Phase segregation during spin-coating was used to fabricate hexagonal close packed Qdots/organic double heterojunction devices. They also introduced an innovative solvent drying process [212,249] during fabrication of QLED device in order to get phase separation and consequently to achieve two distinct layers of polymer and Qdot-monolayer from the mixture. The organic film served as a hole transport layer (HTL) in the QLED and produced the spectra shown inFigure 15. A broad NIR emission (1200 nm- >1700 nm) [220] with an EQE of 0.02% (maximum light output 150 nW/mm^2 at 50 mA and 2.5 V) was also achieved from a device with the structure shown in Figure 15 (ITO//poly(3,4-ethylene-dioxy-thiophene) poly(styrene-sulfonate) (PEDOT:PSS)// HgTe (4.6 nm)//Al).

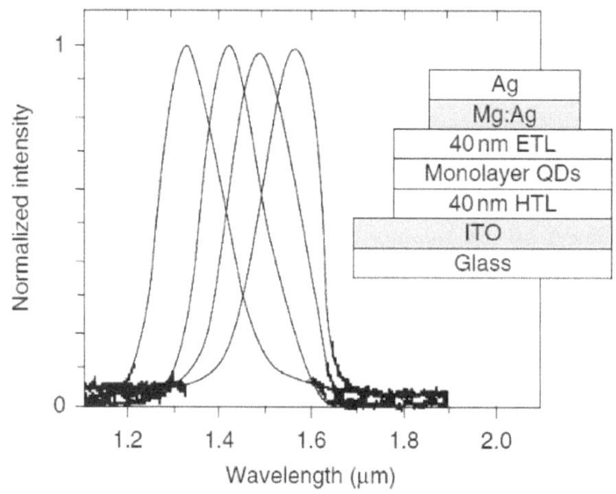

Figure 15. PbSe Qdots based Qdot-LED device structure and NIR-EL spectra using different sized Qdots (ETL: electron transport layer; QDs: quantum dots, HTL: hole transport layer, ITO: indium tin oxide) [Reproduced with permission from [165] Copyright Wiley-VCH Verlag GmbH & Co. KGaA].

In another study of QLEDs, *Coe-Sullivan et al.* [249] used thermal evaporation to deposit 30 nm of 3-(4-Bi-phenylyl)-4-phenyl-5-tert-butylphenyl-1,2,4-tru-azole (TAZ) as a hole blocking layer and a 30 nm thick Alq3 layer onto a monolayer of CdSe/ZnS Qdots followed by a Mg:Ag/Ag cathode. Thermal evaporation of the organic layers offsets some of the advantages of QLEDs and has discouraged commercialization of QLEDs. For instance, Alq3 as an ETL and calcium as an electron injection layer was used to obtain an efficient color QLED [252]. Unfortunately, the device was unstable due to degradation of the organic layer and oxidation of the reactive metal cathode [265,266]. In addition, this device fabrication requires a costly batch vacuum deposition process [254]. Furthermore, defects are always present at the organic-inorganic interface which reduces the efficiency. Also, the electron transport rate and intrinsic electron density of an organic thin film is low, leading to an electron density much lower than the hole density and therefore charge imbalanced and again lower efficiencies. Inorganic transport layers were employed to overcome some of these disadvantages. For examples, a NiO layer was used for HTL in an EL device [243,257], because it is chemically and electrically stable and compatible with CdSe Qdots. *Zhao et al.* [251] demonstrated a solution processed TPD based polymer composite HTL where the composite avoided TPD's tendency to recrystallize during device operation. TPBI (1,3,5-tris(N-phenylbenzimidazole-2-yl)-benzene) ETL was deposited on a monolayer of 4.6 nm CdSe/CdS Qdots using a thermal evaporation method. A monolayer of CdS/ZnS Qdots (QY: 20–30%) exhibited blue EL from a device structure of ITO// 4,4'-N,N"-dicarbazolyl-biphenyl (CBP)// CdS/ZnS Qdots monolayer// 3-(4-biphenylyl)-4-phenyl-5-*tert*-butylphenyl-1,2,4-triazole (TAZ)// Alq3//Mg:Ag//Ag. But the quantum efficiency was only 0.1%. At low current, emission was dominated by the CdS/ZnS Qdots with a FWHM of 30 nm and a peak at 468 nm. At higher currents, EL emission from the organic layers dominated. Using Alq3 as the ETL, *Sun et al.* [252] recently reported red (600 nm), orange (589 nm), yellow (546 nm) and green (517 nm) emitting QLEDs from an organic inorganic hybrid structure using highly luminescent (QY>70%) CdSe/ZnS and CdSe/CdS/ZnS core/shell Qdots. The maximum luminance values for red, orange, yellow and green were 9064, 3200, 4470 and 3700 cd/m^2, respectively. *Caruge et al.* [257] developed a fully inorganic QLED with good long term stability by using a sputtered deposited ZnO:SnO_2 amorphous layer ETL and alloyed ZnCdSe Qdots instead of core/shell Qdots. Although the QLED showed a maximum luminescence value of ~2000 cd/m^2 with an EQE of 0.1%, the charge transporting layers were fabricated using vacuum sputter deposition. A blue emitting solution processed QLED device with a maximum brightness 1600 cd/m^2 and efficiency 0.5 cd/A was reported by *Tan et al.* [267]. *Cho et al.* suggested [250] that the high valance band energy (>6.5 eV) for II-VI Qdots *versus* ITO (4.5–5.1 eV) was the cause of the low efficiency of QLED devices. Using solution-based fabrication steps (except for the Al cathode), they demonstrated a QLED-based display (as shown in Figure 16) having a crosslinked Qdot emitting layer, a HTL and a continuous TiO_2 thin film ETL. They reported a turn-on voltage of these devices of 1.9 V, lower than the band-gap of the Qdots (2.1 eV). *Bae et al.* [268] recently reported a brightness of more than 10,000 cd/m^2 from a green emitting QLED (shown in Figure 17) with a turn-on of 3.5V and luminous efficiency of 5.2 cd/A. They also reported a blue emitting QLED using CdZnS alloy Qdots (Figure 18). Finally, we have fabricated hybrid and completely inorganic QLEDs using solution processing (except for the Al cathode) and a unique ZnO nanoparticle as an ETL. We have achieved remarkable luminance and efficacy values for red green and blue emitting devices (yet to be published). Nevertheless, we have tabulated some of the research results in Table 7.

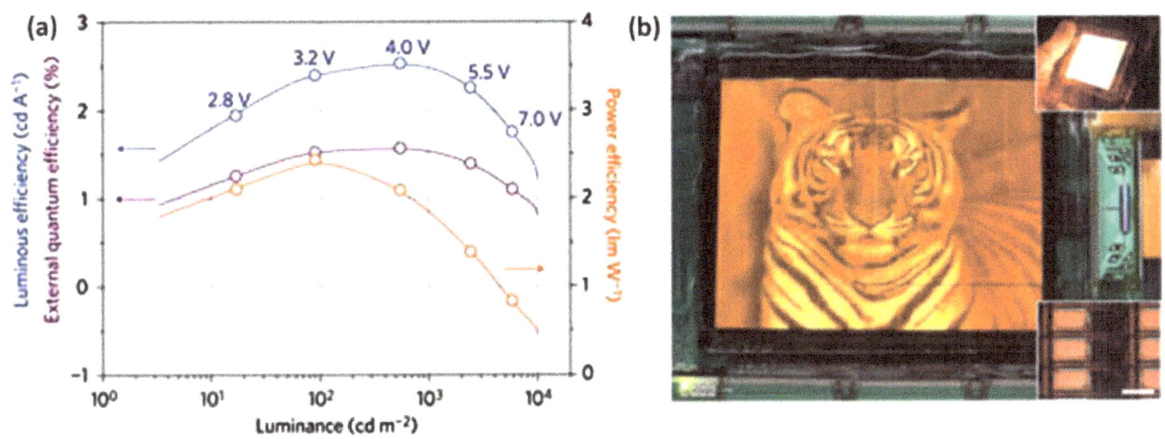

Figure 16. (a) Efficiencies *vs.* luminance data from red emitting QLED devices fabricated using CdSe Qdots; (device exhibited maximum luminance of 12380 cd/m^2, turn-on voltage of 1.9V and power efficiency of 2.41 lm/W) (b) four inch active matrix QLED display (320 ° 240 pixel array) using a-Si TFT [Reprinted by permission from Macmillan Publishers Ltd. [Nature Photonics] [250], Copyright 2009].

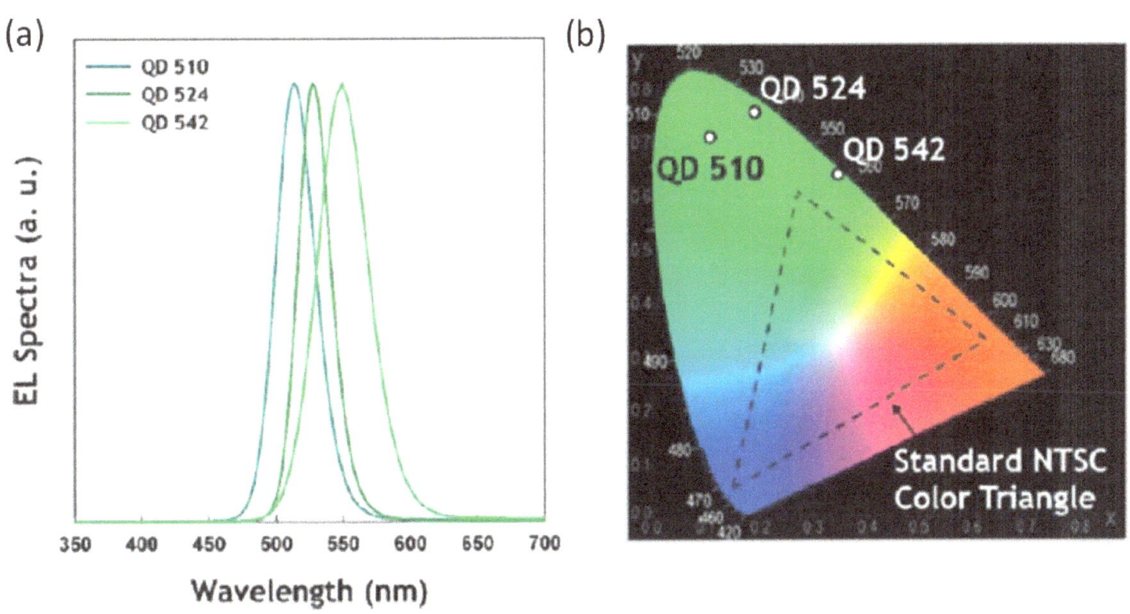

Figure 17. (a) Electroluminescence spectra at 510 nm, 524nm and 542 nm from three different sized CdSe/ZnS Qdots (QD 510: 6.7 nm; QD 524: 7.4 nm; QD 542: 7.8 nm; Maximum brightness obtained from the green emitting device was above 10,000 cd/m^2. (b) Positions of emitted green colors on the color coordinate (CIE) diagram [Reproduced with permission from [268]. Copyright Wiley-VCH Verlag GmbH & Co. KGaA].

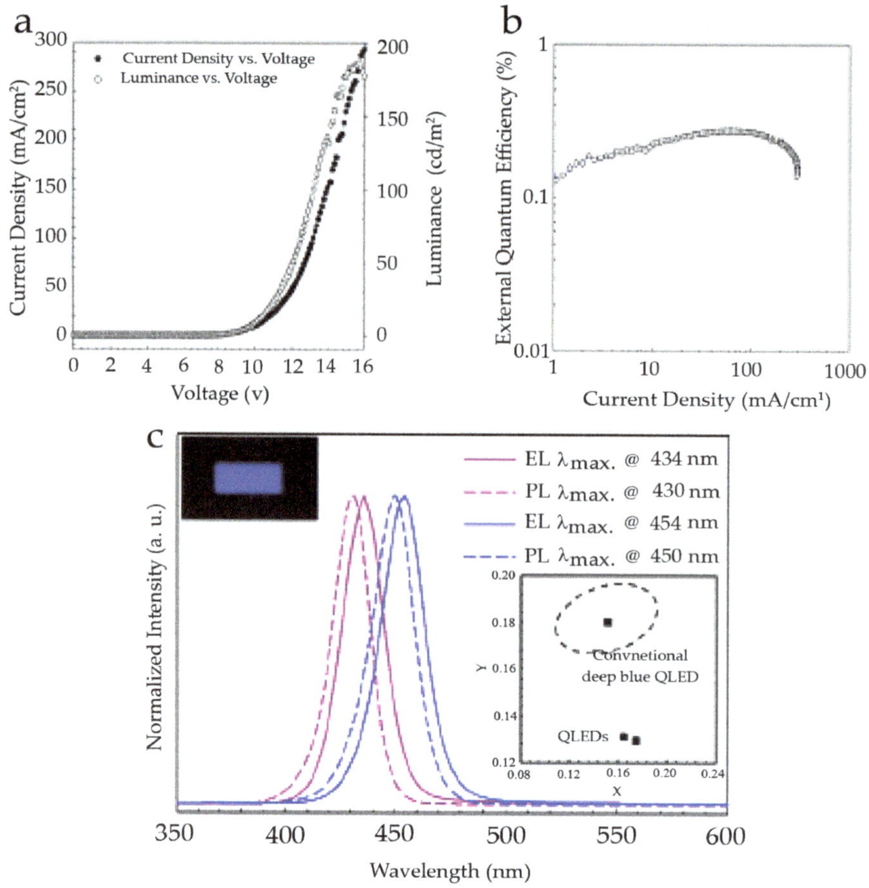

Figure 18. Characteristics of blue emitting QLED using CdZnS/ZnS Qdots: (a) L-I-V characteristics of QLED (b) EQE *vs.*current density and (c) EL and PL spectra from devices. Insets in (c) show a blue emitting electroluminescence device pixel and CIE color coordinate of the QLED [244] (Reproduced with permission from IOP Publishing Ltd).

Table 7. Literature reported QLED device structures and their efficacy values.

Emitting Wavelength (nm)	$V_{turn-on}$ (V)	L_{max} (nit)	LE (cd/A)	PE (lm/W)	EQE (%)	Device Structure	Ref.
~610	4	100	--	--	0.001–0.01%	*ITO//CdSe//PPV//Mg & ITO//PPV//CdSe//Mg*	[256]
520, 550, 610	--	--	--	--	0.0005%	*ITO//(mixture PVK – t-Bu-PBD-CdSe)//Al*	[262]
560	4	600			0.2%	*ITO//PPV//CdSe/CdS// MgAg*	[253]
600	3–3.5				0.1%	*ITO//PPV (multilayer) //CdSe//Al*	[247]
620	3.5				0.01–0.2%	*ITO//PPV//(CdSe in block co-polymer)//Al*	[248]
550–650	2.3–3.5				0.1%	*ITO//(mixture of PDDA & CdTe)//Al*	[269]
560	~3.5	2000	1.9	--	0.52%	*ITO//TPD//CdSe/ZnS//Alq₃//Mg: Ag ITO//TPD//CdSe/ZnS//TAZ//Alq₃/ /Mg:Ag*	[242]
~580		13			0.005%	*ITO//(mixture of CdSeS & PDBD)//TAZ//Alq₃//LiF//Al*	[26]
1300–1400	15				0.5%	*ITO//PEDOT//(Mixture of InAs/ZnSe & MEHPPV or F6BT)//Ca:Al*	[264]
540–635	3.5	--	--	--	1.1%	*ITO//TPD//CdSe/ZnS(monolayer) //TAZ//Alq₃//Mg:Ag//Ag*	[249]

1330–1560	~3				0.001%	*ITO//TPD or αNPD//PbSe//*Alq_3*//BCP// Mg:Ag//Ag*	[165]
1000–1500					1.2%*	*ITO//(mixture of PbS & MEHPPV or CNPPV)//Mg//Ag*	[197]
~610	5	500			0.2%	*ITO//PVK//CdSe/ZnS//bu-PBD//Al*	[245]
615	3	7000	2.0	1.0	2%	*ITO//TPD//CdSe//*Alq_3*//Mg:Ag//Ag*	[212]
573–619					0.001–0.1%	*Au//pGaN//CdSe/ZnS//nGaN//In*	[254]
1590	~1.2				0.02%	*ITO//PEDOT:PSS//HgTe// Al*	[220]
610	4	1000			1.2%	*ITO//PS-TPD-PFCB//TCTA-BVB//CdSe/CdS//TPBI//Ca//Ag*	[251]
625		3000			0.18%	*ITO//NiO//CdSe/ZnS// Alq3//Ag:Mg//Ag*	[243]
520	2.5				0.5%	*ITO//CBP//CdZnSe/CdZnS//TAZ// Alq3//Mg:Ag//Ag*	[27]
~440 (B) ~545 (G) ~610 (R)	2–3	830 (W)	0.9 (W)	0.57 (W)	0.35% (B) 0.65% (G) 1.6% (R) 0.36% (W)	*ITO//PEDOT:PSS//TPD//Qdots// TAZ//Alq3//Mg/Ag//Ag*	[167]
517 (G) 546 (Y) 589 (O) 600 (R)	4 (G), 5 (Y), 3 (O), 3 (R)	3700 (G), 4470 (Y), 3200 (O), 9064 (R)	1.1–2 (G-R)	<1.1		*ITO//PEDOT-PSS//poly-TPD//CdSe/ZnS or CdSe/CdS/ZnS//*Alq_3*//Ca/Al*	[252]
460	2.5	1,600	0.5	0.5	0.06	*ITO//PEDOT:PSS//poly-TPD//CdS/ZnS//Al*	[267]
638	3.8	1950			0.1%	*ITO//NiO//ZnCdSe//ZnO:*SnO_2*//Ag*	[257]
~600	1.9	12,380	1.67			*ITO//PEDOT:PSS//TFB//CdSe/CdS/ZnS//*TiO_2*//Al*	[250]
434–450	5	150	--	--	0.1–0.3	*ITO//PEDOT:PSS//poly-TPD:CBP//CdZnS/ZnS// TPBI//LiF//Al*	[244]
510, 524, 542	3.5	16,000	6.0	--	1.4	*ITO//PEDOT:PSS//poly-TPD//CdSe/ZnS//TPBI//LiF //Al*	[268]

*Internal quantum efficiency value reported $V_{turn-on}$: turn-on voltage of the device; L_{max}: maximum luminance value reported; LE: luminous efficiency; PE: power efficiency; EQE: external quantum efficiency. Alq3: tris-(8-hydroxyquinoline) aluminum; BCP: bathocuproine; CBP: 4,4'-N,N'-dicarbazolylbiphenyl; CNPPP: poly(2-(6-cyano-6'-methylheptyloxy)- 1,4- phenylene); F6BT: poly[(9,9-dihexylfluorenyl-2,7-diyl)-co-(1,4-{benzo-[2,1',3] thiadiazole})]; ITO: indium tin oxide; MEHPPV: poly[2-methoxy-5-(2-ethylhexyloxy)-1,4-phenylenevinylene]; αNPD: 4,4-bis[N-(1-naphyl)-N-phenylamino]biphenyl; t-Bu-PBD: 2-(4-biphenylyl)-5-(4-tertbutylphenyl)-1,3,4 oxadiazole; PEDOT: poly~3,4-ethylenedioxythiophene; PFBD: poly(9,9'-dioctylfluorene-co-N-(4-butylphenyl) diphenylamine; PFCB: perfluorocyclobutane; PPV: poly(phenylene vinylene); PS: polystyrene; PSS: polystyrene sulfonate; PVK: polyvinyl carbazole; TAZ: 3-(4-Biphenylyl)-4-phenyl-5-tertbutylphenyl-1, 2, 4-triazole; TFB: Poly[(9,9-dioctylfluorenyl-2,70diyl)-co-(4-4'-(N-(4-secbutylphenyl)) diphenylamine)]; TPBI: 1,3,5-tris(N-phenylbenzimidazole-2-yl)-benzene; TPD: N, N'-diphenyl-N, N'-bis(3-methylphenyl)-(1, 1'-biphenyl)-4, 4'-diamine; B: blue; G: green; Y: yellow; O: orange; R: red; W; white

Downconversion of Blue or Ultraviolet Light

Qdots are being used in downconversion of high energy to lower energy light. The advantages of Qdots phosphors over conventional inorganic phosphor and/or organic dyes include the (i) high QY of Qdots, (ii) emission range from UV to visible to NIR, (iii) better stability compared to organics, (iv) narrow FWHM ~30 nm (higher color saturation compared to conventional phosphor with typical FWHM: 50–100 nm), and (v) a large absorption window (band-gap to UV, enabling simultaneous excitation of different size Qdots). The main disadvantage of Qdots is self-absorption resulting from overlap of the absorption spectra from larger Qdots with the emission spectra and emission spectra from smaller Qdots. In 2000, *Lee et al.* [161] reported downconversion of 425 nm blue (from GaN based commercial LED) and UV light (from Hg lamp) using 2.0, 2.6, 4.6 and 5.6 nm CdSe/ZnS Qdots dispersed in a poly-aurylmethacrylate polymer matrix. The composite was prepared at 70–75 °C for 2 hrs by mixing as-synthesized TOP capped CdSe/ZnS Qdots and monomer of 1-aurylmethacrylate, then adding ethyleneglycol dimethacrylate and a radical initiator azo-bis-isobutyronitrile for crosslinking. Recently, organics-capped ZnSe Qdots [196] were excited with a near-UV InGaN LED to produce white light with *Commission Internationale de l'Eclairage*(CIE) x, y chromaticity coordinates of 0.38, 0.41. Downconversion of blue light (455 nm) from an InGaN LED resulted in white light emission with a color rendering index (CRI) of > 90 and CIE of 0.33, 0.33, when a mixture of green and red emitting CdSe/ZnSe Qdots was used [270], as shown inFigure 19. We also summarized some of the literature reports on downconversion of Qdots in Table 8.

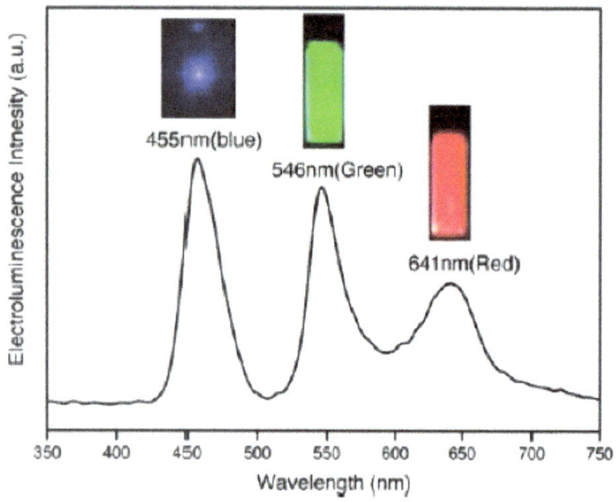

Figure 19. Downconversion of 455 nm blue emission from an InGaN light emitting diode to green and red by size-tuned CdSe/ZnSe Qdots in a silicone matrix (color rendering index: 91; chormacity coordinate: 0.33, 0.33) [reprinted with permission from [270] (© 2006 IEEE)].

Table 8. Selected reports on the use of Qdots to downconvert blue or UV light from inorganic LEDs.

Year	Source light	Qdots	Matrix	Emitted light	Ref.
2000	UV (Hg lamp), Blue GaN Commercial LED	CdSe/ZnS (2.0, 2.6, 4.6, 5.6 nm)	Polyauryl-methacrylate	UV: Blue, orange, red; Blue: red (590 nm)	[161]
2005	InGaN (near UV)	ZnSe (TOPO & Stearic acid coated)	organics coated ZnSe (10 wt%) dispersed in epoxy resin	White, CIE (0.38, 0.41) Conversion efficiency: 30% relative to RGB commercial phosphors	[196]
2006	InGaN (455 nm)	CdSe/ZnSe (G); CdSe/ZnSe (R) CdSe/ZnSe (Y)	TOPO-coated CdSe/ZnSe dispersed in silicone	White, CIE: (0.33, 0.33), CRI: 91 with R& G; White, CIE: (0.32, 0.33), CRI: 50 with Y; Efficiency: 15-30 lm/W	[270]
2007	390 nm UV LED	CdSe/CdS/ZnS	2wt% Qdot in chloroform & epoxy resin at 1:1 (vol); Thermally cured	Red (620 nm)	[168]
2007	InGaN/GaN (440 nm, 452 nm)	CdSe/ZnS (440-452	Qdots blended with resin; 400 -1700 μm (Qdot density: 3.04-140 nanomoles/1ml resin)	White; with 453 nm & CdSe/ZnS (540, 500, 580 & 520 nm): CIE (0.24, 0.33), CRI: 71	[271]
2008	InGaN/GaN (blue/green)	CdSe/ZnS (620nm, R) & Au particles (for surface Plasmon enhanced emission)	5 wt% Qdots and 0.05 wt% Au in toluene spin-coated on LED (thickness ~200 nm)	White: (0.27, 0.24); Conversion efficiency ~53%	[272]

R: red, G: green, B; blue, O: orange; Y: yellow; W; white; CIE: International Commission on illumination; CRI: color rendering index; vol.: by volume

Quantum Dots in Solar Cell Device Fabrication

Increasing demands of energy for human activity warrant immediate attention, and solar energy is clean and plentiful. Approximately, a 9° 10^{22} J of energy reaches the earth everyday from the sun, compared to daily consumption by mankind of about 9 ° 10^{18} J [273]. Although the maximum thermodynamic efficiency (*Shockley-Queisser* limitation) from a single junction silicon solar cell is only about 31% [274], a multi-junction cell, that uses multiple subcell band-gaps to divide the broad solar spectrum into smaller segments, can exhibit higher for theoretical conversion efficiencies: e.g., ~40% for three and >58% for four or more subcells [275]. In fact, *King et al.* reported an efficiency of 40.7% from a three-junction GaInP/GaInAs/Ge cell under the standard spectrum for terrestrial concentrator solar cells at 240 suns [275]. Below, the benefits of using Qdots to fabricate inorganic-organic hybrid solar cells are discussed.

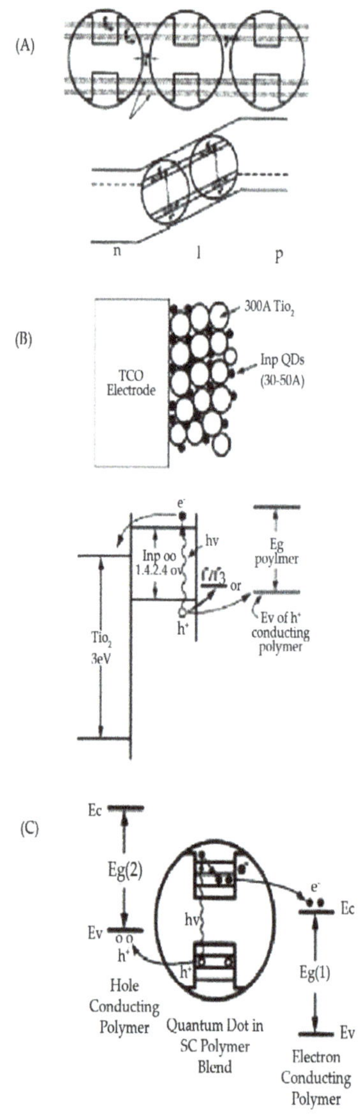

Figure 20. Solar cell with Qdots can be broadly classified in three categories: (A) P-I-N solar cell with a Qdots array, (B) Qdot sensitized and (C) Qdots dispersed [reprinted from publication [290], Copyright 2002, with permission from Elsevier].

A major factor leading to the superior efficiency of inorganic devices compared to organic or hybrid devices is the difference in carrier mobilities. Organic semiconductors possess relatively high absorption coefficients (usually $\geq 10^5$ cm^{-1}) [276]. In a solar cell, four consecutive processes occur: (1) absorption of light and exciton formation, (2) exciton diffusion, (3) charge separation, and (4) charge transportation. Due to poor mobility and a short exciton lifetime in conductive polymers, organics possess low exciton diffusion length (10–20 nm). In other words, excitons that form far from the electrode or carrier transport layer recombine and conversion efficiency decreases. For example, single and multiwall carbon nanotubes (SWNT and MWNT) were linked to CdTe Qdots capped with thioglycolic acid to produce solar cells [277]. The nanotubes provided a good hole transport route to the electrode, and CdTe acted as an excited-state electron donor. The highest monochromatic photon quantum efficiency was 2.3% from a hybrid cell consisting of single SWNT/pyrene+/red-emitting CdTe Qdots. Therefore, the combination of organic and inorganic materials improved the conversion efficiency of the solar cell. Hybrid inorganic/organic hybrid solar cells have been reported using CdSe [278], TiO_2 [279], ZnO [280], PbS [281], PbSe [282], $CuInS_2$ [283] and $CuInSe_2$ [284]. Use of Qdots in solar cell is advantageous because: (i) quantum confinement enables size-tuned tunable band-gaps so that a multi-junction device is possible using the same Qdot composition to absorb entire gamut of sunlight from UV to visible to IR (0.5 eV to 3.5 eV); (ii) hot carrier relaxation dynamics can be significantly reduced due to confinement of exciton; (iii) multiple exciton generation or MEG [105,182,285,286,287,288,289] (e.g., IV-VI, Si, InAs Qdots) with a single photon is possible (*i.e.,* a QY >100%); (iv) good heterojunction with hole conductors are possible; (v) they are more stable and resistant to oxygen, moisture and UV radiation as compared to polymer (vi) lower-cost solution based fabrication processes can be used; (vii) substrate can be flexible. In general, Qdot based solar cells can be classified into three categories [285,290], as shown in Figure 20.

Quantum Dot Sensitized Solar Cell

For several decades, dyes have been used as sensitizers for harvesting energy. In the 1990s, the dye-sensitized solar cell (DSC) was extensively developed by *Graetzel* [291,292]. In a DSC, light is absorbed by the dye followed by an electron transfer from an excited state of the dye molecule into the conduction band of a wide band-gap semiconductors [292,293], such as TiO_2, ZnO, TiO_2, NbO_2or Ta_2O_5. Figure 21 shows the band position of some of the semiconductors that are used in photochemical cells. The hole on the dye is scavenged by a redox couple in solution. For better efficiency and charge separation, less than a monolayer of dye is required on the wide band-gap semiconductors. Usually dyes have a strong absorption band in the visible region but very low absorption in the UV and NIR regions. In addition, dyes in DSC suffer from photodegradation. Therefore, tunable band-gap semiconductors are of interest for the fabrication of sensitized solar cell. For example, the absorption of Qdots can be tuned from UV through the visible into the NIR, and band-edge absorption is favorable for effective light harvesting [293]. Finally, surface passivation can enhance photostability of Qdots. Selected results for Qdots sensitized solar cell are summarized in Table 9.

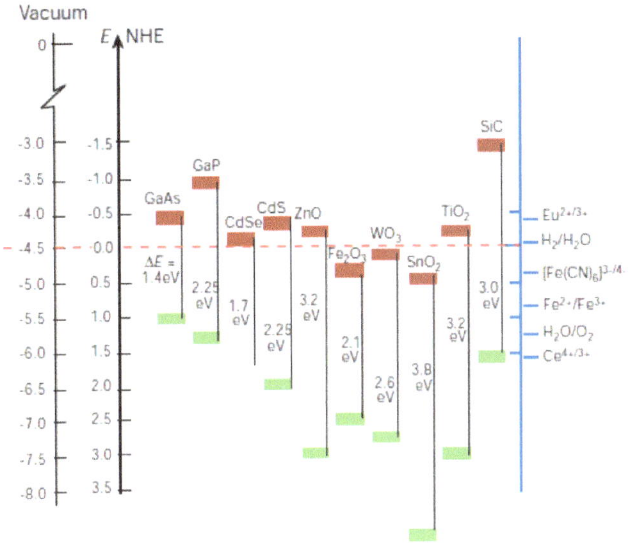

Figure 21. Valence and conduction band energies of various semiconductors, which have been used in solar cells. The outer left ordinate shows the energies referenced to the vacuum level, while the inner scale shows the normal hydrogen electrode scale (both in eV). (Reprinted by permission from Macmillan Publishers Ltd. [Nature (London)] [292], Copyright 2001).

Sensitization of highly porous TiO_2 electrode by *in situ* synthesized CdS Qdots (4–20 nm) was demonstrated in 1990 [294,295]. After further development, a photocurrent QY of 80% with V_{OC} up to 1 V was measured using PbS, CdS, Ag_2S, Sb_2S_3 and BiS_3 Qdots as sensitizers [293]. Recently, organic hole conductor layers are introduced have replaced solution in electrochemical cells. In other words, research on electrochemical solar cells is moving towards heterojuction-based solar cell. *Plass et al.* [296] demonstrated a high surface area *p-n* heterojunction solar cell, where the heterojunction was sensitized to visible light by PbS Qdots. By functionalizing the surface, self assembled Qdots [297,298,299,300] on titania were used to improve the efficiency of solar cell.*Kamat's group* [299] used different bifunctional linker molecules to assemble Qdots on TiO_2particles. *Niitsoo et al.* [300] reported an efficiency of 2.8% from a self-assembled monolayer of CdSe or CdS Qdot (6–8 nm) on titania using chemical bath deposition technique. A similar method was used by *Lin et al.* [298] and they achieved an efficiency of ~1.4%. Using TiO_2 inverse opal and Qdots, *Diguna et al.* [301] achieved a power conversion efficiency of 2.7%. *Sun et al.* [302] used TiO_2nanotubes arrays and CdS Qdots for photoelectrochemical solar cell. Under AM 1.5 G illumination, the conversion efficiency was found to be 4.15%. Titania nanotubes were produced by anodic oxidation [303]. Recently, a conversion efficiency of > 4% was reported [304] when a 'cascade' energy level structure of TiO_2/CdS/CdSe/ZnS was used.

Table 9. Selected data for Qdot sensitized solar-cell.

Year	Qdots	Results	Conversion Efficiency %	Device Structure	Ref
1994	CdS: 4nm PbS: 5nm	J_{SC}: ~1.9 mA/cm^2 V_{OC}: 1 V FF: 0.45 EQE: 80% (@ 460 nm)		ITO//TiO2//CdS or PbS	[293]
1998	InP: 6.5 nm	FF: 0.685	1.5% (400–800 nm)	TiO2//InP	[305]
2002	PbS: 7 nm	V_{OC}: 0.24V EQE: 45%	(at AM 1.5) 0.49%	SnO2:F//TiO2//PbS//Spiro-OMeTAD, Au	[296]
2006	CdSe CdS ~5nm	J_{SC}: 10.5 mA/cm^2 V_{OC}: 0.66 V FF: 39.5	2.8%	TiO2//CdSe or CdS (self-assembled)	[300]
2007	CdS (4–6 nm)	J_{SC}: 3.44 mA/cm^2 V_{OC}: 0.657 V FF: 0.6	1.35%	TiO2//CdS (Self-assembled)	[298]
2007	CdSe	J_{SC}: 7.51 mA/cm^2 V_{OC}: 0.71 V FF: 0.5	2.7%	TiO2//F//CdSe//F//ZnS	[301]
2008	CdS	J_{SC}: 7.82 mA/cm^2 V_{OC}: 1.27 V FF: 0.578	2.8%	TiO2 (nanotubes)//CdS	[302]
2009	CdS CdSe	FF: 0.49 V_{OC}: 0.5137 V J_{sc}: 16.8	4.22% (AM 1.5)	TiO_2//CdS(3)//CdSe(4)//ZnS, Au	[304]

AM: Air mass; VOC: open circuit voltage; ISC: short circuit current; FF: Fill factor; OMeTAD: 2,2',7,7'-tetrakis(N,N-di-p-methoxyphenyl-amine)9,9'-spirobifluorene.

Quantum Dot Dispersed Solar Cell

In 1992, *Wang and Herron* measured the photoconductivity effect of Qdots using *in situ* synthesized CdS Qdots in poly(vinyl-carbazole) matrix with Al electrode [306]. Greenham *et al.* fabricated an inorganic-organic hybrid solar cell using a mixture of 5 nm CdSe Qdot in MEH-PPV polymer that was spin-coated onto an ITO/glass substrate and an Al cathode was deposited to complete the device. The energy conversion efficiency under monochromatic illumination at 514 nm (5 W/m^2) was 0.2% [307]. Because organic semiconductor materials generally exhibit low electron mobilities (below 10^{-4} cm^{2}V^{-1}s^{-1}), mixing them with inorganic nanostructured semiconductors can result in improved device performance [308]. *Huynh et al.* [309] improved the device performance by incorporating Qdots and nanorod in poly-3(hexylthiophene) (also known as P3HT). Incorporation of Qdots resulted in a maximum EQE of 55% with a power conversion efficiency of 1.7%. ZnO Qdots have been used since they are *n*-type semiconductors with high carrier concentrations and electron mobilities [310]. Hybrid solar cells based on ZnO Qdots under an AM1.5 illumination exhibited an energy conversion efficiency of 0.9% for P3HT, 1.6% for poly[2-methoxy-5-(3',7'-dimethyloctyloxy)-1,4-phenylene vinylene] (MDMO-PPV), and EQE of 27% at 480 nm for P3HT and 40% for MDMO-PPV. The device structure was ITO/ PEDOT:PSS/ZnO: polymer (26 vol%, thickness ~200nm)/Al. Thermal annealing of the spin cast ZnO nanoparticles/polymer blend improved charge transport and was crucial to achieving the efficiencies reported above. Note that the above discussion emphasizes results mainly from spherical semiconducting nanoparticles. There are reports on the use of nanorod [311,312], tetrapod [313,314,315] and highly brunched [316] nanoparticles in solar cells. Selected results from Qdots-based solar cells are summarized in Table 10.

Table 10. Selected results for solar cells using Qdot dispersed in a conducting polymer.

Year	Qdot Size	Results	Conversion Efficiency % (at AM 1.5 G)	Device Structure	Ref
1997	5nm	V_{OC}: 0.5 V FF: 0.26 EQE: 12%	0.2 (monochromatic illumination at 514 nm)	ITO// MEHPPV: CdSe or CdS//Al	[307]
2002	CdSe or CdS Dot: 7nm Rod: 7 X 60 nm	V_{OC}: 0.7V FF: 0.4 EQE: 55% (20% Qdots)	1.7	ITO//CdSe: P3HT//Al	[309]
2004	CdSe: Rod: 7 x 30 nm		1.5	ITO//PEDOT:PSS//P3HT-functionalized CdSe //Al	[317]
2006	PbS: 4nm	V_{OC}: 1 I_{SC}: -0.13 FF: 0.28	0.7	ITO//PEDOT:PSS//PbS: MEHPPV//Al	[318]

FF: Fill factor; I_{SC}: short circuit current; ITO: indium tin oxide; MEHPPV: poly[2-methoxy-5-(2-ethylhexyloxy)-1,4-phenylenevinylene]; PEDOT:PSS: poly(3,4-ethylene-dioxy-thiophene) poly(styrene-sulfonate); P3HT: poly-3(hexylthiophene); V_{OC}: open circuit voltage.

Quantum Dots in Other Optoelectronic Devices

Qdot photonic devices, including optical amplifiers and lasers operating at room temperature, have been studied [19,319,320]. A major difficulty in achieving lasing is the very efficient nonradiative *Auger* recombination processes. Several strategies have been proposed to improve the lasing, including increased Qdot concentrations in the active layer, use of improved optical feedback structures, and optimized optical waveguides. Two dimensional waveguides loaded with luminescent colloidal CdTe Qdots grown with a layer-by-layer deposition technique have been investigated [321]. The waveguides exhibited propagation loss coefficients of <1 cm^{-1}. The losses depended weakly on the width of the waveguides but depend strongly on the surface roughness. High optical gain of ~230 cm^{-1} was demonstrated using femtosecond pulses [321], showing that this kind of Qdot waveguides could be suitable for lasers and optical amplifiers. Qdots embedded composites were also used to demonstrate photorefractivity and other nonlinear optoelectronic properties [222,322]. PbSe Qdots in a polymeric host showed large photorefraction, photoconductivity and optical gain when excited with a low-power, continuous-wave laser at 1550 nm [323]. Dynamic-photorefractive holographic gratings were written in the composite. The net optical gain and significant diffraction efficiency achieved with the low power laser makes these nanocomposites potential choices for infrared imaging and optical communication applications.

III-V semiconductors, such as InGaAlN, GaP, GaAs, InP and InAs, are very important for the development of optoelectronic devices as it is possible to engineer III-V Qdots simply by tuning the Qdot size and composition to emit anywhere from the IR to the UV [324,325]. Nitride-based Qdots have attracted enormous research interest because of their large built-in electric fields [324]. These large fields originate from both spontaneous polarization in the wurtzite crystal structure, as well as the lattice-mismatch strain acting through their large piezoelectric constants. Emission from InAs/InP Qdots can be controlled by size as well as the As and P concentrations, and are important for fiber optic telecommunication systems [326]. InP-based material can emit over the range of wavelengths necessary for photonic devices, especially near 1.55 μm. InAsP Qdots are suitable for monolithically integrated photonic devices on a single chip, including lasers, phase modulators, detectors, and passive waveguides.

Application of Quantum Dots in Bioimaging Applications

Currently, magnetic resonance imaging (MRI), optical imaging, and nuclear imaging are emerging as key imaging techniques in biological systems [327]. They differ in terms of sensitivity, resolution, complexity, acquisition time and operational cost. However, these techniques are complementary to each other most of the times. There are several reviews on the physical basis of these techniques [327,328], the instrumentation [329,330] and the issues that affect their performance [331,332]. Currently, a significant amount of research is aimed at using the unique optical properties of Qdots in biological imaging [23,58,59,92,332,333,334,335,336,337,338,339,340, 341,342]. Much of optical bioimaging is based on traditional dyes [343,344], but there are several drawbacks associated with their use. It is well-known that cell autofluorescence in the visible spectrum [345] leads to the following five effects. (1) The autofluorescence can mask signals from labeled organic dye molecules. (2) Instability of organic dye under photo-irradiation is well known in bioimaging which results in only short observation times. (3) In general, conventional dye molecules have a narrow excitation window which makes simultaneous excitation of multiple dyes difficult. (4) Dyes are sensitive to the environmental conditions, such as variation in pH. (5) Most of the organic dyes have a broad emission spectrum with a long tail at red wavelengths which creates spectral cross talk between different detection channels and makes it difficult to quantitate the amounts of different probes. Qdots, on the other hand, are of interest in biology for several reasons, including (1) higher extinction coefficients; (2) higher QYs; (3) less photobleaching; (4) absorbance and emissions can be tuned with size; (5) generally broad excitation windows but narrow emission peaks; (6) multiple Qdots can be used in the same assay with minimal interference with each other; (7) toxicity may be less than conventional organic dyes, and (8) the Qdots may be functionalized with different bio-active agents. In addition, NIR-emitting Qdots can be used to avoid interference from the autofluorescence, since cell, hemoglobin and water have lower absorption coefficient and scattering effects in the NIR region (650 – 900 nm) (see Figure 22). Light is routinely used for intravital microscopy, but imaging of deeper tissue (500 μm – 1 cm) requires the use of NIR light [346]. Inorganic Qdots are more photostable under ultraviolet excitation than organic molecules, and their fluorescence is more saturated.

Among nanostructured materials, Qdot-based size-tuned emission color offers the potential to develop a multicolor optical coding technique, e.g., by functionalizing different sized CdSe Qdots with different molecules. Researchers have used Qdots for *in vivo* and *in vitro* imaging and diagnostic of live cell as a complement to or replacement of conventional organic dyes [353,354,355]. Some of the studies are tabulated in Table 11. Exploiting the properties of Qdots, such as (1) sharp and UV-NIR tunable fluorescence, (2) charge transfer through fluorescence resonance energy transfer (FRET), (3) surface enhanced Raman spectroscopy (SERS) (4) Radio opacity and paramagnetic properties, and/or (5) MRI contrast agent, it has been shown that Qdots can be used for bioimaging purpose. In the following section, we briefly discuss these imaging techniques that are reported in literature.

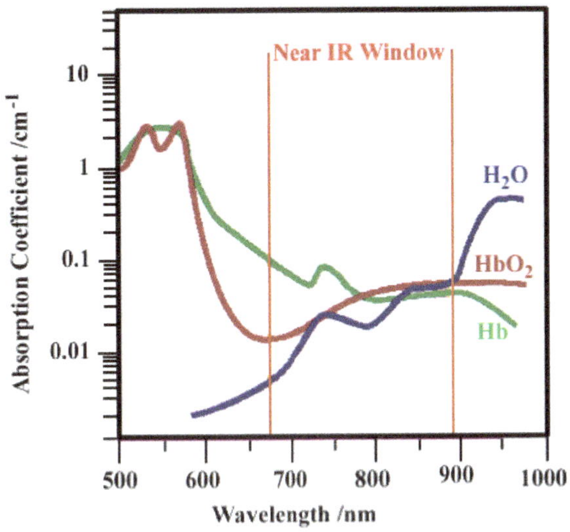

Figure 22. Absorption *versus* wavelength for hemoglobin and water showing the near infrared (NIR) window for *in vivo*imaging to minimize absorption and scattering [346].

Table 11. Selected *in vitro* and *in vivo* bioimaging studies using Qdots.

Qdots	Purpose	Imaging Techniques	Emission/ Size of Qdots	Ref
$CdSe/CdS/SiO_2$	Mouse fibroblast cell imaging	*In vitro* Fluorescence	550 nm & 630 nm	[347]
CdSe/ZnS	Biological detection/ sensing	*In vitro* fluorescence	1–4 nm	[348]
$CdSe/ZnS/SiO_2$	Phagokinetic track imaging	*In vitro* fluorescence	554 nm & 626 nm	[349]
CdSe/ZnS	Tumor vasculature and lung endothelium imaging	*In vitro* and *in vivo* fluorescence	<10 nm	[350]
CdTe/CdSe	Cancer cell lymph nodes imaging	*In vivo* Fluorescence	NIR	[351]
CdSe/ZnS	Maltose binding Protein	*In vitro* FRET	560 nm	[352]

FRET: Fluorescence resonance energy transfer; NIR; near infrared.

Fluorescence for Bioimaging

Qdots fluorescence-based bioimaging [356,357,358] can be broadly classified into four types of modes: intensity, spectrum, lifetime and time-gated. All of these modes can be used at the same time for multimodality imaging. Generally, a high QY from Qdots is required for intensity-based imaging. On the other hand, their narrow emitting spectra make Qdots suitable for multiple colors imaging. The longer fluorescence lifetime of Qdots compared with that of tissue avoid the noise from autofluorescence. Therefore, there is an advantage to use both lifetime and time-gated modes simultaneously (Figure 23). PL from Qdots has been a widely used tool in biology. In 1998,*Alivisatos* and his group showed that the Qdots were potential candidates for biological applications [347]. To establish the use of Qdots, biotin was covalently bound to the Qdot surface and used to label fibroblasts, which was incubated in phalloidin-biotin and streptavidin.

For biological and medical applications, it is of importance to study the photophysical properties of Qdots in living cells [359], particularly photo-induced optical properties of the intracellular Qdots. After injecting thiol-capped CdTe Qdots into living cells, the PL intensity increased with time and the emission peak blue-shifted [359]. De-oxygenation prevented the PL blue shift, suggesting that photoactivated oxygen was responsible. The activated oxygen is presumably formed from the oxygen that intercalates the thiol layer at the Qdot core surface. When Qdots are used as fluorescence probes for cellular imaging, the effects of the PL blue shift and photobleaching must be considered.

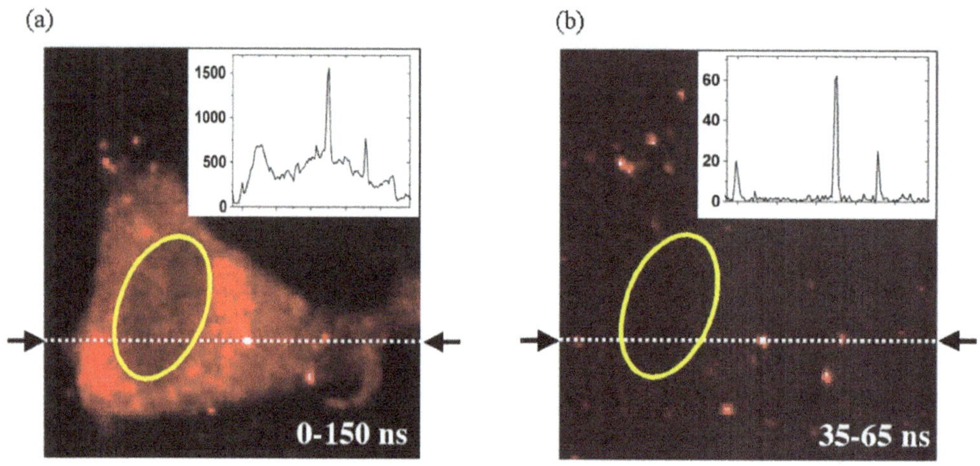

Figure 23. Time-resolved confocal optical micrography of 3T3 cell. (a) A micrograph acquired from all the detected photons; (b) Time-gated micrograph constructed from only photons that arrived 35–65 ns after the laser pulse (laser intensity: 0.1 kW.cm-2; integration time per pixel: 25 ms. Inset figures show cross-sections along the same horizontal line (indicated by the black arrows) for (a) and (b) [360] (Copyright Optical Society of America).

Spectral encoding Qdot technology [338,361] is expected to open new opportunities in gene expression studies, high-throughput screening, and medical diagnostics. The broad absorption spectra of the Qdots allow single wavelength excitation of emission from different-sized Qdots. Multicolor optical coding for biological assays has been achieved by using different sizes of CdSe quantum dots with precisely controlled ratios. The use of 10 intensity levels and 6 colors could theoretically encode one million nucleic acid or protein sequences. *Nie et al.* embedded different-sized Qdots into highly uniform and reproducible polymeric microbeads, which yielded bead identification accuracies as high as 99.99% [362].

The luminescent lifetime of CdSe Qdots (several tens of nanoseconds) is longer than that of cell autofluorescence (~1 ns), which permits measurement of marker spectra and location without high backgrounds through the use of time-gated fluorescent spectroscopy and/or microscopy. In addition, the photostability of CdSe is much better than that of conventional organic dyes [363], allowing data acquisition over long times with continuous excitation. Compared with conventional organic dyes, Qdots have longer lifetime allowing acquisition of low background PL images by using time-gated fluorescent microscopy, as shown in Figure 22 [360]. In an another study [364] CdSe Qdots-based deep tissue imaging of the vasculature system was carried out highlighting various internal structures. The mice used in this study showed no ill effects from the Cd-containing labels.

In order to enhance the lifetime of the emission, some transition or rare earth elements are intentionally incorporated into the Qdots. These activators create local quantum states that lie within the band-gap and provide states for excited electrons or traps for charge carriers and result in radiative relaxations towards the ground state. For transition metal ions such as Mn^{2+}, the lifetime of the luminescence [46,76,180] is on the order of milliseconds due to the forbidden *d-d*transition. *Santra et al.* [93,95,365] demonstrated *in vivo* bioimaging capability using amine modified Mn-doped CdS/ZnS core/shell Qdots conjugated to a TAT peptide (a cell penetrating peptide). Transmission optical and fluorescence micrographs (Figures 24(a) and (b)) of a cross-section of fixed brain tissue clearly showed the blood capillaries (broken white circle in Figure 24(a)) and surrounding brain cells. It was also shown that the TAT conjugated Qdots reached the nucleus of the brain cells (green-circled brown spots in Figure 24(a)). It is well known that the TAT peptide can rapidly translocate through the plasma membrane and accumulate in the cell nucleus [93]. The histological analysis of the brain tissue supports the fact that TAT-conjugated Qdots crossed the blood-brain-barrier, migrated to brain parenchyma and reached the cell nuclei. Endothelial cells in the blood capillaries were found heavily loaded with CdS:Mn/ZnS Qdots and appeared as bright yellow lines in Figure 24(b).

The use of NIR photons is promising for biomedical imaging in living tissue due to longer attenuation distances and lack of autofluorescence in the IR region. This technology often requires exogenous contrast agents with combinations of hydrodynamic diameter, absorption, QY and stability that are not possible with conventional organic dyes [351]. Qdots-based contrast agent offers these properties. In addition, the emission can be tuned to the NIR window (see Figure 22) either by controlling the size of the Qdots or by incorporating rare-earth activators. The emission of CdTe/CdSe Qdots can be tuned into the NIR while preserving the absorption cross-section. It was shown that a polydentate phosphine coating onto the Qdots made the Qdots water soluble, allowing them to be dispersed in serum. Injection of only 400 pmol of NIR emitting Qdots permitted real time imaging of sentinel lymph nodes that were 1 cm below the surface using an excitation power density of only 5 mW/cm^2 [351].

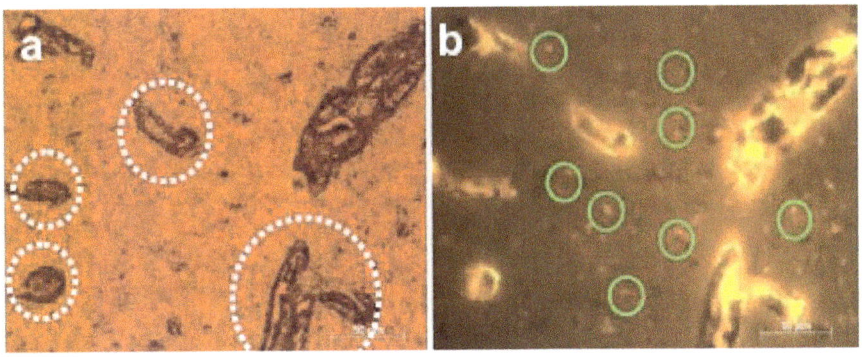

Figure 24. (a) Transmission optical (b) fluorescence microscopy images (40X) of cross-section of fixed brain tissue showing luminescence in (b) from CdS:Mn/ZnS core/shell Qdots [93] – Reproduced by permission of The Royal Society of Chemistry (hyperlink to the article: http://dx.doi.org/10.1039/b503234b).

Use of Fluorescence Resonance Energy Transfer in Bioimaging

Fluorescence resonance energy transfer (FRET) is a phenomenon in which photo-excitation energy is transferred from a donor fluorophor to an acceptor molecule. Based on *Förster* theory, the rate of this energy transfer depends on the spectral overlap of donor emission and acceptor absorption and the donor-acceptor spatial arrangement [366].The ability of Qdots to participate in FRET provides a mechanism for signal transduction in optical sensing schemes. CdSe Qdots can be used to build on-off switches by utilizing *Förster* resonance energy transfer between the Qdot donor and an organic acceptor. In such an optical sensing scheme, Qdots could act both as a donor and an acceptor. *Mattoussi and co-works* [367,368] have studied such an energy transfer between donor Qdots and acceptor dye molecules. In the blend of water-soluble CdSe/ZnS Qdots and maltose binding protein (MBP), MBP was assembled onto the surface of Qdots by both electrostatic self-assembly and metal affinity coordination (Figure 25). With increased fraction of MBPs, emission from the dye increased while that from Qdots decreased. In addition, the emission intensity from dye-labeled MBPs was dependent on the emitted color from different sized Qdots and the spectral overlap (as expected from *Förster* theory). *Mattoussi et al.* also investigated Qdots as an acceptor in a FRET process [369]. This on-off switch has the potential to be used as a sensor in many important applications, including healthcare, environmental monitoring and biodefense systems.

All of these experiments confirmed that water-soluble Qdots have potential applications in biosensor or bioimaging. FRET has been utilized to probe biological activity. *Patolsky et al.* reported that telomerization and DNA replication can be monitored with CdSe/ZnS Qdots [370]. As telomerzation proceeded, the emission from Texas-Red-labeled dUTP increased while that from quantum dots decreased *via* FRET. In replication studies, the Texas-Red dUTP was brought into close proximity of the nanocrystal, resulting in FRET from Qdots to organic dye. These results suggested the possibility of using Qdots in the detection of cancer cells or in amplification of DNA on chip arrays.

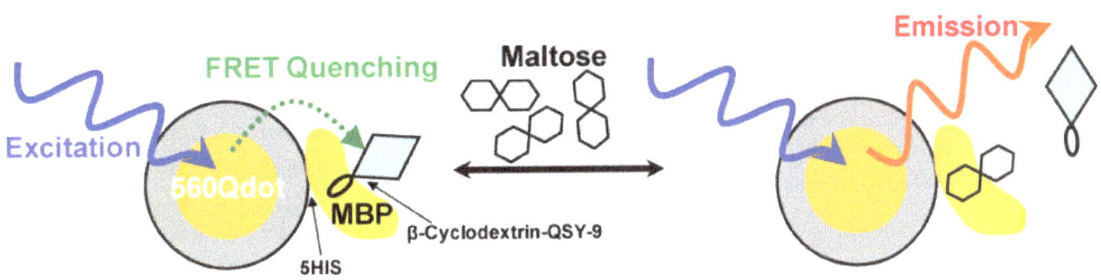

Figure 25. A schematic showing on the function of 560nm Qdot-MBP nanosensor. Each 560 nm emitting Qdot is surrounded by an average of ~10 MBP moieties (a single MBP is shown for simplicity). Formation of Qdot-MBP-β-CD-QSY9 (maximum absorption ~565 nm) results in quenching of Qdot emission. Added maltose displaces β-CD-QSY9 from the sensor assembly, resulting in an increase in direct Qdot emission (Reprinted by permission from Macmillan Publishers Ltd. [Nature Materials] [352] Copyright 2003).

FRET coupled to quenching provided alternative path for sensing. The luminescence *via* FRET is turned on by the appearance of analyte, which displaces a quencher or a terminal energy acceptor.*Mattoussi et al.* have developed a sensor for maltose by adapting their CdSe-MBP conjugates for analyte displacement strategies [352]. First, a β-cyclodextrin conjugated to a non-fluorescent QSY9 quencher dye was docked to the MBP saccharide binding site of the CdSe/ZnS-MBP. Second, maltose displaced the β-cyclodextrin-QSY9 conjugate to restore quantum dots emission. This approach is general and the concept of an antibody fragment bound to a Qdot surface through noncovalent self-assembly should find wide use for other analytes of interest.

Surface Enhanced Raman Spectroscopy

Surface enhanced Raman Spectroscopy (SERS) is a near-field probe which is sensitive to local environments. Qdot-based SERS can be used two ways in biomedical application, first as single molecule-based SERS that measures the unique fingerprint spectra of pure analytes on a Qdot. In a second approach, Qdots are covered by a monolayer of analyte, and SERS spectra are obtained from an ensemble of nanoparticles. In this case, population-average data are determined and one can get robust data from complex milieu or whole blood circulation.

By using the single molecule SERS technique along with Au/silica or core/shell Qdots, biomarker can be very sensitive. Currently, bioconjugated SERS have also been developed to identify protein biomakers on the surfaces of living cancer cells. For example, targeted gold nanoparticles are prepared by using a mixture of thiol–polyethylene glycol (PEG) and a heterofunctional PEG for live cancer cell detection, which binds to the epidermal growth factor receptor (EGFR) with high specificity and affinity [371]. Human head and neck carcinoma cells are EGFR-positive and can give strong SERS signals [372].

In contrast, the human non-small cell lung carcinoma does not express EGFR receptors, showing little or no SERS signals. Single cell profiling studies are of great clinical significance because EGFR is a validated protein target for monoclonal antibody and protein-kinase based therapies. In addition, Qdots-based SERS can be used for *in vivo* tumor targeting and detection. *Qian et al.*reported that they injected small dosage of nanoparticles into subcutaneous and deep muscular sites in live animals and highly resolved SERS signals were obtained [373]. It is estimated that the achievable penetration depth is about 1–2 cm for *in vivo* SERS tumor detection.

Radio-Opacity and Paramagnetic Properties

CdS:Mn/ZnS core/shell Qdots were characterized for radio-opacity and magnetic hysteresis [94,95,365] for possible use as contrast agent in computer tomography (CT) and MRI due to electron dense Cd and paramagnetic Mn, respectively [94]. For radio-opacity, the Qdots sample was compared with a conventional radio-opaque dye, *Omnipaque*, used for CT scans and angiography. It was found that the X-ray absorption of Qdots was less than that of *Omnipaque*. In this respect, Qdot may not provide sufficient contrast for current radiographic practice. A superconductor quantum interface device magnetometer was used to measure the magnetization of CdS:Mn/ZnS Qdots. A typical room temperature hysteresis curve for paramagnetic CdS:Mn was observed, but it is too small for MRI imaging.

Magnetic Resonance-based Bioimaging

MRI is essentially proton nuclear magnetic resonance (NMR) [332,374]. Protons are excited with short pulses of radio frequency radiation and the free induction decay as they relax is measured and deconvoluted by a *Fourier* transform, which provides an image of the tissue. Areas of high proton densities, e.g., water or lipid molecules, have a strong signal and appear bright. Areas of bone or tendon, which have a low proton density, have a weak signal and appear dark. A major limitation of MRI is its inability to distinguish between various types of soft tissues where the relative proton densities can be very similar. Regions having air pockets and fecal matter, such as the bowel, are hard to image because of inconsistent proton density. Therefore, various contrast agents such as perfluorochemicals, oils, fats and nanomaterials, have been studied to circumvent these imaging problems. Unlike organic molecules, nanomaterials-based contrast agents are miscible in aqueous systems which allow them to be used intravenously. Therefore, they are well suited for *in vivo* applications such as tracking blood flow in the brain. Another advantage of Qdots over other nanoparticles is that they offer multimodal imaging capabilities [83]. However, appropriate functionalization of the Qdots is needed in order to make the Qdot a suitable contrast agent for MRI.

Qdot-based contrast agents change the strength of the MRI signal at a desired location. For example, paramagnetic contrast agents change the rate at which protons decay from their excited state to the ground state, allowing more rapid decay through energy transfer to a neighboring nucleus [374]. As a result, regions containing the paramagnetic contrast agent appear darker in an MRI than regions without the agent. When paramagnetic Qdots are delivered to the liver, the uptake rate of Qdots by healthy liver cells is much higher that by diseased cells. Consequently, the healthy regions are darker than the diseased regions. Several experimental reports in the literature demonstrate the usefulness and multi-modal use of Qdots in MRI applications [69,375,376,377,378]. In these reports, the Qdots are often coated with a water soluble paramagnetic coating to enhance contrast. *Yang et al.* [69] synthesized a water soluble Gd-functionalized silica coated CdS:Mn/ZnS Qdots and studied them a for MRI contrast agent. Longitudinal (T_1) and traverse (T_2) proton relaxation times were measured with a single slice, spin-echo image sequence at 4.7 Tesla. Increased magnetic resonance (MR) signal intensity, as shown in Figure 26(a), was observed [69,95,365] with increasing Gd concentrations due to the shorter water relaxation time T_1. In T_2weighted images, the MR signal intensity was substantially decreased by the effects of increased Gd on the T_2 of water (Figure 26(b)). For control experiments, T_1 and T_2 weighted images of serial dilutions of Qdots without Gd^{3+} ions were recorded and could not be distinguished from those of deionized (DI) water.

Normalized T_1 and T_2 weighted intensities *versus* repetition time (T_R) and echo time (T_E), respectively, for DI water and a series of diluted Gd-functionalized Qdots (from 0.36 to 0.0012 mM of Gd) showed increasingly faster recovery of longitudinal magnetization and faster decay of transverse magnetization for increased Gd concentrations [69,95,365], as shown in Figure 26(c) andFigure 26(d) . The efficacy of a contrast agent is generally expressed by its relaxivity (R_i), that is defined by $1/T_i = 1/T^o + R_i[\mathrm{Gd}]$, [379], where T_i is the relaxation time for a contrast agent solution concentration of [Gd], and T^o is the relaxation time in the absence of a contrast agent. The relaxivities R_1 and R_2 were found to be 20.5 and 151 $mM^{-1}s^{-1}$, respectively. Compared to commercially available contrast agents, Gd-functionalized Qdots exhibited higher R_1 and R_2 values under the same magnetic field strength of 4.7 T [380]. High relaxivities were attributed to a reduced tumbling rate of the Gd^{3+}-based contrast agents by grafting the contrast agent to rigid macromolecules and avoiding free rotation of the chelate [381,382]. Although the Gd-Qdots can serve as either a T_1 or T_2 contrast agent, the R_2/R_1 ratio of ~7.4 indicates that they may be most effective as a T_2 contrast agent.

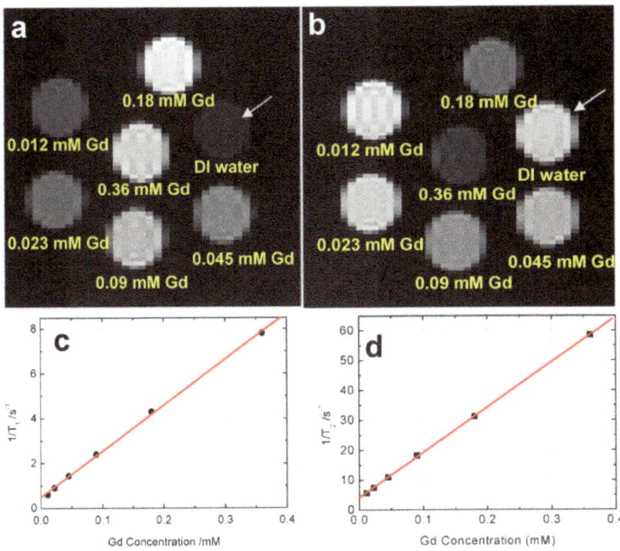

Figure 26. Magnetic resonance data from Gd-functionalized silica-coated CdS:Mn/ZnS Qdots: (a) T_1-weighted (repetition time (T_R) = 11000 ms, echo time (T_E) = 4.2 ms), (b) T_2-weighted (T_R = 11000 ms, T_E = 24 ms) images of deionized (DI) water and serial dilutions of Gd-functionalized Qdots (0.36, 0.18, 0.09, 0.045, 0.023, 0.012 mM of Gd). Linear plots of Gd concentration*versus* $1/T_1$ (c) and $1/T_2$ (d) to obtain ionic relaxivities of R_1 and R_2 of Gd-Qdots [reproduced with permission from [69]. Copyright Wiley-VCH Verlag GmbH & Co. KGaA].

Recent advances in use of Qdots in biology studies support the promise of a quantum leap in the extensive use of Qdots in future biological applications. It is predicted that Qdots will be able to provide unprecedented sensitivity and selectivity over the traditional practices for molecular imaging. The use of Qdots emitting in the NIR region will provide greater sensitivity and the longer lifetime of their excited states (as compared to organic fluorophores and proteins) will lead to for improved bioimaging. Despite the advantages for Qdots-based bioimaging, several issues related to Qdots need to be addressed before *in vivo* use, especially their toxicity. *Son et al.* [383] reported an ion exchange at the surface of CdSe Qdots that suppressed their PL intensity and led to the release of Cd^{2+}, which is known to be toxic to human. The only way to partially recover the PL emission was to add excess Cd^{2+}, which is unacceptable for biological application. In addition, bare Qdots were reported to be cytotoxic [384]. Some of the Qdots properties are limiting, such as size of Qdots, which sometimes is larger than the traditional organic marker dyes. As research on nanoparticles with novel properties continues, it should be possible to overcome these drawbacks and to develop multifunctional, multimodal Qdot-based systems for better biological imaging.

PERSPECTIVE

In this review, selected aspects of the structure, properties, application and performance of Qdots have been discussed. Among the various branches in nanotechnology, these zero dimension nanostructures have paved the way for numerous advances in both fundamental and applied sciences. This is due to the fact that the Qdot exhibit significantly different optical, electronic and physical properties as compared to bulk materials. With respect to synthesis of Qdots, significant progress has been made in studies of the growth kinetics through both theoretical models and experimental data. Procedures ranging from simple wet chemical methods to very sophisticated and equipment-intensive atomic layer deposition techniques are being used to synthesize Qdots. The 'bottom-up' approaches are being widely explored resulting in a wider variety of methods to generate Qdots. Despite the large amount of research, there is still a lot to understand about the use of Qdots in large scale biological and solid-state optical applications, as we discussed throughout the article. For example, Qdot-based LED showed EQE ~1.5% (see Figure 27). It was also demonstrated that a QLED-based display is possible. Although preliminary experiments at the laboratory level have been successful, scaling up the production and retaining the properties of Qdots is not trivial. Continued research and development on Qdots will provide further improvements in quantum efficiency, better device fabrication and new materials which extend the emission into the near IR region.

Nanobiotechnology is predicted to be a major R&D area for this next century. However several stumbling blocks need to be overcome to ensure a close marriage of biology and nanotechnology. Bio-functionalizing the Qdots and interface engineering to control when electronic can pass are major challenges. Several applications, like displays, lighting, selective sensors, bio-imaging, MRI contrast agents, and bio-labels, need attentions to improve further and to answer questions about Qdot synthesis, properties, ageing and toxicity from the scientific and engineering community.

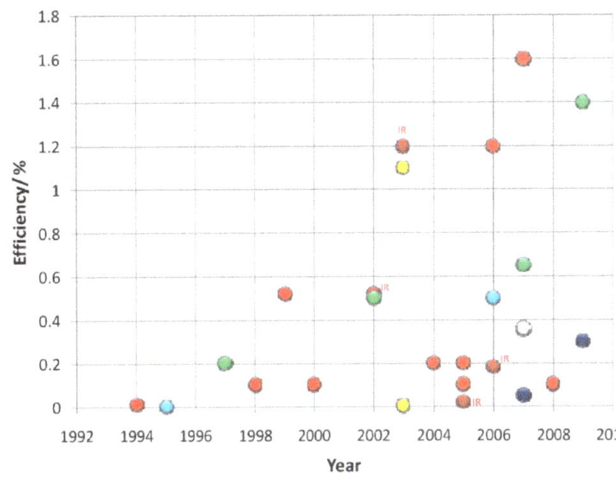

Figure 27. Efficiency of Qdot-based light emitting diode (QLED) *vs.* year reported in the literature; Colors of the legends represent the emitted color from the QLED (red: 601–700 nm; yellow: 561–600; green: 531–560 nm; sky blue: 491–530 nm; blue: 440–490 nm; IR: infrared emitting QLEDs are marked with IR; white: white emitting QLED).

ACKNOWLEDGEMENT

This work was supported by ARO Grant W911NF-07-1-0545.

REFERENCES

1. Henglein, A. Small-particle research–physicochemical properties of extremely small colloidal metal and semiconductor particles. *Chem. Rev.* 1989, *89*, 1861–1873.

2. Trindade, T.; O'Brien, P.; Pickett, N.L. Nanocrystalline semiconductors: Synthesis, properties, and perspectives. *Chem. Mater.* 2001, *13*, 3843–3858.

3. Kuchibhatla, S.; Karakoti, A.S.; Bera, D.; Seal, S. One dimensional nanostructured materials.*Prog. Mater. Sci.* 2007, *52*, 699–913.

4. Bera, D.; Kuiry, S.C.; Seal, S. Synthesis of nanostructured materials using template-assisted electrodeposition. *JOM* 2004, *56*, 49–53.

5. Alivisatos, A.P. Perspectives on the physical chemistry of semiconductor nanocrystals. *J. Phys. Chem.* 1996, *100*, 13226–13239.

6. Bera, D.; Qian, L.; Holloway, P.H. *Phosphor Quantum Dots*; John WIley & Sons, Ltd: West Sussex, UK, 2008.

7. Walter, P.; Welcomme, E.; Hallegot, P.; Zaluzec, N.J.; Deeb, C.; Castaing, J.; Veyssiere, P.; Breniaux, R.; Leveque, J.L.; Tsoucaris, G. Early use of PbS nanotechnology for an ancient hair dyeing formula. *Nano Lett.* 2006, *6*, 2215–2219.

8. Rocksby, H.P. Color of selenium ruby glasses. *J. Soc. Glass Technol.* 1932, *16*, 171.

9. Ekimov, A.I.; Onushchenko, A.A. Quantum size effect in 3-dimensional microscopic semiconductor crystals. *JETP Lett.* 1981, *34*, 345–349.

10. Efros, A.L.; Efros, A.L. Interband absorption of light in a semiconductor sphere. *Sov. Phys. Semicond.* 1982, *16*, 772–775.

11. Rossetti, R.; Ellison, J.L.; Gibson, J.M.; Brus, L.E. Size effects in the excited electronic states of small colloidal CdS crystallites. *J. Chem. Phys.* 1984, *80*, 4464–4469.

12. Spanhel, L.; Haase, M.; Weller, H.; Henglein, A. Photochemistry of colloidal semiconductors .20. Surface modification and stability of strong luminescing CdS particles. *J. Am. Chem. Soc.*1987, *109*, 5649–5655.

13. Kortan, A.R.; Hull, R.; Opila, R.L.; Bawendi, M.G.; Steigerwald, M.L.; Carroll, P.J.; Brus, L.E. Nucleation and growth of CdSe on ZnS quantum crystallite seeds, and vice versa, in inverse micelle media. *J. Am. Chem. Soc.* 1990, *112*, 1327–1332.

14. Murray, C.B.; Norris, D.J.; Bawendi, M.G. Synthesis and characterization of nearly monodisperse CdE (E = S, Se, Te) semiconductor nanocrystallites. *J. Am. Chem. Soc.* 1993, *115*, 8706–8715.

15. Grieser, F.; Furlong, D.N.; Scoberg, D.; Ichinose, I.; Kimizuka, N.; Kunitake, T. Size-quantized semiconductor cadmium chalcogenide particles in Langmuir-Blodgett-films. *J. Chem. Soc. Faraday Trans.* 1992, *88*, 2207–2214.

16. Sundaram, M.; Chalmers, S.A.; Hopkins, P.F.; Gossard, A.C. New quantum sructures. *Science*1991, *254*, 1326–1335.

17. Cingolani, R.; Ploog, K. Frequency and density dependent radiative recombination processes in III-V semiconductor quantum-wells and superlattices. *Adv. Phys.* 1991, *40*, 535–623.

18. Alivisatos, A.P. Semiconductor clusters, nanocrystals, and quantum dots. *Science* 1996, *271*, 933–937.

19. Klimov, V.I.; Mikhailovsky, A.A.; Xu, S.; Malko, A.; Hollingsworth, J.A.; Leatherdale, C.A.; Eisler, H.J.; Bawendi, M.G. Optical gain and stimulated emission in nanocrystal quantum dots. *Science* 2000, *290*, 314–317.

20. Artemyev, M.V.; Woggon, U.; Wannemacher, R.; Jaschinski, H.; Langbein, W. Light trapped in a photonic dot: Microspheres act as a cavity for quantum dot emission. *Nano Lett.* 2001, *1*, 309–314.

21. Graham-Rowe, D. From dots to devices. *Nat. Photonics* 2009, *3*, 307–309.

22. Coe-Sullivan, S. Quantum dot developments. *Nat. Photonics* 2009, *3*, 315–316.

23. Bera, D.; Qian, L.; Holloway, P.H. *Semiconducting Quantum Dots for Bioimaging*; Informa Heathcare: New York, NY, USA, 2009; Vol. 191.

24. Yoffe, A.D. Low-dimensional systems–quantum-size effects and electronic-properties of semiconductor microcrystallites (zero-dimensional systems) and some quasi-2-dimensional systems. *Adv. Phys.* 1993, *42*, 173–266.

25. Yoffe, A.D. Semiconductor quantum dots and related systems: Electronic, optical, luminescence and related properties of low dimensional systems. *Adv. Phys.* 2001, *50*, 1–208.

26. Jang, E.; Jun, S.; Pu, L. High quality CdSeS nanocrystals synthesized by facile single injection process and their electroluminescence. *Chem. Comm.* 2003, 2964–2965.

27. Steckel, J.S.; Snee, P.; Coe-Sullivan, S.; Zimmer, J.R.; Halpert, J.E.; Anikeeva, P.; Kim, L.A.; Bulovic, V.; Bawendi, M.G. Color-saturated green-emitting QD-LEDs. *Angew. Chem. Int. Ed.*2006, *45*, 5796–5799.

28. Yang, H.S.; Holloway, P.H.; Ratna, B.B. Photoluminescent and electroluminescent properties of Mn-doped ZnS nanocrystals. *J. Appl. Phys.* 2003, *93*, 586–592.

29. Tolbert, S.H.; Herhold, A.B.; Johnson, C.S.; Alivisatos, A.P. Comparison of quantum confinement effects on the electronic absorption-spectra of direct and indirect gap semiconductor nanocrystals. *Phys. Rev. Lett.* 1994, *73*, 3266–3269.

30. Tolbert, S.H.; Alivisatos, A.P. The wurtzite to rock-salt structural transformation in CdSe nanocrystals under high-pressure. *J. Chem. Phys.* 1995, *102*, 4642–4656.

31. Woggon, U. *Optical Properties of Semiconductor Quantum Dots*; Springer-Verlag: Berlin, Germay, 1997; Volume 136.

32. Bryan, J.D.; Gamelin, D.R. Doped semiconductor nanocrystals: Synthesis, characterization, physical properties, and applications. *Prog. Inorg. Chem.* 2005, *Volume 54*, 47–126.

33. Erwin, S.C.; Zu, L.J.; Haftel, M.I.; Efros, A.L.; Kennedy, T.A.; Norris, D.J. Doping semiconductor nanocrystals. *Nature (London)* 2005, *436*, 91–94.

34. Norris, D.J.; Efros, A.L.; Erwin, S.C. Doped nanocrystals. *Science* 2008, *319*, 1776–1779.

35. Pradhan, N.; Goorskey, D.; Thessing, J.; Peng, X.G. An alternative of CdSe nanocrystal emitters: Pure and tunable impurity emissions in ZnSe nanocrystals. *J. Am. Chem. Soc.* 2005,*127*, 17586–17587.

36. Zheng, Y.H.; Zhao, J.H.; Bi, J.F.; Wang, W.Z.; Ji, Y.; Wu, X. G.; Xia, J.B. Cr-doped InAs self-organized diluted magnetic quantum dots with room-temperature ferromagnetism. *Chin. Phys. Lett.* 2007, *24*, 2118–2121.
37. Bhargava, R.N. Doped nanocrystalline materials–Physics and applications. *J. Lumin.* 1996, *70*, 85–94.
38. Stowell, C.A.; Wiacek, R.J.; Saunders, A.E.; Korgel, B.A. Synthesis and characterization of dilute magnetic semiconductor manganese-doped indium arsenide nanocrystals. *Nano Lett.*2003, *3*, 1441–1447.
39. Beaulac, R.; Schneider, L.; Archer, P.I.; Bacher, G.; Gamelin, D.R. Light-induced spontaneous magnetization in doped colloidal quantum dots. *Science* 2009, *325*, 973–976.
40. Radovanovic, P.V.; Gamelin, D.R. Electronic absorption spectroscopy of cobalt ions in diluted magnetic semiconductor quantum dots: Demonstration of an isocrystalline core/shell synthetic method. *J. Am. Chem. Soc.* 2001, *123*, 12207–12214.
41. Hou, S.L.; Zhang, X.Q.; Mao, H.B.; Wang, J.Q.; Zhu, Z.Q.; Jing, W.P. Photoluminescence and XPS investigations of Cu^{2+}-doped ZnS quantum dots capped with polyvinylpyrrolidone.*Phys. Status Solidi B* 2009, *246*, 2333–2336.
42. Kim, J.U.; Kim, Y.K.; Yang, H. Reverse micelle-derived Cu-doped Zn1-xCdxS quantum dots and their core/shell structure. *J. Colloid. Interface Sci.* 2010, *341*, 59–63.
43. Murugadoss, A.; Chattopadhyay, A. Tuning photoluminescence of ZnS nanoparticles by silver. *Bull. Mater. Sci.* 2008, *31*, 533–539.
44. Fujii, M.; Yamaguchi, Y.; Takase, Y.; Ninomiya, K.; Hayashi, S. Photoluminescence from impurity codoped and compensated Si nanocrystals. *Appl. Phys. Lett.* 2005, *87*, 211919 1–3.
45. Orlinskii, S.B.; Schmidt, J.; Baranov, P.G.; Hofmann, D.M.; Donega, C.D.; Meijerink, A. Probing the wave function of shallow Li and Na donors in ZnO nanoparticles. *Phys. Rev. Lett.*2004, *92*, 047603 1–4.
46. Yang, H.S.; Santra, S.; Holloway, P.H. Syntheses and applications of Mn-doped II-VI semiconductor nanocrystals. *J. Nanosci. Nanotechnology* 2005, *5*, 1364–1375.
47. Yang, Y.A.; Chen, O.; Angerhofer, A.; Cao, Y.C. Radial-position-controlled doping in CdS/ZnS core/shell nanocrystals. *J. Am. Chem. Soc.* 2006, *128*, 12428–12429.
48. Kulkarni, S.K.; Winkler, U.; Deshmukh, N.; Borse, P.H.; Fink, R.; Umbach, E. Investigations on chemically capped CdS, ZnS and ZnCdS nanoparticles. *Appl. Surf. Sci.* 2001, *169*, 438–446.
49. Wang, W.Z.; Germanenko, I.; El-Shall, M.S. Room-temperature synthesis and characterization of nanocrystalline CdS, ZnS, and CdxZn1-xS. *Chem. Mater.* 2002, *14*, 3028–3033.
50. Petrov, D.V.; Santos, B.S.; Pereira, G.A.L.; Donega, C.D. Size and band-gap dependences of the first hyperpolarizability of CdxZn1-xS nanocrystals. *J. Phys. Chem. B* 2002, *106*, 5325–5334.
51. Bailey, R.E.; Nie, S.M. Alloyed semiconductor quantum dots: Tuning the optical properties without changing the particle size. *J. Am. Chem. Soc.* 2003, *125*, 7100–7106.
52. Zhong, X.H.; Han, M.Y.; Dong, Z.L.; White, T.J.; Knoll, W. Composition-tunable ZnxCd1-xSe nanocrystals with high luminescence and stability. *J. Am. Chem. Soc.* 2003, *125*, 8589–8594.
53. Gurusinghe, N.P.; Hewa-Kasakarage, N.N.; Zamkov, M. Composition-tunable properties of CdSxTe1-x alloy nanocrystals. *J. Phys. Chem. C* 2008, *112*, 12795–12800.
54. Korgel, B.A.; Monbouquette, H.G. Controlled synthesis of mixed core and layered (Zn,Cd)S and (Hg,Cd) S nanocrystals within phosphatidylcholine vesicles. *Langmuir* 2000, *16*, 3588–3594.
55. Harrison, M.T.; Kershaw, S.V.; Burt, M.G.; Eychmuller, A.; Weller, H.; Rogach, A.L. Wet chemical synthesis and spectroscopic study of CdHgTe nanocrystals with strong near-infrared luminescence. *Mater. Sci. Eng. B* 2000, *69*, 355–360.
56. Lee, H.; Holloway, P.H.; Yang, H. Synthesis and characterization of colloidal ternary ZnCdSe semiconductor nanorods. *J. Chem. Phys.* 2006, *125*, 2363181–2363189.
57. Zhong, X.H.; Feng, Y.Y.; Knoll, W.; Han, M.Y. Alloyed ZnxCd1-xS nanocrystals with highly narrow luminescence spectral width. *J. Am. Chem. Soc.* 2003, *125*, 13559–13563.
58. Smith, A.M.; Gao, X.H.; Nie, S.M. Quantum dot nanocrystals for *in vivo* molecular and cellular imaging. *Photochem. Photobiol.* 2004, *80*, 377–385.
59. Bailey, R.E.; Smith, A.M.; Nie, S.M. Quantum dots in biology and medicine. *Physica E* 2004,*25*, 1–12.
60. Wang, Y.; Herron, N. Nanometer-sized semiconductor clusters–materials synthesis, quantum size effects, and photophysical properties. *J. Phys. Chem.* 1991, *95*, 525–532.
61. Bang, J.; Yang, H.; Holloway, P.H. Enhanced and stable green emission of ZnO nanoparticles by surface segregation of Mg. *Nanotechnology* 2006, *17*, 973–978.
62. Kucur, E.; Bucking, W.; Giernoth, R.; Nann, T. Determination of defect states in semiconductor nanocrystals by cyclic voltammetry. *J. Phys. Chem. B* 2005, *109*, 20355–20360.
63. Colvin, V.L.; Goldstein, A.N.; Alivisatos, A.P. Semiconductor nanocrystals covalently bound to metal-surfaces with self-assembled monolayers. *J. Am. Chem. Soc.* 1992, *114*, 5221–5230.
64. Dabbousi, B.O.; Murray, C.B.; Rubner, M.F.; Bawendi, M.G. Langmuir-Blodgett manipulation of size-selected CdSe nanocrystallites. *Chem. Mater.* 1994, *6*, 216–219.
65. Murray, C.B.; Kagan, C.R.; Bawendi, M.G. Self-organization of CdSe nanocrystallites into 3-dimensional quantum-dot superlattices. *Science* 1995, *270*, 1335–1338.
66. Wang, Q.; Kuo, Y.C.; Wang, Y.W.; Shin, G.; Ruengruglikit, C.; Huang, Q.R. Luminescent properties of water-soluble denatured bovine serum albumin-

coated CdTe quantum dots. *J. Phys. Chem. B* 2006, *110*, 16860–16866.

67. Ma, N.; Yang, J.; Stewart, K.M.; Kelley, S.O. DNA-passivated CdS nanocrystals: Luminescence, bioimaging, and toxicity profiles. *Langmuir* 2007, *23*, 12783–12787.

68. Chen, X.B.; Lou, Y.B.; Samia, A.C.; Burda, C. Coherency strain effects on the optical response of core/shell heteronanostructures. *Nano Lett.* 2003, *3*, 799–803.

69. Yang, H.S.; Santra, S.; Walter, G.A.; Holloway, P.H. Gd-III-functionalized fluorescent quantum dots as multimodal imaging probes. *Adv. Mater.* 2006, *18*, 2890–2894.

70. Peng, X.G.; Schlamp, M.C.; Kadavanich, A.V.; Alivisatos, A.P. Epitaxial growth of highly luminescent CdSe/CdS core/shell nanocrystals with photostability and electronic accessibility. *J. Am. Chem. Soc.* 1997, *119*, 7019–7029.

71. Reiss, P.; Bleuse, J.; Pron, A. Highly luminescent CdSe/ZnSe core/shell nanocrystals of low size dispersion. *Nano Lett.* 2002, *2*, 781–784.

72. Qian, L.; Bera, D.; Tseng, T.K.; Holloway, P.H. High efficiency photoluminescence from silica-coated CdSe quantum dots. *Appl. Phys. Lett.* 2009, *94*, 073112.

73. Geng, B.Y.; Du, Q.B.; Liu, X.W.; Ma, J.Z.; Wei, X.W.; Zhang, L.D. One-step synthesis and enhanced blue emission of carbon-encapsulated single-crystalline ZnSe nanoparticles. *Appl. Phys. Lett.* 2006, *89*, 033115.

74. Ebenstein, Y.; Mokari, T.; Banin, U. Fluorescence quantum yield of CdSe/ZnS nanocrystals investigated by correlated atomic-force and single-particle fluorescence microscopy. *Appl. Phys. Lett.* 2002, *80*, 4033–4035.

75. Nazzal, A.Y.; Wang, X.Y.; Qu, L.H.; Yu, W.; Wang, Y.J.; Peng, X.G.; Xiao, M. Environmental effects on photoluminescence of highly luminescent CdSe and CdSe/ZnS core/shell nanocrystals in polymer thin films. *J. Phys. Chem. B* 2004, *108*, 5507–5515.

76. Yang, H.; Holloway, P.H.; Cunningham, G.; Schanze, K.S. CdS:Mn nanocrystals passivated by ZnS: Synthesis and luminescent properties. *J. Chem. Phys.* 2004, *121*, 10233–10240.

77. Nirmal, M.; Dabbousi, B.O.; Bawendi, M.G.; Macklin, J.J.; Trautman, J.K.; Harris, T.D.; Brus, L.E. Fluorescence intermittency in single cadmium selenide nanocrystals. *Nature (London)*1996, *383*, 802–804.

78. Balet, L.P.; Ivanov, S.A.; Piryatinski, A.; Achermann, M.; Klimov, V.I. Inverted core/shell nanocrystals continuously tunable between type-I and type-II localization regimes. *Nano Lett.*2004, *4*, 1485–1488.

79. Kim, S.; Fisher, B.; Eisler, H.J.; Bawendi, M. Type-II quantum dots: CdTe/CdSe (core/shell) and CdSe/ZinTe (core/shell) heterostructures. *J. Am. Chem. Soc.* 2003, *125*, 11466–11467.

80. Yang, H.S.; Holloway, P.H.; Santra, S. Water-soluble silica-overcoated CdS:Mn/ZnS semiconductor quantum dots. *J. Chem. Phys.* 2004, *121*, 7421–7426.

81. Boldt, K.; Bruns, O.T.; Gaponik, N.; Eychmuller, A. Comparative examination of the stability of semiconductor quantum dots in various biochemical buffers. *J. Phys. Chem. B* 2006, *110*, 1959–1963.

82. Wolcott, A.; Gerion, D.; Visconte, M.; Sun, J.; Schwartzberg, A.; Chen, S.W.; Zhang, J.Z. Silica-coated CdTe quantum dots functionalized with thiols for bioconjugation to IgG proteins. *J. Phys. Chem. B* 2006, *110*, 5779–5789.

83. Santra, S.; Bagwe, R.P.; Dutta, D.; Stanley, J.T.; Walter, G.A.; Tan, W.; Moudgil, B.M.; Mericle, R.A. Synthesis and characterization of fluorescent, radio-opaque, and paramagnetic silica nanoparticles for multimodal bioimaging applications. *Adv. Mater.* 2005, *17*, 2165–2169.

84. Santra, S.; Dutta, D.; Moudgil, B.M. Functional dye-doped silica nanoparticles for bioimaging, diagnostics and therapeutics. *Food Bioprod. Process.* 2005, *83*, 136–140.

85. Empedocles, S.A.; Bawendi, M.G. Quantum-confined stark effect in single CdSe nanocrystallite quantum dots. *Science* 1997, *278*, 2114–2117.

86. Jeang, E.H.; Lee, J.H.; Je, K.C.; Yim, S.Y.; Park, S.H.; Choi, Y.S.; Choi, J.G.; Treguer, M.; Cardinal, T. Fabrication and optical characteristics of CdS/Ag metal-semiconductor composite quantum dots. *Bull. Korean Chem. Soc.* 2004, *25*, 934–936.

87. Je, K.C.; Ju, H.; Treguer, M.; Cardinal, T.; Park, S.H. Local field-induced optical properties of Ag-coated CdS quantum dots. *Opt. Express* 2006, *14*, 7994–8000.

88. Talapin, D.V.; Mekis, I.; Gotzinger, S.; Kornowski, A.; Benson, O.; Weller, H. CdSe/CdS/ZnS and CdSe/ZnSe/ZnS core-shell-shell nanocrystals. *J. Phys. Chem. B* 2004, *108*, 18826–18831.

89. Xie, R.G.; Kolb, U.; Li, J.X.; Basche, T.; Mews, A. Synthesis and characterization of highly luminescent CdSe-Core CdS/Zn0.5Cd0.5S/ZnS multishell nanocrystals. *J. Am. Chem. Soc.*2005, *127*, 7480–7488.

90. Xu, J.F.; Xiao, M. Lasing action in colloidal CdS/CdSe/CdS quantum wells. *Appl. Phys. Lett.*2005, *87*, 173117 1.

91. Qian, L.; Bera, D.; Holloway, P. Photoluminescence from ZnS/CdS:Mn/ZnS quantum well quantum dots. *Appl. Phys. Lett.* 2008, *92*, 093103.

92. Santra, S.; Xu, J.S.; Wang, K.M.; Tan, W.H. Luminescent nanoparticle probes for bioimaging.*J. Nanosci. Nanotech.* 2004, *4*, 590–599.

93. Santra, S.; Yang, H.; Stanley, J.T.; Holloway, P.H.; Moudgil, B.M.; Walter, G.; Mericle, R.A. Rapid and effective labeling of brain tissue using TAT-conjugated CdS:Mn/ZnS quantum dots. *Chem. Comm.* 2005, 3144–3146.

94. Santra, S.; Yang, H.S.; Holloway, P.H.; Stanley, J.T.; Mericle, R.A. Synthesis of water-dispersible fluorescent, radio-opaque, and paramagnetic CdS:Mn/ZnS quantum dots: A multifunctional probe for bioimaging. *J. Am. Chem. Soc.* 2005, *127*, 1656–1657.

95. Holloway, P.H.; Yang, H.; Lee, H.; Seo, S.; Santra, S.; Qian, L.; Bera, D. Nanophosphor: PL, EL, and Biological

Markers; Proc. Int. Display Workshops, IDW'06. Otsu, Japan, 6–8 December 2006.

96. Yu, Z.H.; Guo, L.; Du, H.; Krauss, T.; Silcox, J. Shell distribution on colloidal CdSe/ZnS quantum dots. *Nano Lett.* 2005, *5*, 565–570.

97. Yang, H.; Holloway, P.H. Efficient and Photostable ZnS-passivated CdS:Mn luminescent nanocrystals. *Adv. Func. Mater.* 2004, *14*, 152–156.

98. Rosenthal, S.J.; McBride, J.; Pennycook, S.J.; Feldman, L.C. Synthesis, surface studies, composition and structural characterization of CdSe, core/shell and biologically active nanocrystals. *Surf. Sci. Rep.* 2007, *62*, 111–157.

99. McBride, J.R.; Kippeny, T.C.; Pennycook, S.J.; Rosenthal, S.J. Aberration-corrected Z-contrast scanning transmission electron microscopy of CdSe nanocrystals. *Nano Lett.* 2004, *4*, 1279–1283.

100. McBride, J.; Treadway, J.; Feldman, L.C.; Pennycook, S.J.; Rosenthal, S.J. Structural basis for near unity quantum yield core/shell nanostructures. *Nano Lett.* 2006, *6*, 1496–1501.

101. Katari, J.E.B.; Colvin, V.L.; Alivisatos, A.P. X-Ray photoelectron-spectroscopy of CdSe nanocrystals with applications to studies of the nanocrystal surface. *J. Phys. Chem.* 1994, *98*, 4109–4117.

102. Bera, D.; Qian, L.; Sabui, S.; Santra, S.; Holloway, P.H. Photoluminescence of ZnO quantum dots produced by a sol-gel process. *Opt. Mater.* 2008, *30*, 1233–1239.

103. Hines, M.A.; Guyot-Sionnest, P. Synthesis and characterization of strongly luminescing ZnS-capped CdSe nanocrystals. *J. Phys. Chem.* 1996, *100*, 468–471.

104. Dabbousi, B.O.; RodriguezViejo, J.; Mikulec, F.V.; Heine, J.R.; Mattoussi, H.; Ober, R.; Jensen, K.F.; Bawendi, M.G. (CdSe)ZnS core-shell quantum dots: Synthesis and characterization of a size series of highly luminescent nanocrystallites. *J. Phys. Chem. B* 1997, *101*, 9463–9475.

105. Klimov, V.I. Mechanisms for photogeneration and recombination of multiexcitons in semiconductor nanocrystals: Implications for lasing and solar energy conversion. *J. Phys. Chem. B* 2006, *110*, 16827–16845.

106. Brus, L.E. A simple-model for the ionization-potential, electron-affinity, and aqueous redox potentials of small semiconductor crystallites. *J. Chem. Phys.* 1983, *79*, 5566–5571.

107. Steigerwald, M.L.; Brus, L.E. Semiconductor crystallites–a class of large molecules. *Acc. Chem. Res.* 1990, *23*, 183–188.

108. Ogawa, S.; Hu, K.; Fan, F.R.F.; Bard, A.J. Photoelectrochemistry of films of quantum size lead sulfide particles incorporated in self-assembled monolayers on gold. *J. Phys. Chem. B* 1997,*101*, 5707–5711.

109. Haram, S.K.; Quinn, B.M.; Bard, A.J. Electrochemistry of CdS nanoparticles: A correlation between optical and electrochemical band gaps. *J. Am. Chem. Soc.* 2001, *123*, 8860–8861.

110. Kucur, E.; Riegler, J.; Urban, G.A.; Nann, T. Determination of quantum confinement in CdSe nanocrystals by cyclic voltammetry. *J. Chem. Phys.* 2003, *119*, 2333–2337.

111. Bae, Y.; Myung, N.; Bard, A.J. Electrochemistry and electrogenerated chemiluminescence of CdTe nanoparticles. *Nano Lett.* 2004, *4*, 1153–1161.

112. Poznyak, S.K.; Osipovich, N.P.; Shavel, A.; Talapin, D.V.; Gao, M.Y.; Eychmuller, A.; Gaponik, N. Size-dependent electrochemical behavior of thiol-capped CdTe nanocrystals in aqueous solution. *J. Phys. Chem. B* 2005, *109*, 1094–1100.

113. Inamdar, S.N.; Ingole, P.P.; Haram, S.K. Determination of band structure parameters and the quasi-particle gap of CdSe quantum dots by cyclic voltammetry. *Chem. Phys. Chem.* 2008, *9*, 2574–2579.

114. Reiss, P.; Quemard, G.; Carayon, S.; Bleuse, J.; Chandezon, F.; Pron, A. Luminescent ZnSe nanocrystals of high color purity. *Mater. Chem. Phys.* 2004, *84*, 10–13.

115. Karanikolos, G.N.; Alexandridis, P.; Itskos, G.; Petrou, A.; Mountziaris, T.J. Synthesis and size control of luminescent ZnSe nanocrystals by a microemulsion-gas contacting technique.*Langmuir* 2004, *20*, 550–553.

116. Bawendi, M.G.; Wilson, W.L.; Rothberg, L.; Carroll, P.J.; Jedju, T.M.; Steigerwald, M.L.; Brus, L.E. Electronic-structure and photoexcited-carrier dynamics in nanometer-size CdSe clusters.*Phys. Rev. Lett.* 1990, *65*, 1623–1626.

117. Efros, A.L.; Rosen, M.; Kuno, M.; Nirmal, M.; Norris, D.J.; Bawendi, M. Band-edge exciton in quantum dots of semiconductors with a degenerate valence band: Dark and bright exciton states. *Phys. Rev. B* 1996, *54*, 4843–4856.

118. Efros, A.L.; Rosen, M. The electronic structure of semiconductor nanocrystals. *Annu. Rev. Mater. Sci.* 2000, *30*, 475–521.

119. Efros, A.L.; Rosen, M. Random telegraph signal in the photoluminescence intensity of a single quantum dot. *Phys. Rev. Lett.* 1997, *78*, 1110–1113.

120. Stefani, F.D.; Knoll, W.; Kreiter, M.; Zhong, X.; Han, M.Y. Quantification of photoinduced and spontaneous quantum-dot luminescence blinking. *Phys. Rev. B* 2005, *72*, 125304 1–7.

121. Kuno, M.; Fromm, D.P.; Hamann, H.F.; Gallagher, A.; Nesbitt, D.J. Nonexponential "blinking" kinetics of single CdSe quantum dots: A universal power law behavior. *J. Chem. Phys.* 2000, *112*, 3117–3120.

122. van Sark, W.; Frederix, P.; van den Heuvel, D.J.; Bol, A.A.; van Lingen, J.N.J.; Donega, C.D.; Gerritsen, H.C.; Meijerink, A. Time-resolved fluorescence spectroscopy study on the photophysical behavior of quantum dots. *J. Fluoresc.* 2002, *12*, 69–76.

123. Stefani, F.D.; Zhong, X.H.; Knoll, W.; Han, M.Y.; Kreiter, M. Memory in quantum-dot photoluminescence blinking. *New J. Phys.* 2005, *7*, 197.

124. Issac, A.; von Borczyskowski, C.; Cichos, F. Correlation between photoluminescence intermittency of CdSe

quantum dots and self-trapped states in dielectric media. *Phys. Rev. B*2005, *71*, 161302 1.

125. Gfroerer, T.H. *Photoluminescence in analysis of surface and interfaces*; John Wiley & Sons Ltd.: Chichster, UK, 2000; pp. 9209–9231.

126. Djurisic, A.B.; Leung, Y.H.; Choy, W.C.H.; Cheah, K.W.; Chan, W.K. Visible photoluminescence in ZnO tetrapod and multipod structures. *Appl. Phys. Lett.* 2004, *84*, 2635–2637.

127. Norberg, N.S.; Gamelin, D.R. Influence of surface modification on the luminescence of colloidal ZnO nanocrystals. *J. Phys. Chem. B* 2005, *109*, 20810–20816.

128. Djurisic, A.B.; Leung, Y.H. Optical properties of ZnO nanostructures. *Small* 2006, *2*, 944–961.

129. Cheng, H.M.; Lin, K.F.; Hsu, H.C.; Hsieh, W.F. Size dependence of photoluminescence and resonant Raman scattering from ZnO quantum dots. *Appl. Phys. Lett.* 2006, *88*, 261909.

130. Lu, J.G.; Ye, Z.Z.; Zhang, Y.Z.; Liang, Q.L.; Fujita, S.; Wang, Z.L. Self-assembled ZnO quantum dots with tunable optical properties. *Appl. Phys. Lett.* 2006, *89*, 023122.

131. Xu, P.S.; Sun, Y.M.; Shi, C.S.; Xu, F.Q.; Pan, H.B. The electronic structure and spectral properties of ZnO and its defects. *Nucl. Instrum. Methods Phys. Res. Sect. B* 2003, *199*, 286–290.

132. Aleksandra, B.D.; Leung, Y. H. Optical properties of ZnO nanostructures. *Small* 2006, *2*, 944–961.

133. Spanhel, L.; Anderson, M.A. Semiconductor clusters in the sol-gel process−quantized aggregation, gelation, and crystal-growth in concentrated ZnO colloids. *J. Am. Chem. Soc.*1991, *113*, 2826–2833.

134. Lee, J. D. *Concise Inorganic Chemistry*; Blackwell Science: Noida, India, 2005; p. 1032.

135. Xiu, F.X.; Yang, Z.; Mandalapu, L.J.; Liu, J.L.; Beyermann, W.P. p-Type ZnO films with solid-source phosphorus doping by molecular-beam epitaxy. *Appl. Phys. Lett.* 2006, *88*, 052106.

136. Viswanatha, R.; Chakraborty, S.; Basu, S.; Sarma, D.D. Blue-emitting copper-doped zinc oxide nanocrystals. *J. Phys. Chem. B* 2006, *110*, 22310–22312.

137. Wang, Y.S.; Thomas, P.J.; O'Brien, P. Optical properties of ZnO nanocrystals doped with Cd, Mg, Mn, and Fe ions. *J. Phys. Chem. B* 2006, *110*, 21412–21415.

138. Hirate, T.; Sasaki, S.; Li, W.C.; Miyashita, H.; Kimpara, T.; Satoh, T. Effects of laser-ablated impurity on aligned ZnO nanorods grown by chemical vapor deposition. *Thin Solid Films*2005, *487*, 35–39.

139. Liu, S.M.; Liu, F.Q.; Wang, Z.G. Relaxation of carriers in terbium-doped ZnO nanoparticles.*Chem. Phys. Lett.* 2001, *343*, 489–492.

140. Cheng, B.C.; Xiao, Y.H.; Wu, G.S.; Zhang, L.D. Controlled growth and properties of one-dimensional ZnO nanostructures with Ce as activator/dopant. *Adv. Func. Mater.* 2004, *14*, 913–919.

141. Ishizumi, A.; Kanemitsu, Y. Structural and luminescence properties of Eu-doped ZnO nanorods fabricated by a microemulsion method. *Appl. Phys. Lett.* 2005, *86*, 253106.

142. Wu, G.S.; Zhuang, Y.L.; Lin, Z.Q.; Yuan, X.Y.; Xie, T.; Zhang, L.D. Synthesis and photoluminescence of dy-doped ZnO nanowires. *Physica E* 2006, *31*, 5–8.

143. Zhang, X.T.; Liu, Y.C.; Zhang, J.Y.; Lu, Y.M.; Shen, D.Z.; Fan, X.W.; Kong, X.G. Structure and photoluminescence of Mn-passivated nanocrystalline ZnO thin films. *J. Cryst. Growth* 2003,*254*, 80–85.

144. Xu, C.X.; Sun, X.W.; Dong, Z.L.; Tan, S.T.; Cui, Y.P.; Wang, B.P. Manganese-doped zinc oxide tetratubes and their photoluminescent properties. *J. Appl. Phys.* 2005, *98*, 113513.

145. Yang, L.W.; Wu, X.L.; Huang, G.S.; Qiu, T.; Yang, Y.M. *In situ* synthesis of Mn-doped ZnO multileg nanostructures and Mn-related Raman vibration. *J. Appl. Phys.* 2005, *97*, 014308.

146. Ronning, C.; Gao, P.X.; Ding, Y.; Wang, Z.L.; Schwen, D. Manganese-doped ZnO nanobelts for spintronics. *Appl. Phys. Lett.* 2004, *84*, 783–785.

147. Shen, G.Z.; Cho, J.H.; Yoo, J.K.; Yi, G.C.; Lee, C.J. Synthesis and optical properties of S-doped ZnO nanostructures: Nanonails and nanowires. *J. Phys. Chem. B* 2005, *109*, 5491–5496.

148. Foreman, J.V.; Li, J.Y.; Peng, H.Y.; Choi, S.J.; Everitt, H.O.; Liu, J. Time-resolved investigation of bright visible wavelength luminescence from sulfur-doped ZnO nanowires and micropowders. *Nano Lett.* 2006, *6*, 1126–1130.

149. Yang, P.; Lu, M.K.; Xu, D.; Yuan, D.L.; Zhou, G.J. ZnS nanocrystals co-activated by transition metals and rare-earth metals–a new class of luminescent materials. *J. Lumin.* 2001, *93*, 101–105.

150. Qu, S.C.; Zhou, W.H.; Liu, F.Q.; Chen, N.F.; Wang, Z.G.; Pan, H.Y.; Yu, D.P. Photoluminescence properties of Eu3+-doped ZnS nanocrystals prepared in a water/methanol solution. *Appl. Phys. Lett.* 2002, *80*, 3605–3607.

151. Suyver, J.F.; Wuister, S.F.; Kelly, J.J.; Meijerink, A. Synthesis and photoluminescence of nanocrystalline ZnS : Mn^{2+}. *Nano Lett.* 2001, *1*, 429–433.

152. Bhargava, R.N.; Gallagher, D.; Hong, X.; Nurmikko, A. Optical-Properties of Manganese-Doped Nanocrystals of Zns. *Phys. Rev. Lett.* 1994, *72*, 416–419.

153. Bol, A.A.; Meijerink, A. Long-lived Mn2+ emission in nanocrystalline ZnS : Mn^{2+}. *Phys. Rev. B* 1998, *58*, R15997–R16000.

154. Su, F.H.; Fang, Z.L.; Ma, B.S.; Ding, K.; Li, G.H.; Chen, W. Pressure dependence of Mn^{2+}luminescence in differently sized ZnS : Mn nanoparticles. *J. Phys. Chem. B* 2003, *107*, 6991–6996.

155. Khosravi, A.A.; Kundu, M.; Kuruvilla, B.A.; Shekhawat, G.S.; Gupta, R.P.; Sharma, A.K.; Vyas, P.D.; Kulkarni, S.K. Manganese-doped zinc-sulfide nanoparticles by aqueous method.*Appl. Phys. Lett.* 1995, *67*, 2506–2508.

156. Beermann, P.A.G.; McGarvey, B.R.; Muralidharan, S.; Sung, R.C.W. EPR spectra of Mn^{2+}-doped ZnS quantum dots. *Chem. Mater.* 2004, *16*, 915–918.

157. Eaton, D.F. Reference materials for fluorescence measurement. *Pure Appl. Chem.* 1988, *60*, 1107–1114.

158. Qian, L.; Bera, D.; Holloway, P.H. Temporal evolution of white light emission from CdSe quantum dots. *Nanotechnology* 2008, *19*, 285702.

159. Bera, D.; Qian, L.; Holloway, P.H. Time-evolution of photoluminescence properties of ZnO/MgO core/shell quantum dots. *J. Phys. D* 2008, *41*, 182002.

160. Hines, M.A.; Guyot-Sionnest, P. Bright UV-blue luminescent colloidal ZnSe nanocrystals. *J. Phys. Chem. B* 1998, *102*, 3655–3657.

161. Lee, J.; Sundar, V.C.; Heine, J.R.; Bawendi, M.G.; Jensen, K.F. Full color emission from II-VI semiconductor quantum dot-polymer composites. *Adv. Mater.* 2000, *12*, 1102.

162. Talapin, D.V.; Rogach, A.L.; Kornowski, A.; Haase, M.; Weller, H. Highly luminescent monodisperse CdSe and CdSe/ZnS nanocrystals synthesized in a hexadecylamine-trioctylphosphine oxide-trioctylphospine mixture. *Nano Lett.* 2001, *1*, 207–211.

163. Norris, D.J.; Yao, N.; Charnock, F.T.; Kennedy, T.A. High-quality manganese-doped ZnSe nanocrystals. *Nano Lett.* 2001, *1*, 3–7.

164. Qu, L.H.; Peng, X.G. Control of photoluminescence properties of CdSe nanocrystals in growth. *J. Am. Chem. Soc.* 2002, *124*, 2049–2055.

165. Steckel, J.S.; Coe-Sullivan, S.; Bulovic, V.; Bawendi, M.G. 1.3 mu m to 1.55 mu m tunable electroluminescence from PbSe quantum dots embedded within an organic device. *Adv. Mater.* 2003, *15*, 1862–1866.

166. Donega, C.D.; Hickey, S.G.; Wuister, S.F.; Vanmaekelbergh, D.; Meijerink, A. Single-step synthesis to control the photoluminescence quantum yield and size dispersion of CdSe nanocrystals. *J. Phys. Chem. B* 2003, *107*, 489–496.

167. Anikeeva, P.O.; Halpert, J.E.; Bawendi, M.G.; Bulovic, V. Electroluminescence from a mixed red-green-blue colloidal quantum dot monolayer. *Nano Lett.* 2007, *7*, 2196–2200.

168. Lim, J.; Jun, S.; Jang, E.; Baik, H.; Kim, H.; Cho, J. Preparation of highly luminescent nanocrystals and their application to light-emitting diodes. *Adv. Mater.* 2007, *19*, 1927–1932.

169. van Dijken, A.; Makkinje, J.; Meijerink, A. The influence of particle size on the luminescence quantum efficiency of nanocrystalline ZnO particles. *J. Lumin.* 2001, *92*, 323–328.

170. Chen, H.S.; Lo, B.; Hwang, J.Y.; Chang, G.Y.; Chen, C.M.; Tasi, S.J.; Wang, S.J.J. Colloidal ZnSe, ZnSe/ZnS, and ZnSe/ZnSeS quantum dots synthesized from ZnO. *J. Phys. Chem. B*2004, *108*, 17119–17123.

171. Bol, A.A.; Meijerink, A. Luminescence quantum efficiency of nanocrystalline ZnS : Mn^{2+}. 1. Surface passivation and Mn^{2+} concentration. *J. Phys. Chem. B* 2001, *105*, 10197–10202.

172. Hattori, Y.; Isobe, T.; Takahashi, H.; Itoh, S. Luminescent properties of ZnS:Mn^{2+}nanocrystals/SiO2 hybrid phosphor synthesized by *in situ* surface modification co-precipitation. *J. Lumin.* 2005, *113*, 69–78.

173. Becker, W.G.; Bard, A.J. Photo-luminescence and photoinduced oxygen-adsorption of colloidal zinc-sulfide dispersions. *J. Phys. Chem.* 1983, *87*, 4888–4893.

174. Baral, S.; Fojtik, A.; Weller, H.; Henglein, A. Photochemistry and radiation-chemistry of colloidal semiconductors .12. Intermediates of the oxidation of extremely small particles of CdS, ZnS, and Cd3P2 and size quantization effects (a pulse-radiolysis study). *J. Am. Chem. Soc.* 1986, *108*, 375–378.

175. Dunstan, D.E.; Hagfeldt, A.; Almgren, M.; Siegbahn, H.O.G.; Mukhtar, E. Importance of surface-reactions in the photochemistry of ZnS colloids. *J. Phys. Chem.* 1990, *94*, 6797–6804.

176. Bol, A.A.; Meijerink, A. Luminescence quantum efficiency of nanocrystalline ZnS:Mn^{2+}. 2. Enhancement by UV irradiation. *J. Phys. Chem. B* 2001, *105*, 10203–10209.

177. Cao, L.X.; Zhang, J.H.; Ren, S.L.; Huang, S.H. Luminescence enhancement of core-shell ZnS : Mn/ZnS nanoparticles. *Appl. Phys. Lett.* 2002, *80*, 4300–4302.

178. Jin, C.M.; Yu, J.Q.; Sun, L.D.; Dou, K.; Hou, S.G.; Zhao, J.L.; Chen, Y.M.; Huang, S.H. Luminescence of Mn^{2+} doped ZnS nanocrystallites. *J. Lumin.* 1995, *66–7*, 315–318.

179. Shavel, A.; Gaponik, N.; Eychmuller, A. Efficient UV-blue photoluminescing thiol-stabilized water-soluble alloyed ZnSe(S) nanocrystals. *J. Phys. Chem. B* 2004, *108*, 5905–5908.

180. Bera, D.; Holloway, P.H.; Yang, H. Efficient and photostable ZnS-passivated CdS:Mn quantum dots. *ECS Trans.* 2006, *3*, 5–17.

181. Williams, E.W.; Hall, R. *Luminescence and the light emitting diode*, 1 Ed. ed; Pergomon Press: New York, NY, USA, 1977; p. 237.

182. Klimov, V.I.; Mikhailovsky, A.A.; McBranch, D.W.; Leatherdale, C.A.; Bawendi, M.G. Quantization of multiparticle Auger rates in semiconductor quantum dots. *Science* 2000, *287*, 1011–1013.

183. Scherer, A.; Craighead, H.G.; Beebe, E.D. Gallium-arsenide and aluminum gallium-arsenide reactive ion etching in boron-trichloride argon mixtures. *J. Vac. Sci. Technol. B* 1987, *5*, 1599–1605.

184. Tsutsui, K.; Hu, E.L.; Wilkinson, C.D.W. Reactive ion etched II-VI quantum dots–dependence of etched profile on pattern geometry. *Jpn. J. Appl. Phys. Part 1* 1993, *32*, 6233–6236.

185. Chason, E.; Picraux, S.T.; Poate, J.M.; Borland, J.O.; Current, M.I.; delaRubia, T.D.; Eaglesham, D.J.; Holland, O.W.; Law, M.E.; Magee, C.W.; Mayer, J.W.; Melngailis, J.; Tasch, A.F. Ion beams in silicon processing and characterization. *J. Appl. Phys.* 1997, *81*, 6513–6561.

186. Burda, C.; Chen, X.B.; Narayanan, R.; El-Sayed, M.A. Chemistry and properties of nanocrystals of different shapes. *Chem. Rev.* 2005, *105*, 1025–1102.

187. Sashchiuk, A.; Lifshitz, E.; Reisfeld, R.; Saraidarov, T.; Zelner, M.; Willenz, A. Optical and conductivity properties of PbS nanocrystals in amorphous zirconia sol-gel films. *J. Sol-Gel Sci. Technol.* 2002, *24*, 31–38.

188. Hoener, C.F.; Allan, K.A.; Bard, A.J.; Campion, A.; Fox, M.A.; Mallouk, T.E.; Webber, S.E.; White, J.M. Demonstration of a shell core structure in layered CdSe-ZnSe small particles by x-ray photoelectron and Auger spectroscopies. *J. Phys. Chem.* 1992, *96*, 3812–3817.

189. Yang, H. Syntheses and application of Mn-doped II-VI semiconductor nanocrystals. Dissertation, Univ. Florida, Gainesville, 2003.

190. Yang, H.; Holloway, P.H. Enhanced photoluminescence from CdS:Mn/ZnS core/shell quantum dots. *Appl. Phys. Lett.* 2003, *82*, 1965–1967.

191. Qu, L.H.; Peng, Z.A.; Peng, X.G. Alternative routes toward high quality CdSe nanocrystals.*Nano Lett.* 2001, *1*, 333–337.

192. Yu, W.W.; Wang, Y.A.; Peng, X.G. Formation and stability of size-, shape-, and structure-controlled CdTe nanocrystals: Ligand effects on monomers and nanocrystals. *Chem. Mater.*2003, *15*, 4300–4308.

193. Talapin, D.V.; Haubold, S.; Rogach, A.L.; Kornowski, A.; Haase, M.; Weller, H. A novel organometallic synthesis of highly luminescent CdTe nanocrystals. *J. Phys. Chem. B* 2001, *105*, 2260–2263.

194. Singh, S.; Rama, N.; Rao, M.S.R. Influence of d-d transition bands on electrical resistivity in Ni doped polycrystalline ZnO. *Appl. Phys. Lett.* 2006, *88*, 222111 1–3.

195. Yang, H.S.; Lee, H.; Holloway, P.H. Anisotropic growth of luminescent Eu3+- or Er3+-doped Gd2O3 nanocrystals. *Nanotechnology* 2005, *16*, 2794–2798.

196. Chen, H.S.; Wang, S.J.J.; Lo, C.J.; Chi, J.Y. White-light emission from organics-capped ZnSe quantum dots and application in white-light-emitting diodes. *Appl. Phys. Lett.* 2005, *86*, 131905.

197. Bakueva, L.; Musikhin, S.; Hines, M.A.; Chang, T.W.F.; Tzolov, M.; Scholes, G.D.; Sargent, E.H. Size-tunable infrared (1000–1600 nm) electroluminescence from PbS quantum-dot nanocrystals in a semiconducting polymer. *Appl. Phys. Lett.* 2003, *82*, 2895–2897.

198. Battaglia, D.; Peng, X.G. Formation of high quality InP and InAs nanocrystals in a noncoordinating solvent. *Nano Lett.* 2002, *2*, 1027–1030.

199. Bae, W.K.; Nam, M.K.; Char, K.; Lee, S. Gram-scale one-pot synthesis of highly luminescent blue emitting Cd1-xZnxS/ZnS nanocrystals. *Chem. Mater.* 2008, *20*, 5307–5313.

200. Olshavsky, M.A.; Goldstein, A.N.; Alivisatos, A.P. Organometallic synthesis of GaAs crystallites exhibiting quantum confinement. *J. Am. Chem. Soc.* 1990, *112*, 9438–9439.

201. Brennan, J.G.; Siegrist, T.; Carroll, P.J.; Stuczynski, S.M.; Reynders, P.; Brus, L.E.; Steigerwald, M.L. Bulk and nanostructure group-II-VI compounds from molecular organometallic precursors. *Chem. Mater.* 1990, *2*, 403–409.

202. Peng, X.G.; Wickham, J.; Alivisatos, A.P. Kinetics of II-VI and III-V colloidal semiconductor nanocrystal growth: "Focusing" of size distributions. *J. Am. Chem. Soc.* 1998, *120*, 5343–5344.

203. Kher, S.S.; Wells, R.L. A straightforward, new method for the synthesis of nanocrystalline GaAs and GaP. *Chem. Mater.* 1994, *6*, 2056–2062.

204. Micic, O.I.; Sprague, J.R.; Curtis, C.J.; Jones, K.M.; Machol, J.L.; Nozik, A.J.; Giessen, H.; Fluegel, B.; Mohs, G.; Peyghambarian, N. Synthesis and characterization of InP, GaP, and GaInP2 quantum dots. *J. Phys. Chem.* 1995, *99*, 7754–7759.

205. Guzelian, A.A.; Katari, J.E.B.; Kadavanich, A.V.; Banin, U.; Hamad, K.; Juban, E.; Alivisatos, A.P.; Wolters, R.H.; Arnold, C.C.; Heath, J.R. Synthesis of size-selected, surface-passivated InP nanocrystals. *J. Phys. Chem.* 1996, *100*, 7212–7219.

206. Guzelian, A.A.; Banin, U.; Kadavanich, A.V.; Peng, X.; Alivisatos, A.P. Colloidal chemical synthesis and characterization of InAs nanocrystal quantum dots. *Appl. Phys. Lett.* 1996, *69*, 1432–1434.

207. Cao, Y.W.; Banin, U. Synthesis and characterization of InAs/InP and InAs/CdSe core/shell nanocrystals. *Angew. Chem. Int. Ed.* 1999, *38*, 3692–3694.

208. Peng, X.G.; Manna, L.; Yang, W.D.; Wickham, J.; Scher, E.; Kadavanich, A.; Alivisatos, A.P. Shape control of CdSe nanocrystals. *Nature (London)* 2000, *404*, 59–61.

209. Manna, L.; Scher, E.C.; Alivisatos, A.P. Synthesis of soluble and processable rod-, arrow-, teardrop-, and tetrapod-shaped CdSe nanocrystals. *J. Am. Chem. Soc.* 2000, *122*, 12700–12706.

210. Hikmet, R.A.M.; Talapin, D.V.; Weller, H. Study of conduction mechanism and electroluminescence in CdSe/ZnS quantum dot composites. *J. Appl. Phys.* 2003, *93*, 3509–3514.

211. Peng, Z.A.; Peng, X.G. Formation of high-quality CdTe, CdSe, and CdS nanocrystals using CdO as precursor. *J. Am. Chem. Soc.* 2001, *123*, 183–184.

212. Coe-Sullivan, S.; Steckel, J.S.; Woo, W.K.; Bawendi, M.G.; Bulovic, V. Large-area ordered quantum-dot monolayers *via* phase separation during spin-casting. *Adv. Func. Mater.* 2005,*15*, 1117–1124.

213. Jun, S.; Jang, E.J.; Chung, Y.S. Alkyl thiols as a sulfur precursor for the preparation of monodisperse metal sulfide nanostructures. *Nanotechnology* 2006, *17*, 4806–4810.

214. Zhu, J.J.; Koltypin, Y.; Gedanken, A. General sonochemical method far the preparation of nanophasic selenides: Synthesis of ZnSe nanoparticles. *Chem. Mater.* 2000, *12*, 73–78.

215. Qian, H.F.; Li, L.; Ren, J.C. One-step and rapid synthesis of high quality alloyed quantum dots (CdSe-CdS) in aqueous phase by microwave irradiation with controllable temperature.*Mater. Res. Bull.* 2005, *40*, 1726–1736.

216. Wang, L.C.; Chen, L.Y.; Luo, T.; Qian, Y.T. A hydrothermal method to prepare the spherical ZnS and

flower-like CdS microcrystallites. *Mater. Lett.* 2006, *60*, 3627–3630.

217. Yang, H.Q.; Yin, W.Y.; Zhao, H.; Yang, R.L.; Song, Y.Z. A complexant-assisted hydrothermal procedure for growing well-dispersed InP nanocrystals. *J. Phys. Chem. Solids* 2008, *69*, 1017–1022.

218. Xie, Y.; Qian, Y.T.; Wang, W.Z.; Zhang, S.Y.; Zhang, Y.H. A benzene-thermal synthetic route to nanocrystalline GaN. *Science* 1996, *272*, 1926–1927.

219. Rogach, A.L.; Katsikas, L.; Kornowski, A.; Su, D.S.; Eychmuller, A.; Weller, H. Synthesis and characterization of thiol-stabilized CdTe nanocrystals. *Ber.Bunsen Ges. Phys. Chem. Chem. Phys.* 1996, *100*, 1772–1778.

220. O'Connor, E.; O'Riordan, A.; Doyle, H.; Moynihan, S.; Cuddihy, A.; Redmond, G. Near-infrared electroluminescent devices based on colloidal HgTe quantum dot arrays. *Appl. Phys. Lett.* 2005, *86*, 201114.

221. Sankaran, V.; Cummins, C.C.; Schrock, R.R.; Cohen, R.E.; Silbey, R.J. Small PbS clusters prepared *via* romp block copolymer technology. *J. Am. Chem. Soc.* 1990, *112*, 6858–6859.

222. Winiarz, J.G.; Zhang, L.M.; Park, J.; Prasad, P.N. Inorganic: Organic hybrid nanocomposites for photorefractivity at communication wavelengths. *J. Phys. Chem. B* 2002, *106*, 967–970.

223. Leonard, D.; Krishnamurthy, M.; Reaves, C.M.; Denbaars, S.P.; Petroff, P.M. Direct formation of quantum-sized dots from uniform coherent islands of InGaAs on GaAs-surfaces. *Appl. Phys. Lett.* 1993, *63*, 3203–3205.

224. Xin, S.H.; Wang, P.D.; Yin, A.; Kim, C.; Dobrowolska, M.; Merz, J.L.; Furdyna, J.K. Formation of self-assembling CdSe quantum dots on ZnSe by molecular beam epitaxy. *Appl. Phys. Lett.*1996, *69*, 3884–3886.

225. Leonardi, K.; Selke, H.; Heinke, H.; Ohkawa, K.; Hommel, D.; Gindele, F.; Woggon, U. Formation of self-assembling II-VI semiconductor nanostructures during migration enhanced epitaxy. *J. Cryst. Growth* 1998, *184*, 259–263.

226. Kurtz, E.; Shen, J.; Schmidt, M.; Grun, M.; Hong, S.K.; Litvinov, D.; Gerthsen, D.; Oka, T.; Yao, T.; Klingshirn, C. Formation and properties of self-organized II-VI quantum islands.*Thin Solid Films* 2000, *367*, 68–74.

227. Swihart, M.T. Vapor-phase synthesis of nanoparticles. *Curr. Opin. Colloid Interface Sci.* 2003, *8*, 127–133.

228. Frank, F.C.; van der Merwe, J.H. One-dimentional dislocations. I. Static theory. *Proc. Roy. Soc. London A* 1949, *198*, 205.

229. Volmer, M.; Weber, A. Keimbildung in ubersettigten gebilden. *Z. Phys. Chem.* 1926, *119*, 277.

230. Stranski, I.N.; Krastanow, V.L. Sitzungberichte der akademie der wissenschaften in wien.*Akad. Wiss. Lit. Mainz Math.-Natur. Kl. IIb* 1939, *146*, 797.

231. Eaglesham, D.J.; Cerullo, M. Dislocation-free stranski-krastanow growth of Ge on Si(100).*Phys. Rev. Lett.* 1990, *64*, 1943–1946.

232. Kim, Y.G.; Joh, Y.S.; Song, J.H.; Baek, K.S.; Chang, S.K.; Sim, E.D. Temperature-dependent photoluminescence of ZnSe/ZnS quantum dots fabricated under the Stranski-Krastanov mode. *Appl. Phys. Lett.* 2003, *83*, 2656–2658.

233. Tsukamoto, S.; Bell, G.R.; Arakawa, Y. Heteroepitaxial growth of InAs on GaAS(001) by *in situ* STM located inside MBE growth chamber. *Microelectron. J.* 2006, *37*, 1498–1504.

234. Jiao, Y.H.; Wu, J.; Xu, B.; Jin, P.; Hu, L.J.; Liang, L.Y.; Wang, Z.G. MBE InAs quantum dots grown on metamorphic InGaAs for long wavelength emitting. *Physica E* 2006, *35*, 194–198.

235. Nakamura, S.; Kitamura, K.; Umeya, H.; Jia, A.; Kobayashi, M.; Yoshikawa, A.; Shimotomai, M.; Nakamura, S.; Takahashi, K. Bright electroluminescence from CdS quantum dot LED structures. *Electron. Lett.* 1998, *34*, 2435–2436.

236. Lobo, C.; Leon, R. InGaAs island shapes and adatom migration behavior on (100), (110), (111), and (311) GaAs surfaces. *J. Appl. Phys.* 1998, *83*, 4168–4172.

237. Muller, C.D.; Falcou, A.; Reckefuss, N.; Rojahn, M.; Wiederhirn, V.; Rudati, P.; Frohne, H.; Nuyken, O.; Becker, H.; Meerholz, K. Multi-colour organic light-emitting displays by solution processing. *Nature (London)* 2003, *421*, 829–833.

238. Meerheim, R.W.K.; He, G.; Pfeiffer, M.; Leo, K. Highly efficient organic light emitting diodes (OLED) for displays and lighting. *Proc. SPIE Int. Soc. Opt. Eng.* 2006, *6192*, 1–16.

239. Adachi, C.; Baldo, M.A.; Thompson, M.E.; Forrest, S.R. Nearly 100% internal phosphorescence efficiency in an organic light-emitting device. *J. Appl. Phys.* 2001, *90*, 5048–5051.

240. Gu, G.; Garbuzov, D.Z.; Burrows, P.E.; Venkatesh, S.; Forrest, S.R.; Thompson, M.E. High-external-quantum-efficiency organic light-emitting devices. *Opt. Lett.* 1997, *22*, 396–398.

241. Xu, J.; Cui, D.H.; Lewis, B.A.; Wang, A.Y.; Xu, S.Y.; Gerhold, M. Microcavity light-emitting devices based on colloidal semiconductor nanocrystal quantum dots. *IEEE Photonics Technol. Lett.* 2005, *17*, 2008–2010.

242. Coe, S.; Woo, W.K.; Bawendi, M.; Bulovic, V. Electroluminescence from single monolayers of nanocrystals in molecular organic devices. *Nature (London)* 2002, *420*, 800–803.

243. Caruge, J.M.; Halpert, J.E.; Bulovic, V.; Bawendi, M.G. NiO as an inorganic hole-transporting layer in quantum-dot light-emitting devices. *Nano Lett.* 2006, *6*, 2991–2994.

244. Bae, W.K.; Kwak, J.; Lim, J.; Lee, D.; Nam, M.K.; Char, K.; Lee, C.; Lee, S. Deep blue light-emitting diodes based on Cd1-xZnxS@ZnS quantum dots. *Nanotechnology* 2009, *20*, 075202.

245. Chaudhary, S.; Ozkan, M.; Chan, W.C.W. Trilayer hybrid polymer-quantum dot light-emitting diodes. *Appl. Phys. Lett.* 2004, *84*, 2925–2927.

246. Irwin, M.D.; Buchholz, B.; Hains, A.W.; Chang, R.P.H.; Marks, T.J. p-Type semiconducting nickel oxide as an efficiency-enhancing anode interfacial layer in polymer

bulk-heterojunction solar cells. *Proc. Natl. Acad. Sci. USA* 2008, *105*, 2783–2787.

247. Mattoussi, H.; Radzilowski, L.H.; Dabbousi, B.O.; Thomas, E.L.; Bawendi, M.G.; Rubner, M.F. Electroluminescence from heterostructures of poly(phenylene vinylene) and inorganic CdSe nanocrystals. *J. Appl. Phys.* 1998, *83*, 7965–7974.

248. Mattoussi, H.; Radzilowski, L.H.; Dabbousi, B.O.; Fogg, D.E.; Schrock, R.R.; Thomas, E.L.; Rubner, M.F.; Bawendi, M.G. Composite thin films of CdSe nanocrystals and a surface passivating/electron transporting block copolymer: Correlations between film microstructure by transmission electron microscopy and electroluminescence. *J. Appl. Phys.* 1999, *86*, 4390–4399.

249. Coe-Sullivan, S.; Woo, W.K.; Steckel, J.S.; Bawendi, M.; Bulovic, V. Tuning the performance of hybrid organic/inorganic quantum dot light-emitting devices. *Org. Electron.* 2003, *4*, 123–130.

250. Cho, K.S.; Lee, E.K.; Joo, W.J.; Jang, E.; Kim, T.H.; Lee, S.J.; Kwon, S.J.; Han, J.Y.; Kim, B.K.; Choi, B.L.; Kim, J.M. High-performance crosslinked colloidal quantum-dot light-emitting diodes. *Nat. Photonics* 2009, *3*, 341–345.

251. Zhao, J.L.; Bardecker, J.A.; Munro, A.M.; Liu, M.S.; Niu, Y.H.; Ding, I.K.; Luo, J.D.; Chen, B.Q.; Jen, A.K.Y.; Ginger, D.S. Efficient CdSe/CdS quantum dot light-emitting diodes using a thermally polymerized hole transport layer. *Nano Lett.* 2006, *6*, 463–467.

252. Sun, Q.; Wang, Y.A.; Li, L.S.; Wang, D.Y.; Zhu, T.; Xu, J.; Yang, C.H.; Li, Y.F. Bright, multicoloured light-emitting diodes based on quantum dots. *Nat. Photonics* 2007, *1*, 717–722.

253. Schlamp, M.C.; Peng, X.G.; Alivisatos, A.P. Improved efficiencies in light emitting diodes made with CdSe(CdS) core/shell type nanocrystals and a semiconducting polymer. *J. Appl. Phys.* 1997, *82*, 5837–5842.

254. Mueller, A.H.; Petruska, M.A.; Achermann, M.; Werder, D.J.; Akhadov, E.A.; Koleske, D.D.; Hoffbauer, M.A.; Klimov, V.I. Multicolor light-emitting diodes based on semiconductor nanocrystals encapsulated in GaN charge injection layers. *Nano Lett.* 2005, *5*, 1039–1044.

255. Murray, C.B.; Kagan, C.R.; Bawendi, M.G. Synthesis and characterization of monodisperse nanocrystals and close-packed nanocrystal assemblies. *Annu. Rev. Mater. Sci.* 2000, *30*, 545–610.

256. Colvin, V.L.; Schlamp, M.C.; Alivisatos, A.P. Light-emitting-diodes made from cadmium selenide nanocrystals and a semiconducting polymer. *Nature (London)* 1994, *370*, 354–357.

257. Caruge, J.M.; Halpert, J.E.; Wood, V.; Bulovic, V.; Bawendi, M.G. Colloidal quantum-dot light-emitting diodes with metal-oxide charge transport layers. *Nat. Photonics* 2008, *2*, 247–250.

258. Zhu, T.; Shanmugasundaram, K.; Price, S.C.; Ruzyllo, J.; Zhang, F.; Xu, J.; Mohney, S.E.; Zhang, Q.; Wang, A.Y. Mist fabrication of light emitting diodes with colloidal nanocrystal quantum dots. *Appl. Phys. Lett.* 2008, *92*, 023111 1–3.

259. Kobayashi, S.; Tani, Y.; Kawazoe, H. Quantum dot activated all-inorganic electroluminescent device fabricated using solution-synthesized CdSe/ZnS nanocrystals. *Jpn. J. Appl. Phys. Part 2*2007, *46*, L966–L969.

260. Taylor, R.M.; Church, K.H.; Sluch, M.I. Red light emission from hybrid organic/inorganic quantum dot AC light emitting displays. *Displays* 2007, *28*, 92–96.

261. Steckel, J.S.; Zimmer, J.P.; Coe-Sullivan, S.; Stott, N.E.; Bulovic, V.; Bawendi, M.G. Blue luminescence from (CdS)ZnS core-shell nanocrystals. *Angew. Chem. Int. Ed.* 2004, *43*, 2154–2158.

262. Dabbousi, B.O.; Bawendi, M.G.; Onitsuka, O.; Rubner, M.F. Electroluminescence from CdSe quantum-dot polymer composites. *Appl. Phys. Lett.* 1995, *66*, 1316–1318.

263. Yang, H.S.; Holloway, P.H. Electroluminescence from hybrid conjugated polymer– CdS:Mn/ZnS core/shell nanocrystals devices. *J. Phys. Chem. B* 2003, *107*, 9705–9710.

264. Tessler, N.; Medvedev, V.; Kazes, M.; Kan, S.H.; Banin, U. Efficient near-infrared polymer nanocrystat light-emitting diodes. *Science* 2002, *295*, 1506–1508.

265. Ghosh, A.P.; Gerenser, L.J.; Jarman, C.M.; Fornalik, J.E. Thin-film encapsulation of organic light-emitting devices. *Appl. Phys. Lett.* 2005, *86*.

266. McElvain, J.; Antoniadis, H.; Hueschen, M.R.; Miller, J.N.; Roitman, D.M.; Sheats, J.R.; Moon, R.L. Formation and growth of black spots in organic light-emitting diodes. *J. Appl. Phys.*1996, *80*, 6002–6007.

267. Tan, Z.N.; Zhang, F.; Zhu, T.; Xu, J.; Wang, A.Y.; Dixon, D.; Li, L.S.; Zhang, Q.; Mohney, S.E.; Ruzyllo, J. Bright and color-saturated emission from blue light-emitting diodes based on solution-processed colloidal nanocrystal quantum dots. *Nano Lett.* 2007, *7*, 3803–3807.

268. Bae, W.K.; Kwak, J.; Park, J.W.; Char, K.; Lee, C.; Lee, S. Highly efficient green-light-emitting diodes based on CdSe@ZnS quantum dots with a chemical-composition gradient. *Adv. Mater.*2009, *21*, 1690–1694.

269. Gao, M.Y.; Lesser, C.; Kirstein, S.; Mohwald, H.; Rogach, A.L.; Weller, H. Electroluminescence of different colors from polycation/CdTe nanocrystal self-assembled films. *J. Appl. Phys.* 2000, *87*, 2297–2302.

270. Chen, H.S.; Hsu, C.K.; Hong, H.Y. InGaN-CdSe-ZnSe quantum dots white LEDs. *IEEE Photonics Technol. Lett.* 2006, *18*, 193–195.

271. Nizamoglu, S.; Ozel, T.; Sari, E.; Demir, H.V. White light generation using CdSe/ZnS core-shell nanocrystals hybridized with InGaN/GaN light emitting diodes. *Nanotechnology* 2007,*18*, 065709.

272. Yeh, D.M.; Huang, C.F.; Lu, Y.C.; Yang, C.C. White-light light-emitting device based on surface plasmon-enhanced CdSe/ZnS nanocrystal wavelength conversion on a blue/green two-color light-emitting diode. *Appl. Phys. Lett.* 2008, *92*, 09112 1–3.

273. Gratzel, M. From space to earth: The story of solar electricity. *Nature (London)* 2000, *403*, 363.

274. Shockley, W.; Queisser, H.J. Detailed balance limit of efficiency of p-n junction solar cells. *J. Appl. Phys.* 1961, *32*, 510–519.

275. King, R.R.; Law, D.C.; Edmondson, K.M.; Fetzer, C.M.; Kinsey, G.S.; Yoon, H.; Sherif, R.A.; Karam, N.H. 40% efficient metamorphic GaInP/GaInAs/Ge multijunction solar cells. *Appl. Phys. Lett.* 2007, *90*, 183516.

276. Hoppe, H.; Sariciftci, N.S. Organic solar cells: An overview. *J. Mater. Res.* 2004, *19*, 1924–1945.

277. Guldi, D.M.; Rahman, G.M.A.; Sgobba, V.; Kotov, N.A.; Bonifazi, D.; Prato, M. CNT-CdTe versatile donor-acceptor nanohybrids. *J. Am. Chem. Soc.* 2006, *128*, 2315–2323.

278. Han, L.L.; Qin, D.H.; Jiang, X.; Liu, Y.S.; Wang, L.; Chen, J.W.; Cao, Y. Synthesis of high quality zinc-blende CdSe nanocrystals and their application in hybrid solar cells.*Nanotechnology* 2006, *17*, 4736–4742.

279. Oey, C.C.; Djurisic, A.B.; Wang, H.; Man, K.K. Y.; Chan, W.K.; Xie, M.H.; Leung, Y.H.; Pandey, A.; Nunzi, J.M.; Chui, P.C. Polymer-TiO2 solar cells: TiO2 interconnected network for improved cell performance. *Nanotechnology* 2006, *17*, 706–713.

280. Olson, D.C.; Piris, J.; Collins, R.T.; Shaheen, S.E.; Ginley, D.S. Hybrid photovoltaic devices of polymer and ZnO nanofiber composites. *Thin Solid Films* 2006, *496*, 26–29.

281. Michler, P.; Imamoglu, A.; Mason, M.D.; Carson, P.J.; Strouse, G.F.; Buratto, S.K. Quantum correlation among photons from a single quantum dot at room temperature. *Nature (London)*2000, *406*, 968–970.

282. Cui, D.H.; Xu, J.; Zhu, T.; Paradee, G.; Ashok, S.; Gerhold, M. Harvest of near infrared light in PbSe nanocrystal-polymer hybrid photovoltaic cells. *Appl. Phys. Lett.* 2006, *88*, 183111.

283. O'Regan, B.; Schwartz, D.T.; Zakeeruddin, S.M.; Gratzel, M. Electrodeposited nanocomposite n-p heterojunctions for solid-state dye-sensitized photovoltaics. *Adv. Mater.* 2000, *12*, 1263–1267.

284. Bereznev, S.; Konovalov, I.; Opik, A.; Kois, J. Hybrid CuInS2/polypyrrole and CuInS2 /poly(3,4-ethylenedioxythiophene) photovoltaic structures. *Synth. Met.* 2005, *152*, 81–84.

285. Nozik, A.J. Exciton multiplication and relaxation dynamics in quantum dots: Applications to ultrahigh-efficiency solar photon conversion. *Inorg. Chem.* 2005, *44*, 6893–6899.

286. Pijpers, J.J.H.; Ulbricht, R.; Tielrooij, K.J.; Osherov, A.; Golan, Y.; Delerue, C.; Allan, G.; Bonn, M. Assessment of carrier-multiplication efficiency in bulk PbSe and PbS. *Nat. Physics* 2009, *5*, 811–814.

287. Beard, M.C.; Ellingson, R.J. Multiple exciton generation in semiconductor nanocrystals: Toward efficient solar energy conversion. *Laser Photonics Rev.* 2008, *2*, 377–399.

288. Allan, G.; Delerue, C. Influence of electronic structure and multiexciton spectral density on multiple-exciton generation in semiconductor nanocrystals: Tight-binding calculations. *Phys. Rev. B* 2008, *77*, 125340.

289. Beard, M.C.; Knutsen, K.P.; Yu, P.R.; Luther, J.M.; Song, Q.; Metzger, W.K.; Ellingson, R.J.; Nozik, A.J. Multiple exciton generation in colloidal silicon nanocrystals. *Nano Lett.* 2007, *7*, 2506–2512.

290. Nozik, A.J. Quantum dot solar cells. *Physica E* 2002, *14*, 115–120.

291. Oregan, B.; Gratzel, M. A low-cost, high-efficiency solar-cell based on dye-sensitized colloidal TiO2 films. *Nature (London)* 1991, *353*, 737–740.

292. Gratzel, M. Photoelectrochemical cells. *Nature (London)* 2001, *414*, 338–344.

293. Vogel, R.; Hoyer, P.; Weller, H. Quantum-sized PbS, CdS, Ag2S, Sb2S3, and Bi2S3 particles as sensitizers for various nanoporous wide-bandgap semiconductors. *J. Phys. Chem.* 1994, *98*, 3183–3188.

294. Vogel, R.; Pohl, K.; Weller, H. Sensitization of highly porous, polycrystalline TiO2 electrodes by quantum sized CdS. *Chem. Phys. Lett.* 1990, *174*, 241–246.

295. Gopidas, K.R.; Bohorquez, M.; Kamat, P.V. Photoelectrochemistry in semiconductor particulate systems .16. Photophysical and photochemical aspects of coupled semiconductors−Charge-transfer processes in colloidal CdS-TiO2 and CdS-AgI systems. *J. Phys. Chem.* 1990, *94*, 6435–6440.

296. Plass, R.; Pelet, S.; Krueger, J.; Gratzel, M.; Bach, U. Quantum dot sensitization of organic-inorganic hybrid solar cells. *J. Phys. Chem. B* 2002, *106*, 7578–7580.

297. Peter, L.M.; Riley, D.J.; Tull, E.J.; Wijayantha, K.G.U. Photosensitization of nanocrystalline TiO2 by self-assembled layers of CdS quantum dots. *Chem. Commun.* 2002, 1030–1031.

298. Lin, S.C.; Lee, Y.L.; Chang, C.H.; Shen, Y.J.; Yang, Y.M. Quantum-dot-sensitized solar cells: Assembly of CdS-quantum-dots coupling techniques of self-assembled monolayer and chemical bath deposition. *Appl. Phys. Lett.* 2007, *90*, 143517.

299. Robel, I.; Subramanian, V.; Kuno, M.; Kamat, P.V. Quantum dot solar cells. Harvesting light energy with CdSe nanocrystals molecularly linked to mesoscopic TiO2 films. *J. Am. Chem. Soc.* 2006, *128*, 2385–2393.

300. Niitsoo, O.; Sarkar, S.K.; Pejoux, C.; Ruhle, S.; Cahen, D.; Hodes, G. Chemical bath deposited CdS/CdSe-sensitized porous TiO2 solar cells. *J. Photochem. Photobiol. A* 2006, *181*, 306–313.

301. Diguna, L.J.; Shen, Q.; Kobayashi, J.; Toyoda, T. High efficiency of CdSe quantum-dot-sensitized TiO2 inverse opal solar cells. *Appl. Phys. Lett.* 2007, *91*, 023116 1–3.

302. Sun, W.T.; Yu, Y.; Pan, H.Y.; Gao, X.F.; Chen, Q.; Peng, L.M. CdS quantum dots sensitized TiO2 nanotube-array photoelectrodes. *J. Am. Chem. Soc.* 2008, *130*, 1124–1125.

303. Karakoti, A.S.; Filmalter, R.; Bera, D.; Kuchibhatla, S.; Vincent, A.; Seal, S. Spiral growth of one dimensional titania nanostructures using anodic oxidation. *J. Nanosci. Nanotech.* 2006, *6*, 2084–2089.

304. Lee, Y.L.; Lo, Y.S. Highly efficient quantum-dot-sensitized solar cell based on co-sensitization of CdS/CdSe. *Adv. Func. Mater.* 2009, *19*, 604–609.

305. Zaban, A.; Micic, O.I.; Gregg, B.A.; Nozik, A.J. Photosensitization of nanoporous TiO2 electrodes with InP quantum dots. *Langmuir* 1998, *14*, 3153–3156.

306. Wang, Y.; Herron, N. Photoconductivity of CdS nanocluster-doped polymers. *Chem. Phys. Lett.* 1992, *200*, 71–75.

307. Greenham, N.C.; Peng, X.G.; Alivisatos, A.P. Charge separation and transport in conjugated polymer cadmium selenide nanocrystal composites studied by photoluminescence quenching and photoconductivity. *Synth. Met.* 1997, *84*, 545–546.

308. Arango, A.C.; Carter, S.A.; Brock, P.J. Charge transfer in photovoltaics consisting of interpenetrating networks of conjugated polymer and TiO2 nanoparticles. *Appl. Phys. Lett.*1999, *74*, 1698–1700.

309. Huynh, W.U.; Dittmer, J.J.; Alivisatos, A.P. Hybrid nanorod-polymer solar cells. *Science* 2002,*295*, 2425–2427.

310. Beek, W.J.E.; Wienk, M.M.; Janssen, R.A.J. Efficient hybrid solar cells from zinc oxide nanoparticles and a conjugated polymer. *Adv. Mater.* 2004, *16*, 1009–1013.

311. Sun, B.Q.; Greenham, N.C. Improved effciency of photovoltaics based on CdSe nanorods and poly(3-hexylthiophene) nanofibers. *Phys. Chem. Chem. Phys.* 2006, *8*, 3557–3560.

312. Wang, P.; Abrusci, A.; Wong, H.M.P.; Svensson, M.; Andersson, M.R.; Greenham, N.C. Photoinduced charge transfer and efficient solar energy conversion in a blend of a red polyfluorene copolymer with CdSe nanoparticles. *Nano Lett.* 2006, *6*, 1789–1793.

313. Sun, B.Q.; Snaith, H.J.; Dhoot, A.S.; Westenhoff, S.; Greenham, N.C. Vertically segregated hybrid blends for photovoltaic devices with improved efficiency. *J. Appl. Phys.* 2005, *97*, 014914.

314. Sun, B.Q.; Marx, E.; Greenham, N.C. Photovoltaic devices using blends of branched CdSe nanoparticles and conjugated polymers. *Nano Lett.* 2003, *3*, 961–963.

315. Dayal, S.; Kopidakis, N.; Olson, D.C.; Ginley, D.S.; Rumbles, G. Photovoltaic devices with a low band gap polymer and CdSe nanostructures exceeding 3% efficiency. *Nano Lett.* 2010, *10*, 239–242.

316. Gur, I.; Fromer, N.A.; Chen, C.P.; Kanaras, A.G.; Alivisatos, A.P. Hybrid solar cells with prescribed nanoscale morphologies based on hyperbranched semiconductor nanocrystals.*Nano Lett.* 2007, *7*, 409–414.

317. Liu, J.S.; Tanaka, T.; Sivula, K.; Alivisatos, A.P.; Frechet, J.M.J. Employing end-functional polythiophene to control the morphology of nanocrystal-polymer composites in hybrid solar cells. *J. Am. Chem. Soc.* 2004, *126*, 6550–6551.

318. Watt, A.A.R.; Blake, D.; Warner, J.H.; Thomsen, E.A.; Tavenner, E.L.; Rubinsztein-Dunlop, H.; Meredith, P. Lead sulfide nanocrystal: conducting polymer solar cells. *J. Phys. D* 2005, *38*, 2006–2012.

319. Klimov, V.I.; Ivanov, S.A.; Nanda, J.; Achermann, M.; Bezel, I.; McGuire, J.A.; Piryatinski, A. Single-exciton optical gain in semiconductor nanocrystals. *Nature (London)* 2007, *447*, 441–446.

320. Wang, C.J.; Wehrenberg, B.L.; Woo, C.Y.; Guyot-Sionnest, P. Light emission and amplification in charged CdSe quantum dots. *J. Phys. Chem. B* 2004, *108*, 9027–9031.

321. Roither, J.; Pichler, S.; Kovalenko, M.V.; Heiss, W.; Feychuk, P.; Panchuk, O.; Allam, J.; Murdin, B.N. Two- and one-dimensional light propagations and gain in layer-by-layer-deposited colloidal nanocrystal waveguides. *Appl. Phys. Lett.* 2006, *89*, 111120.

322. Liu, B.; Li, H.P.; Chew, C.H.; Que, W.X.; Lam, Y.L.; Kam, C.H.; Gan, L.M.; Xu, G.Q. PbS-polymer nanocomposite with third-order nonlinear optical response in femtosecond regime.*Mater. Lett.* 2001, *51*, 461–469.

323. Choudhury, K.R.; Sahoo, Y.; Prasad, P.N. Hybrid quantum-dot-polymer nanocomposites for infrared photorefractivity at an optical communication wavelength. *Adv. Mater.* 2005, *17*, 2877–2881.

324. Williams, D.P.; Andreev, A.D.; O'Reilly, E.P. Dependence of exciton energy on dot size in GaN/AlN quantum dots. *Phys. Rev. B* 2006, *73*, 241301.

325. Pan, G.Q.; Kordesch, M.E.; Van Patten, P.G. New pyrolysis route to GaN quantum dots.*Chem. Mater.* 2006, *18*, 3915–3917.

326. Barik, S.; Tan, H.H.; Jagadish, C.; Vukmirovic, N.; Harrison, P. Selective wavelength tuning of self-assembled InAs quantum dots grown on InP. *Appl. Phys. Lett.* 2006, *88*.

327. Cassidy, P.J.; Radda, G.K. Molecular imaging perspectives. *J. R. Soc. Interface* 2005, *2*, 133–144.

328. Schillaci, O.; Danieli, R.; Padovano, F.; Testa, A.; Simonetti, G. Molecular imaging of atheroslerotic plaque with nuclear medicine techniques (Review). *Int. J. Mol. Med.* 2008, *22*, 3–7.

329. Lecchi, M.; Ottobrini, L.; Martelli, C.; Del Sole, A.; Lucignani, G. Instrumentation and probes for molecular and cellular imaging. *Q. J. Nucl. Med. Mol. Imaging* 2007, *51*, 111–126.

330. Pichler, B.J.; Wehrl, H.F.; Judenhofer, M.S. Latest advances in molecular imaging instrumentation. *J. Nucl. Med.* 2008, *49*, S5–S23.

331. Levenson, R.M.; Lynch, D.T.; Kobayashi, H.; Backer, J.M.; Backer, M.V. Multiplexing with multispectral imaging: From mice to microscopy. *Ilar J.* 2008, *49*, 78–88.

332. Sharma, P.; Brown, S.; Walter, G.; Santra, S.; Moudgil, B. Nanoparticles for bioimaging. *Adv. Colloid Interface Sci.* 2006, *123*, 471–485.

333. Wang, Y.; Deng, Y.L.; Qing, H.; Xie, H.Y. Advance in real-time and dynamic biotracking and bioimaging based on quantum dots. *Chem. J. Chin. Univer.-Chin.* 2008, *29*, 661–668.

334. Parak, W.J.; Pellegrino, T.; Plank, C. Labelling of cells with quantum dots. *Nanotechnology*2005, *16*, R9–R25.

335. Li, Z.B.; Cai, W.; Chen, X.Y. Semiconductor quantum dots for *in vivo* imaging. *J. Nanosci. Nanotech.* 2007, *7*, 2567–2581.

336. Kumar, S.; Richards-Kortum, R. Optical molecular imaging agents for cancer diagnostics and therapeutics. *Nanomedicine* 2006, *1*, 23–30.

337. Smith, A.M.; Dave, S.; Nie, S.M.; True, L.; Gao, X.H. Multicolor quantum dots for molecular diagnostics of cancer. *Expert Rev. Mol. Diagn.* 2006, *6*, 231–244.

338. Gao, X.H.; Yang, L.L.; Petros, J.A.; Marshal, F.F.; Simons, J.W.; Nie, S.M. *In vivo* molecular and cellular imaging with quantum dots. *Curr. Opin. Biotechnol.* 2005, *16*, 63–72.

339. Jaiswal, J.K.; Goldman, E.R.; Mattoussi, H.; Simon, S.M. Use of quantum dots for live cell imaging. *Nat. Methods* 2004, *1*, 73–78.

340. Smith, A.M.; Duan, H.W.; Mohs, A.M.; Nie, S.M. Bioconjugated quantum dots for *in vivo*molecular and cellular imaging. *Adv. Drug Delivery Rev.* 2008, *60*, 1226–1240.

341. Gao, X.H.; Dave, S.R. Quantum dots for cancer molecular imaging. In *Bio-Applications of Nanoparticles*; 2007; Volume 620, pp. 57–73.

342. Zrazhevskiy, P.; Gao, X. Quantum dots for cancer molecular imaging. *Minerva Biotecnologica*2009, *21*, 37–52.

343. Hillman, E.M.C. Optical brain imaging *in vivo*: Techniques and applications from animal to man. *J. Biomed. Optics* 2007, *12*, 051402.

344. Luker, G.D.; Luker, K.E. Optical imaging: Current applications and future directions. *J. Nucl. Med.* 2008, *49*, 1–4.

345. Wu, W.W.; Li, A.D. Optically switchable nanoparticles for biological imaging. *Nanomedicine*2007, *2*, 523–531.

346. Weissleder, R. A clearer vision for *in vivo* imaging. *Nat. Biotechnol.* 2001, *19*, 316–317.

347. Bruchez, M.; Moronne, M.; Gin, P.; Weiss, S.; Alivisatos, A.P. Semiconductor nanocrystals as fluorescent biological labels. *Science* 1998, *281*, 2013–2016.

348. Chan, W.C.W.; Nie, S.M. Quantum dot bioconjugates for ultrasensitive nonisotopic detection. *Science* 1998, *281*, 2016–2018.

349. Parak, W.J.; Boudreau, R.; Le Gros, M.; Gerion, D.; Zanchet, D.; Micheel, C.M.; Williams, S.C.; Alivisatos, A.P.; Larabell, C. Cell motility and metastatic potential studies based on quantum dot imaging of phagokinetic tracks. *Adv. Mater.* 2002, *14*, 882–885.

350. Akerman, M.E.; Chan, W.C.W.; Laakkonen, P.; Bhatia, S.N.; Ruoslahti, E. Nanocrystal targeting *in vivo*. *Proc. Natl. Acad. Sci. USA* 2002, *99*, 12617–12621.

351. Kim, S.; Lim, Y.T.; Soltesz, E.G.; De Grand, A.M.; Lee, J.; Nakayama, A.; Parker, J.A.; Mihaljevic, T.; Laurence, R.G.; Dor, D.M.; Cohn, L.H.; Bawendi, M.G.; Frangioni, J.V. Near-infrared fluorescent type II quantum dots for sentinel lymph node mapping. *Nat. Biotechnol.*2004, *22*, 93–97.

352. Medintz, I.L.; Clapp, A.R.; Mattoussi, H.; Goldman, E.R.; Fisher, B.; Mauro, J.M. Self-assembled nanoscale biosensors based on quantum dot FRET donors. *Nat. Mater.* 2003, *2*, 630–638.

353. Sapsford, K.E.; Pons, T.; Medintz, I.L.; Mattoussi, H. Biosensing with luminescent semiconductor quantum dots. *Sensors* 2006, *6*, 925–953.

354. Sukhanova, A.; Devy, M.; Venteo, L.; Kaplan, H.; Artemyev, M.; Oleinikov, V.; Klinov, D.; Pluot, M.; Cohen, J.H.M.; Nabiev, I. Biocompatible fluorescent nanocrystals for immunolabeling of membrane proteins and cells. *Anal. Biochem.* 2004, *324*, 60–67.

355. Dahan, M.; Levi, S.; Luccardini, C.; Rostaing, P.; Riveau, B.; Triller, A. Diffusion dynamics of glycine receptors revealed by single-quantum dot tracking. *Science* 2003, *302*, 442–445.

356. Michalet, X.; Pinaud, F.F.; Bentolila, L.A.; Tsay, J.M.; Doose, S.; Li, J.J.; Sundaresan, G.; Wu, A.M.; Gambhir, S.S.; Weiss, S. Quantum dots for live cells, *in vivo* imaging, and diagnostics.*Science* 2005, *307*, 538–544.

357. Hassan, M.; Klaunberg, B.A. Biomedical applications of fluorescence imaging *in vivo*. *Comp. Med.* 2004, *54*, 635–644.

358. Medintz, I.L.; Uyeda, H.T.; Goldman, E.R.; Mattoussi, H. Quantum dot bioconjugates for imaging, labelling and sensing. *Nat. Mater.* 2005, *4*, 435–446.

359. Zhang, Y.; He, J.; Wang, P.N.; Chen, J.Y.; Lu, Z.J.; Lu, D.R.; Guo, J.; Wang, C.C.; Yang, W.L. Time-dependent photoluminescence blue shift of the quantum dots in living cells: Effect of oxidation by singlet oxygen. *J. Am. Chem. Soc.* 2006, *128*, 13396–13401.

360. Dahan, M.; Laurence, T.; Pinaud, F.; Chemla, D.S.; Alivisatos, A.P.; Sauer, M.; Weiss, S. Time-gated biological imaging by use of colloidal quantum dots. *Opt. Lett.* 2001, *26*, 825–827.

361. Wu, C.L.; Hong, J.Q.; Guo, X.Q.; Huang, C.B.; Lai, J.P.; Zheng, J.S.; Chen, J.B.; Mu, X.; Zhao, Y.B. Fluorescent core-shell silica nanoparticles as tunable precursors: Towards encoding and multifunctional nano-probes. *Chem. Comm.* 2008, 750–752.

362. Han, M.Y.; Gao, X.H.; Su, J.Z.; Nie, S. Quantum-dot-tagged microbeads for multiplexed optical coding of biomolecules. *Nat. Biotechnol.* 2001, *19*, 631–635.

363. Dubertret, B.; Skourides, P.; Norris, D.J.; Noireaux, V.; Brivanlou, A.H.; Libchaber, A. *In vivo*imaging of quantum dots encapsulated in phospholipid micelles. *Science* 2002, *298*, 1759–1762.

364. Larson, D.R.; Zipfel, W.R.; Williams, R.M.; Clark, S.W.; Bruchez, M.P.; Wise, F.W.; Webb, W.W. Water-soluble quantum dots for multiphoton fluorescence imaging *in vivo*. *Science*2003, *300*, 1434–1436.

365. Holloway, P.H.; Yang, H.; Lee, H.; Seo, S.; Santra, S.; Qian, L.; Bera, D. *Nanophosphor: PL, EL, and Biological Markers* Proc. Int. Symp. Display lighting Phosphor Mater, ISDLPMW'06. 2006.

366. Lakowicz, J.R. *Principles of Fluorescence Spectroscopy,*, 2nd ed.; Academic Publishers: New York, NY, USA, 1999.

367. Tran, P.T.; Goldman, E.R.; Anderson, G.P.; Mauro, J.M.; Mattoussi, H. Use of luminescent CdSe-ZnS nanocrystal bioconjugates in quantum dot-based nanosensors. *Phys. Status Solidi B*2002, *229*, 427–432.

368. Clapp, A.R.; Medintz, I.L.; Mauro, J.M.; Fisher, B.R.; Bawendi, M.G.; Mattoussi, H. Fluorescence resonance energy transfer between quantum dot donors and dye-labeled protein acceptors. *J. Am. Chem. Soc.* 2004, *126*, 301–310.

369. Clapp, A.R.; Medintz, I.L.; Fisher, B.R.; Anderson, G.P.; Mattoussi, H. Can luminescent quantum dots be efficient energy acceptors with organic dye donors? *J. Am. Chem. Soc.* 2005,*127*, 1242–1250.

370. Patolsky, F.; Gill, R.; Weizmann, Y.; Mokari, T.; Banin, U.; Willner, I. Lighting-up the dynamics of telomerization and DNA replication by CdSe-ZnS quantum dots. *J. Am. Chem. Soc.* 2003, *125*, 13918–13919.

371. Herbst, R.S.; Shin, D.M. Monoclonal antibodies to target epidermal growth factor receptor-positive tumors–A new paradigm for cancer therapy. *Cancer* 2002, *94*, 1593–1611.

372. Reuter, C.W.M.; Morgan, M.A.; Eckardt, A. Targeting EGF-receptor-signalling in squamous cell carcinomas of the head and neck. *Br. J. Cancer* 2007, *96*, 408–416.

373. Qian, X.; Peng, X.-H.; Ansari, D.O.; Yin-Goen, Q.; Chen, G.Z.; Shin, D.M.; Yang, L.; Young, A.N.; Wang, M.D.; Nie, S. *In vivo* tumor targeting and spectroscopic detection with surface-enhanced Raman nanoparticle tags. *Nat. Biotech.* 2008, *26*, 83–90.

374. Martin, C.R.; Mitchell, D.T. Nanomaterials in analytical chemistry. *Anal. Chem.* 1998, *70*, A322–A327.

375. Tan, W.B.; Zhang, Y. Multi-functional chitosan nanoparticles encapsulating quantum dots and Gd-DTPA as imaging probes for bio-applications. *J. Nanosci. Nanotech.* 2007, *7*, 2389–2393.

376. Mulder, W.J.M.; Koole, R.; Brandwijk, R.J.; Storm, G.; Chin, P.T.K.; Strijkers, G.J.; Donega, C.D.; Nicolay, K.; Griffioen, A.W. Quantum dots with a paramagnetic coating as a bimodal molecular imaging probe. *Nano Lett.* 2006, *6*, 1–6.

377. van Tilborg, G.A.F.; Mulder, W.J.M.; Chin, P.T.K.; Storm, G.; Reutelingsperger, C.P.; Nicolay, K.; Strijkers, G.J. Annexin A5-conjugated quantum dots with a paramagnetic lipidic coating for the multimodal detection of apoptotic cells. *Bioconjug. Chem.* 2006, *17*, 865–868.

378. Bakalova, R.; Zhelev, Z.; Aoki, I.; Ohba, H.; Imai, Y.; Kanno, I. Silica-shelled single quantum dot micelles as imaging probes with dual or multimodality. *Anal. Chem.* 2006, *78*, 5925–5932.

379. Toth, E.; Bolskar, R.D.; Borel, A.; Gonzalez, G.; Helm, L.; Merbach, A.E.; Sitharaman, B.; Wilson, L.J. Water-soluble gadofullerenes: Toward high-relaxivity, pH-responsive MRI contrast agents. *J. Am. Chem. Soc.* 2005, *127*, 799–805.

380. Rohrer, M.; Bauer, H.; Mintorovitch, J.; Requardt, M.; Weinmann, H.J. Comparison of magnetic properties of MRI contrast media solutions at different magnetic field strengths.*Invest. Radiol.* 2005, *40*, 715–724.

381. Pierre, V.C.; Botta, M.; Raymond, K.N. Dendrimeric gadolinium chelate with fast water exchange and high relaxivity at high magnetic field strength. *J. Am. Chem. Soc.* 2005, *127*, 504–505.

382. Tan, W.B.; Zhang, Y. Multifunctional quantum-dot-based magnetic chitosan nanobeads. *Adv. Mater.* 2005, *17*, 2375–2380.

383. Son, D.H.; Hughes, S.M.; Yin, Y.D.; Alivisatos, A.P. Cation exchange reactions-in ionic nanocrystals. *Science* 2004, *306*, 1009–1012.

384. Green, M. Semiconductor quantum dots as biological imaging agents. *Angew. Chem., Int. Ed.*2004, *43*, 4129–4131.

CHAPTER 2

QUANTUM DOTS PREPARED BY DROPLET EPITAXIAL METHOD

Ákos Nemcsics

Keywords: Droplet epitaxy, GaAs, AlGaAs, AlAs, QD, MBE

INTRODUCTION

The electrons "boxed up" or confined in quantum confinement and the calculation of their energy levels, which are well known from textbooks, were regarded for decades as idealistic concepts far from practical reality. The molecular beam epitaxial (MBE) technology, developed for compound semiconductors, made the introduction of the low-dimension structures into the everyday scientific research possible. The structures confined in all three dimensions are called quantum dots (QDs).

For the fabrication of QDs or any zero-dimensional structures, various methods were developed. For a long time, the only known method for the production of epitaxially grown zero-dimensional system was the strain-induced method, based on lattice mismatch. Presently, the most used technique is the lattice-mismatched technique by the Stranski–Krastanov mode [1–5]. QDs, such as In As nanocluster prepared on GaAs surface, are the oldest known system. The droplet epitaxy (DE), however, is one of the latest methods for QD preparation. This method of preparation is not only an alternative way to the conventional method but also a production method for a number of zero-dimensional quantum structures such as ringlike QDs, double ringlike QDs, inverse QDs (filled nanoholes), or QD molecules. This technique is also fully compatible with MBE technology.

Citation: Akos Nemcsics (2015). Quantum Dots Prepared by Droplet Epitaxial Method, Quantum Dots - Theory and Applications, Dr. Vasilios Stavrou (Ed.), InTech, DOI: 10.5772/60823.

The subjects of this paper are the technologies used, growth kinetics, and some properties of the zero-dimensional nanostructures of different shapes, grown by DE. The fundamentals of DE were first demonstrated by Koguchi and his coworkers in the early 1990s [6–9]. One advantage is that it is not restricted by mismatch conditions; in addition to that is the possible fabrication of strain-free QDs and similar nanostructures by DE, against the Stranski–Krastanov-based growth method. The shape diversity makes it a preferable process for that application. The size, shape, and elementary distribution of the developed structures are dependent on the developing technology used. It is obvious that the physical parameters of the structures are important from both the theoretical and practical point of view.

In this paper, QDs grown by DE will be described, selected from III and V class materials. Here, the QDs are mostly GaAs with the use of AlGaAs as barrier material. This process consists of the following two basic steps: first, the metallic (e.g., Ga) nanosized droplets are generated on the surface by the Volmer–Weber growth mode, followed by the second step of droplet crystallization that is the transformation into GaAs QDs in arsenic atmosphere.

For the control of the process, the kinetics of the growth process knowledge is necessary, which is so far lacking the full theoretical description.

The DE technology for QD preparation has superiority over the nowadays widely used strain-induced method. The greatest advantage is that both lattice matched and mismatched can be used for QD fabrication. At the same time, the size and density of the QD's can be controlled independently, and the distribution is more uniform. However, the shapes of the QDs show a wide spectrum. This spectrum includes the conventionally shaped QDs, ringlike QDs, and complex multicomponent QD structures [9–13]. To avoid intermixing, low temperature can be applied at preparation. A wide variety substrate orientation has beneficial effect on the application.

Although the technique is quite recent, it is already successfully applied in a number of cases. QDs had significantly improved the general performance of the optoelectronic devices, like semiconductor lasers and optical amplifiers. DE fabricated QDs introduced further improvements for the beneficial reasons described previously [10].

Future quantum devices, based on quantum mechanical and electromagnetic interactions, require lateral QD configurations. With the modified DE method, aligned QD pairs and QD molecules can be produced without involving lithographic technology [14, 15].

THE DROPLET EPITAXIAL TECHNIQUE

Technological Background and Preliminaries of the Droplet Epitaxial Technique

It is well known in our profession that the grown structure can be classified in three growth modes depending on the interaction among the constituents. These modes are the layer-by-layer growth or Frank–van der Merve mode, the island growth or Volmer–Weber mode, and the intermediate mode (layer-plus-island growth) or Stranski–Krastanov mode (Figure 1) [16]. The morphology of the interface is determined by the interplay between deposition, desorption, and surface diffusion. If an atom or molecule from the environment arrives at a random position on the surface, the deposition process bonds with the surface atoms and sticks.

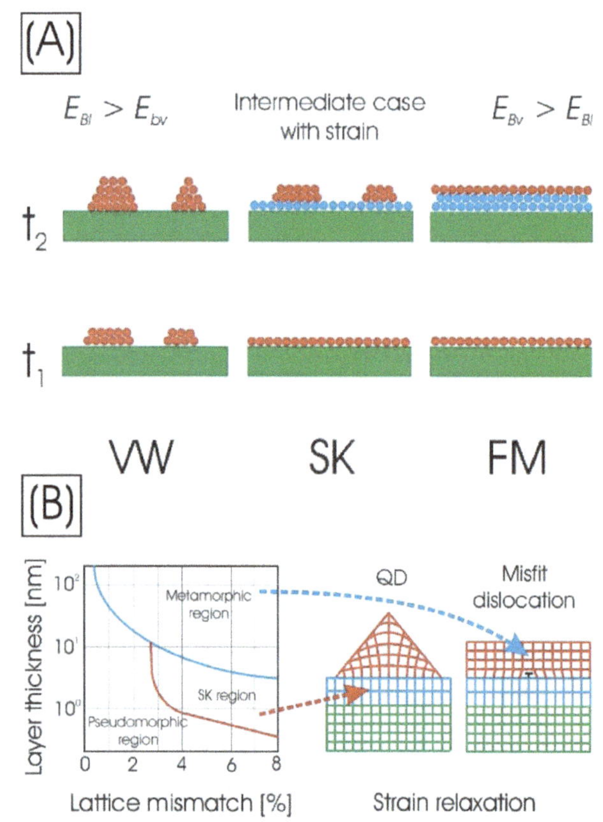

Figure 1. (A) Schematic illustration of the three growth modes. (B) Strain relaxation via QD formation and via generation of misfit dislocation.

The process competing with deposition is desorption. During the desorption process, some atoms deposited on the surface leave the interface. The probability of the desorption depends on how strongly the atom is bonded to the surface. The strength of the bonds depends either on the type of atom or on the local atom arrangement of the surface where the atom sticks. The desorption can be negligible for many materials under typical MBE conditions. The deposited atoms diffuse on the crystal surface, searching for the energetically most favorable position. The diffusion length can be quite large and depends on the temperature and the binding energies to the substrate (E_{Bv}, binding in vertical direction).

In the Frank–van der Merve mode, the interaction between the neighboring atoms (E_{Bl}, lateral directions) in the layer is weaker than that with the substrate atoms ($E_{Bv} > E_{Bl}$) [17]. In this growth mode, islands of monolayer (ML) height coalesce before a new layer can nucleate on top of them. In the Volmer–Weber mode, the situation is just the opposite, that is, the binding energy among the deposited atoms is stronger than to the substrate surface ($E_{Bl} > E_{Bv}$). In this case, the growth proceeds to many atomic layers at discrete islands before these islands merge.

In the Stranski–Krastanov mode, the process is initiated in a layer-by-layer fashion, but islanding commences after the growth of a certain layer thickness. In certain situations, layer-by-layer growth is desirable because of the need for multilayered structures with exact layer thickness and flat interfaces. This requires that the nucleation takes place as a single event on the substrate. In this growth mode, the lattice mismatch between the grown layer and the substrate material is the most common case.

The deformation takes place by the even layer to match the lattice of the substrate. In this case, elastic energy is generated in the structure, which deforms mainly the lattice of the grown layer. When this deformation energy overcomes a critical value, it can be relaxed in two ways. One possible way is to nucleate nanosized islands on the surface of the even layer, which is called the wetting layer. The absence of the lateral layer allows the atomic planes to laterally relax, reducing the elastic energy in the structure. According the first mentioned way, the so-called strain-induced QD can be grown. The other way is to generate misfit dislocations at the interface [18–21]. The dislocations in the substrate and at the interface can overgrow into the grown layer [23–25]. These defects are called threading dislocations.

Figure 1 shows the following two out of three growth modes, Volmer–Weber and Stranski–Krastanov for nanostructure production. One may ask why only the Stranski–Krastanov growth mode is widely used for QD production. The answer lies in the development of the MBE technology. The application of III–V materials opened the way for the production of heteroepitaxial structures, specifically the GaAs–AlGaAs–AlAs system, for its identical lattice parameter. The growth of InAs and InGaAs created particular technological difficulties due to the lattice mismatch. In the early 1990s, the success of InGaAs growth on GaAs, with differing indium content [26–27], brought about a big advance in this field. Initially, the thickness of the layers stayed under a critical value, but this limitation was successfully overcome. The strain-induced QD production has superseded the DE, dominating the field earlier.

The archetypal system of the clustered nanostructures is the lattice-mismatched system such as InAs on GaAs, where the strain-induced process leads to the formation of QD. The lattice-mismatched structure can relax either by the generation of misfit dislocation or by the nucleation of nanosized islands on the even layer (Figure 1). The essential driving force for coherent QD formation, after an under layer formed, is the strain relaxation, whereby the energy gained from the increase in surface area via QD formation more than compensates the increase in interfacial free energy. It is generally accepted that one of the essential driving forces, for coherent lattice-mismatched QD formation, is strain relaxation.

The field of self-organized strain-induced QDs is dominated by two material system, InAs/GaAs (001) and Ge/Si (001). For instance, the MBE growth of about 1.5 MLs of InAs on GaAs results in the spontaneous formation of InAs QDs driven by the strain between the deposited InAs and the GaAs substrate. The driving force for the self-organized QD formation is the strain energy induced by the lattice mismatch of about 7%, in which the condition restricts the material choice. Two families of the shape such as pyramids and domes can be created during the defect-free QD transformations [28].

Principles of DE

The QD preparation in a self-assembled manner by using the DE technique is an advantageous alternative and extension to the strain-induced QD creation. The basic idea of the DE originated from Koguchi and his coworkers [29, 30]. In comparison with the technology of the strain-induced Stranski–Krastanov QD production, the DE is more flexible regarding the choice of the QD material and also regarding the shape and distribution of the resulted QDs and other unconventionally-shaped QD structures. For example, it is possible to create not only InAs QD on GaAs but also strain-free GaAs QD on AlGaAs surface. Furthermore, this technology of DE allows us to produce ringlike QD and further special-shaped nanosized structures and structure complexes [31–35].

In the case of the DE, the clustering on the surface is carried out with the help of Volmer–Weber like growth mode. The basic idea based on the separation of the III-column and V-column materials supply during the MBE growth (Figure 2). Here, the QD preparation consists of two main parts such as the formation of metallic nanosized droplet on the surface and its crystallization with the help of the non-metallic component of the compound semiconductor. It is very important to mention that this DE technique is entirely compatible with the MBE technology. This circumstance allows us to combine the DE method with the other conventional MBE processes.

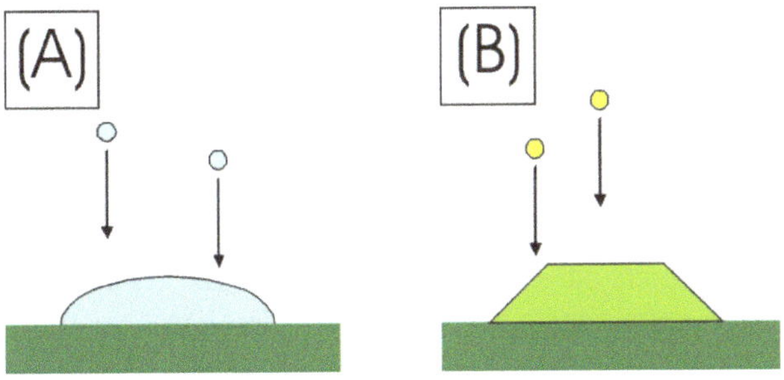

Figure 2. (A) First step of DE: generation of metallic droplet on the surface. (B) Second step: crystallization to create QD.

First, the III-column elements such as Ga, In, Al, etc., are deposited. These metallic components form clusters that are nanosized droplets on the substrate surface. At the droplet formation, the main driving forces are the diffusion, the minimalization of the surface tension, and the Ostwald ripening [36–42]. The dominance of these driving forces depends on the technological circumstances. The second step of the DE technique is the crystallization of the metallic cluster by the molecular beam irradiation of the nonmetallic V-column element. It is important to suppress or to control the lateral diffusion of the group III elements from the cluster.

QD Engineering

As known, the electronic and optoelectronic properties of the nanostructure depend strongly on the shape of the nanostructure. The technology used for growth can be controlled by the size, shape, and elementary distribution of the developing structure. These technological and physical parameters are very important from both theoretical and practical points of view. In the DE technique, because of the shape and distribution diversity, the key point is to discover growth mechanism and its relation with the parameters of the developed structures. Since the discovery of the DE, several scientific knowledge are gathered. However, until now, no full theoretical description of the underlying growth kinetic is available.

An interesting correlation between the GaAs QD shapes and their volume was observed by Heyn and coworkers [43, 44]. As a result of this discovery, they developed the first growth model for the DE of GaAs QDs [43, 44]. The experiment was carried out with the cystallization of Ga droplets on AlGaAs surface. Here, growth temperature ranged between 140°C and 300°C. On the arsenic terminated surface, the deposition of Ga with flux 0.025–0.79 ML/s was executed. It was clearly visible that growth temperature strongly influenced QD density.

The QD density dependence on temperature shows the scaling law [44]. Under 200°C, the slope of the temperature shows E_a = 0.235 eV. Over 200°C, data did not follow the scaling law, but QD density decreased. At this temperature, coarsening of the cluster begins. This effect goes back to the Ostwald ripening, which means the growth of large clusters at the expense of smaller ones, hence causing a decrease in the total cluster density.

Higher temperature combined with lover arsenic flux will result in a ringlike nanostructure. At longer waiting time and lower arsenic flux, these rings fall below the original substrate surface and the encircling surround. After the heat treatment of the sample with the clusters formed, the total lack of arsenic results in nanoholes. The creation of nanoholes can lead back to the thermal etching of the substrate by the liquid metallic component. At a typical process temperature of 570°C, etching starts approximately 20 s later than the beginning of the annealing step, followed by desorption approximately 100 s after. The depth of the nanoholes can be found by multiplying the etching time by the etching rate.

These nanoholes can be filled with low band-gap material to create the so-called inverted QD. Furthermore, the nanoholes can serve as templates for QD preparation. In the utilization of surface anisotropy, QD pairs can be grown. The nanohole and the nanomound can initialize QD molecules, where the QD per molecules ranged from two to six. In the case of InAs QDs on

GaAs (001) surface, the number of QDs per GaAs mound can be effectively controlled by varying InAs ML coverage [45].

DROPLET FORMATION

The Initial Surface Status

During the QD formation, the status of the substrate surface is very important. It is well known that surface reconstruction influences strongly the growing layer and the forming structure [16]. Surface reconstruction depends on the temperature and the ambient environment environment, too. As mentioned before, the DE technique is compatible with the MBE technology. It means that the growing surface can be continuously monitored with reflection high-energy electron diffraction (RHEED) to determine the status of the surface, e.g., the surface reconstruction. During the growth, the RHEED pattern and its specular spot intensity allows a real-time tracking of the surface status changes. Here, we are using the technological most important (001) surface orientation. The bipolar (001) surface of GaAs can be terminated by either Ga or As atoms with rich variety of reconstructions are formed depending on surface stoichiometries [16].

Before the droplet creation, we have to grow a layer from a wide band-gap material for barrier. This layer growth happened with a conventional MBE technology, which is carried out in arsenic ambient pressure. Hence, the initial status of the surface is arsenic terminated. After the layer growth, we have to change the temperature depending on the planned nanostructure shape. The deposited MLs are not equal with the droplet volume. Assume that Ga is deposited on AlGaAs surface at 200°C. Due to the strong binding energy of As to Ga and substrate, the first Ga ML is consumed for the formation of Ga terminated surface. This first ML volume does not contribute to the formation of the Ga droplets. It means the coverage of Ga located in the droplet is Ft-A. The total Ga surface coverage is $\Theta = Ft$, where it is resulted with flux F for a time t, and A is the incorporated value of ML.

Although no complete phase diagram of GaAs (001) has been mapped, certain surface phases are generally observed during growth or after growth and annealing. For example, the less As-rich surface exhibits a (2 × 4) / c(2 × 8) pattern and arises from 0.75 ML of As. This reconstruction is normally present during the MBE of GaAs. Areas with (2 × 4) and c(2 × 8) symmetry can coexist. With larger coverage of As, a c(4 × 4) reconstruction with As coverage of 1.75 ML is observed [46, 47]. Depending on the substrate temperature and the Ga-to-As ratio during MBE, the GaAs (001) surface is known to display various surface reconstructions ranging from the As-rich c(4 × 4)β, c(4 × 4)α, c(2 × 8), and (2 × 4) to the Ga-rich (6 × 6), c(8 × 2), (4 × 6), and (4 × 2) surface. The surface symmetry critically depends on the preparation conditions [46, 47].

In situ investigations of this large set of reconstructions with RHEED are well established. Using scanning tunneling microscopy (STM), most of these surface reconstructions could also be imaged in real space, and the structure and the structural models were refined, too. Most of the previous studies were based on the diffraction techniques and no direct observation for the (4 × 6) phase exists. The STM study reveals that (4 × 6) symmetry arises from the coexistence of (2 × 4) and (4 × 2) units [48]. We have seen before that the change in temperature and other parameters results in the continuous change in surface reconstruction. During the change of the temperature, not only the reconstruction but also the specular spot intensity changes. The results show that at lower temperature, the specular spot intensity is high, and at higher temperature, the intensity is low. Furthermore, in the up and down and back directions, the temperature versus intensity function shows hysteretic behavior (Figure 3). The explanation of this phenomenon can lead us to the understanding of the intricate surface structures [49,50].

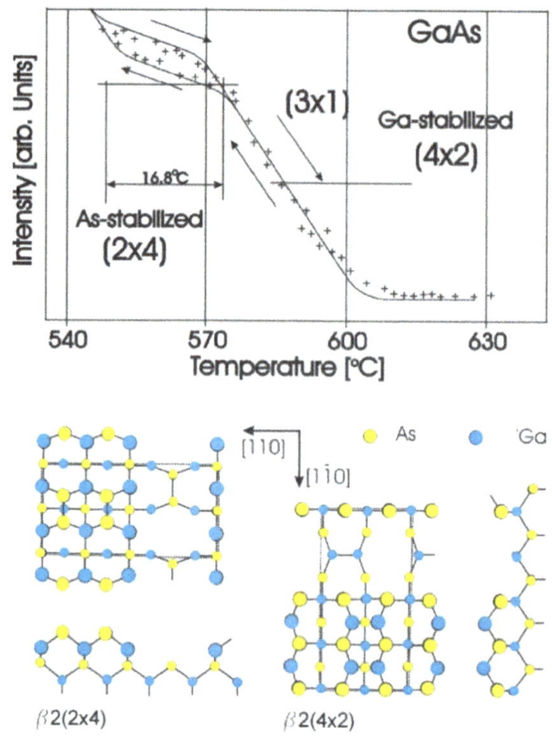

Figure 3. Ga and As stabilized GaAs (001) surface depending on the substrate temperature.

The initial step of the DE is the droplets creation on the surface. It is already well documented that the droplet nucleation requires Ga-terminated surface, which means that a part of Ga coverage incorporate into the surface. During the initial stage of Ga deposition, fundamental change happens in the surface reconstruction, which strongly influences the droplet formation. Before the Ga deposition, the surface reconstruction can be usually c(4 × 4), (2 × 4), or (4 × 6), where the values of As coverage are 1.75, 0.75, or 1.12 ML, respectively. The previous paragraphs show the complicity of the surface reconstruction and other difficult unsolved problems associated with the process. The part from the metallic surface deposition migrating into the droplet depends largely on the initial state of the surface.

Size and Density of the Droplets

The size and the density of the metallic droplets depend on the substrate temperature and on the deposited MLs. In the case of Ga droplets on AlGaAs surface under 200°C, the cluster density versus temperature follows scaling law. Over 200°C, the scaling law is broken. Under this temperature, the clusters begin to coalesce, which drastically reduce the cluster number. This decrease of cluster number is governed by the Ostwald ripening [36–42].

The reduction of the cluster number is affected by the diffusion length of the constituents. The condition required for the coalescence is the overlap of the diffusion length and the mean distance of the neighboring clusters. It characterized by a local interaction between neighboring clusters of slightly different size. The smaller cluster starts to decompose to maintain the gradient of the radius equilibrium concentration toward the larger cluster due to the Gibbs–Thomson effect [51]. The appearance of such processes is critically dependent on the initial conditions.

The Ostwald ripening describes the driving force, which determines the size distribution of cluster ensemble on the surface [38–40, 42, 52]. This description supposes a driving force that depends on the mean radius of the islands. It is supposed that atoms detach more readily from smaller islands and condense rather to larger ones. Therefore, for a given amount of material on the surface, larger islands grow at the expense of smaller ones. The temporal evolution of the island radius (r) can be described by a differential equation: $dr/dt = b(T) / r^2(1/r - 1/r_c)$, where r_c is the critical cluster radius and b(T) is a temperature-dependent parameter.

Droplet–Surface Interaction

In this section, we will discuss the interaction between the droplet and the substrate surface. Here, the dependence of the droplet shape and the so-called thermal etching are discussed. It is known that the droplet shape is determined by the wettability of the substrate surface, and this property can be described by contact angle. There are three types of situations: no wetting, partial wetting, and complete wetting, where these situations correspond somewhat with the three growth modes. The contact angle depends on some circumstances, such as the status of the carrier surface, the droplet size, and the temperature. This is because the nuclei may wet the substrate, strongly changing their geometry. The deposited material will not wet the substrate because this would be accompanied by an overall increase in free energy of the system. The liquid deposit will spread out to maximize the area of the interface. For all the other situations, a partial wetting with a contact angle θ should be considered, that is, $\gamma_s = \gamma_i + \gamma_e \cos\theta$, where γ_e and γ_s represent the surface free energies of deposited cluster and substrate, respectively, and γ_i is the interfacial free energy (Figure 4) [53].

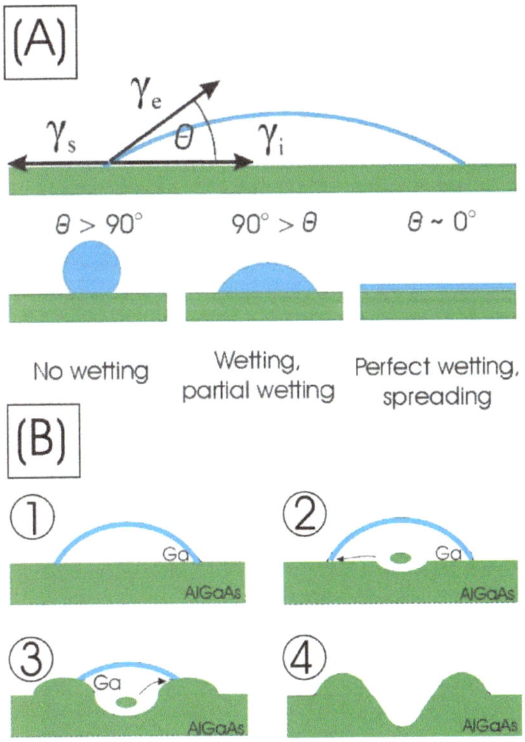

Figure 4. (A) The contact angle dependence on the surface tension and wettability. (B) Localized thermal etching of the compound semiconductor surface by a metallic droplet.

In our DE case, we have to discuss a further interaction between the droplet and the carrier surface. This interaction is called thermal etching, droplet etching, or local droplet etching. The explanation of the process will be discussed with the help of Ga/GaAs/AlGaAs system. Where Ga droplet forms on AlGaAs surface.

It is known from the liquid phase epitaxy that thermal etching takes place at the interface Ga melt and AlGaAs surface [54–57]. In the first step in DE technology, Ga droplets are generated on the surface. Due to thermal etching, this droplet-shaped Ga melt can solve the arsenide from the substrate material. At the interface, the metallic components of the substrate mix with the Ga atoms of the droplet. During the QD formation, this thermal etching takes place, too. In the initial stage of the QD formation, enough time is available for the mixing of the atoms at the interface. In the case of GaAs QD on AlGaAs substrate, Al content can be observed in the GaAs QD. This phenomenon was confirmed by the author and coworkers [58]. This thermal etching effect can be observed also at evolution of the ringlike QDs.

Our observations show that the middle part of ringlike QDs is below the surface of the substrate surrounding the QDs. This phenomenon is also the consequence of thermal etching. At slow crystallization, etching can act longer; therefore, it results in deeper holes in the center. At high temperature, quite deep nanohole can be achieved [85].

Thermal etching can be utilized also for nanostructure preparation. With the help of droplets created on the substrate surface, the surface can locally be etched. At higher temperature, this etching phenomenon is more significant. After the local thermal etching, the remained deep craters are the so-called nanoholes. During the DE QD fabrication, the Ga deposition temperature is ranged between 100°C and 350°C typically. These temperatures are very low compared to usual MBE growth conditions. In this case, the droplet thermal etching is less significant. At significant higher temperature, the situation is different.

If the temperature ranged between 450°C and 620°C, the local thermal etching under the droplet is already considerable. After an annealing time, the droplet is replaced in a nanohole surrounded with lobes. This local droplet etching shows a very promising technique for the structuring of the substrate surface without any lithographic processes. Furthermore, this patterning is also fully compatible with the MBE technology. In this way, nanohole was first presented by Wang and coworkers [59].

This finding opened a new way for the fabrication of another type of QD. This system was based on the filling of the nanohole, where the fill of the hole is carried out with lower band-gap material than the substrate. In this way created, nanohole filled inverted QDs are demonstrated by Heyn and coworkers [60–66]. In the process of nanohole filling, we utilize the Ga diffusion on the substrate surface. The temperature and orientation dependence of the Ga (and other metallic component) diffusion can be utilized at further nanostructure preparation.

FORMATION OF QD

Preparation Technology

The initialization of the alternatively grown conventionally shaped QD preparation, the basic idea of DE, originated from Koguchi and his coworkers [6, 30]. In this way, it is possible to create strain-free GaAs QD on AlGaAs surface. Along this discovery, further unconventionally shaped QD types were discovered [9, 32–35]. In the following, we will detail the DE-grown single hump-shaped QD in the case of GaAs/AlGaAs system. As discussed above, this process consists roughly of two main steps. First, metal (in our case Ga) droplets are generated on the surface in a Volmer–Weber-like growth mode. In the second step, crystallization takes place. Under arsenic pressure, the droplets transform into crystalline nanostructures. The shape of the nanostructures and their distribution depend on the initial droplet size and distribution and on the further technological parameters such as substrate temperature, arsenic pressure, waiting time, etc.

Here, typical experimental parameters for conventional QD preparation are described [43]. The GaAs QDs are grown on AlGaAs (001) surface. The growth experiments are performed in a solid source MBE system equipped with effusion cells for Ga and Al evaporation and valved cracker cell for arsenic ambient pressure. The evolution of growth front is in situ monitored with RHEED. First, on GaAs (001) wafer, pure GaAs layer is grown, and it is followed by an AlGaAs layer with Al content of 0.3. After the AlGaAs layer preparation, the sample is cooled down to 200°C. The $\theta = 3.75$ ML Ga is deposited with the flux of 0.75 ML/s without any As flux. After the Ga deposition, a 60-s waiting time is performed. The annealing is carried out at a temperature of 350°C and at an As pressure of 5×10^{-5}Torr. The growth process are tracked continuously in the direction of [11¯1¯0] with the help of RHEED [67]. In this direction, the RHEED pattern is more informative for the status of the QD evolution compared with the perpendicular [110] direction. After the growth process, the quantum objects can be ex situ investigated (AFM and TEM and PL methods), which can provide us further information to understand the growth kinetics.

During the QD preparation, the sequence of the RHEED pictures in the main growth stages is depicted in Figure 5. In the initial stage, the RHEED pattern showed sharp streaks (stage at t_0). After the Ga deposition, the pattern diffused on the RHEED screen (stage at t_1). Almost at the same time with the offering of arsenic pressure, the RHEED pattern changed suddenly from diffused to spotty (stage at t_2). During the annealing phase, the pattern changed slowly (some minutes) from spotty to spots with tails (stage at t_3). The density of QDs is 3.6×10^{10} cm^{-2}. It is shown that the characteristic RHEED pattern of QD is still recognizable even if the dot density is one order of magnitude less [43].

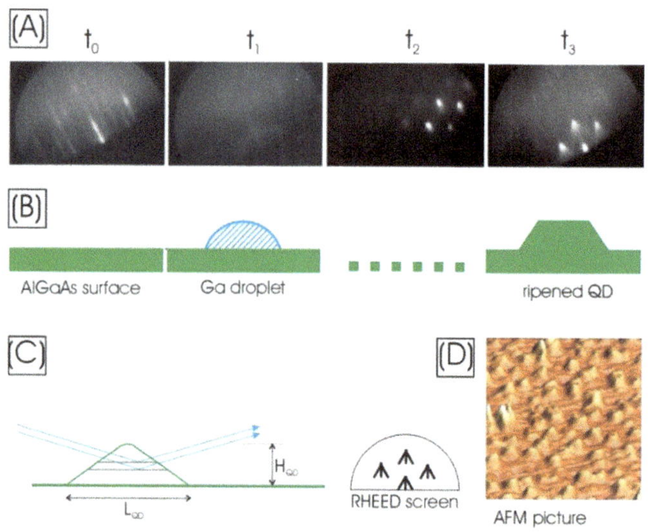

Figure 5. (A) During the QD evolution, the changing of the RHEED pattern and (B) its temporal evaluation. (C) Interaction of QD and electron beam. (D) AFM picture of DE-grown QDs.

Geometry and Electronic Structure of QD

As we demonstrated earlier, the shape of the QDs depends on the volume and also the density of QDs. According the shape, there are two types of QDs. The larger type of QD is a truncated pyramid-like shape with side facets of 55°. The smaller QD is a pyramid-like form with a side facet of 25° [43, 44]. In the following, the geometry of the QD will be discussed in the case of the smaller type of QD.

The atomic resolution structure of a typical DE-grown QD is shown in Figure 6. (The image was taken with the electron beam parallel to the [110] zone axis of the AlGaAs single crystal substrate [43,67–68].) The typical dimensions of the QD shown in Figure 6 are 54 nm base width and 5.5 nm height. The steepness of the QD side wall measured by TEM as well as by AFM shows about 25°. Figure 6 also shows that all the lattice fringes of the AlGaAs substrate are continued in the GaAs QD without any distortion. It is well known that the lattice parameters of these two substances are practically the same. No crystal defects were observed within the QD or at the interface with the host crystal [68, 69].

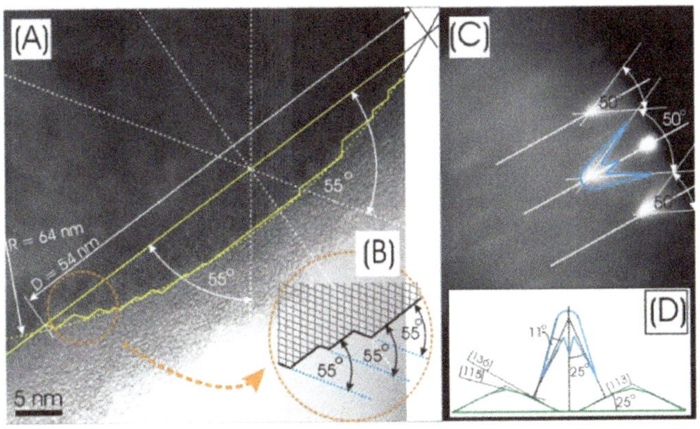

Figure 6. (A) Cross-sectional TEM image of a DE-grown QD, and (B) its stepped surface. (C) Broadened chevron tails and (D) their explanation.

The feature mentioned is shown in a darker contrast of the QD, also a few atomic surface layer of AlGaAs crystal between the QDs. The darker contrast on the AlGaAs surface can be interpreted as follows. The AlGaAs layer, at low temperature, shows As-rich c(4×4) surface. Nominally, θ = 3.75 ML Ga is deposited on the surface without arsenic flux. In the duration the Ga supply, a few ML of Ga is combined with excess arsenic surface atoms. Thus, a thin GaAs layer forms on the surface while the rest of the Ga forms nanodroplets. During the crystallization, the droplet and the surface layer go into similar composition. As a result, the QD and the surface layer shows similar darker contrast compared to the host material [69]. The existence of this GaAs surface layer was predicted earlier from the comparison of the PL measurement and the energy level calculation [32].

The chevron tails are connected with the faceting of QDs as it was verified in the case of diverse-shaped DE QDs [10]. In case of DE, it can be shown from the AFM measurement and from the tilted TEM picture that the shape and size of QDs are very uniform [69], and their side angle is about 25° [43, 68], which corresponds to the half opening angle of the chevron tails (Figure 6). Despite the same shape of the QDs, the chevron tails are not sharp but broad [68]. The observed side angle near to 25° corresponds to the (113) crystallographic plane [67, 68].

A cross-sectional TEM image shows that the side of the QD is not a single crystalline plane but has stepped shape (faceting). The steps consist of planes parallel with interface (parallel with {002} crystal planes) and planes with 55° to the interface, corresponding to {111} planes. The envelope curve of the QD cross section is a circle segment with a radius of R = 64 nm [68].

The electronic structure of the DE-grown GaAs QDs was investigated by PL spectroscopy [70, 71]. Furthermore, lasing was demonstrated in such structure [72]. After postgrown annealing, improved optical performance was demonstrated [72]. In the case of DE-grown QDs, the better optical properties compared to strain-induced QDs originated from the technology. The DE-grown QDs are prepared at very low substrate temperatures, which are unfavorable in view of the incorporation of undesired background impurities and crystal defects.

Some Aspects to the Kinetics of QD Growth

The growth kinetic can primarily be investigated by the in situ RHEED tracking. In the annealing phase of QD production, the diffused RHEED pattern becomes spotty nearly simultaneously with the opening of arsenic source. The sufficient arsenic quantity and the low temperature make the build in (infiltration) of As in the Ga phase (i.e., crystallization) possible [73]. This process of infiltration takes about 2–3 min to the sharp chevron image. A crystallized shell comes into being on the Ga droplets. The spotty RHEED pattern originates from the transmission electron diffraction. The electron beam goes through the crystalline GaAs shell layers over the droplet. If there are crystallite formation or droplets on the surface, It is observed that bulk scattering of the grazing beam can occur and the RHEED pattern may become dominated by spots rather than streaks due to transmission electron diffraction (Figure 5) [67].

The scattering from several planes strongly modulates the intensity along the reciprocal lattice rod. Thus, the streaks observed from two-dimensional surface are not observed when transmission dominates [67]. For the transmission case, the reciprocal lattice is an array of points each broadened owing to the finite size of the scattering region. During the annealing, the As diffuses inside of droplets, while excess As builds in (infiltrates) in the shell [73, 74]. Thus, the droplet crystallizes slowly. In the given moment, the rounded shell will be broken by the cornered crystallite grown inside of the structure. At the same time, a chevron-shaped spot develops on the RHEED screen [43, 48, 67, 68].

The angle between two RHEED streaks starting from same reciprocal lattice point is about 55°, as shown in final stage of the QD evolution. These so-called chevron tails are attributed perpendicularly to the facets of the QD [67, 69]. The volume of the QDs is large enough to receive transmission pattern during the electron scattering. The main lateral expansion L_{QD} and height H_{QD} of QD – according to the AFM measurement – are 50 and 5 nm, respectively (see Figure 6). The mean free path of the electrons Λ in GaAs between the crystal planes without collision at the typical incidence angle of RHEED (about 2°) is less than 20 nm. Thus, in our case, there are several (ca. 9) lattice planes to receive transmission character [67].

The analytical TEM investigation provides further contribution to the evolution kinetics. In our TEM investigations, we detected Al content in GaAs QD [48]. Here, this study shows the high-resolution micrograph of a similar QD together with Ga and Al elemental maps of the same area. These images clearly show that the QDs contain both Ga and Al. The presence of Al within the QDs is supported by the explicit protrusion of bright contrast on the Al map at regions corresponding to the QDs. This feature originated from a dissolution process [67]. This process is very important for QR production. This phenomenon will be discussed in the next section.

On the basis of the measurements, the following kinetics explanation seems plausible [68, 69]. The process of GaAs crystallization starts at the edge of the droplet, initialized by the three-phase line (TPL) at this point, serving as discontinuity for the seeding [100]. Although in principle interaction can take place at any point of the droplet, due to the thermal movement, the species, arriving to the edge, will start the seeding of the crystallization process. The described mechanism for this process has been accepted by several authors too. [11, 67]. (Otherwise, it would be difficult to explain the formation of the ringlike QD. In the case of a dot, the seed grows inward, while in the case of ring, it tends to grow outward, which is maintained by the Ga migration.) Since we are dealing with dot shape, the dominant process is arsenic diffusion.

The crystal seed grows inward into the droplet, and also upward simultaneously. This process of growth can only be explained quantitatively because, in the case of nanosizes, the observed bulk processes and properties like diffusion and binding energy cannot be applied. Although similar crystallization processes have been observed, no attempt has been made to explain them until now (see Figure 7) [68]. The growth of the GaAs crystal occurs in direction opposite to the penetration of Al. The crystallization of GaAs can start only after the opening of arsenic cell, while the dissolution of AlAs species occurs immediately upon the deposition of droplets. The process at the surface is quicker than process at the interface. Thus, the process at the surface is the dominant during the processing time of the QD production.

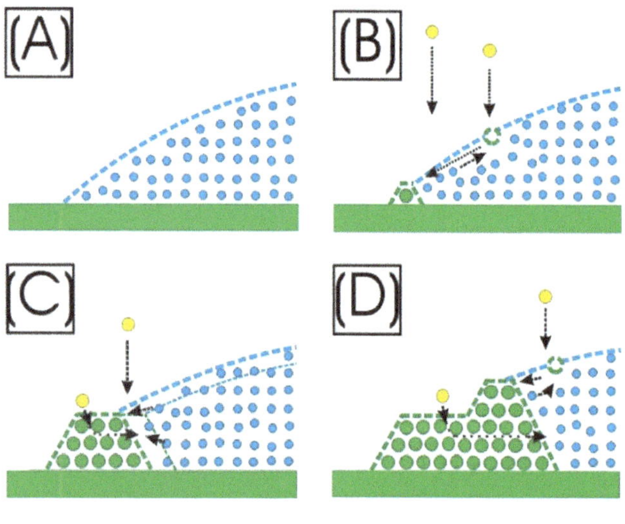

Figure 7. Explanation of the stepped surface evolution.

We explained the evolution kinetics of the stepped outer surface of the QD [68]. The summary of the kinetic is as follows. The intersection of the crystal surface with the droplet is the TPL, which serves as an initial place of crystallization [68]. When a Ga atom of the droplet meets an arsenic atom, they form a GaAs molecule. These GaAs species, making a Brown-like movement over the droplet surface, can reach the TPL, where the crystallization starts (Figure 6). The crystal seed at the TPL will grow on the account of further arriving arsenic species. The outer facet angle of the crystallization center at the edge will be the favorable 55° because it has enough time to find the optimal position (the low index facet). The crystallization seed grows partly upward and also partly in the direction of the droplet inside. During the process of solidification, a circular monocrystalline phase is formed at the droplet edge inheriting the orientation of the perfect substrate. During the process of solidification, the amount of Ga atoms in the droplets decreases so the droplet size decreases too. The outer side of the QD consists of steps of few MLs, where the front panel and the terrace of the step are (111) and (001) planes, respectively.

It is known in fcc crystals that the surface energy of (111) facet is less than that of (001) facet; thus, the latter grows predominantly during crystallization. This takes place by the lateral shift of the low energy (111) step facets with the simultaneous areal growth of (001) facets. With the size decrease of the droplet, its edge moves inward, thus creating a new TPL or crystallization seed at the new place, and the whole process continues as before with the original droplet and substrate. The only difference is that the crystallization takes place along a circle of less and less diameter.

RINGLIKE QD PREPARATION

Growth Technology of Ringlike QD

The preparation process responsible for the DE formation of ringlike QDs is described in the case of GaAs nanostructures on AlGaAs (001) surface. The growth experiments are performed similarly as described in the previous section, but the technological parameters somewhat differs. On GaAs (001) wafers, first pure GaAs layer is grown, and it is followed by an AlGaAs layer with Al content of 0.3.

Then the GaAs sample with AlGaAs layer must be cooled down to 300°C. On the surface, Ga is deposited as described in former section. Then $\theta = 3.75$ ML gallium is deposited with the flux of 0.19 ML/s without any arsenic flux. During the annealing, the temperature remained the same (300 °C), but the As pressure was 4×10^{-6} Torr. The production of the quantum objects was tracked continuously in the direction of $[11\bar{1}\bar{0}]$ with the help of RHEED [67].

Further, two types of ringlike nanostructures were generated with different amounts of deposited Ga. One of them was generated at 570°C on AlGaAs (001) surface applying 6.4 ML Ga. In this case, the AFM measurement shows ringlike QD with deep hole in the middle of the structure and very large clusters [95]. The other type of the nanostructures was prepared similarly, but the deposited Ga was less than that in the former case. The Ga coverage was 3.2 ML. In this case, the AFM picture shows special-shaped nanostructure with very deep crater in the middle surrounded by ringlike bulge formations and also sallow nanocraters, with plane rims (without bulge) [75, 76].

After the growth process, quantum objects are investigated using the AFM method, and the first mentioned types of nanostructure are studied with PL technique, too. Temperature-dependent PL spectra measured on the GaAs ringlike QD samples are compared with the conventionally shaped QDs.

The technology can continuously be tracked by RHEED. The initial stage of the surfaces is the same as in the case of conventional QD. The RHEED pattern of the initial surface shows sharp streaks (stage $t_0$on Figure 8). After the Ga deposition, the pattern diffuses on the RHEED screen similarly as in the case of QDs (stage t_1). The deposited Ga is in liquid phase. The disappearance of the RHEED pattern originates from the appearance of the liquid phase of Ga droplet on the surface. However, after the Ga deposition, the change of the observed RHEED pattern is quite different. After the offering of arsenic background of 4×10^{-6} Torr, the RHEED pattern develops very slowly, over 5 min. Contrary to the case of conventionally shaped QDs, which is almost at the same time of the arsenic cell opening, the RHEED pattern changed suddenly. The developed pattern contains in the middle a streak with a small spot and around elongated larger spots. According the AFM measurement, the density of the ringlike structures is 1.5×10^9 cm^{-2}. In this case, the effect of open surface on RHEED is larger than in the case of QD [69]. It is shown that the characteristic RHEED pattern of QD is still recognizable even if QD density is one order of magnitude less [43].

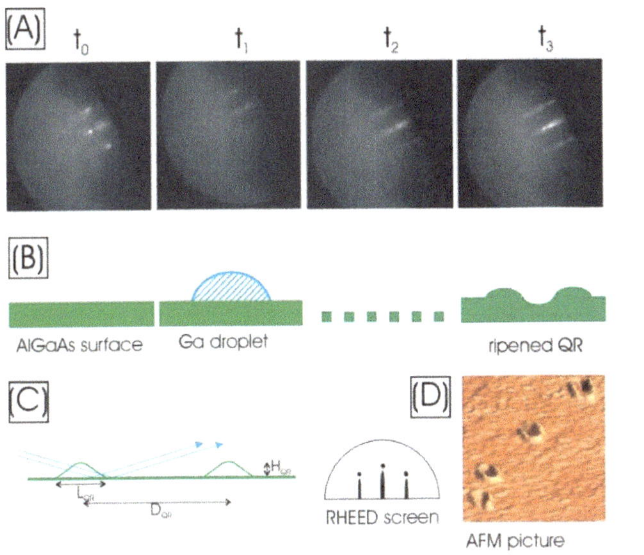

Figure 8. (A) During the evolution of ringlike DE-grown QD, the changing of the RHEED pattern and (B) its temporal evaluation. (C) Interaction of ringlike QD and electron beam. (D) AFM picture of DE-grown ripened nanostructures.

Geometry and Electronic Structure of the Ringlike QD

The shapes of ringlike QDs are various. The main parameters are the following: ring-middle diameter, width of the ring base, and height. As we will show, the facet angle is determined by the volume. The further shape parameter is the level of the ring middle. The ring middle can lie higher or deeper than the original substrate surface. At lower annealing temperature and at less arsenic ambient pressure, the droplet metamorphosis results in ringlike nanostructure with smaller crater in the middle [78–81]. For example, at 4.75 ML Ga deposition, under 1×10^{-5} Torr arsenic pressure and 350°C annealing for 10 min, the resulted structures are ring with shallow crater [79].

In another case, if the ring center lies deeper [75, 77], the structure is called a nanohole. In this case, thermal etching is dominant. If the annealing temperature is higher and the arsenic ambient pressure is smaller, then we receive ring structure with low-lying center. For example, at 8 ML deposited Ga, at 520°C annealing and under various arsenic pressure, we can get nanostructures with various deepness in the middle. If the arsenic pressures are 3×10^{-6}, 2×10^{-6}, 1×10^{-6}, and 1×10^{-7} Torr, then the holes are ca. 3, ca. 7, ca. 8.5, and 9.5 nm deep, respectively [62, 75, 76, 82].

The electronic structure is strongly governed by the shape of the nanostructure. Here, the advantages of the ringlike QD compared to the conventionally shaped QD are described using temperature-dependent PL spectra. Under measurement, the temperature ranges between 4 K and 300 K. The samples are excited by Ar^+ ion laser at 488 nm wavelength. The resolution was better than 0.5 nm [83].

The PL spectra of the uncovered GaAs ringlike QD grown on AlGaAs surface are shown in Figure 9. At 4 K, the spectrum has five peaks. They can be explained as bound exciton (1.5129 eV), exciton bound to acceptor (1.4892 eV), and its longitudinal optical phonon replica (1.4577 eV). Further on at higher energies, two peaks appear (1.5308 and 1.5602 eV). The PL spectra were also recorded as a function of temperature. To verify the identification of the peaks, the temperature dependence of the band-gap energy of GaAs and that of the bound exciton were compared. At 4 K in the case of conventionally shaped QD with identical volume, three peaks can only be seen at lower energy range. It is visible that the quantum confinement in ringlike QD is larger than that in QD at same nanostructure volume. In the following, this phenomenon will be explained [83].

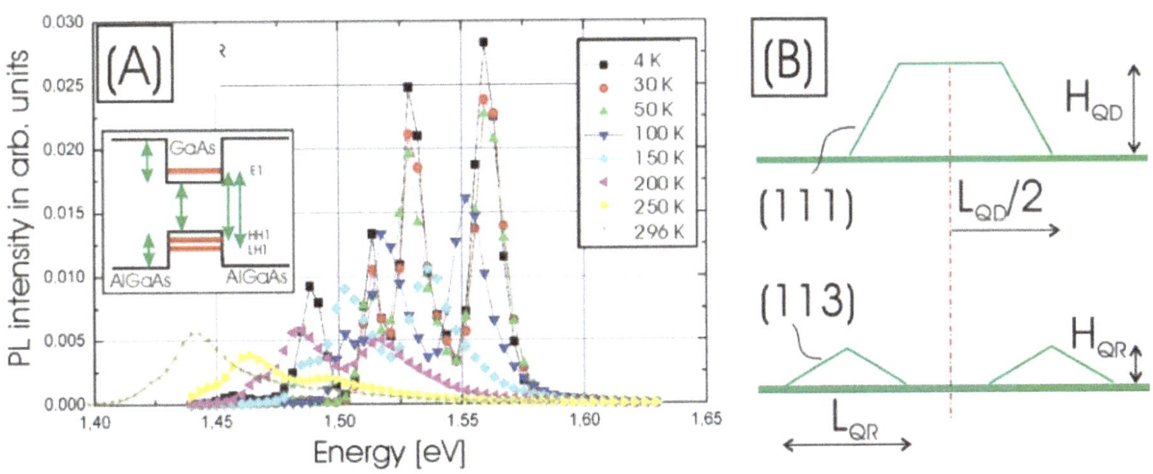

Figure 9. Temperature-dependent PL spectra of ringlike QD. (B) The explanation, why the ringlike QD has significantly sharper PL peaks compared to conventionally shaped QD.

The base diameter of the QD (L_{QD}) investigated is much larger, and its height (H_{QD}) is also larger but comparable. Thus, here no quantum effects are expected. Therefore, it was impossible to detect a peak shift or peak broadening for QD. At the same volume, for ringlike structure, the laterally wide (L_{QR}) nanostructure had a height (H_{QR}) significantly less than 7 nm. Consequently, the supposition of quantum confinement produces proper result [83].

The PL spectra show characteristic peak width, depending on the temperature and the size distribution of these structures. At the same temperature, a broader size variation results in a broader PL peak. Under similar growth conditions, the PL peaks of ringlike structures have significantly narrower full width at half maximum (FWHM) (less than one third) than that of the conventionally shaped QDs [78,84].

The sharper PL peaks can be explained by follows. As it is known, both kinds of QDs is formed from a gallium droplet. Assume that the volume and its variation of the initial droplets are the same in both cases. The facet of the nanostructures is size dependent and cannot be arbitrarily sized [43]. For the QD, let's start from the greater (111) facet (the approx. diameter ($2r = L_{QD}$) is 100 nm [43]). For the ringlike QD, only the (113) facet can be taken into account because of the smaller volume (the approximate width (L_{QR}) is 60 nm [43, 44, 83], dedicated to a circle segment). Due to the crystallographic constraints, the geometry of the formed QDs is determined. The height-to-diameter ratio cannot be arbitrary; it can be defined by a single parameter.

If r ($r = L_{QD}/2$) is the radius of the base circle belonging to the initial droplet, the volume of the developing QD, as a function of r, is given by $V = 1.58r^3$. For the ringlike nanostructure, from the equality of the volumes, the w ($w = L_{QR}$) parameter can be calculated as $w = 0.71r$. (This means that the nano-object with a parameter of $2r = 100$ nm has (111) facets, and the other one with a parameter of $w = 0.71 \times 100 \sim 70$ nm or less has (113) facets. It corresponds to the above presented measured data.) It means, that the height of the nanostructure influences commonly the quantum behavior. The heights of the QD and QR structures can be expressed as a function of their volumes: $m_{QD} = V^{1/3}/1.34$ and $m_{QR} = V^{1/3}/6.56$, respectively. Assuming the same variations of the volumes, it can be seen that the variation of the height for QR is much smaller than for QD; consequently, the corresponding FWHM of the PL peak is smaller [83].

Some Aspects to the Formation of Ringlike Nanostructure

The primer information to discover the formation kinetic originates from RHEED tracking [67]. After the deposition of Ga on AlGaAs (001) surface, the RHEED picture is becoming diffused due to the amorphous nature of the phase present on the surface. The annealing phase begins after the offering of arsenic component ($p_{As} = 4 \times 10^{-6}$ Torr, $T_{sub} = 300°C$). After releasing the arsenic, some time is needed for the development of the characteristic sharp pattern, representative of the crystalline structure. This is an indication that the liquid state on the surface stays longer and that the material transfer processes helps the formation of the ringlike structures. For detailed technological parameters of the different DE nanostructures, we refer the readers to the literature [32–35, 43–45, 62].

After the completion of the growth, the QR structures were investigated with AFM. The perspective AFM image and the top view with line scans are shown in Figure 10. The dimensions of QRs were determined from individual line scans. The density of the nanostructures determined from AFM pictures was 1.5×10^9 cm^{-2}. It is visible that the middles of the nanostructures are deeper than the original surface level [85]. It can be shown, from the AFM measurement that although the shapes and sizes of QRs are fairly uniform, we can observe small deviations from these averages. It is often observed that the smaller diameter rings have deeper depressions in the center and the larger diameter rings have shallower ones in the middle. (In the illustrations, the smaller and larger objects are labeled with "S" and "L", respectively.) The QRs have slightly elongated shapes due to the different binding properties in [110] and $[11\bar{1}\bar{0}]$ directions [86].

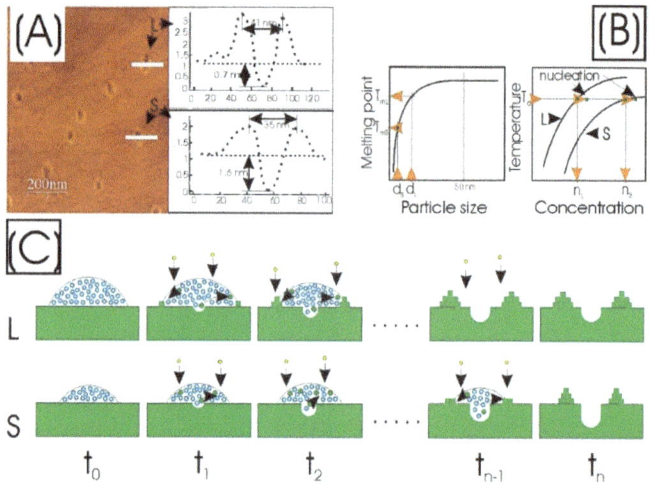

Figure 10. (A) Cavity dependence on the droplet size. (B) Functions of melting point and solubility. (C) Temporal evaluation of the smaller and larger ringlike QDs.

These facts are contradictory to the aspect ratios of ringlike QDs, and the explanation can help to understand the kinetics of the DE quantum structure formation. The explanation to this extraordinary behavior is as follows [85]. It is an obvious assumption that the larger droplets are leading to the development of rings of larger diameter and equally the smaller droplets to smaller rings. The intersection of the crystal surface with the droplet edge is the so-called TPL, which is the initial place of crystallization [85]. The TPL of larger diameter initiates a larger droplet, and equally, the smaller diameter forms from a smaller droplet. It is known from the liquid phase epitaxy that thermal etching takes place at the interface of the Ga melt and AlGaAs surface. This phenomenon was confirmed by analytical TEM for DE QD [69]. The Ga melt can solve the arsenide molecules (e.g., GaAs). These arsenide molecules originate partly from the thermal etching of the AlGaAs substrate surface and partly from the reaction of the external (from effusion source) arsenic atoms. (When a Ga atom of the droplet meets an arsenic atom from the environment, they form a GaAs molecule.) These molecules, due to thermal movement in the droplet, can reach the TPL, where the crystallization takes place. During the process of solidification, the material migrates from the middle to the edge of the nanostructure. A circular crystalline phase is formed at the droplet edge. The proposed kinetics for the formation is shown schematically in Figure 10.

It is known that the melting point decreases with the reduction of the particle size. The normalized melting curve versus the diameter of the particle (Figure 10) shows that when the particle size is less than 50 nm, then the melting point depends very strongly on size. In the nanorange, this dependence on the size is stronger than for the bulk. What makes it more complicated is the fact that the melting point depends on the particle shape as well. In this case, the Ga droplet's shape is a segment of a sphere. Its width in the middle falls in the range of 10–20 nm. Here, the change in the melting point is particularly sharp [85].

Around the edges the structure is thin. The experimentally obtained melting curves for near spherical metal nanoparticles show similarly. We use these curves for the qualitative assessment. This indicates that the melting point of the large and that of the small Ga droplets can differ considerably. The solubility curves for different particle sizes (Figure 10) show that at the same temperature, the larger droplet has lower saturating concentration than the smaller one. The meaning of this is that crystallization in the larger droplet will take place earlier, at lower arsenide concentration than in a smaller one. The smaller droplet will crystallize later, during arsenide concentration.

The temporal evaluation of the smaller and larger ringlike QDs is shown in lower part of the figure onset (Figure 10) [85]. In other words, in the larger droplets, the probability of the formation of the crystallization seeds is higher; therefore, the crystallization takes place earlier so less time is spent on material transportation, causing the development of the depression in the middle. When the droplets are small, the probability is less, and crystallization starts later, leaving more time for the formation of deeper depression in the center of the ring. This process is influenced by other factors as well. The melting temperature of the nanostructure is dropping with its diminishing size, staying longer in liquid state at the same temperature, spending more time on the formation of deeper in the center [85].

The above-described finding can be proven by further experiments (in Section 5.1, ringlike nanostructure production is described where the amounts of the deposited Ga were 3.2 and 6.4 ML, respectively). The explanation is as follows [85, 87, 88]: At a given temperature, there is a critical droplet size (CDS) under which the solution begins. After the Ga deposition, droplets form, followed by the growth of the larger droplets at the expense of smaller ones according the Ostwald ripening. When the droplet is smaller than the CDS, the substrate solution begins. In the case of Ga 6.4 ML, the formation of small holes and large clusters can be observed. The sizes of the droplets formed are above the CDS. After the deposition, the differentiation of the droplets begins. The smaller droplets reach critical size and start solving the substrate. This state is frozen via opening the arsenic cell.

When the deposited Ga is 3.2 ML, we can observe shallow holes with plane rims. The explanation is as follows [85, 87, 88]. In this case, the quantity of the deposited Ga is small. The formed droplets are under CDS. Therefore, the solution starts under the droplets immediately after the droplet formation. Under the small droplets, the solution is faster, but the material of liquid Ga is used up in a short time. The reduction in material is due to, first, the solution and, second, the migration of materials toward the larger droplets. After a short time, at the smaller droplets, the solution stops, while it carriers on further under the larger ones. The larger droplets will not be spent, and therefore the surrounding ring will freeze after the opening of the arsenic cell.

SPECULAR-SHAPED QDS AND COMPLEXES

The Inverted QDs

The ringlike QD is an unconventional QD, which has advantages over the conventional QD. As we have shown, the DE technique is a many-sided tool in this field. It allows us to fabricate further interesting nanostructures. Here, we will further discuss unconventional-shaped QDs and their complexes. In the present subsection, we will focus on an alternative QD preparation technique. This is not only an alternative preparation but the QD with inverted technology can also have advantages at special application.

A further recent method for the fabrication of strain-free QD is the nanohole filling. The nanohole is created by localized thermal etching, and it is filled subsequently [95]. This is a QD with inverted technology. The localized thermal etching takes place at conventional MBE growth temperatures, and we expect only very low level of crystal defects. The nanoholes are created in a self-organized fashion by local material removal [89]. For the inverted QD fabrication, nanoholes are generated using Al droplets on AlAs surface. Subsequently, the holes are filled with GaAs layer to form QDs of controlled height [90]. The nanoholes are filled with GaAs in pulsed mode.

Here, we show a cross-sectional study of an inverted QD [66]. The technology of this QD is as follows. The structure is grown on GaAs (001) surface. On the surface, AlGaAs and AlAs layer sequence is grown. The AlGaAs layer has 0.23 Al content. The AlAs and the AlGaAs layers are 5.5 and 19.5 nm, respectively. After the growth of every AlAs layer, 3.2 ML Al is deposited at 650°C without any arsenic flux. The deposited Al MLs form droplets on the surface. This is followed by an annealing step of 180 s, during which the droplets transform into nanoholes.

Following that, the holes are partially filled by the deposition of 0.6 nm thick GaAs in a growth-interrupted fashion. During the hole filling, the substrate temperature is 600°C. Depending on the foreseen high of the QD, some growth of 0.5 s and pause of 30 s sequences are carried out. Finally, the GaAs QD can be covered with AlGaAs. The scheme and the TEM image of the inverted QD structure are shown in Figure 11. The above-described pulsed technique allows us to create not only single QDs but also vertically aligned QD pairs [91].

The distances on the TEM image correspond with the intended layer thicknesses. The picture shows GaAs QD embedded in AlGaAs barrier material. Furthermore, the TEM image shows a thickening of the AlAs layer in the surrounding the nanohole. Elementary maps taken using electron energy loss spectroscopy confirm that the thickened region consists of AlAs. We identify the additional AlAs on top of the flat AlAs layer as the wall that surrounds the nanohole opening. Furthermore, the TEM image onFigure 11 shows that the next AlAs layer grown on top of the inverted QD layer is bended upward at the location of the QD, resulting in the formation of a kind of hill [66].

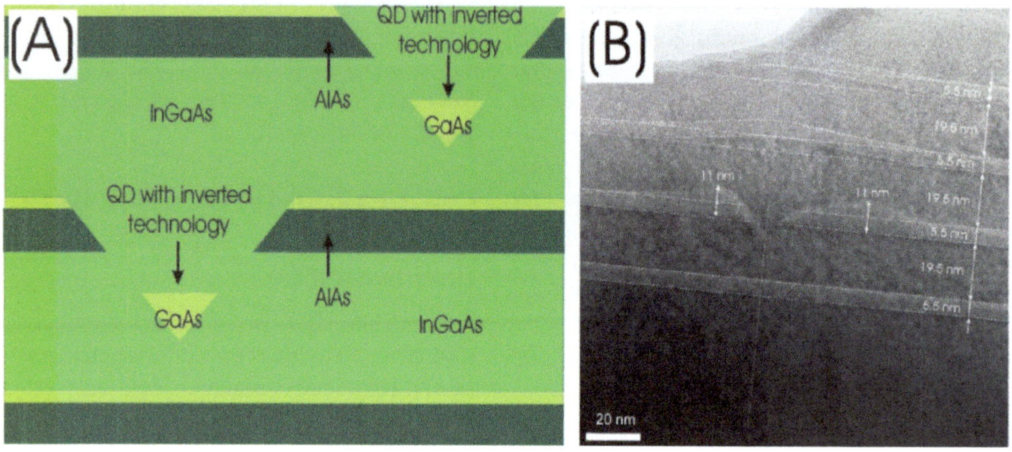

Figure 11. (A) Scheme of the sample with inverted QDs. (B) Cross-sectional TEM image of a GaAs QD embedded in InGaAs layer.

The Laterally Aligned QD Pairs

Two semiconductor QDs in close proximity, which can each spatially confine an individual charge carrier in a discrete energy level, interact quantum mechanically with each other. In particular, the wave functions of the charge carriers confined in each QDs of the pair begin to overlap, resulting in an efficient tunneling. Furthermore, the wave functions may become mixed to develop molecular orbital. Moreover, resonance in the optical range leads to the formation coupled QD pairs with the help of dipole–dipole interaction. These research leads toward quantum information processing. The QD pairs and their systems offer, at least conceptually, the potential of implementing scalable arrays of qubits.

In this part, we are dealing GaAs QD pairs prepared on AlGaAs surface with the usage of the anisotropy of the (001) oriented surface. We will show two preparational series. One of them is carried out under lower temperature, at less amount of deposited MLs. The other ones is prepared under higher temperature at higher amount of deposited Ga.

In the first case, AlGaAs with Al content of 0.27 is grown on the GaAs (001) surface. After this, Ga droplets are created at 330°C temperature on the substrate. The crystallization happens at 200°C, under accurate control of the arsenic flux [104]. The ripened structure basically consists of two QDs aligned in the [0 1¯1¯ 1] crystalographic direction. The average base size and height of each QD are 45 and 10 nm, respectively. The QDs are separated by an average distance of 39 nm between their apexes, as measured by AFM. The density of the structure is 2×10^8 cm^{-2} [104]. For the study of its optical properties, the QD pairs are embedded in AlGaAs barrier layer.

The second technology is carried out on GaAs surface. First, AlGaAs layer is grown with 0.3 Al amount. At 550°C substrate temperature, a large amount of Ga is deposited to create droplets on the surface. The structure is "arsenized" by fine control of the flux [99]. The resulting dots are rather large. The individual pairs have an interdot distance of about 130 nm and are aligned along the [0 1¯1¯ 1] direction. The dots with a height of 10 nm reside on a shallow base with a diameter of 300 nm. The density of the structure is 2.3×10^8 cm^{-2} [99]. For optical characterization, the QD pairs are buried by the deposition of AlGaAs layer.

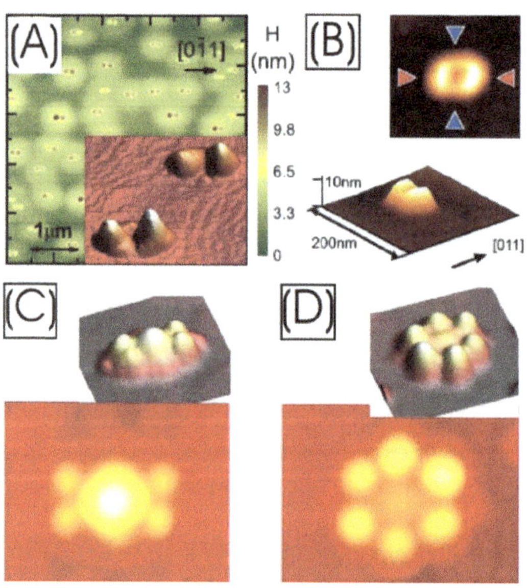

Figure 12. QD pairs grown at high substrate temperature with large amounts of Ga MLs. (Illustration originated from R. Pomraenke et al.; Phys. Rev. B 77 (2008) 075314) [105].) (B) (B) QD pairs grown at low substrate temperature with low amounts of Ga MLs. (Illustration originated from M. Yamagiwa et al.; Appl. Phys. Lett. 89 (2006) 113115) [104].) (C and D) Growth of quad- and hexa-QD molecules initiated by GaAs mound. (Illustrations originated from J.H. Lee et al.; Appl. Phys. Lett. 89 (2006) 202101) [110].)

In the first case, the micro-PL spectra of a single QD pair show the ensemble of emissions, which may indicate the existence of a tunnel coupling between the members of the pair. However, it must be mentioned that these emission lines are still not fully understood [104]. At the second experiment, the PL study is performed at a rather larger inter-dot separation. Here the formation of coherently coupled molecular QD states is suppressed, and excitonic interactions between neighboring QDs within each members are weak [105]. The process of the QD pair evolution is also not fully understood. Instead of the hemispherical shape of the initial droplet, the ripened structure is anisotropic. The explanation is based on the anisotropic surface potential of the GaAs (001) surface (Figure 12).

Configuration of QD Ensembles

Until now, the most widely studied system of interacting self-assembled QDs is a vertically stacked QD molecule. However, this system is limited by its one-dimensional nature and is not easily scalable. Alternatively, a variety of techniques can be used to modify the substrate to nucleate laterally coupled QD molecules in two dimensions by MBE technology. Various shape droplets can subsequently serve as templates for further growth, without the need for further surface preparation.

The DE-like created nanohole and also the homo-DE-grown island can serve as an initialization place for the growth of further nanostructures. The nanohole as a template for QD molecules is demonstrated by Songmuang and coworkers [106]. The nanohole can serve as a template for QD pair too [107]. Salamo and coworkers proposed a new way for the preparation of QD molecules [92-94, 110]. They adopt a hybrid growth approach, utilizing both DE and strain-induced growth to overcome some limitations of the Stranski–Krastanov QD growth mode alone on a planar GaAs surface [102]. Using MBE, a self-assembled InGaAs QD molecule is realized around GaAs mound formed by DE on GaAs (001) surface [102].

The number of QDs per GaAs mound can be effectively controlled by varying the InAs ML coverage. The number of QDs per template ranges from two to six. The technology is as follows. On GaAs (001) surface, 3 ML Ga is deposited at 500°C without any arsenic flux to form Ga droplets. Subsequently, 80 s of annealing occurs, and the substrate temperature is decreased to 150°C. The Ga droplet is crystallized under 1.3×10^{-5} Torr equivalent pressure for 100 s. During this time, nanoscale GaAs mound forms as template. The substrate temperature is raised again to 500°C, and InAs deposition follows. If the deposited InAs are 1.4, 1.6, and 2.0 ML, then the created nanostructures are bi-QD molecule, quad-QD molecule, and hexa-QD molecule [108].

During the QD molecule preparation, the height of the GaAs mound decreases from the original value, and also the diameter decreases. It appears that the InAs growth started mixing with the Ga atoms from the GaAs mounds, resulting in InGaAs shoulders on initial templates. This technique allows to fabricate the so-called quantum-rod pair and lip-shaped structures too [108]. The technology is extended with incident angle-controlled molecular beam technique [109].

ACKNOWLEDGEMENTS

In alphabetical order, the author would like to thank J. Balázs, J. Bozsik, A. Csík, M. Csutorás, L. Dobos, Z. Hajnal, Ch. Heyn, W. Hansen, P. Kucsera, J. Makai, C. Manzoni, B. Pécz, B. Pődör, I. Réti, K. Sakoda, S. Sanguinetti, A. Stemmann, J. Takács, G. Tényi, L. Tóth, and A. Ürmös for their fruitful cooperation and to the "Concert-Japan ERA-NET FemtoTera" Project for financial support via OTKA-NN 114457 Found.

REFERENCES

1. D. Leonard, M. Krisnamurthy, C. M. Reaves, S. P. Denbaas, P. M. Petroff; Appl. Phys. Lett. 63 (1993) 3203.
2. V. Bressler-Hill, S. Varma, A. Lorke, B. Z. Nosho, P. M. Petroff, W. H. Weinberg; Phys. Rev. Lett. 74 (1995) 3209.
3. W. Yang, H. Lee, T. J. Johnson, P. C. Sercel, A. G. Norman; Phys. Rev. B 61 (2000) 2784.
4. O. Stier, M. Grundmann, D. Bimberg; Phys. Rev. B 59 (1999) 5688.
5. Ch Heyn, A. Bolz, T. Maltezopoulos, R. L. Johnson, W. Hansen; J. Cryst. Growth 278 (2005) 46.
6. N. Koguchi, S. Takahashi, T. Chikyow; J. Cryst. Growth 111 (1991) 688.
7. T. Mano, K. Watanabe, S. Tsukamoto, H. Fujikoa, M. Oshima, N. Koguchi; Jpn. J. Appl. Phys. 38 (1999) L1009.
8. J M Lee, D H Kim, H Kong, J C Woo, S J Park; J. Cryst. Growth 212 (2000) 67.
9. T Mano, T Kuroda, K Mitsuishi, M Yamagiwa X J Guo, K Furuya, K Sakoda, N Koguchi; J. Cryst. Growth 301–302 (2007) 740.
10. T. Mano, S. Tsukamoto, H. Fujioka, M. Oshima, N. Koguchi; J. Cryst. Growth 227–228 (2001) 1069.
11. Z. Gong, Z. C. Nin, S. S. Huang, Z. D. Fang, B. Q. Sun, J. B. Xia; Appl. Phys. Lett. 87 (2005) 093116.
12. T. Kuroda, T. Mano, T. Ochiai, S. Sanguinetti, K. Sakoad, G. Kigo, N. Koguchi; Phys. Rev. Lett. 72 (2005) 205301.
13. S. Hwang, Z. Nin, Z. Fang, H. Ni, Z. Gong, J. B. Xia; Appl. Phys. Lett. 89 (2006) 031921.
14. M. Yamagiwa, T. Mano, T. Kuroda, T. Tateno, K. Sakoda, G. Kido, N. Koguchi; Appl. Phys. Lett. 89 (2006) 113115.
15. J. H. Lee, Zh. Wang, N. W. Strom, Yu. I. Mazur, G. J. Salamo; Appl. Phys. Lett. 89 (2006) 202101.
16. H. Lüth: Surfaces and Interfaces of Solid Materils; Springer Verlag, Berlin (1998).
17. J. Tersoff, F. K. LeGoues; Phys. Rev. Lett. 72 (1994) 3570.
18. J. M. Gérard, O. Cabrol, J. Y. Marzin, N. Lebouché, J. M. Moison; Mat. Sci. Eng. B 37 (1996) 8.
19. D. Bimberg, M. Grundmann, N. N. Ledentsov: Quantum Dot Heterstructures, John Wiley, New York (1999).
20. B. A. Joyce, D. D. Vvedensky: Growth modes and morphologies on GaAs surfaces, (in Thin Films: Heteroepitaxial Systems, Eds.: W. K. Lin, M. B. Santos) World Sci., Singapore (1999) p 368.
21. V. A. Shcukin, D. Bimberg; Rev. Mod. Phys. 71 (1999) 1125.

22. P.Ribas, V. Krishnamoorthy, R.M. Park; Appl. Phys. Lett. 57 (1990) 1040.
23. E.A. Fitzgerald; Mater. Sci. Rep. 7 (1991) 621.
24. Á. Nemcsics, F. Riesz, L. Dobos; Thin Solid Films 343 (1989) 520.
25. Á. Nemcsics, F. Riesz; Phys. Stat. Sol. C 0 (2003) 893.
26. R. Schnurpfeil, K. Restöft, A. Müller, Á. Nemcsics, R. Manzke, M. Skibowski; J. Electr. Spectr. Rel. Phen. 68 (1994) 175.
27. Á. Nemcsics, J. Olde, M. Geyer, R. Schnurfpeil, R. Manzke, M. Skibowski; Phys. Stat. Sol. A 155 (1996) 427.
28. S. Lee, I. Daruka, C.S. Kim, A-L Barabási, J.L. Metz, J.K. Furdyna; Phys. Rev. Lett. 81 (1998) 3479.
29. N. Koguchi, S. Takahashi, T. Chikyow; J. Cryst. Growth 111 (1991) 688.
30. N. Koguchi, K. Ishige; Jpn. J. Appl. Phys. 32 (1993) 2052.
31. T. Mano, T. Kuroda, K. Mitsuishi, M. Yamagiwa, X. J. Guo, K. Furuya, K. Sakoda, N. Koguchi; J. Cryst. Growth 301–302 (2007) 740.
32. T. Mano, S. Tsukamoto, H. Fujioka, M. Oshima, N. Koguchi; J. Cryst. Growth 227–228 (2001) 1069.
33. Z. Gong, Z. C. Nin, S. S. Huang, Z. D. Fang, B. Q. Sun, J. B. Xia; Appl. Phys. Lett. 87 (2005) 093116.
34. T. Kuroda, T. Mano, T. Ochiai, S. Sanguinetti, K. Sakoad, G. Kigo, N. Koguchi; Phys. Rev. Lett. 72 (2005) 205301.
35. S. Hwang, Z. Nin, Z. Fang, H. Ni, Z. Gong, J. B. Xia; Appl. Phys. Lett. 89 (2006) 031921.
36. W.Z. Ostwald; Z. Phys. Chem. 34 (1900) 975.
37. M. Zinke-Allmang, L.C. Feldman, S. Nakahara; Appl. Phys. Lett. 51 (1987) 975.
38. M. Zinke-Allmang, L.C. Feldman, W. van Saarloos; Phys. Rev. Lett. 68 (1992) 2358.
39. G.Z. Pan, K.N. Tu; Appl. Phys. Lett. 68 (1996) 1654.
40. G.R. Carlow, M. Zinke-Allmang; Phys. Rev. Lett. 78 (1997) 4601.
41. K. Shorlin, S. Krylov, M. Zinke-Allmang; Physica A 261 (1998) 248.
42. Raab, G. Springholz; Appl. Phys. Lett. 77 (2000) 2991.
43. Ch. Heyn, A. Stemmann, A. Schramm, H. Welsch, W. Hansen, Á. Nemcsics; Appl. Phys. Lett. 90 (2007) 203105.
44. Ch. Heyn, A. Stemmann, A. Schramm, H. Welsch, W. Hansen, Á. Nemcsics; Phys. Rev. B 76 (2007) 075317.
45. J.H. Lee, Z.M. Wang, N.W. Strom, Y.I. Mazur, G.J. Salamo; Appl. Phys. Lett. 89 (2006) 202101.
46. Ohtake; Surf. Sci. Rep. 63 (2008) 295.
47. Q.K. Xue, T. Hashizume, T. Sakurai; Prog. Surf. Sci. 56 (1997) 1.
48. D.K. Biegelen, R.D. Brinans, J.E. Northrup, L.E. Swartz; Phys. Rev. B 41 (1990) 5701.
49. Á. Nemcsics, J. Takacs; Acta Polytechn. Hung. 7 (2010) 109.
50. Á. Nemcsics, J Takacs; Semiconductors 45 (2011) 91.
51. M. Zinke-Allang, L.C. Feldman, M.H. Grabow; Surf. Sci. Rep. 16 (1992) 377.
52. V. A. Shcukin, D. Bimberg; Rev. Mod. Phys. 71 (1999) 1125.
53. M. Wautelet, D. B. Beljonne, J.-L. Brédas, J. Cornil, R. Lazzaroni, P. Lecére, M. Alexander, P. Dubois, P. Gills, Y. Gossuin, R. Muller, A. Ouakssim, A. Roch, D. Duviver, J. Robert, R. Gouttebaron, M. Hecq, F. Monteverde: Nanotechnologie; Oldenbourg Verlag, München (2003).
54. Grossy, M.B. Small; J. Cryst. Growth 11 (1971) 157.
55. H. Ijmin, S. Gonda; J. Cryst. Growth 33 (1976) 215.
56. N.S. Peev; J. Cryst. Growth 98 (1989) 499.
57. E. Lendvay, T. Görög, V. Rakovics; J. Cryst. Growth 72 (1985) 616.
58. Á. Nemcsics, L. Tóth, L. Dobos, Ch. Heyn, A. Stemmann, A. Schramm, W. Hansen; Superlatt. Microstr. 48 (2010) 351.
59. Zh. M Wang, B.L. Liang, K A Sablon, G.J. Salamo; Appl. Phys. Lett. 90 (2007) 113120.
60. Ch. Heyn, A. Stemmann, W Hansen; Appl. Phys. Lett. 95 (2009) 1731100.
61. Ch. Heyn, A. Stemmann T Köppen Ch Strelow, T Kipp, M Grave, S Mendach, W Hansen; Appl. Phys. Lett. 94 (2099) 183113.
62. Ch Heyn, A Stemmann, R Eiselt, W Hansen; J. Appl. Phys. 105 (2009) 054316.
63. Stemmann, Ch. Heyn, T. Köppen, T. Kipp, W. Hansen; Appl. Phys. Lett. 93 (2008) 123108.
64. Ch. Heyn, A. Stemmann, W. Hansen; J. Cryst. Growth 311 (2009) 1839.
65. Ch Heyn, A. Stemmann, T. Köppen, Ch. Strelow, T. Kipp, M. Grave, S. Mendach, W. Hansen; Nanoscale Res. Lett. 5 (2010) 576.
66. Á. Nemcsics, Ch. Heyn, L. Tóth, L. Dobos, A. Stemmann, W. Hansen; J. Cryst. Growth 335 (2011) 58.
67. Á. Nemcsics, Ch. Heyn, A. Stemmann, A. Schramm, H. Welsch, W. Hansen; Mat. Sci. Eng. B 165 (2009) 118.
68. Á. Nemcsics, L. Tóth, L. Dobos, A. Stemmann; Microel. Reliab. 51 (2011) 927.
69. D. Choi, M. Warusawithana, C.O. Chui, J. Chen, W. Tsai, D.G. Schlom, J.S. Harris; Mat. Res. Symp. Proc 996 (2007) H5–H31.
70. H. Lee, R. Lowe-Webb, W. Yang, P.C. Sercel; Appl. Phys. Lett. 72 (1998) 812.
71. K. Watanabe, N. Koguchi, Y. Gotoh; Jpn J Appl. Phys 39 (2000) L79.
72. T. Mano, T. Kuroda, M. Yamagiwa, G. Kido, K. Sakoda, N. Koguchi; Appl. Phys. Lett. 89 (2006) 183102.
73. Á. Nemcsics; Semiconductor 39 (2005) 1352.
74. Mojzes, T. Sebestyén, P. B. Barna, G. Gergely, D. Szigethy; Thin Solid Films 61 (1979) 27.
75. N. Koguchi, S. Takahashi, T. Chikyow; J. Cryst. Growth 111 (1991) 688.
76. N. Koguchi, K. Ishige; Jpn. J. Appl. Phys. 32 (1993) 2052.

77. J.M. Garcia, D. Grandos, J.P. Silveira, F Briones; Microel. Journal 35 (2004) 7.
78. T. Mano, T. Kuroda, M. Yamagiwa, G. Kido, K. Sakoda, N. Koguchi; Appl. Phys. Lett. 89 (2006) 183102.
79. C. Zhao, Y.H. Chen, B. Xu, P. Jin, Z.G. Wang; Appl. Phys. Lett. 91 (2007) 033112.
80. S. Sanguinetti, T. Mano, A. Gerosa, C. Somaschini, S. Bietti, N. Koguchi, E. Grilli, M. Guzzi, M. Gurioli, M. Abbarachi; J. Appl. Phys. 104 (2008) 113519.
81. C. Somaschi, S. Bietti, N. Koguchi, S. Sanguinetti; Appl. Phys. Lett. 97 (2010) 203109.
82. A Stemmann, T Köppen, M Grave, S Wildfang, S Mendach, W Hansen, Ch Heyn; J. Appl. Phys. 106 (2009) 064315.
83. Á. Nemcsics, J. Balázs, B. Pődör, J. Makai, A. Szemann; Phys. Stat. Sol. C 8 (2011) 2826.
84. K. Watanabe, S. Tsukamoto, Y. Gotoh, N. Koguchi; J. Cryst. Growth 227 (2001) 1073.
85. Á. Nemcsics, A. Stemmann, J. Takács; Microel. Reliab. 52 (2012) 430.
86. Á. Nemcsics; J. Cryst. Growth 217 (2000) 223.
87. Á. Nemcsics; AIP Conf. Proc. 1598 (2014) 79.
88. Á. Nemcsics; Acta Politechn. Hung. 8 (2011) 5.
89. Ch. Heyn, A. Stemmann, T. Köppen, Ch. Strelow, T. Kipp, M. Grave, S. Mendach, W. Hansen; Appl. Phys. Lett. 94 (2009) 183113.
90. Z.M. Wang, B.L. Liang, K.A. Sablon, G.J. Salamo; Appl. Phys. Lett. 90 (2007) 113120.
91. V. Pollojärvi, A. Stemmann, M. Guina, A. Stemmann, Ch. Heyn; Nanotechnology 22 (2011) 105603.
92. J.H. Lee, S. Sablon, Z.M. Wang, G.J. Salamo; J. Appl. Phys. 103 (2008) 054301.
93. M. Hanke, M. Dubslaff, M. Schmidbauer, Z.M. Wang, Y.I. Mazur, P.M. Lytvyn; Appl. Phys. Lett. 95 (2009) 023103.
94. M. Dubslaff, M. Hanke, M. Burghammer, S. Schröder, R. Hoppe, C.G. Schroeder, Y.I. Mazur, Z.M. Wang, J.H. Lee, G.J. Salamo; Appl. Phys. Lett.98 (2011) 213105.
95. Ch. Heyn; Phys. Rev. B 83 (2011) 165302.
96. R. Viswanatha, S. Saha-Dasgupta, D.D. Sarma; Phys. Rev. B 72 (2005) 045333.
97. G. Bastarg: Wave Mechanics Applied to Semiconductor Heterostructures, Les Éditions, Les Ulis (1990).
98. T. Mano. T. Kuroda, S. Sanquinetti, T. Ochiai, T. Tateno, J. S. Kim, T. Noda, M. Kawabe, K. Sakoda, G. Kido, N. Koguchi; Nano Lett. 5 (2005) 425.
99. H. Lee, R. Lowe-Webb, W. Yang, P.C. Sercel; Appl. Phys. Lett. 72 (1998) 812.
100. K. Watanabe, N. Koguchi, Y. Gotoh; Jpn J Appl. Phys. 39 (2000) L79.
101. T. Mano, T. Kuroda, M. Yamagiwa, G. Kido, K. Sakoda, N. Koguchi; Appl. Phys. Lett. 89 (2006) 183102.
102. G.A. Satunkin; J. Cryst. Growth 255 (2003) 170.
103. T. Mano, K. Watanabe, S. Tsukamoto, H. Fujikoa, M. Oshima, N. Koguchi; J. Cryst. Growth 209 (2000) 504.
104. M. Yamagiwa, T. Mano, T. Kuroda, T. Tateno, K. Sakoda, G. Kido, N. Koguchi; Appl. Phys. Lett. 89 (2006) 113115.
105. R. Pomraenke, C. Lienau, Y.I. Mazur, Z.H. Wang, B. Liang, G.G. Tarasov, G.J. Salamo; Phys. Rev. B 77 (2008) 075314.
106. R. Sonmuang, S. Kiravittaya, O.G. Schmidt; Appl. Phys. Lett. 82 (2003) 2892.
107. P.A. Gonzalez, J.M. Sanchez, Y. Gonalez, B. Alen, D. Fuster, L. Gonzalez; Cryst. Growth Des. 9 (2009) 2525.
108. K.A. Sablon, J.H. Lee, Z.M. Wang, J.H. Scultz, G.J. Salamo; Appl. Ph ys. Lett. 92 (2008) 203106.
109. M.K. Yakes, C.D. Cress, J.G. Tiscler, A.S. Bracker; Nano 4 (2010) 3877.
110. J.H. Lee, Zh.M. Wang, Yu, I. Mazur, G.J. Salamo; Appl. Phys. Lett. 89 (2006) 202101.

CHAPTER 3

QUANTUM DOTS SOLAR CELLS

Khalil Ebrahim Jasim

INTRODUCTION

One of the foremost challenges in solar cells industry is reducing the cost/watt of delivered solar electricity. In conventional microstructures (bulk) single junction solar cells, photons with energies less than semiconductor bandgap are not harvested while those with energies much larger than the bandgap produce hot-carries and upon cooling down (thermalization) the excess energy get wasted as heat. Therefore, novel materials or structures with tunable bandgap or intermediate band that can be tuned to match the spectral distribution of solar spectrum are crucial. Quantum dots (QDs) have the advantage of tunable bandgap as a result of size variation as well as formation of intermediate bands. In contrast to traditional semiconductor materials that are crystalline or amorphous, quantum dots can be molded into a variety of different types, in two-dimensional (sheets) or three-dimensional arrays. They can be processed to create junctions on inexpensive substrates such as plastics, glass or metal sheets. They can easily be combined with organic polymers and dyes.

Quantum dots are a special class of semiconductors, which are nanocrystals, composed of periodic groups of II-VI, III-V, or IV-VI materials and can confine electrons (quantum confinement). When the size of a QD approaches the size of the material's exciton Bohr radius, quantum confinement effect becomes prominent and electron energy levels can no longer be treated as continuous band, they must be treated as discrete energy levels. Hence, QD can be considered as an artificial molecule with energy gap and energy levels spacing dependent on its size (radius). The energy band gap increases with a decrease in size of the quantum dot, as shown in Figure 4. As the size of a QD increases its absorption peak is red shifted due to shrinkage of its bandgap (see Figure 5). The adjustable bandgap of quantum dots allow the construction of nanostructured solar cell that is able to harvest more of the solar spectrum. QDs have large intrinsic dipole moments, which may lead to rapid charge separation. Quantum dots have been found to emit up to three electrons per photon dueto multiple exciton generation (MEG), as opposed to only one for standard crystalline silicon solar cell. Theoretically, this could boost solar power efficiency from 20 % to as high as 65 %.

Generally speaking, there are three important parameters that characterize the performance of a photovoltaic cell. These are the open-circuit voltage (V_{oc}), the short circuit current (I_{sc}), and the fill factor (F F). However, the fill factor is also a function of V_{oc} and I_{sc}. Therefore, these last two parameters are the key factors for determining the cell's power conversion efficiency. Under ideal conditions, each photon incident on the cell with energy greater than the band gap will produce an electron flowing in the external circuit. The fill factor is determined from the maximum area of the I-V characteristics under illumination and the short circuit current and open circuit voltage, or

$$FF = \frac{V_{mp}\, I_{mp}}{V_{oc}\, I_{sc}} \quad (1)$$

where V_{mp} and I_{mp} are the operating point that will maximize the power output. In this case, the energy conversion efficiency is given by:

Citation: Khalil Ebrahim Jasim (2015). Quantum Dots Solar Cells, Solar Cells - New Approaches and Reviews, Prof. Leonid A. Kosyachenko (Ed.), InTech, DOI: 10.5772/59159.

$$\eta = \frac{V_{oc}\, I_{sc}\, FF}{P_{in}} \quad (2)$$

where P_{in} is the input power.

This chapter main objective is to give an introductory coverage of a more sophisticated subject. After we review the physics, designs, structures, and some growth/synthesis techniques of quantum dots. We will give a comprehensive description of some architectures of QD solar cells (e.g., Schottky cell, p-i-n configuration, depleted heterojunction, and quantum dots sensitized solar cell. Also, challenges and opportunities of quantum dots solar cells will be discussed.

BRIEF HISTORY OF QUANTUM DOTS

Since the early days of 1960s colloidal semiconductor crystallites or quantum dots concept has been suggested as a new structure of semiconductor materials. In 1981 Ekimove [1] reported the existence of semiconductor crystallites in a glass matrix. In 1985 Louis Brus developed a quantum model of spherical quantum dots based on effective mass model [2]. Quantum dot term was coined in 1988 by Reed's group [3]. Smith and his coworkers [4] successfully reported the growth of quantum film by depositing 3D silver Ag islands on gallium arsenide GaAs substrate in a two-step process. By the end of the 1990s, commercial productions of colloidal quantum dots become feasible. In 2004 a research group at Los Alamos [5] reported that a quantum dot is capable of emitting up to three electrons per photon, as opposed to only one for standard single crystalline layer semiconductors. It is now well established that absorption of high energy photon ($8\times E_g$) quantum dot is capable of generation seven photo-generated charge carriers due to inverse Auger recombination [6]. Now a days quantum dots are promising nanostructure materials in photonics and biomedical applications.

PHYSICS OF QUANTUM DOTS

In a bulk semiconductor, electrons and holes are free to move and there is no confinement and hence they have continuous energy values, where energy levels are so close to each other and packed such that energy bands are formed. Occupied bands called valance band and empty ones called conduction bands. The highest occupied band (valance band) and the lowest unoccupied band (conduction band) are separated by what is called energy bandgap E_g. Exciton is formed when electron hole pair are generated. The bond electron-hole system (exciton) form a hydrogen like atom. The separation of between the electron-hole is called Bohr's radius. Table 1 presents examples of exciton Bohr radius for some semiconductors.

Table 1. Exciton Bohr radius and bandgap energy of some common semiconductors.

Semiconductor Structure	**Exciton Bohr Radius (nm)**	**Bandgap Energy (eV)**
PbS	40.0	0.41
GaAs	28.0	1.43
CdTe	15.0	1.50
CdSe	10.6	1.74
ZnSe	8.4	2.58
CdS	5.6	2.53

Dimensionality of a material specifies how many dimensions do the carriers of thematerial act as free carriers. In bulk semiconductor continuous density of states results in both conduction and valence bands. However, when the number of atoms in the lattice is very few, the density of states becomes discrete, and loses the continuous 'band' like feature. Generally speaking, when a material has one or more dimensions small enough to affect its electronic density of state as illustrated in Figure 1, then the material is said to be confined. Accordingly we can have quantum wells, quantum wires, and quantum dots. Bulk semiconductor materials are example of three dimensional systems where density of states is proportional to $(E-E_{c/v})^{1/2}$. Quantum well system is a two dimensional system where electrons are confined in one dimension and therefore possess step like density of states. Quantum wire system is a one dimensional systemwhere electrons are confined in two dimensions and therefore possess density of states proportional to $(E-E_{c/v})^{-1/2}$. Quantum dot is a zero dimensional system where electron motion is confined in three dimensions. Therefore, a quantum dot possess atomic like density of states that is described mathematically by a delta function $\delta(E-E_{c/v})$.

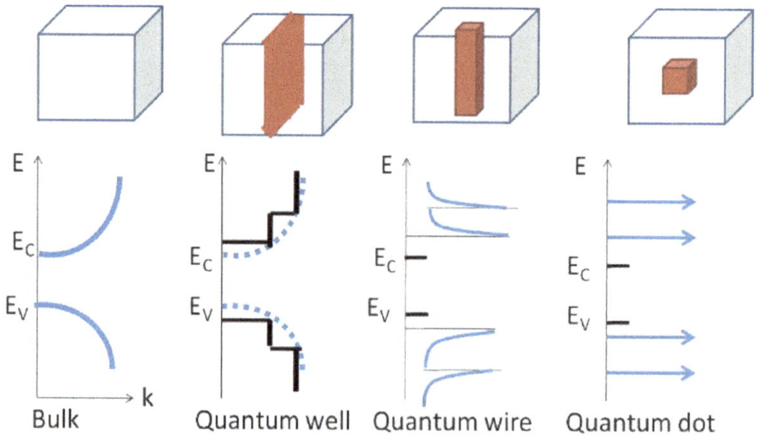

Figure 1. Schematic of density of states as system dimensionality is reduced. The density of states in different confinement configurations: (a) bulk; (b) quantum well; (c) quantum wire; (d) quantum dot. The conduction and valence bands split into overlapping subbands that become successively narrower as the electron motion is restricted in more dimensions. dimension. Adopted from [7].

In fact, quantum confinement is mainly because of relatively few atoms present in a quantum dot (see Figure 2), where excitons get confined to a much smaller space, on the order of the material's exciton Bohr radius. This pronounced confinement leads to discrete, quantized energy levels more like those of an atom than the continuous bands of a bulk semiconductor. For this reason in some literatures, quantum dots have sometimes been referred to as "artificial atoms."

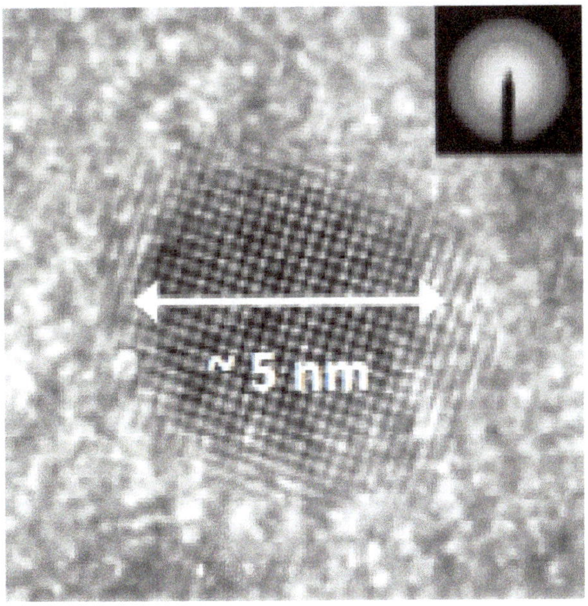

Figure 2. TEM of colloidal lead selenide PbSe quantum dot, from [8].

In order to specify energy eigenvalues of electrons and holes in a quantum dot, a good approximation is the infinite potential well illustrated in Figure 3. Therefore, a quantum dot can be imagined as quantum box. Assuming $\Psi_n(r)$ is the wave function of the nth state. Schrödinger equation for an electron confined in one-dimensional infinite square potential well of size L is:

$$-\frac{\hbar^2}{2m}\frac{d^2\psi_n}{dx^2} + V(x)\psi_n = E_n\psi_n \tag{3}$$

The energy eigenvalues E_n and eigen-wavefunction $\Psi_n(r)$ of the Schrödinger equation are given

$$\psi_n(x) = \sqrt{\frac{2}{L}}\sin\left(\frac{n\pi}{L}x\right) + \sqrt{\frac{2}{L}}\cos\left(\frac{n\pi}{L}x\right) \tag{4}$$

$$E_n = \frac{\pi^2\hbar^2}{2mL^2}n^2,\ n = 1,2,3,.... \tag{5}$$

Now if we extend the confinements of electron in three-dimensional potential well (box with dimensions L_x, L_y, and L_z) its momentum and energy will be quantized in all dimensions and we have:

$$E_{n_{x,y,z}} = \frac{\pi^2\hbar^2}{2m}\left(\frac{n_x^2}{L_x^2} + \frac{n_y^2}{L_y^2} + \frac{n_z^2}{L_z^2}\right) \tag{6}$$

With quantum dot (cubic box) of side dimension $L_x = L_y = L_z = L$, then E_n is written as:

$$E_{n_{x,y,z}} = \frac{\pi^2\hbar^2}{2mL^2}\left(n_x^2 + n_y^2 + n_z^2\right) = \frac{\pi^2\hbar^2}{2mL^2}n^2 \tag{7}$$

Similar energy eigenvalues can be written for holes. One must specify the m_e for electron and m_h for hole. Considering spherical shape of quantum dot with radius R, based on the effective mass model developed by Louis Brus [2] for colloidal quantum dots. The band gap E^{QD} can be approximated by:

$$E^{QD} = E_g^{bulk} + \frac{\hbar^2\pi^2}{2R^2}\left(\frac{1}{m_e} + \frac{1}{m_h}\right) - \frac{1.8e^2}{4\pi\varepsilon\varepsilon_0 R} \tag{8}$$

where ε is the relative permittivity, and $\varepsilon_0 = 8.85410^{-14}$ F.cm^{-1} the permittivity of free space.

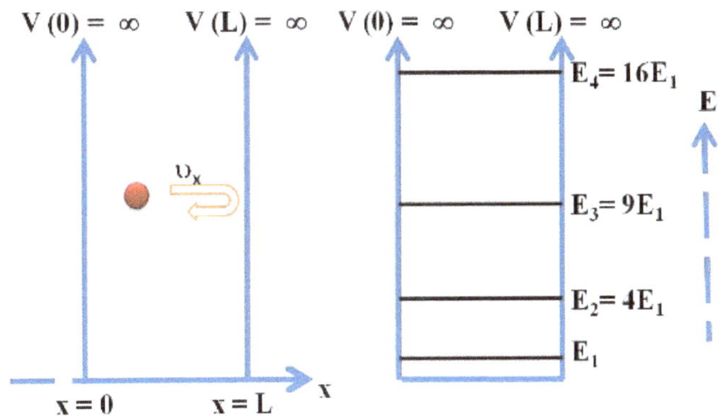

Figure 3. Infinite quantum well model.

In Equation (8), the first term on the right hand side describes the energy bandgap value of the bulk, the second term represents particle in a box quantum confinement model, and the third term details the Coulomb attraction between electron and hole (exciton). As the radius of the quantum dot decreases the Coulomb attraction term could be neglected compared to the second term in calculations. Therefore, Equation (8) indicates that bandgap energy eigenvalues increases as the quantum dot size decreases.

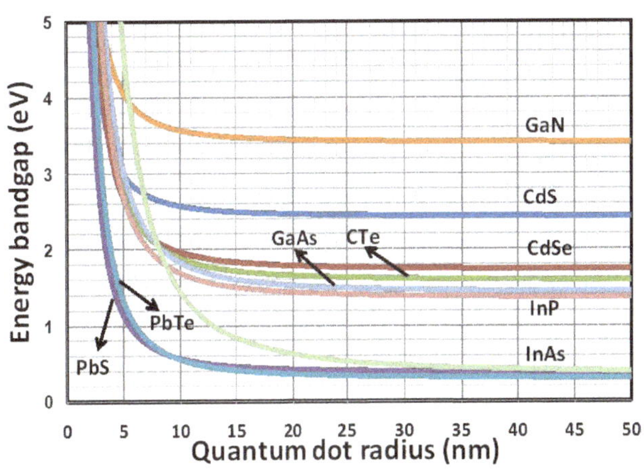

Figure 4. Variation of quantum dot energy bandgap vs. dot size for some common semiconductors. From [9].

Figure 4 presents few examples of well-known quantum dots, only the first and the second terms have been considered. Other detailed models such as the strong confinement model [10] have been adopted in determining the quantum dot sizes such as CdS.

Figure 5 shows the measured absorbance of three different sizes of lead sulfide (PbS) quantum dots suspended in toluene using dual beam spectrophotometer. Since a quantum dot bandgap is tunable depending on its size, the smaller the quantum dot the higher energy is required to confine excitons into its volume. Also, energy levels increase in magnitude and spread out more. Therefore, exciton characteristic peak is blue shifted.

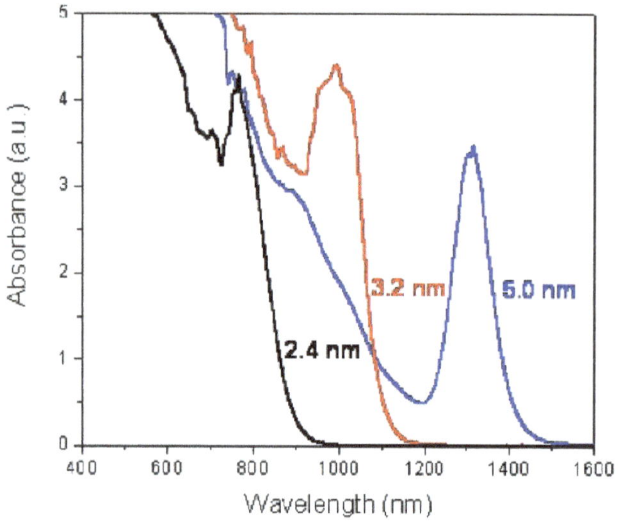

Figure 5. Measured absorbance of three different sizes PbS quantum dot suspended in toluene.

Quantum dot structured materials can be used to create the so called intermediate band. The large intrinsic dipole moments possessed by quantum dots, lead to rapid charge separation. Solar cells based on intermediate bandgap materials expected to achieve maximum theoretical efficiency as high as 65%.

Nozik and his coworkwers [11-13] invesitigated Multiple exciton generation in semiconductor quantum dots. As illustrated in Figure 6, unlike bulk semiconductors such as crystalline silicon, quantum dots can generate multiple exciton (electron-hole pairs) after collision with one photon of energy exceeding the bandgap. In bulk semiconductor absorption of photon with energy exceeding the bandgap promotes an electron from the valance band to higher level in the conduction band these electrons are called hot carrier. The excited electron (hot carrier) undergoes many nonradiative relaxation (thermalization: multi-phonon emission) before reaching the bottom of the conduction band. However, in a quantum dot the hot carrier undergoes impact ionization process (carrier multiplication). Therefore, absorption of a single photon generates multiple electron-hole pairs.This phenomena is called multiple exciton generation MEG. Therefore, absorption of UV photons in quantum dots produces more electrons than near infrared photons.

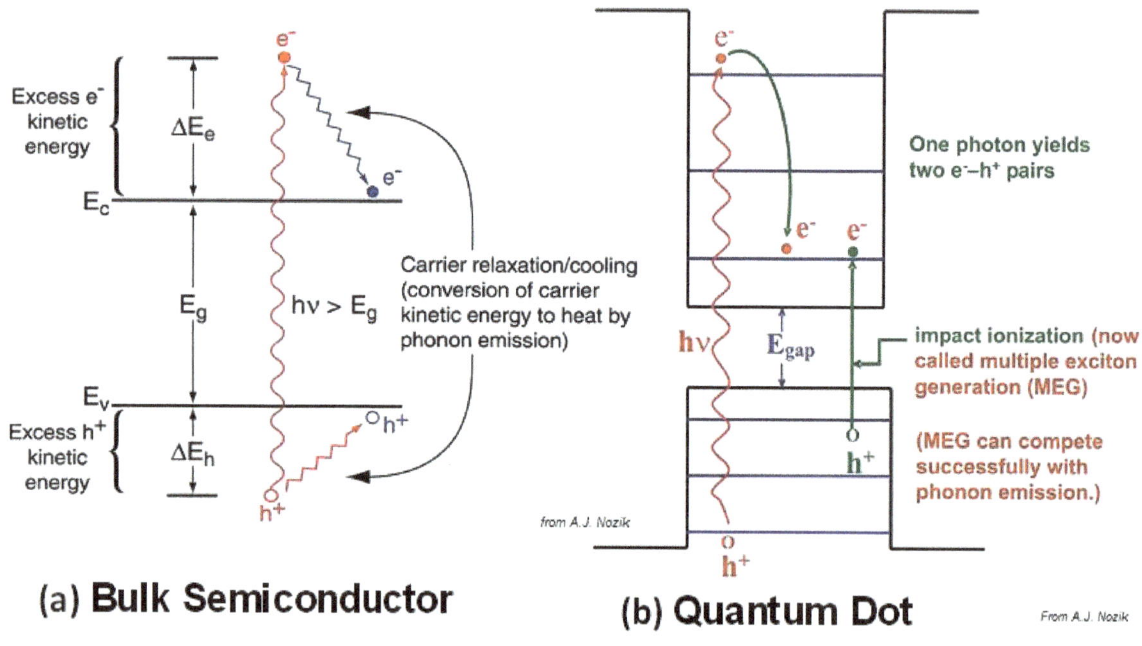

Figure 6. Thermalization of hot carriers in a bulk semiconductor (a) and a quantum dot (b), from [12].

After excitation of charge carriers the timing of the processes leading to generation of multiple charge carriers are detailed in [14] and schematically illustrated in Figure 7. After absorption of photon it takes hot carries 0.1 ps to go for impact ionization, then after 2 ps carries cool down. It takes around 20 ps for Auger recombination and finally 2ps are needed for carries to cool down and become ready for new excitation.

Experimentally carrier multiplication process-one photon generates more than one exciton via impact ionization or inverse of Auger recombination-in quantum dot has been investigated and quantified using the transient absorption spectroscopy technique. In short, [5] laser pulses are directed on the sample one for excitation and another for absorption. Measuring time of relaxation indicates whether there are single excitons recombining radiatively or biexcitons recombining via Auger recombination.

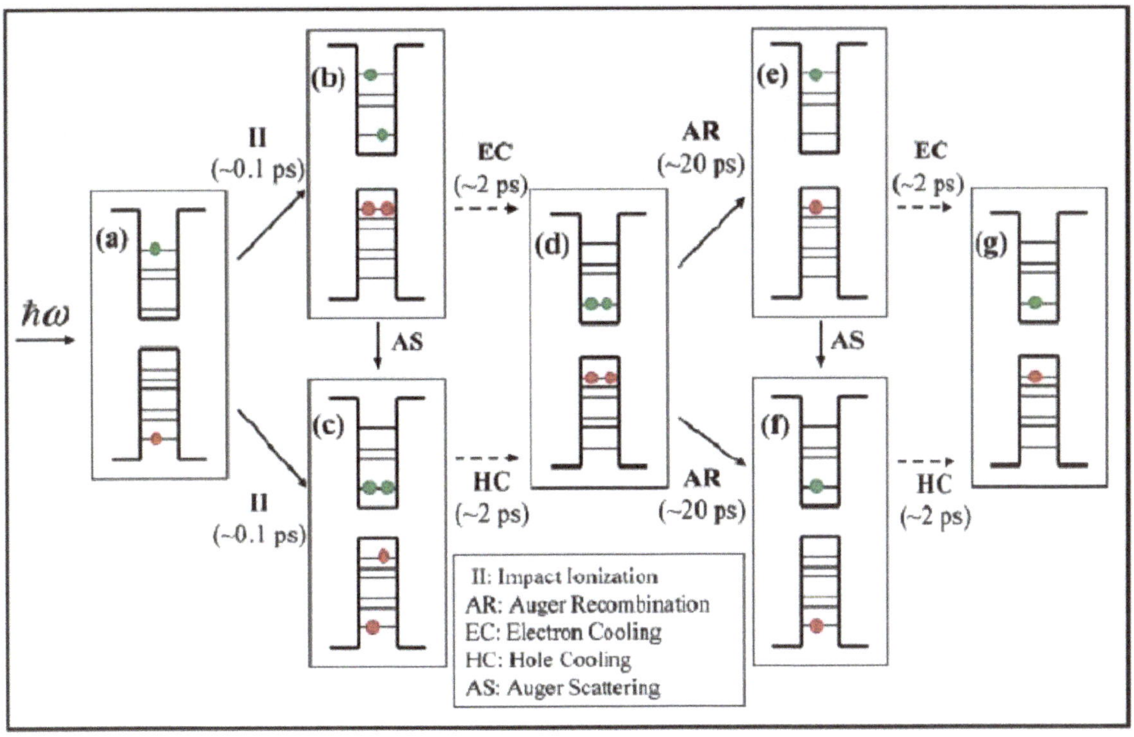

Figure 7. Illustration of the time of the processes leading to generation of multiple charge carriers in a quantum dot, from [14].

It has been discovered that there is linear proportionality between absorption change for number of generated electron-hole pairs less than 3 and number of generated electron-hole pairs. For example, investigations using PbSe and PbS QDs; confirmed 300% photo-generated carrier density (quantum yield QY)in PbSe QDs by a photon with energy of 4 times the energy spacing between the HOMO and LUMO (E_g) of the quantum dot [5, 15] as shown in Figure 8-a.Further investigations by Schaller and his coworkers [6] reported higher carrier multiplication efficiencies of 700% of photo-generated carrier density at higher photon energy to bandgap (E_g) ratios (photon energy $8\times E_g$).Multiplication onset starts at photon with energy equals to 3 times E_g as shown in Figure 8-b.

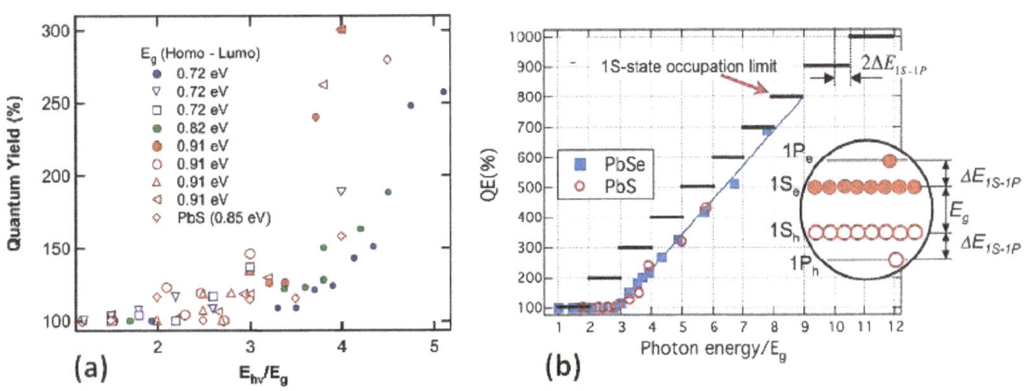

Figure 8. Experimental confirmation of carrier multiplication efficiencies. From(a) [5] and (b) [6].

SYNTHESIS OF QUANTUM DOTS

There are many methods or techniques used in fabrication or synthesis of quantum dots that can be classified as top-down (carving large piece of a material to the desired nanostructures) or bottom-up (assembling atoms to form the desired nanostructures) methods. Generally speaking, some of these techniques are practical in solar cells applications. A summary of each synthesis method is given below.

Electron-Beam Lithography

Electron-beam lithography is one of the top-down methods. Electron beam system is used to etch well defined patterns on a bulk semiconductor and or super lattice, then a semiconductor layers are grown according to a well-established protocol in micro-technology.

Self-Assembly Growth

Self-assembly growth is one of the bottom-up methods. Molecular Beam Epitaxy (MBE) and Metal Organic Chemical

Vapor Deposition (MOCVD) are widely used in the growth of the super lattice with different materials of different lattice constants. Semiconducting compound with a smaller lattice constant is grown (deposited atom by atom) on the surface of a compound with a larger lattice constant. The relaxation of the grown layer after specific growth thickness due to lattice mismatch results in nucleation of islands of random shapes and controllable size. This growth mode is known as Stranski-Krastanov growth (see Figure 9).MBE method, beside it is sophisticated and slow, it is expensive. However, growth control of quantum dots is precise and multi-layers of quantum dots is possible. MOCVD is used in mass production of sample wafers and in contrast to MBE the growth of crystals is by chemical reaction and not physical deposition. In this approach high vacuum and temperature are required.

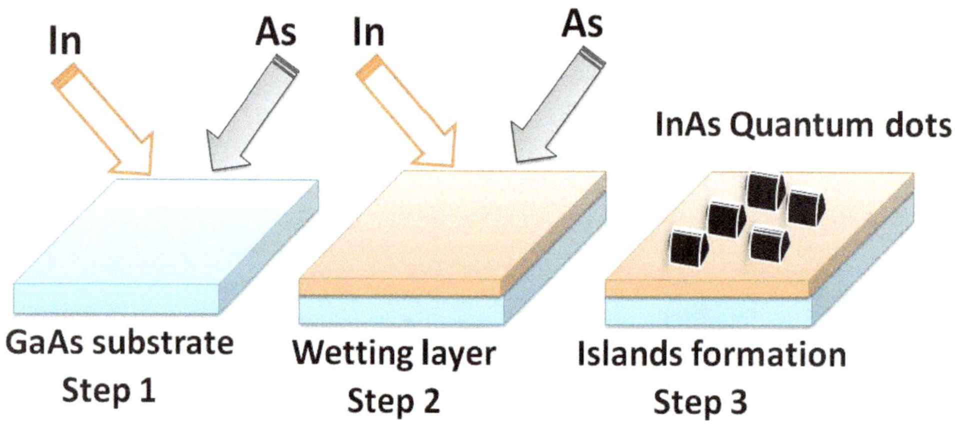

Figure 9. Schematic of InAs island formation on GaAs surface by Stranski-Krastanow growth.

Organometallic or Colloidal (Wet Chemistry) Synthesis Method

Organometallic or colloidal (Wet Chemistry) synthesis method is one of bottom-up methods. It's also a self-assembly technique. It's relatively inexpensive and fast growth method to synthesis large quantity of quantum dots all at once. The growth could occur at non-extreme conditions. Figure 10 show a schematic of an example of the setup used in synthesis of colloidal quantum dots.

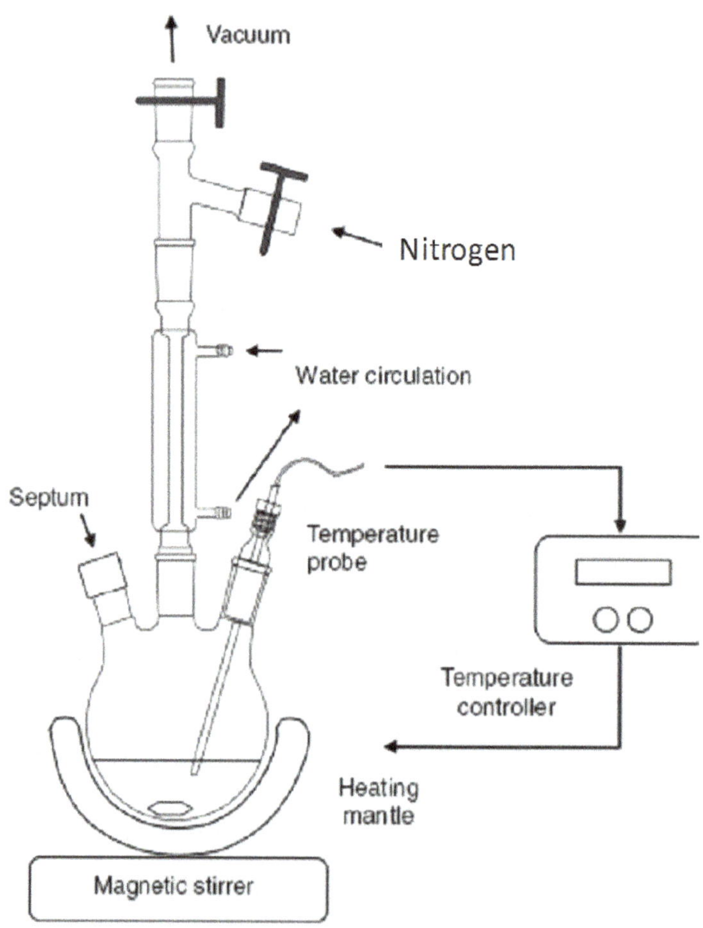

Figure 10. Schematic of a common setup used in synthesis of colloidal quantum dots.

As an example recipe given by [16] to synthesis colloidal Lead Sulfide PdS quantum dots: Heat together in a flask at 150 °C for 1 hr under N_2 flow (to make lead oleate):

- 220 mg lead(II) oxide
- 0.25 ml oleic acid (tech grade)
- 9.75 ml 1-octadecene (ODE, tech grade)

After cooling to 90-100 °C, quickly inject into the reaction flask 5 ml of 0.1 M hexamethyldisilathiane (0.5 mmol) dissolved in ODE. Then immediately cool with an ice bath.

Precipitate (methanol/butanol mixture), centrifuge and resuspend in nonpolar solvent (e.g., toluene, tetrachloroethylene, or hexanes).

$$\mathrm{PbO} + 2\mathrm{C}_{18}H_{34}O_2 - (N_2\, ODE\langle 122\,°\,C) \rightarrow Pb(^{et)}_{ol\,ae} + H_2O$$

$$Pb(^{et)}_{ol\,ae} + (^{\;\;i2}_{(\,)_{(_{C}H_3S)}} S - (N_2\, ODE\langle 122\,°\,C) \rightarrow PbS\ quantum\ dots\ capped\ with\ OA$$

To manipulate functionality, alter the charge and reactivity of the surface, synthesized quantum dots usually caped with a shell from different composition. Also, the shell can enhance the stability and dispersability of the colloidal core. Magnetic, optical, or catalytic functions may be readily imparted to the dispersed colloidal core. In fact, encasing colloids in a shell of different composition may perhaps protect the core from extraneous physical and chemical changes.

Generally speaking, colloidal quantum dots could be categorized as Type-I (e.g., CdSe/ZnS) and Type-II QDs (e.g., CdTe/CdSe) as shown in Figure 11:

- In Type-I QDs, all charge carriers are confined in the core material in which radiative recombination occurs (see Figure 11-a).
- In Type-II QDs, charge carriers are segregated in the core and shell; radiative recombination occurs across the material interface (see Figure 11-b).

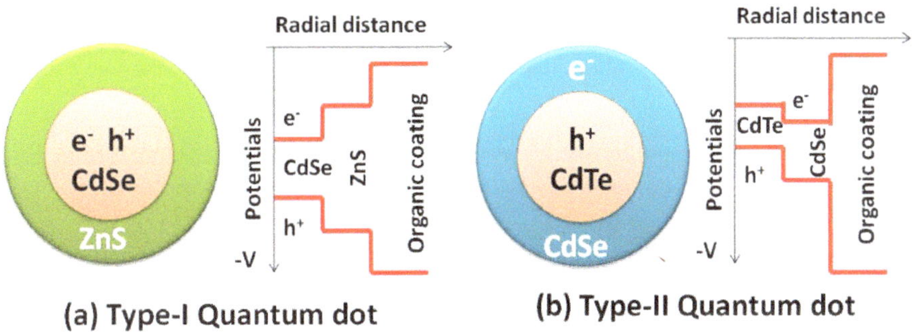

Figure 11. Illustration of (a) Type-I quantum dots (b) and Type-II quantum dots.

In fact, many researchers investigating quantum dot sensitized solar cells adopt other growth strategies to assemble quantum dots on electrode surface. We used the most direct and easiest one, the drop casting (electrode is socked with quantum dots solution drop by drop). Chemical bath deposition CBD has been used by [17], Electrophoritic deposition has been adopted by [18], successive ionic layer adsorption and reaction (SILAR) has been used by [19], and the used of bifunctional linker was demonstrated by Subramanian and coworkers [20].

QUANTUM DOTS SOLAR CELLS

There are many realistic quantum dots based solar cells. In this section we classify them in seven strategies just as an approach to fulfill the main objective of this chapter that is introducing quantum dot solar cells. For example, [21] metal-semiconductor junction, polymer-semiconductor, and semiconductor-semiconductor systems are well know strategies of quantum dots based solar cell. Moreover; quantum dots attached to n-C_{60} composite clusters see [18] or single and multi-walled carbon nanotubes based solar cells have been investigated [22,23].

Metal-Semiconductor Junction Solar Cell

Figure 12-a schematically shows the structure of the metal-semiconductor junctionwhich also called Schottky barrier quantum dotsbased solar cell. Itis basically fabricated from quantum dots layers (Nanocrystals film) sandwiched between metallic electrode and ITO counter electrode deposited on transparent glass substrate to act as photo-electrode.In the band diagram shown in Figure 12-b, a depletion region is due tocharge transfer to QD film. Because of high electron density in metal (~ 10^{22}cm^{-3}), the depletion is negligible on its side of the cell.

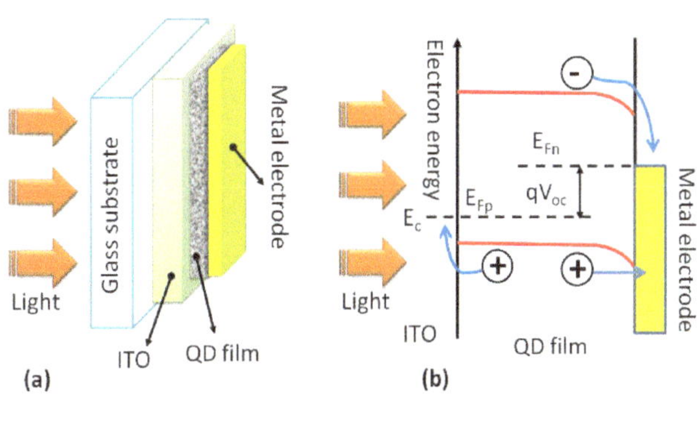

Figure 12. (a) Schematic of Schottky barrier quantum dots based solar cell, (b) band diagram of Schottky solar cell. Adopted from [24].

Layer-by-layer fabrication of quantum dots film is prepared by layer-by-layer dip coating of the substrate. For example, in layer by layer fabrication of PbSe quantum dots films, quantum dots film prepared by dip-coating, alternating between PbSe NCs in hexane and then 0.1 M EDT in anhydrous a cetonitrile, allowing the film to dry between each layer [25]. Device with 140 nm-thick QD film achieved an efficiency of ~2.2 % [26] heterojunction solar cell has been investigated by sputtering 150 nm layer of ZnO and growing layers of PbS quantum dots fabricated in air (4.1 mg/mL concentration of PbS 820 nm NC in hexanes;1 mM solution of 1,2-ethanedithiol in a cetonitrile layer-by-layer deposition for 20 cycles (~ 120nm PbS film thickness). Tested devices showed good air stability over ~1,000 hrs with cell power efficiency of 1.53 %.

It has been found [24] that the open circuit voltage in Schottky cell is constrained due to pinning of Fermi level resulted from the formed defect states at the metal-semiconductor interface. Moreover, as illustrated in Figure 12-a, photoelectrons are firstly generated near the ITO (ohmic contact) and need to pass through long pathway to the metallic electrode. Hence, many of the photoelectrons are encountering high recombination probability. Also on the onset of photovoltaic action the hole current injection to metallic contact is not reduce because the barrier to hole injection become less effective as shown in Figure 12-b. Gao J. et. al. [27] reported a method to eliminate such disadvantage by inserting a layer of a transition metal oxide (TMO) between the quantum dot (QD) layer and metal electrode. An n-Type transition metal oxide such asmolybdenum oxide (MoO_x) and vanadium oxide (V_2O_x) work as a hole extraction layer in PbS quantum dot solar cells resulted in power conversion efficiency =4.4%. The formation of dipole at the interface of MoO_x and PbS enhances the band binding and hence allowing efficient hole extraction from the PbS film valance band by the MoO_x.

Hybrid Silicon/QD Film (NCS) Solar Cells

Kilimov and his coworkers [28] demonstrated the feasibility of hybrid PV devices that combine advantages of mature silicon fabrication technologies with the unique electronic properties of semiconductor QD (see Figure 13).The hybrid silicon/PbS QD film (Nanocrystals) solar cells show external quantum efficiencies of 7%at infrared energies and 50% in the visible and a power conversion efficiency of up to 0.9%.

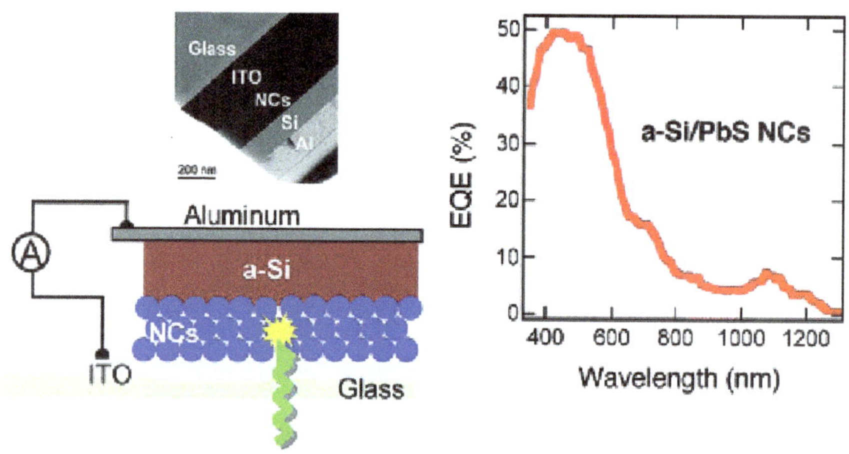

Figure 13. Hybrid silicon/PbS QD film solar cells. From [28].

P-I-N Structure Solar Cell

The advances in multiple quantum well solar cells [29], see Figure 14, and remarkable efficiency improvement of such p-i--n structure configurations due efficient management of photo-generated carries have encouraged researchers to replace the III-V intrinsic region of semiconductors quantum wells structures with quantum dots.

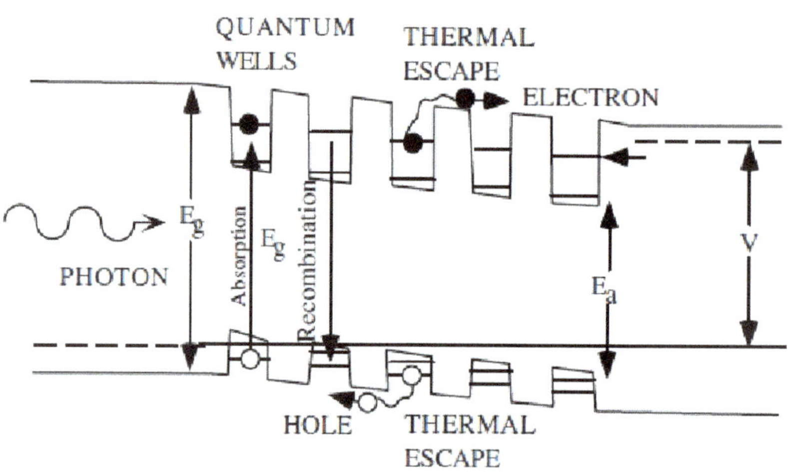

Figure 14. Multiple quantum well solar cell band structure. From [29].

Nozik and his coworkers [11,13] suggested a general p-i-n junction configuration for incorporating array of QDs into solar cells. QDs array forms the active medium of the cell (the intrinsic region). In this arrangement quantum size effects should not beeradicated due to cell architecture, excitons must be separated prior to Auger recombination, and upon absorption of photons excitons ought to be separated into free-charge carriers and transported to appropriate electrodes. Figure 15, shows quantum dot layers forming the i-region of a solar cell. The QDs are electronically coupled to each other to sustain electron and hole transport. Also, mini-bands are formed and hence could facilitate intermediate band formation [30]. Intermediate energy band is energy absorbing band which can accept electron excitations from the valance band and then allow transitions to the conduction band. In quantum dots solar cells with intermediate bands photon of energies lower than the bandgap are basically absorbed and high energy photons produce hot-carries (see Figure 15). Collecting charge-carries while they are hot enhances cell voltage. On the other hand, photocurrent enhancement could be achieved by getting more from the hot-carriers via inverse Auger recombination (impact ionization) leading to multiple exciton generation MEG. One needs to bear in mind electron tunneling mechanism between quantum dots.

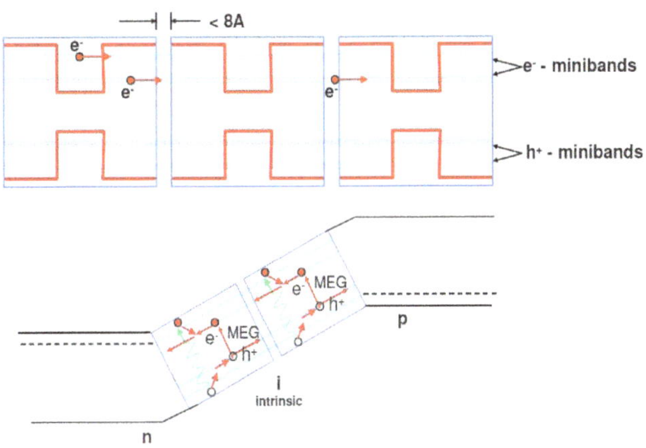

Figure 15. Possible p-i-n QD arrangements for solar cells.From [11].

Investigation done by Nozik group found that [31] by comparing the performance of GaAs base solar cell to one incorporating InAs/GaAs QDs intrinsic region sandwiched between p and n GaAs emitters (grown by MBE system in the Stranski–Krastanov growth mod), the cell exhibits an extended response for photon energies lower than the GaAs bandgap. The contribution to the total current of the cell came as a result of harvesting below-bandgap-energy photons is minute (1%).

Polymer-Semiconductor Structure Configuration

In 2002 Huynh and his coworkers [32] investigated photovoltaic action in hybrid nanorods-polymer solar cells and under Air Mass (A.M.) 1.5 Global solar conditions, a power conversion efficiency of 1.7 % was obtained. This works and others encouraged many groups to investigate incorporation of quantum dots in polymers. Quantum dots are dispersed in organic-semiconductor polymer matrices. For example, in a hole-conducting polymer such as MEH-PPV (poly(2-methoxy,5-(2-ethyl)-hexyloxy-pphenylenevinylene) disordered arrays of CdSe quantum dots are synthesized. When the cell illuminated, the absorbed photons results in photo-generated carriers. As illustrated in Figure 16 [31], the photo-generated holes are injected into the MEH-PPV polymer phase which is in contact with metallic electrode to facilitate their collection. Sun et al., in 2003 reported that solar power conversion efficiencies of 1.8 % were achieved under AM1.5 illumination fora device containing 86 wt % of quantum dots [33].

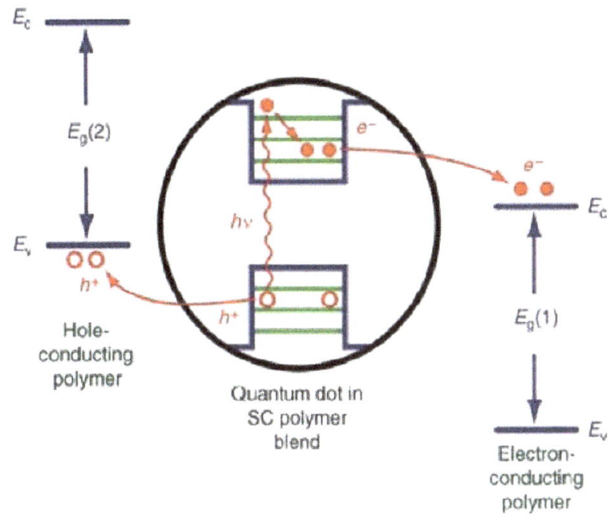

Figure 16. Shows how photo-generated carriers from quantum dots injected in conducting polymers. From [31].

Depleted Heterojunction Quantum Dots Solar Cells

In depleted heterojunction colloidal quantum dot solar cells as detailed in ref. [24] a nanostructured wide bandgap semiconductor such as TiO_2 and quantum dot film are sandwiched between conductive transparent electrode (glass coated with Tin Oxide SnO_2F) and metallic (such as gold) coated electrode (see Figure 17-a).Figure 17-b illustrate the energy band diagram. Although TiO_2 has low carrier density (~10^{16} cm^{-3}) compared to metal, a depletion region in the cells forms due to charge transfer to QD film. And because of high electron density in metal (~ 10^{22} cm^{-3}), the depletion is negligible on its side of the cell. Depleted heterojunction cell overcome the disadvantages encountered with Schottky cell.

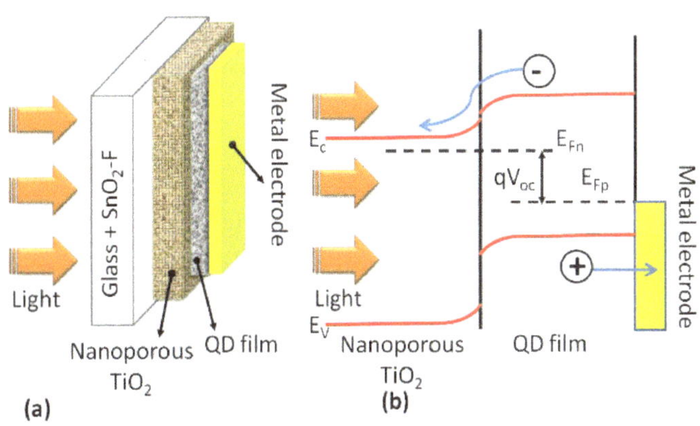

Figure 17. (a) Schematic diagram of Depleted-heterojunction Colloidal Quantum Dots Solar Cells, (b) energy band diagram. From [24].

In their work Grätzel and coworkers [24] have reported that the resultant depleted heterojunction solar cells offer power conversion efficiency of 5.1% at AM1.5 illumination. The devices were capable of harvesting broadband of the solar spectrum as a result of employing infrared-bandgap size-effect-tuned PbS CQD.

Quantum Dot Sensitized Solar Cells

The structure and operation principle of QD sensitized photovoltaic cell is almost identical to dye sensitized cells [9, 34,35] with the exception that now the QDs are the source of current injection. Quantum dots can be produced in situ or more without difficulty adsorbed from a colloidal QD solution. The structure of the photovoltaic cell is shown schematically in Figure 18. In this figure, we distinguish four essential elements of the cell, namely, the conducting and counter conducting electrodes, the nanostructured TiO_2 layer, the quantum dot energy levels, and the electrolyte.

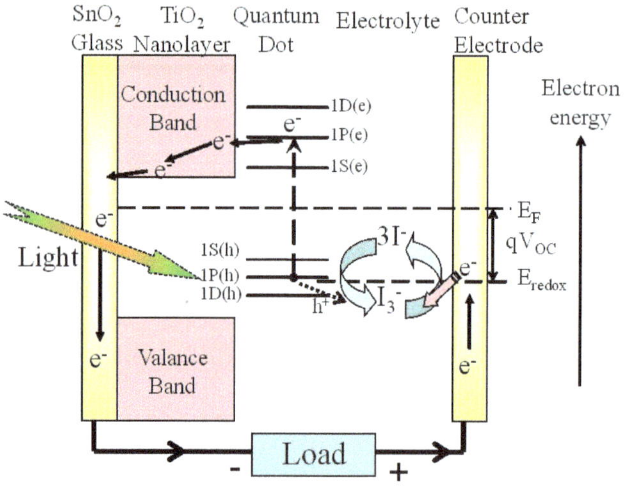

Figure 18. Schematic diagram illustrating the structure and operation of quantum dots-sensitized photovoltaic cell.

The operation of the cell can be described by the following steps and the corresponding process equations:

Upon absorption of a photon, a quantum dot is excited from the ground state (QD_S) to a higher energy state (QD_S^*), as illustrated by Eq.(1) below.

$$\textit{Excitation process}: QD_S + h\nu \rightarrow QD_S^* \tag{9}$$

Where QDS and QD_S^* is the quantum dot in its ground state and excited state respectively.

The absorption process results in the creation of electron-hole pair in the form of exciton. Dissociation of the exciton occurs if the thermal energy exceeds its binding energy.

$$\textit{Exciton dissociation}: QD_S^* \rightarrow e^{-*} + h^{+*} (\text{free carriers}) \tag{10}$$

The excited electron is then injected in the conduction band of the wide bandgap semiconductor nanostructured TiO_2 thin film. This process will cause the oxidation of the photosensitizer (The QDs).

$$\textit{Injection process}: QD_S^* + TiO_2 \rightarrow TiO_2(e^{-*}) + QD_S^+ \tag{11}$$

The injected electron is transported between the TiO_2 nanoparticles, and then gets extracted to a load where the work done is delivered as electrical energy.

$$\textit{Energy generation}: TiO_2(e^{-*}) + \text{C.E.} \rightarrow TiO_2 + e^{-*}(\text{C.E.}) + \text{elecrical energy} \tag{12}$$

Where C.E. stands for counter electrode. The counter electrode is identical to the photoelectrode where the nanostructured TiO_2 is deposited. The counter electrode is usually coated with a catalyst (graphite).

Most common electrolytes used in QDSCs are aqueous polysulfide and organic electrolyte with I^-/I_3^--redox couple. In some works the liquid electrolyte has been replacedwith solid-state hole conductors such as spiro-OMeDAT andCuSCN [36]. Assuming electrolyte used in the cell contains I^-/I_3^-redox ions, that play the role of electron mediator between the TiO_2 photoelectrode and the counter electrode [9]. Therefore, the oxidized photosensitizer states (QD_S^+) are regenerated by receiving an electron from the oxidized I^-ion redox mediator, regenerating the ground state (QD_S), and I^-becomes oxidized to the oxidized state I_3^-(triodide ions).

$$\textit{Regeneration of QDs}: QD_S^+ + \frac{3}{2}I^- \rightarrow QD_S + \frac{1}{2}I_3^- \tag{13}$$

The I_3^-diffuses to the counter electrode and substitutes the internally donated electron with that from the external load and gets reduced back to I^-ion.

$$\textit{Electron capture reaction}: \frac{1}{2}I_3^- + e^-(\text{C.E}) \rightarrow \frac{3}{2}I^- + \text{C.E.} \tag{14}$$

Overall, generation of electric power in this type of cells causes no permanent chemical transformation.

To enhance electron injection into the conduction band of the TiO_2 thin film, one must choose a sensitizer with a proper matching energy gap. Quantum dots can fulfill the necessary energy gap requirement by choosing the ones with the proper size. It is interesting to note that for the QD to effectively accept the donated electron from the redox mediator. Finally, the maximum potential produced by the cell is determined by the energy separation between the electrolyte chemical potential (E_{redox}) and the Fermi level (E_F) of the TiO_2 layer, as shown in Figure 18.

The bulk of many research works done on QD synthesized solar cells focused on CdS, CdSe, and CdTe QD as sensitizers [37-39] The choice of these materials follows the success of earlier studies on identifying the morphological and electrolyte effects on their performance and stability [40-41]. Incident Photon to carriers conversion efficiency IPCE is directly related to the product of light harvesting efficiency = $1\text{-}10^{-A(\lambda)}$ (where $A(\lambda)$ is the spectral absorbance of the quantum dots sample, electron injection efficiency Φ_{inj} (how efficient are electrons injected from excited quantum dot into TiO_2 conduction band), and electron collection efficiency Φ_{coll} (how efficent are electrons are collected by the photoelectrode), hence we have:

$$IPCE = LHE \times \Phi_{inj} \times \Phi_{coll} \tag{15}$$

Light harvesting efficiency could be affected by both the type and size of quantum dots. For example PdS based quantum dots have broader spectral absorbance than CdS quantum dots. Kamat reported that charge injection from excited CdSe quantum dots into nanostructured TiO_2 film can be controlled by varying solution pH as illustrated in Figure 19 "At increasing solution pH, the conduction band of TiO_2shifts 59 mV/pH unit to a more negative potential, thereby decreasing the driving force and thus decreasing the rate of nonradiative electron transfer from excited CdSe. The emission yield and the average emission lifetime increase with increasing pH, thus providing a way to monitor the variation in medium pH." [42].

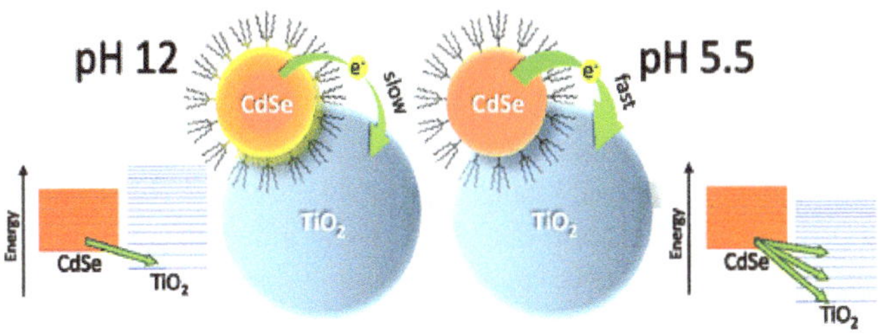

Figure 19. Illustration of how charge injection from excited CdSe quantum dots into nanostructured TiO_2 film can be controlled by varying solution pH. From [42].

Kongkanand and co-workers has investigated the effect of quantum dots size on charge injection rate [43]. They found that smaller-sized CdSe quantum dots show greater charge injection rates and also higher IPCE at the excitonic band. Interestingly, Larger particles have better absorption in the visible region, on the other hand, it cannot inject electrons into TiO_2 as effectively as smaller-sized CdSe quantum dots. It has been found that surface treatments can strongly influence charge transfer, recombination, and transport processes of photogenerated electrons and holes in QDSCs [44]. Figure 20 schematically shows that transport of electrons through the nonporous electrode is dominated by diffusion [45]. A path taken by an electron is not simply a straight line. An electron undergoes through a crisscross path before reaching the photoelectrode. The electron diffusion length, L_n is related to electron lifetime (τ_n) and diffusion coefficient (D_n) as:

$$L_n = \sqrt{D_n \tau_n} \tag{16}$$

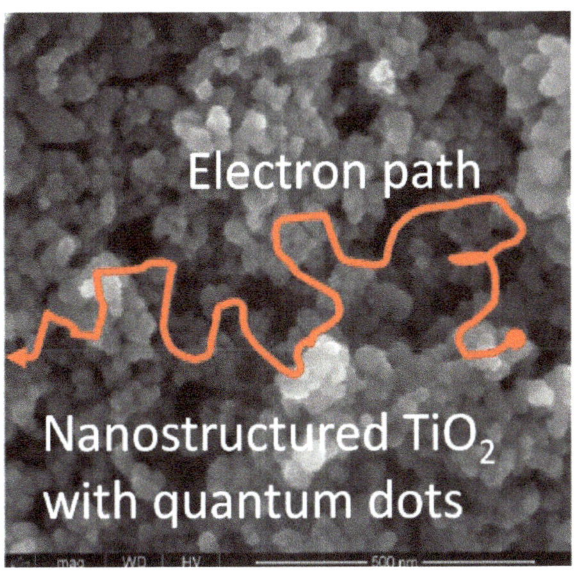

Figure 20. Illustrate a path taken by an electron after being injected into nanoporous TiO_2 layer (Not to scale). The nanoporous TiO_2 layer has been imaged using scanning electron microscope SEM.

Cell conversion efficiency is affect by morphology of the photoelectrode [43]. Tube type is more advantageous to the fast electronic conduction, due to shorter diffusion path compared with the particle type. For example, OTE / TiO_2 (Nanoparticles)/CdSe: 0.6 % versus Ti / TiO_2 Nanotubes / CdSe: 0.7 %.

Interesting results have been reported by some investigators who studied the incorporation of a layer of PbS quantum dots in thin film solar cells, by direct growth of PbS quantum dots on nanostructured TiO_2electrodes [27]. Deposition of a transition metal oxide (n-type) layer on grown layer of PbS quantum dots to act as hole extractor layers [46] or employing a graded recombination layer [47].

Several methods have been employed to prepare TiO_2 thin layer. We prepared nanostructured thin films following the procedure detailed in [9, 34]. In this method, a suspension of TiO_2 is prepared by adding 9 ml of nitric acid solution of PH 3-4 (in ml increment) to 6 g of colloidal P25 TiO_2 powder in mortar and pestle. To get a white free flow-paste, we added 8 ml of distilled water (in 1 ml increment) during the grinding process. Finally, a drop of transparent surfactant is added in 1 ml of distilled water to ensure uniform coating and adhesion to the transparent conducting electrode. It was found that the ratio of the nitric acid solution to the colloidal P25 TiO_2 powder is a critical factor for cell performance. If the ratio exceeds certain threshold value, the resulting film becomes too thick and has a tendency to peel off. On the other hand, a low ratio reduces appreciably the efficiency of light generation.

Doctor blade technique was employed by depositing the TiO_2 suspension uniformly on a cleaned (rinsed with ethanol) conductive plate. The TiO_2 film was allowed to dry for few minutes and then annealed at approximately 450 °C (in a well-ventilated zone) for about 15 minutes to form a nanoporous TiO_2 layer as shown in Figure 20. The conductive plate is then allowed to cool slowly to room temperature. This is a necessary condition to remove stresses and avoid cracking of the glass or peeling off the TiO_2 layer. Once the TiO_2 nanocrystalline layer is prepared, it is coated with colloidal QDs. The counter electrode is coated with graphite that Acts as a catalyst in regenerating quantum dots. Both the photo- and counter electrodes are clamped together and drops of electrolyte are applied to fill the clamped cell. The electrolyte used is iodide electrolyte (0.5 M potassium iodide mixed with 0.05 M iodine in water free ethylene glycol) containing a redox couple (traditionally the iodide/tri-iodide I^-/I^{3-}couple). The measurements of the open circuit voltage and short circuit current have been performed under direct illumination from a solar simulator producing 1 sun illumination. UV or IR cut-off filters have been used to eliminate carrier generation from TiO_2 layer and to impede cell overheating.No antireflection coatings on the photoelectrode have been applied. Figure 21 shows an example of the obtained I-V characteristics of PdS quantum dots (size 3.2 nm) sensitized cell with power conversion efficiency =1.8 %.

As it's the case with dye sensitized solar cells, quantum dots sensitized solar cells light harvesting efficiency could be enhanced via efficient light management [48] by increasing light scattering effect [49]. The distance traveled by transmitted photons in the cell is increases by multiple scattering and hence get highly probable by the sensitizer. Surface plasmons resonance effect in the cell also has been suggested [50]. Another approach that is effective in enhancing photovoltaic effect is the reduction of the charge recombination by controlling transparent-conducting-oxide/electrolyte interface such that injected electrons in photoelectrode are excluded from recombining with the redox couple in electrolyte.

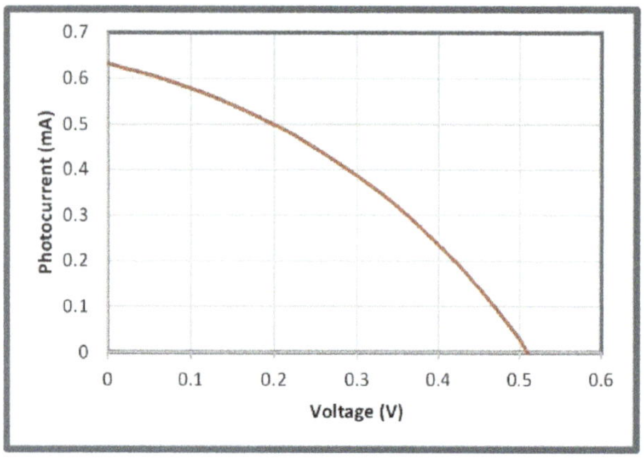

Figure 21. I-V characteristics of PbS quantum dot sensitized solar cell. Quantum dots have on the average radius of 3.2 nm.

Quantum Dot-Dye Sensitized Solar Cells

Incorporation of dye with quantum dots as sensitizer of wide bandgap semiconductor has attracted attention of many research groups. For example, we found that some natural dyes are enhancing power conversion efficiency, while some others are not. Figure 22 shows an example of the I-V characteristics of a first round study of assembled cells illuminated with a collimated beam from a hot filament lamp. In cell preparation we followed the same strategy described in section 5.6 such that after coating the photoelectrode with PbS quantum dots it was soaked in dye for an hour. Then, the electrode was rinsed with deionized water and ethanol. After that the cell is assembled and tested. The dye used was extracted from a pomegranate.

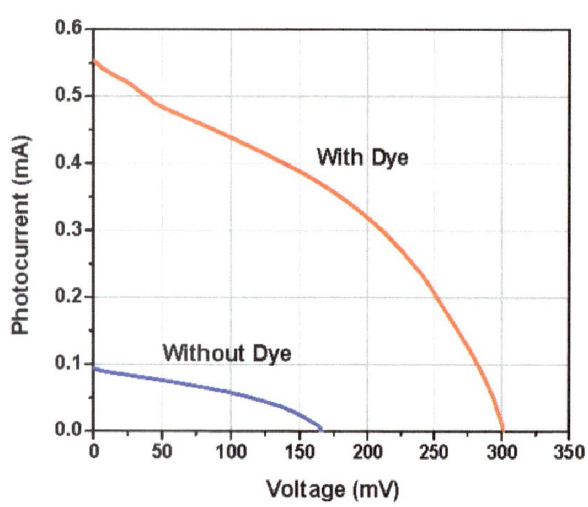

Figure 22. I-V characteristics of typical assembled quantum dot-dye sensitized solar cell. Quantum dots average size of 2.4 nm and pomegranate dye extract used as sensitizers of TiO_2 nanoporous layer.

New configuration based on quantum dot-dye bilayer-sensitized solar cells has been demonstrated by Zaban and co-workers [51]. The bi-sensitizer layer cell is made up of a nanocrystalline TiO_2/CdS QD+amorphous TiO_2/N719 dye. The main aim was to providea configuration having increased charge-separation efficiency by slowing the interfacial charge recombination processes that resulted in 250% increase in cell efficiency compared to a QD monolayer cell.

The configuration investigated by Zaban and co-workers [52] established on making colloidal quantum dots (CdSe/CdS/ZnS core/shell/shell quantum dots) that serve as antennas. Via nonradiative energy transfer, absorbed light is funneled to the charge separating dye molecules (SQ02 dye molecules). The colloidal quantum dot donors are incorporated into the solid TiO_2 photoelectrode resulting in high energy transfer efficiency as well as substantial improvement of the cell stability. In this approach the processes of light absorption can be separated from charge carrier injection. Therefore, this approach enables optimization of each independently. Time resolved luminescence measurements relate the significant contribution of the QDs to the spectral response of the cell in the presence of the dye to Förster resonance energy transfer from the QDs to the dye molecules.

The efficiency of solar cells can be enhanced by combining quantum dots with some dye and used as a sensitizer. We suggest doing further investigations in order to understand QD-dye system. The performance of dye sensitized or quantum dots solar cells can be increased by optimizing preparation technique, using different types of electrolyte, utilizing different nanostructures (e.g., rods, stars), and replacing TiO_2 with other types of wide bandgap semiconductors such as zinc oxide ZnO.

CONCLUSION

Crystalline semiconductor solar cells besides possessing low efficiency due to their band gab limit (Shockley-Queisser limit)[53] they are expensive in terms of manufacturing cost per generated Watt of delivered electric power. In single junction bulky semiconductor solar cells, photon of energies less than the band gap are wasted since none of them are absorbed. Moreover, excess energy of those photons with energies greater than the bandgap is wasted as heat as a result of hot-carriers thermalization. A quantum dot is a crystalline semiconductor nanoparticles. Examples of well investigated quantum dots structures are CdS, CdSe, PbS, and PbSe. The operation principle of quantum dots sensitized solar cell is similar to that of the dye sensitized solar cells DSSCs. In a quantum dot, confinement effect arises from size effect when particle size is smaller or comparable to exciton Bohr radius. As the size of the quantum dot decreases its characteristic excitonic beak get blue shifted.

Growth or synthesis methods of quantum dots are well established. In quantum dots, the rate of energy dissipation is significantly reduced, and the charge carriers are confined within a minute volume, thereby increasing their interactions and enhancing the probability for generating multiple excitons due to hot carries mechanism. There are many proposed quantum dot solar cells configurations.

The functional principle of QD-sensitized solar cell is the same as that of DSSC. The difference is that the dye in DSSC is replaced with quantum dots. This class of third generation solar cell is promising and recently attracting considerable attention. Operation principles, and performance limitations are well understood and many solutions have been proposed to enhance cell efficiency.

ACKNOWLEDGEMENTS

The author is greatly indebted to prof. Dr. Shawqi Aldallal for his kind support and encouragements. His valuable suggestions are highly appreciated. Special thanks to Miss. Fatema Aljaboori for here assistance in obtaining the I-V characteristic presented in Figure 21.

REFERENCES

1. Ekimov A.I., and Onushchenko A.A. Quantum size effect in three-dimensional microscopic semiconductor crystals. JETP Lett. 1981; 34 345–349.
2. Brus L. E. Electronic wave functions in semiconductor clusters: experiment and theory. J. Phys. Chem. 1986; 90 2555-2560.
3. Reed M.A., Randall J.N., Aggarwal RJ, Matyi R.J., Moore T.M., and Wetsel A.E. Observation of discrete electronic states in a zero-dimensional semiconductor nanostructure. Phys Rev Lett. 1988; 60 (6) 535–537.
4. Smith A.R., Chao K.-J., Niu Q., and Shih C.-K. Formation of atomically flat silver film on GaAs with a 'silver-mean' quasiperiodicity. Science 1996; 273 226-228.
5. Schaller R.D., and Klimov V.I. High Efficiency Carrier Multiplication in PbSe Nanocrystals: Implications for Solar Energy Conversion. Phys. Rev. Letts 2004; 92 186601-186614.
6. Schaller R.D., Sykora M., Pietryga J.M., and Klimov V.I. Seven Excitons at a Cost of One: Redefining the Limits for Conversion Efficiency of Photons into Charge Carriers. Nano Letters 2006; 6 424-429.
7. Saleh, B., and Teich, M. Fundamentals of Photonics, ch.15. John Wiley & Sons, Inc. ISBNs: 0-471-83965-5 1991.
8. Sargent E.H. Infrared Quantum Dots. Adv. Mat. 2005; 17 515-522.
9. Jasim, K.E. Dye Sensitized Solar Cells-Working Principles, Challenges and Opportunities. In: Solar Cells-Dye-Sensitized Devices, Leonid A. Kosyachenko, (Ed),, Book Published by InTech Open Access Publisher, ISBN 978-953-307-735-2. 2011. Chapter 8, pp. 171-204.
10. Gaponenko S.V. Strong Confinement Approximation in Optical Properties of Semiconductor Nanocrystals. New York: Cambridge University Press; 1998.
11. Nozik, A.J. Quantum dot solar cells. Physica E 2002;14 115-120.
12. Nozik A.J. Multiple exciton generation in semiconductor quantum dots. Chemical Physics Letters 2008; 457 3-11.
13. Nozik A.J. Nanoscience and Nanostructures for Photovoltaics and Solar Fuels. Nano Lett. 2010; 10 2735-2741.
14. Franceschetti A., An J.M., and Zunger A. Impact ionization can explain carrier multiplication in PbSe quantum dots. Nano Letters 2006; 6 2191-2195.
15. Ellingson R.J., Beard M.C., Johnson J.C., Yu P., Micic O.I., Nozik A.J., Shabaev A., and Efros, Al. L. Highly Efficient Multiple Exciton Generation in Colloidal PbSe and PbS Quantum Dots. Nano Letters 2005; 5 865-871.
16. Hines M. A., Scholes G.D. Colloidal PbS nanocrystals with size-tunable near-infrared emission: observation of post-synthesis. Adv. Mater.2003; 15 1844-1849.
17. Yochelis S., and Hodes G. Nanocrystalline CdSe formation by direct reaction between Cd ions and selenosulfate solution. Chem Mater 2004; 16 2740-2744.
18. Brown P. & Kamat, P.V. Quantum Dot Solar Cells. Electrophoretic Deposition of CdSe-C60 Composite Films and Capture of Photogenerated Electrons with nC60 Cluster Shell. J. Am. Chem. Soc. 2008; 130 8890-8891.
19. Baker D.R. and Kamat P.V. Photosensitization of TiO2 Nanostructures with CdS Quantum Dots: Particulate versus Tubular Support Architectures. Adv. Funct. Mater. 2009; 19 805–811.
20. Subramanian V, Kuno M, Kamat, V.Quantum dot solar cells. harvesting light energy with CdSe nanocrystals molecularly linked to mesoscopic TiO2 films. J Am Chem Soc 2006; 128 (7) 2385–2393.
21. Kamat P.V. Quantum Dot Solar Cells. Semiconductor Nanocrystals as Light Harvesters. J. Phys. Chem. C 2008; 112 18737-18753.
22. Sheeney-Haj-Kia L., Basnar B., Willner I. Efficient Generation of Photocurrents by using CdS/Carbon nanotube Assemblies on Electrodes. Angew. Chem. Int. Ed. 2005; 44 78-83.
23. Huang Q., and Gao L. Synthesis and Characterization of CdS/multiwalled carbon nanotube heterojunctions. Nanotecnol 2004; 15 1855-1860.
24. Pattantyus-Abraham A.G., Kramer I.J., Barkhouse A.R., Wang X., Konstantatos G., Debnath R., Levina L., Raabe I., Nazeeruddin M.K., Gratzel M., and Sargent E.H. Depleted-Heterojunction Colloidal Quantum Dot Solar Cells. ACSNano 4 (6) 3374–3380.
25. Luther J.M., Law M.et al. Structural, Optical, and Electrical Properties of Self-Assembled Films of PbSe Nanocrystals Treated with 1,2-Ethanedithiol. ACS Nano. 2008; 2 271.
26. Luther M., Gao J., Lloyd M.T., Semonin O.E., Beard M.C., and Nozik A.J. Stability... solar cell. Advanced Materials 2010; 22 (33) 3704–3707.
27. Gao J., Perkins C.L, Luther J.M., Hanna M.C., Chen H-Yu, Semonin O.E., Nozik A.J., Ellingson R.J., and Beard, M.C. n-Type Transition Metal Oxide as a Hole Extraction Layer in PbS Quantum Dot Solar Cells. Nano Letters 2011; 11 3263
28. Sun, B., Findikoglu A.T., Sykora M., Werder D.J., and Klimov V.I. Hybrid Photovoltaics Based on Semiconductor Nanocrystals and Amorphous Silicon. Nano Lett. 2009; 9 (3) 1235–1241.
29. Barnham K. and Duggan G. A new approach to high efficiency multibandgap solar cells. J. Appl. Phys. 1990; 67 3490-3493.
30. Luque A., and Marti A. Increasing the Efficiency of Ideal Solar Cells by Photon Induced Transitions at Intermediate Levels Phys. Rev. Lett. 1997; 78 5014-5017.
31. Luque A., Martí A., and Nozik A.J. Solar Cells Based on Quantum Dots: Multiple Exciton Generation and

Intermediate Bands. MRS BULLETIN 2007, 32, pp: 236-241.

32. Huynh W.U. Hybrid Nanorod-Polymer Solar Cells. Science 2002; 295 2425.

33. Sun B., Marx E., and Greenham, N.C. Photovoltaic Devices Using Blends of Branched CdSe Nanoparticles and Conjugated Polymers. Nano-Letters 2003; 3 (7) 961-963.

34. Hara K., and Arakawa H. Handbook of Photovoltaic Science and Engineering, by A. Luque and S. Hegedus (Eds.), Chapter 15, p.663, John Wiley & Sons, Ltd., ISBN: 0-471-49196-9; 2003.

35. O'Regan B.; and Grätzel M. A Low-cost High-Efficiency Solar Cell Based on Dye Sensitized Colloidal TiO2 Films. Nature 1991; 353 737-740.

36. Rhle S., Shalom, M., and Zaban, A. Quantum-Dot-Sensitized Solar Cells. Chem. Phys, Chem. 2010; 11 2290 –2304.

37. Peng A., and Peng X. Formation of high-quality CdTe, CdSe, and CdS nanocrystals using CdO as precursor". Journal of the American Chemical Society 2001; 123 (1) 183–184.

38. Lee H.J., Yum J.H., Leventis H.C. CdSe quantum dot-sensitized solar cells exceeding efficiency 1% at full-sun intensity. Journal of Physical Chemistry C. 2008; 112 (30) 11600 –11608.

39. Bang J. H. and Kamat, P. V. Quantum dot sensitized solar cells. A tale of two semiconductor nanocrystals: CdSe and CdTe,. ACS Nano. 2009; 3 (6) 1467–1476.

40. Wang P., Zakeeruddin S. M., Comte P., Exnar I., and Grätzel M. Gelation of Ionic Liquid-Based Electrolytes with Silica Nanoparticles for Quasi-Solid-State Dye-Sensitized Solar Cells. Chem. Commun. 2002; 24 2972-2973.

41. Wang P., Zakeeruddin S. M., Moser J.E., Nazeeruddin M. K., Sekiguchi T., and Grätzel, M. Molecular-scale interface engineering of TiO2 nanocrystals: Improving the efficiency and stability of dye-sensitized solar cells. Nat. Mater. 2003; 2 402.

42. Chakrapani V., Tvrdy K., and Kamat P.V. Modulation of Electron Injection in CdSe–TiO2 System through Medium Alkalinity", J. Am. Chem. Soc. 2010; 132 (4) 1228-1229.

43. Kongkanand A., Tvrdy K., Takechi K., Kuno M., and Kamat P.V. Quantum Dot Solar Cells. Tuning Photoresponse through Size and Shape Control of CdSe-TiO2 Architecture. J. Am. Chem. Soc. 2008; 130 (12) 4007-4015.

44. Mora-Sero I., and Bisquert J. Breakthroughs in the Development of Semiconductor-Sensitized Solar Cells. J. Phys. Chem. Lett. 2010; 1 3046–3052.

45. Barnes P.R.F, Liu L., Li X., Anderson A.S., Kisserwan H., Ghaddar T.H., James R., Durrant J.S., and O'Regan, B.C. Re-evaluation of Recombination Losses in Dye-Sensitized Cells: The Failure of Dynamic Relaxation Methods to Correctly Predict Diffusion Length in Nanoporous Photoelectrodes. Nano Lett. 2009; 9 (10) 3532–3538.

46. Braga A., Gimenez S., Concina, I., Vomiero, A., and Mora-Sero, I. Panchromatic sensities solar cells based on metal sulfied quantum dots grown directly on nanostructured TiO2 electrode. J. Phys. Chem. Lett. 2011; 2 454-460.

47. W X., Koleilat G.I., Tang J., Liu H., Kramer I.J., Debnath R., Brzozowski L., Barhouse D.A.R., and Levina, L.; and Hoogland, S.; Tandem colloidal quantum dot solar cells employing a graded recombination layer. Nature Photonics 2011 ; 5 480.

48. Choi H., Nahm C., Kim J., Kim C., Kang S., Hwang T., and Park B.Toward highly efficient quantum-dot-and dye-sensitized solar cells. Current Applied Physics 2003; 13 S2-S13.

49. Koo H.-J., Park J., Yoo B., Yoo K. Kim K., and Park N.-G. Size-dependent scattering efficiency in dye-sensitized solar cell. Inorg. Chim. Acta. 2008; 361 677-683.

50. Atwater H.A., and Polman, A. Plasmonics for improved photovoltaic devices. Nat. Mater. 2010; 9 205-213.

51. Shalom M., Albero J., Tachan Z., Martínez-Ferrero E., Zaban A., and Palomares E. Quantum Dot-Dye Bilayer-Sensitized Solar Cells: Breaking the Limits Imposed by the Low Absorbance of Dye Monolayers. J. Phys. Chem. Lett. 2010; 1 1134–1138.

52. Buhbut S., Itzhakov S., Tauber E., Shalom M., Hod I., Geiger T., Garini Y., Oron D., and Zaban, A. Built-in Quantum Dot Antennas in Dye-Sensitized Solar Cells. Acs Nano 2010; 4 (3) 1293–1298.

53. Shockley W., and Queisser H.J. Detailed Balance Limit of Efficiency of p-n Junction Solar Cells", Journal of Applied Physics 1961; 32 510-519.

CHAPTER 4 SEMICONDUCTOR QUANTUM DOTS FOR BIOMEDICIAL APPLICATIONS

Lijia Shao, Yanfang Gao and Feng Yan

Keywords: Quantum dots, bioanalysis, toxicology delivery, photodynamic therapy, cell imaging

INTRODUCTION

Quantum dots (QDs) as colloidal nanocrystalline semiconductors have unique photophysical properties due to quantum confinement effects. They emit different wavelengths over a broad range of the light spectrum from visible to infrared, depending on their sizes and chemical compositions. Compared with the traditional organic fluorophores (e.g., organic dyes and fluorescent proteins), QDs have unique optical and electronic properties, such as larger absorption coefficients, size-tunable light emission, superior signal brightness, resistance to photobleaching and simultaneous excitation of multiple fluorescence colors [1–6]. In addition, the large-surface area of QDs is beneficial to covalently link to biorecognition molecules, such as peptides, antibodies, nucleic acids or small-molecule ligands for further application as fluorescent probes (Figure 1).

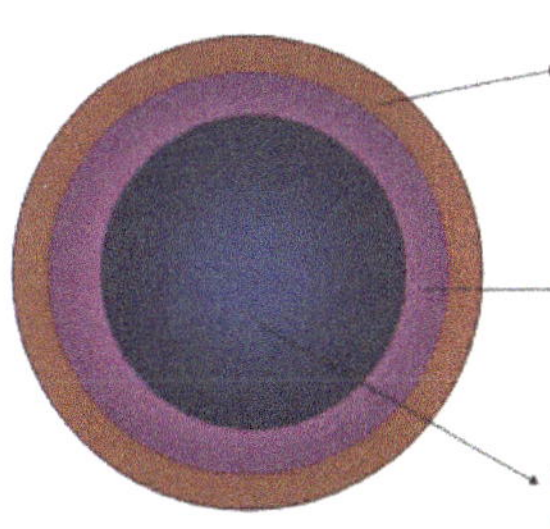

Figure 1. Schematic representation of a quantum dot. QDs are nanocrystals composed of a core of a semiconductor, usually composed of elements from groups II–IV, e.g., CdSe, or groups III–V, e.g., InP. The shell is typically a higher bandgap material such as ZnS. Finally, a capping outer layer such as silica can offer large-surface area for covalently linking to biorecognition molecules such as peptides, antibodies, nucleic acids and small-molecule ligands for further application. The diameter of QDs ranges between 2–10 nm.

These properties of QDs herald a revolution from electronic materials science to biological applications [7]. Current and projected applications of QDs include using fluorescent labels for cellular labeling [1,8,9], intracellular sensors [10,11], deep-tissue and tumor targeting and imaging agents [7,12–17], sensitizers for photodynamic therapy (PDT) [18–21], vectors for gene therapy [22–26], magnetic resonance imaging (MRI) contrast agents [27,28] and so on. This review mainly summarizes the development of synthesis, the surface modification and toxicity of QDs, and briefly focuses on the application developments of QDs in the biomedical field.

THE SURFACE CHEMISTRY AND TOXICITY OF QDS

Early in the 1990s, Bawendi and coworkers first reported a synthesis protocol for QDs with highly monodisperse, regular core structure and tunable particle size [29,30]. Up to now, the most successful and well-developed method to prepare highly luminescent II–VI QDs is the TOP/TOPO synthetic approach [31]. However, these QDs are insoluble in water, which limits their biological applications. Therefore, a number of surface functionalization studies have been developed to make QDs water-soluble and biologically compatible [29–37].

In one common approach, the original hydrophobic coatings are replaced by water-soluble functional molecules (e.g., dithiothreitol [38–40], mercaptocarbonic acids [41–44], 2-aminoethanethiol [33,45], dihydrolipoic acid [34–36,46,47], oligomeric phosphines [37,48], peptides [49–57], and cross-

Citation: Lijia Shao, Yanfang Gao and Feng Yan, Semiconductor Quantum Dots for Biomedicial Applications, doi:10.3390/s111211736.

linked dendrons [58–61]) through the ligand exchange reactions. Because the optical properties of the inorganic core are often very sensitive to the surface, the ligand exchange process may result in poorer performance, particularly in the case of quantum dots [62].

The second approach is to encapsulate QDs in an amphiphile whose hydrophobic ends interleave with, but do not replace, the organic coating on QDs. This improvement for QDs synthesis is significant: (1) protecting the core/shell structure and maintaining the original photophysics of QDs; (2) making QDs water-soluble; (3) providing a biological interface and multiple functions [7]. However these kinds of QDs are not stable in biological settings because of relativelyweak anchoring of the single and double hydrophobic tails to the particle. Additionally, the hydrophilic end groups of even biocompatible surfactants may not protect nanocrystals from nonspecific biomolecular interactions [31]. Scientists have used amphiphilic polymers instead of simple amphiphile because single polymer chains can contain multiple hydrophobic units, their interactions with the native organic coatings on QDs can be numerous, and thus the encapsulant can be bound more strongly than conventional surfactants. However, the range of amphiphilic polymers for creating stable and nonaggregating QDs in biological settings has been relatively limited. Up to now, most of the amphiphilic polymers used are commercial and their hydrophobic/hydrophilic ratios are fixed, hence the cost is high and it may be different to control the process of forming water-soluble QDs and to optimize the forming conditions [31].

Although QDs have great prospects, the toxicity of QDs cannot be overlooked. During the processing of biological applications (e.g., cancer imaging, targeting and PDT treatment), the degradation products of QDs will do harm to the cells which they contact with, or produce immune responses with the components in blood [17]. The toxic degradation production routes are: first, the oxidation of the nanoparticle core/shell material can cause the release of free cadmium or other heavy metals, which will interrupt the normal cell activities [18]; secondly, the photosensitized production of reactive oxygen intermediates (ROI) also plays an important role in mediating the cell damage [63]; thirdly, the toxicity of capping materials should also be considered, several groups in capping materials such as mercaptoacetic acid and tri-*n*-octylphosphine oxide (TOPO) could produce toxicity to cells [12].

To reduce the cytotoxicity of QDs, replacement of the cadmium by nontoxic or less-toxic metals such as indium (In), or encapsulation of the core with a biocompatibile shell should be considered. Though In-based semiconducting dots contain arsenic, another toxin, the cytotoxicity of these dots may be small enough to keep the toxicity low. Fisher and coworkers [64] found that QDs could remain within the body for very long periods. Kim [8] reported that larger QDs generally accumulated in the reticuloendothelial system, such as the liver, spleen and lymphatic system for several months, but the size less than 5 nm could be removed by the kidney quickly. So in order to minimize the toxicity of QDs, QDs can be designed as smaller as they can, which can help them more easily to clean them out from the body.

In spite of the fact many investigators have paid close attention to and observed the side-effect of QDs, the definite metabolism of QDs *in vivo* remains uncertain [65–68]. Thus, it is still a necessary issue to investigate the detailed biochemical and pharmacological mechanism for further application of QDs in the human body.

DELIVERING QDS INTO CELLS

Effective delivery of QDs into the targeted-cell is the primary requirement for the bioapplications of QDs [9,15,17,20,32,54]. It is a major step because if QDs cannot reach their site of action *in vivo*, they is useless. Furthermore, efficient delivery can also allow a reduction in dosage level, avoid non-specific side effects and reduce toxicity risks [66,69,70]. The current methods for delivering QDs into cells mainly include passive delivery, facilitated delivery and active delivery.

The general passive delivery for QDs is endocytosis, which is simple, without further functionalization of the QDs surface with a targeting ligand for uptake [66]. By incubating with the cells at appropriate concentration and exposure time, QDs will enter into cells though the nonspecific cell endocytosis. However, the nonspecific ingestion of this mode caused ineffective endosomal escape, and would impede the delivery of QDs to the cytoplasm or other organelles. Furthermore, high intracellular concentration of QDs can enhance the cytotoxicity in some cases [69].

Facilitated delivery includes four ways: peptide-mediated uptake, protein-mediated delivery, polymer-mediated delivery and small molecule-mediated delivery [66]. Generally, these molecules are noncovalently assembled onto the surface of QDs for bioconjugation. Facilitated delivery could reduce the nonspecific absorption and side effects. However, QDs could also be uptaken by cell through endocytosis, leading to endosomal sequestration during the facilitated delivery strategies (Figure 2). As is well known, the high acidic of endosomes could degrade the QDs conjugates over time, thus free Polyethyleneimine (PEI) was used to encapsule the QDs conjugates to increase the stability [70]. Considering further application of cell imaging, more general endosomal escape strategies need to be developed in order to expand the application of facilitated delivery.

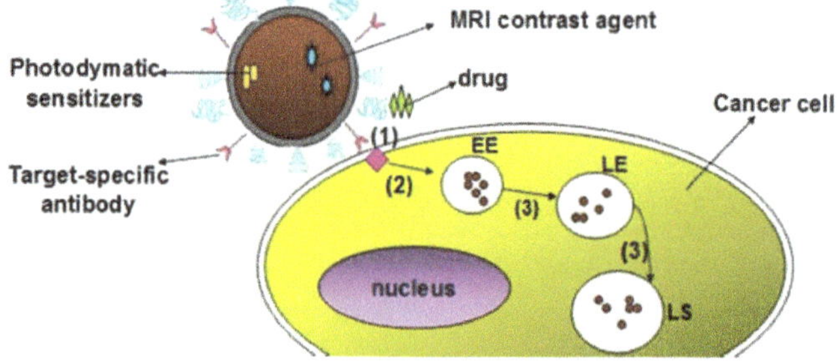

Figure 2. Schematic representation of delivering QDs into cells, the process comprises of three major stages: (1) endocytosis; (2) sequestering in early endosome (EE); (3) translocation to later endosomes (LE) or lysosomes (LS).

Active delivery is a direct physical manipulation of the cell by electroporation and microinjection. In comparison to facilitated delivery, QDs conjugates are delivered directly to the cytoplasm via electroporation by an endocytic pathway, without subsequent endosomal escape. However, the high cellular mortality rate and intracellar aggregation occurring during the delivery should be conquered [71]. Compared with electroporation, microinjection could deliver the QDs directly to the cytoplasm with lower cell death rate, and the rate of microinjection of QDs conjugates to cells depends on the physical constraints of cells, including morphology, membrane thickness, height, *etc.* [66]. Furthermore, this technology is very expensive. Therefore, considering the coexistence of advantages and drawbacks of the mentioned approaches, the appropriate way for delivering QDs into cell should be determined according to the specific experimental requirements. The relationship between the specific examples and the delivery strategies are listed in Table 1 [66].

Table 1. Selected strategies for the intracellular delivery of QDs.

Strategy	Mechanism	Examples	Targeted Cells	References
Passive uptake	Electrostatic interactions	-	HeLa	[15]
			Human macrophages	[27,72]
			Breast cancer (MDA-MB-231)	[73,74]
			Human melanoma cells (LU1205)	[75]
Facilitated delivery	Peptide-mediated	TAT	Human embryonic kidney	[76]
			HeLa	[77]
			Mesenchymal stem cells	[53]
			Jurkat cells	[28]
		Pep-1 (Chariot)	Osteoblast	[78]
			Vascular endothelial cells	[78]
		RGD motify	Fibroblast (NIH 3T3)	[79]
			Epidermoid carcinoma	[80]
	Protein-mediated	Neuropeptide	HeLa	[17]
		Transferrin	Human pancreatic cancer	[2,81,82]
		Antibody	Breast cancer (MCF-7)	[11]
		EGF	Mesenchymal stem cells	[83]
			Chinese hamster ovary	[84-86]
			Medulloblastoma tumors	[87]
			Glioma tumors	[87]
		Cholera toxin B	Fibroblast	[88,89]
		NGF	PC12 neural cells	[90,91]
	Polymer/lipid-mediated	Lipid polymers	Mouse lymphoma	[92]
			HeLa	[93]
			A549 epithelial lung HeLa	[94]
		Polyethylenei-mine	HeLa	[95]
	Drug-mediated	Tiopronin	Fibroblast	[96]
	Small molecule	Glucose/sugar	*S. cerevisiae* (Baker's yeast)	[97,98]
		Folate	Epidermal carcinoma	[99]
		Adenine/AMP	Bacteria (*Bacillus subtilis, E. coli*)	[100,101]
		Dopamin	A9 mouse fibroblast with transfected dopamine receptor	[102]
Active Delivery	Electroporation	-	HeLa	[71]
			Mouse neural stem progenitor	[9]
	Microinjection	-	cells	[71]
			Xenopus embryo	[103]
			HeLa	[104]
			Human embryonic kidney	

QDS-BASED CANCER TARGETING AND IMAGING

The photoluminescence (PL) of QDs is exceptionally bright and stable, making them potential candidates for biomedical imaging and therapeutic interventions. QDs conjugated with cancer specific ligands/antibodies/peptides were found to be effective for detecting and imaging human cancer cells. Gao and coworkers [67] firstly reported the QDs-antibody conjugates for *in vivo* targeting and imaging cancer, in which QDs-antibody conjugates were used as imaging probe for investigating and tracing QDs-PSMA antibody conjugates in mouse bearing subcutaneous human prostate cancer. It was found that the QDs-antibody conjugates were efficiently and uniformly distributed in prostate tumors due to the specific binding between PSMA antigen in prostate cancer cells and PSMA antibody on QDs. Cai and coworkers [105] conjugated NIR QDs with RED peptide, which could bind to the over-expressed αvβ3 integrin on the surface of U87MG glioblastoma cells and MDA-MB-435 human breast cancer cells to target cancer cell *in vivo*. By linking QDs to AFP (alpha-fetoprotein) antibody, an important marker for hepatocellular carcinoma cell lines, a specific immunofluorescent probes was obtained for further detection of AFP antibody in human serum. Yu *et al.* [106] demonstrated that the probe could target the specific hepatocellular carcinoma cells, and the expected results was obtained by investigating distribution of the probes in cancer cells by using a site-by-site measurement. Weng *et al.* [107] functionalized QDs with anti-HER2 scFv to synthesize the immunoliposome-based nanoparticles (QD-ILs). After incubating with HER2-overexpressing SK-BR-3 and MCF-7/HER2 cells, the QD-ILs exhibited efficient receptor-mediated endocytosis. *In vivo* fluorescence imaging showed that QD-ILs had localized prominently in tumors as well as in MPS organs

(Figure 3). Liu *et al.* [68] reported a QDs-based wavelength-resolved spectral imaging for molecular mapping of tumor heterogeneity on human prostate cancer tissue specimens. By conjugating different QDs with specific protein biomarkers, such as E-cadherin, high-molecular-weight cytokeratin, p63, and α-methylacyl CoA racemase, structural distinct prostate glands and single cancer cells could be detected and characterized within the complex microenvironments of radical prostatectomy and needle biopsy tissue specimens using the wavelength-resolved spectral imaging.

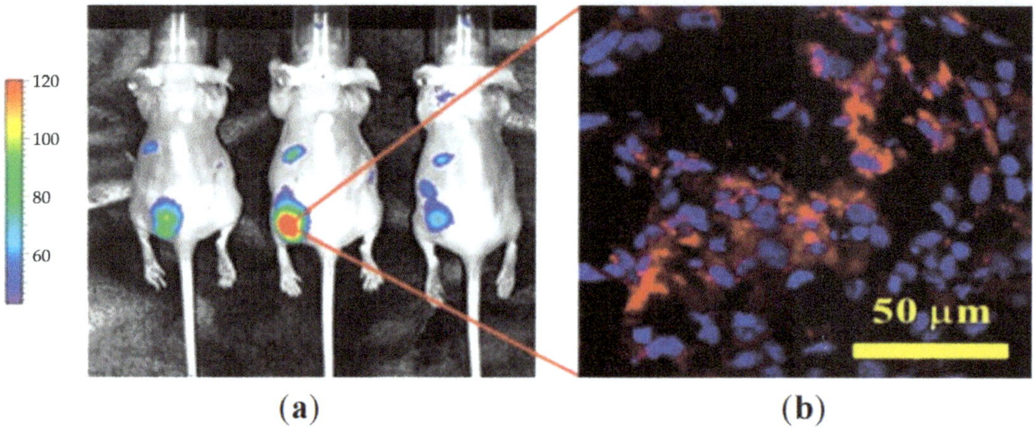

Figure 3. (a) Left panel: *In vivo* fluorescence imaging of three nude mice bearing MCF-7/HER2 xenografts implanted in the lower back 30 h after i.v. injection with anti-HER2 QD-ILs; (b) Right panel: A 5 μm section cut from frozen tumor tissues harvested at 48 h postinjection and examined by confocal microscopy by a 63× oil immersion objective (image size, 146 μm × 146 μm). The tumor section was examined in two-color scanning mode for nuclei stained by DAPI (blue) and QD-ILs (red). (Cited from Weng *et al.* [107]).

The main advantage of QDs imaging is that it is non-ionizing and less hazardous [108]. In recent years, several groups have used QD probes for fluorescence immunostaining of fixed cells and tissue specimens [109–113]. QD-based immunohistochemistry (IHC) can improve both diagnostic sensitivity and specificity. In addition, because multiplexed QD staining can be carried out on intact cells and tissue specimens, it is expected to provide correlated molecular and morphological information, at the same time, this type of integrated biomarker and morphological data are not available from traditional analytical methods such as mass spectrometry, gene chips, protein microarrays, and polymerase chain reactions [109]. However, medical applications of QD-based IHC have achieved only limited success. A major bottleneck is the lack of robust protocols to define the key parameters and steps [109]. For example, there are no consensuses on methods for QD-antibody (QD-Ab) bioconjugation, tissue specimen preparation, multicolor QD staining, image processing and data quantification. So it is necessary to solve these problems, and let the QDs move further.

QDS RELATED PHOTODYNAMIC THERAPY FOR CANCER

Presently, the conventional types of cancer treatment (chemotherapy and radiation therapy), work by destroying fast-growing cells, but other types of fast-growing healthy cells (such as blood and hair cells) also can be damaged along with cancer cells, causing adverse reactions, or side effects. These side effects can range from fatigue and flu-like symptoms to hair loss and blood clotting problems. PDT developed in last century has become an FDA-approved therapy for different malignancies and with potential in other ailments such as coronary heart disease, AIDS and psoriasis [63].

Exploration of the use of light-activated drugs known as photosensitizers (PS) has been one of the most active areas of photomedical research in recent years [18–21,63,114,115]. PDT uses the combination of a photosensitizing drug and light in the presence of oxygen to cause selective damage to the targeting tissue. During PDT, reactive oxygen intermediates (ROI) is generated in the diseased cells by a simple and controllable light-activated process, which involves a photosensitizer that is capable of absorbing light appropriate wavelength and transfers energy or electron to oxygen or other molecules, and creates ROI such as singlet oxygen ($^1O_2$), hydroxyl radical (OH), super oxide anion (O_2^-) and hydrogen peroxide (H_2O_2). Then ROI will immediately react with vital biomolecules in cell organelles, leading to cell damage, mutation, death and photooxidation of cell constituents [19,20,63,114,115]. Singlet oxygen ($^1O_2$) is regarded as the main mediator of photo-induced cytotoxicity in PDT, which causes oxidation and degradation of cellular components, and ultimately cell apoptosis. [20,63,114,115] (Figure 4).

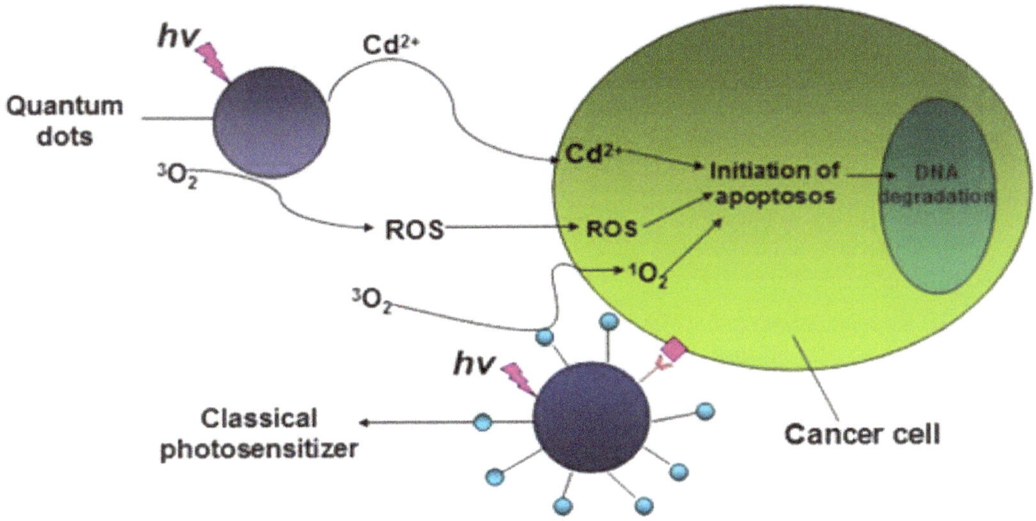

Figure 4. Schematic representation of possible mechanisms for induction of PDT processes by QDs and the classical photosensitizer.

The standard PS drugs for PDT are porphyrin, phthalocyanines and chlorine derivatives. Porphyrin derivatives are the first generation photosensitizer. Despite the clinical success of porphyrin derivatives, some of their disadvantages like prolonged cutaneous photosensitivity, chemical impurity and weak absorption at therapeutic wavelengths have inspired the development of new PDT photosensitizers with improved optical and chemical properties. Phthalocyanines derivatives have favorable photophysical and chemical properties, which include strong absorbance at long wavelengths and chemical tunability through substituent addition on the periphery of the macrocycle or on the axial ligands. However, like most photosensitizing agents, these PS have poor solubility in water and tend to aggregate in aqueous solutions, which can result in loss of photochemical activity and affect their cell penetrating properties [63]. To resolve such issues nanoparticles are currently being explored as potential delivery systems for PDT photosensitizers or directly as PDT agents. The novel QDs-PS conjugates are used as a high ratio of PDT agents and anticancer targeting antibodies, where QDs can act as nanoscaffolds and solubilizers. They can also function as "energy-harvesting antenna" for PDT therapy due to their large one- or two-photon absorption cross-sections. Thus, QDs can be efficiently exited even deep within tissues and sensitized proximal PDT agents via energy transfer from QDs to PDT [21].

The novel QDs-PS conjugates showed many advantages over conventional PS drugs [17–21,63,114,115]: (1) they are species with well-defined size, shape, and composition, and can be synthesized by relatively simple and inexpensive methods; (2) they have been shown to be nontoxic in the absence of light but have the potential to be cytotoxic under irradiation; (3) they have photostability, and tunable and strong absorption, which can be tuned from the UV their composition and size; (4) the surface coating of QDs can be modified to enable them to become water soluble, biocompatible and target-specific.

However, researchers should be further investigated on the basis of predominances of the QDs-PS compared to the convention PS drugs. Despite many desirable properties of QDs for PDT, there still remain several important issues that need to be addressed to fully assess their applicability as PS in PDT. One major issue is the toxicity profile of the QDs inside the cells and their overall photostability once exposed to biological environments [63]. Another important matter that should be carefully investigated is how their surface composition affects the photosensitization process. Still, QDs-PS conjugates for cancer therapy are only suitable to superficial tumours is also need to be resolved [18].

CONCLUSIONS AND OUTLOOK

In the last decade, the unique photophysical properties and functions of QDs have been widely investigated, making them one of the most promising nanomaterials. Their outstanding performances such as high fluorescence yields, stability against photobleaching and the size-dependent luminescence features of QDs provide broad variety of applications for QDs in many fields. By acting as fluorescent, and photoelectrochemical as well as electrochemical probes, various QDs-based optical and electrochemical bioanalysis have already been successfully explored for sensing a wide range of molecules with high sensitivity and specificity. Furthermore, as a biomedical label, QDs can make a worthy contribution to the development of new diagnostic and delivery systems due to their unique optical properties. By combination of functional biomolecule-nanoparticle hybrid systems and the optical imaging and biophysics, QDs have been used as optical reporter units of biocatalytic transformations and can probe intracellular processes *in vitro*. QDs as a novel probe for *in vivo* analysis and clinic therapy such as PDT open an attractive new field with promising prospectives in biomedicine.

ACKNOWLEDGMENTS

This review was financially supported by the National Basic Research Program of China (2010CB732400), the National Natural Science Foundation of China (21075055), the Program for Six Peak Talents of Jiangsu province (66) and the Leading Medical Talents Program from Department of Health of Jiangsu Province.

REFERENCES

1. Bruchez, M.J.; Moronne, M.; Gin, P.; Weiss, S.; Alivisatos, A.P. Semiconductor nanocrystals as

fluorescent biological labels. *Science* 1998, *281*, 2013–2016.

2. Chan, W.C.; Nie, S. Quantum dot bioconjugates for ultrasensitive nonisotopic detection.*Science* 1998, *281*, 2016–2018.
3. Chan, W.C.; Maxwell, D.J.; Gao, X.H.; Bailey, R.E.; Han, M.Y.; Nie, S. Luminescent quantum dots for multiplexed biological detection and imaging. *Curr. Opin. Biotechnol* 2002, *13*, 40–46.
4. Dabbousi, B.O.; Rodriguez-Viejo, J.; Mikulec, F.V.; Heine, J.R.; Mattoussi, H.; Ober, R.; Jensen, K.F.; Bawendi, M.G. (CdSe) ZnS core-shell quantum dots: Synthesis and characterization of a size series of highly luminescent nanocrystallites. *J. Phys. Chem* 1997, *101*, 9463–9475.
5. Lim, Y.T.; Kim, S.; Nakayama, A.; Stott, N.E.; Bawendi, M.G.; Frangioni, J.V. Selection of quantum dot wavelengths for biomedical assays and imaging. *Mol. Imaging* 2003, *2*, 50–64.
6. Mattoussi, H.; Kuno, M.K.; Goldman, E.R.; George, P.; Mauro, J.M. Colloidal semiconductor quantum dot conjugates in biosensing. In *Optical Biosensors: Present and Future*; Ligler, F.S., Rowe Taitt, C.A., Eds.; Elsevier: Amsterdam, The Netherlands, 2002; Chapter 17; pp. 537–569.
7. Michalet, X.; Pinaud, F.F.; Bentolila, L.A. Quantum dots for live cells, *in vivo* imaging, and diagnostics. *Science* 2005, *307*, 538–544.
8. Kim, S.; Lim, Y.T.; Soltesz, E.G.; Grand, A.M.; Lee, J.; Nakayama, A. Near-infrared fluorescent type II quantum dots for sentinel lymph node mapping. *Nat. Biotechnol* 2004, *22*, 93–97.
9. Chen, F.Q.; Gerion, D. Fluorescent CdSe/ZnS nanocrystal-peptide conjugates for long-term, nontoxic imaging and nuclear targeting in living cells. *Nano Lett* 2004, *4*, 1827–1832.
10. Somers, R.C.; Bawendi, M.G.; Nocera, D.G. Nanocrystals as sensors. *Green Chem* 2007, *9*, 403–410.
11. Qian, J.; Yong, K.T.; Roy, I.; Ohulchanskyy, T.Y.; Bergey, E.J.; Lee, H.H.; Tramposch, K.M.; He, S.; Maitra, A.; Prasad, P.N. Imaging pancreatic cancer using surface-functionalized quantum dots. *J. Phys. Chem* 2007, *111*, 6969–6972.
12. Smith, A.M.; Duan, H.; Mohs, A.M.; Nie, S. Bioconjugated quantum dots for *in vivo*molecular and cellular imaging. *Adv. Drug Deliv. Rev* 2008, *60*, 1226–1240.
13. Medintz, I.L.; Uyeda, H.T.; Goldman, E.R.; Mattoussi, H. Quantum dot bioconjugates for imaging, labelling and sensing. *Nat. Mater* 2005, *4*, 435–446.
14. Jaiswal, J.K.; Mattoussi, H.; Mauro, J.M.; Simon, S.M. Long-term multiple color imaging of live cells using quantum dot bioconjugates. *Nat. Biotechnol* 2003, *21*, 47–51.
15. Bakalova, R.; Zhelev, Z.; Aoki, I.; Kanno, I. Designing quantum-dots probes. *Nat. Photonics*2007, *1*, 487–489.
16. Baron, R.; Willner, B.; Willner, I. Biomolecule-nanoparticle hybrids as functional units for nanobiotechnology. *Chem. Commun* 2007, *18*, 323–332.
17. Biju, V.; Muraleedharan, D.; Nakayama, K.; Shinohara, Y.; Itoh, T.; Baba, Y.; Ishikawa, M. Quantum dot-insect neuropeptide conjugates for fluorescence imaging, transfection, and nucleus targeting of living cells. *Langmuir* 2007, *23*, 10254–10261.
18. Juzenas, P.; Chen, W.; Sun, Y.P.; Coelho, M.; Generalov, R.; Generalova, N.; Christensen, I.L. Quantum dots and nanoparticles for photodynamic and radiation therapy of cancer. *Adv. Drug Deliv. Rev* 2008, *60*, 1600–1614.
19. Ochsner, M. Photophysical and photobiological processes in the photodynamic therapy of tumours. *J. Photochem. Photobiol. B* 1997, *39*, 1–18.
20. Oleinick, N.L.; Evans, H.H. The photobiology of photodynamic therapy: Cellular targets and mechanisms. *Radiat. Res* 1998, *150*, 146–156.
21. Tardivo, J.P.; Giglio, A.D.; Oliveira, C.S.; Gabrielli, D.S.; Junqueira, H.C.; Tada, D.B.; Severino, D.; Turchiello, R.F.; Baptista, M.S. Methylene blue in photodynamic therapy: From basic mechanisms to clinical applications. *Photodiagn. Photodyn. Ther* 2005, *2*, 175–191.
22. Jiang, Z.; Li, R.; Todd, N.W.; Stass, S.A.; Jiang, F. Detecting genomic aberrations by fluorescence *in situ* hybridization with quantum dots-labeled probes. *J. Nanosci. Nanotechnol*2007, *7*, 4254–4259.
23. Burke, R.S.; Pun, S.H. Extracellular barriers to *in vivo* PEI and PEGylated PEI polyplex-mediated gene delivery to the liver. *Bioconjug. Chem* 2008, *19*, 693–704.
24. Delehanty, J.B.; Boeneman, K.; Bradburne, C.E.; Robertson, K.; Medintz, I.L. Quantum dots: A powerful tool for understanding the intricacies of nanoparticle-mediated drug delivery.*Expert Opin. Drug Deliv* 2009, *6*, 1091–1112.
25. Zhao, Y.; Zhao, L.; Zhou, L.; Zhi, Y.; Xu, J.; Wei, Z.; Zhang, K.X.; Ouellette, B.F.; Shen, H. Quantum dot conjugates for targeted silencing of bcr/abl gene by RNA interference in human myelogenous leukemia K562 cells. *J. Nanosci. Nanotechnol* 2010, *10*, 5137–5143.
26. Bolhassani, A.; Safaiyan, S.; Rafati, S. Improvement of different vaccine delivery systems for cancer therapy. *Mol. Cancer* 2011. [CrossRef]
27. Nabiev, I.; Mitchell, S.; Davies, A.; Williams, Y.; Kelleher, D.; Moore, R.; Gun'ko, Y.K.; Byrne, S.; Rakovich, Y.P.; Donegan, J.F.; Sukhanova, A.; Conroy, J.; Cottell, D.; Gaponik, N.; Rogach, A.; Volkov, Y. General strategy for designing functionalized magnetic microspheres for different bioapplications. *Nano Lett* 2007, *7*, 3452–3461.
28. Lewin, M.; Carlesso, N.; Tung, C.H.; Tang, X.W.; Cory, D.; Scadden, D.T.; Weissleder, R. Tat peptide-derivatized magnetic nanoparticles allow *in vivo* tracking and recovery of progenitor cells. *Nat. Biotechnol* 2000, *18*, 410–414.
29. Dabbousi, B.O.; Murray, C.B.; Rubner, M.F.; Bawendi, M.G. Langmuir-Blodgett manipulation of size-selected CdSe nanocrystallites. *Chem. Mater* 1994, *6*, 216–219.
30. Danek, M.; Jensen, K.F.; Murray, C.B.; Bawendi, M.G. Electrospray organometallic chemical vapor

deposition-A novel technique for preparation of II–VI quantum dot composites. *Appl. Phys. Lett* 1994, *65*, 2795–2797.

31. Wu, Z.; Zhao, Y.; Qiu, F.; Li, Y.; Wang, S.; Yang, B.; Chen, L.; Sun, J.; Wang, J. Forming water-soluble CdSe/ZnS QDs using amphiphilic polymers, stearyl methacrylate/methylacrylate copolymers with different hydrophobic moiety ratios and their optical properties and stability. *Colloids Surf. A Physicochem. Eng. Asp* 2009, *350*, 121–129.
32. Ghasemi, Y.; Peymani, P.; Afifi, S. Quantum dot: Magic nanoparticle for imaging, detection and targeting. *Acta Biomed* 2009, *80*, 156–165.
33. Lee, C.M.; Jang, D.; Cheong, S.J.; Kim, E.M.; Jeong, M.H.; Kim, S.H.; Kim, D.W.; Lim, S.T.; Sohn, M.H.; Jeong, H.J. Surface engineering of quantum dots for *in vivo* imaging.*Nanotechnology* 2010. [CrossRef]
34. Uyeda, H.T.; Medintz, I.L.; Jaiswal, J.K.; Simon, S.M.; Mattoussi, H. Synthesis of compact multidentate ligands to prepare stable hydrophilic quantum dot fluorophores. *J. Am. Chem. Soc* 2005, *127*, 3870–3878.
35. Pons, T.; Uyeda, H.T.; Medintz, I.L.; Mattoussi, H. Hydrodynamic dimensions, electrophoretic mobility, and stability of hydrophilic quantum dots. *J. Phys. Chem. B* 2006,*110*, 20308–20316.
36. Smith, A.M.; Nie, S. Minimizing the hydrodynamic size of quantum dots with multifunctional multidentate polymer ligands. *J. Am. Chem. Soc* 2008, *130*, 11278–11279.
37. Kim, S.; Bawendi, M.G. Oligomeric ligands for luminescent and stable nanocrystal quantum dots. *J. Am. Chem. Soc* 2003, *125*, 14652–14653.
38. Stewart, M.H.; Susumu, K.; Mei, B.C.; Medintz, I.L.; Delehanty, J.B.; Blanco-Canosa, J.B.; Dawson, P.E.; Mattoussi, H. Multidentate poly(ethylene glycol) ligands provide colloidal stability to semiconductor and metallic nanocrystals in extreme conditions. *J. Am. Chem. Soc*2010, *132*, 9804–9813.
39. Kong, X.L.; Qi, H.; Zhou, H.X.; Ren, L.L.; Deng, C.Y.; Li, F.R. A novel sensitive immunoassay by nucleic acid barcode dot and its application in the detection of prostate-specific antigen.*Clin. Chem. Lab. Med* 2010, *48*, 279–283.
40. Thangadurai, P.; Balaji, S.; Manoharan, P.T. Surface modification of CdS quantum dots using thiols-structural and photophysical studies. *Nanotechnology* 2008, *19*, 435708:1–435708:8.
41. Yuan, Z.; Zhang, A.; Cao, Y.; Yang, J.; Zhu, Y.; Yang, P. Effect of mercaptocarboxylic acids on luminescent properties of CdTe quantum dots. *J. Fluoresc* 2011. [CrossRef]
42. Petkar, K.C.; Chavhan, S.S.; Agatonovik-Kustrin, S.; Sawant, K.K. Nanostructured materials in drug and gene delivery: A review of the state of the art. *Crit. Rev. Ther. Drug Carr. Syst*2011, *28*, 101–164.
43. Zhao, L.; Pang, X.; Adhikary, R.; Petrich, J.W.; Jeffries-El, M.; Lin, Z. Organic-inorganic nanocomposites by placing conjugated polymers in intimate contact with quantum rods.*Adv. Mater* 2011, *23*, 2844–2849.
44. Song, S.; Qin, Y.; He, Y.; Huang, Q.; Fan, C.; Chen, H.Y. Functional nanoprobes for ultrasensitive detection of biomolecules. *Chem. Soc. Rev* 2010, *39*, 4234–4243.
45. Zorn, M.; Bae, W.K.; Kwak, J.; Lee, H.; Lee, C.; Zentel, R.; Char, K. Quantum dot-block copolymer hybrids with improved properties and their application to quantum dot light-emitting devices. *ACS Nano* 2009, *3*, 1063–1068.
46. Goldman, E.R.; Mattoussi, H.; Anderson, G.P.; Medintz, I.L.; Mauro, J.M. Fluoroimmunoassays using antibody-conjugated quantum dots. *Methods Mol. Biol* 2005, *303*, 19–34.
47. Liu, W.; Howarth, M.; Greytak, A.B.; Zheng, Y.; Nocera, D.G.; Ting, A.Y.; Bawendi, M.G. Compact biocompatible quantum dots functionalized for cellular imaging. *J. Am. Chem. Soc*2008, *130*, 1274–1284.
48. Sheng, W.; Kim, S.; Lee, J.; Kim, S.W.; Jensen, K.; Bawendi, M.G. *In situ* encapsulation of quantum dots into polymer microspheres. *Langmuir* 2006, *22*, 3782–3790.
49. Bolhassani, A. Potential efficacy of cell-penetrating peptides for nucleic acid and drug delivery in cancer. *Biochim. Biophys. Acta* 2011, *1816*, 232–246.
50. Wang, J.; Xia, J. Preferential binding of a novel polyhistidine peptide dendrimer ligand on quantum dots probed by capillary electrophoresis. *Anal. Chem* 2011, *83*, 6323–6329.
51. Li, Y.; Zhou, Y.; Wang, H.Y.; Perrett, S.; Zhao, Y.; Tang, Z.; Nie, G. Chirality of glutathione surface coating affects the cytotoxicity of quantum dots. *Angew. Chem. Int. Ed. Engl* 2011, *50*, 5860–5864.
52. Iyer, G.; Pinaud, F.; Xu, J.; Ebenstein, Y.; Li, J.; Chang, J.; Dahan, M.; Weiss, S. Aromatic aldehyde and hydrazine activated Peptide coated quantum dots for easy bioconjugation and live cell imaging. *Bioconjug. Chem* 2011, *22*, 1006–1011.
53. Ranjbarvaziri, S.; Kiani, S.; Akhlaghi, A.; Vosough, A.; Baharvand, H.; Aghdami, N. Quantum dot labeling using positive charged peptides in human hematopoetic and mesenchymal stem cells. *Biomaterials* 2011, *32*, 5195–5205.
54. Kuo, C.W.; Chueh, D.Y.; Singh, N.; Chien, F.C.; Chen, P. Targeted nuclear delivery using peptide-coated quantum dots. *Bioconjug. Chem* 2011, *22*, 1073–1080.
55. Sapsford, K.E.; Granek, J.; Deschamps, J.R.; Boeneman, K.; Blanco-Canosa, J.B.; Dawson, P.E.; Susumu, K.; Stewart, M.H.; Medintz, I.L. Monitoring botulinum neurotoxin a activity with peptide-functionalized quantum dot resonance energy transfer sensors. *ACS Nano* 2011, *5*, 2687–2699.
56. Liu, H.Y.; Gao, X. Engineering monovalent quantum dot-antibody bioconjugates with a hybrid gel system. *Bioconjug. Chem* 2011, *22*, 510–517.
57. Jin, T.; Tiwari, D.K.; Tanaka, S.; Inouye, Y.; Yoshizawa, K.; Watanabe, T.M. Antibody-protein A conjugated quantum dots for multiplexed imaging of surface receptors in living cells. *Mol. Biosyst* 2010, *6*, 2325–2331.

58. Das, A.; Sanjayan, G.J.; Kecskés, M.; Yoo, L.; Gao, Z.G.; Jacobson, K.A. Nucleoside conjugates of quantum dots for characterization of G protein-coupled receptors: Strategies for immobilizing A2A adenosine receptor agonists. *J. Nanobiotechnol* 2010. [CrossRef]

59. Zhao, Y.; Li, Y.; Song, Y.; Jiang, W.; Wu, Z.; Wang, Y.A.; Sun, J.; Wang, J. Architecture of stable and water-soluble CdSe/ZnS core-shell dendron nanocrystals via ligand exchange. *J. Colloid. Interface Sci* 2009, *339*, 336–343.

60. Hwang, S.H.; Moorefield, C.N.; Wang, P.; Jeong, K.U.; Cheng, S.Z.; Kotta, K.K.; Newkome, G.R. Dendron-tethered and templated CdS quantum dots on single-walled carbon nanotubes. *J. Am. Chem. Soc* 2006, *128*, 7505–7509.

61. Advincula, R.C. Hybrid organic-inorganic nanomaterials based on polythiophene dendronized nanoparticles. *Dalton Trans* 2006, *23*, 2778–2784.

62. Rogach, A.L.; Nagesha, D.; Ostrander, J.W.; Giersig, M.; Kotov, N.A. "Raisin bun"-type composite spheres of silica and semiconductor nanocrystals. *Chem. Mater* 2000, *12*, 2676–2685.

63. Samia, A.C.; Dayal, S.; Burda, C. Quantum dot-based energy transfer: Perspectives and potential for applications in photodynamic therapy. *Photochem. Photobiol* 2006, *82*, 617–625.

64. Fischer, H.; Liu, L.; Pang, K.S.; Chan, W. Pharmacokinetics of nanoscale quantum dots: *In vivo* distribution, sequestration, and clearance in the rat. *Adv. Funct. Mater* 2006, *16*, 1299–1305.

65. Robert, F.S. Nanotechnology takes aim at cancer. *Science* 2005, *310*, 1132–1134.

66. Delehanty, J.B.; Mattoussi, H.; Medintz, I.L. Delivering quantum dots into cells: Strategies, progress and remaining issues. *Anal. Bioanal. Chem* 2009, *393*, 1091–1105.

67. Gao, X.H.; Cui, Y.Y.; Levenson, R.M.; Chung, L.W.; Nie, S.M. *In vivo* cancer targeting and imaging with semiconductor quantum dots. *Nat. Biotechnol* 2004, *22*, 969–976.

68. Liu, J.; Stephen, K.L.; Vijay, A.V. Molecular mapping of tumor heterogeneity on clinical tissue specimens with multiplexed quantum dots. *ACS Nano* 2010, *4*, 2755–2765.

69. Hardman, R. A toxicologic review of quantum dots: Toxicity depends on physicohemical and environmental factors. *Environ. Health Perspect* 2006, *114*, 165–172.

70. Ciofani, G.; Raffa, V.; Menciass, A.; Cuschieri, A. Cytocompatibility, interactions, and uptake of polyethyleneimine-coated boron nitride nanotubes by living cells: Confirmation of their potential for biomedical applications. *Biotechnol. Bioeng* 2008, *101*, 850–858.

71. Derfus, A.M.; Chan, W.C.; Bhatia, S.N. Intracellular delivery of quantum dots for live cell labeling of organelle tracking. *Adv. Mater* 2004, *16*, 961–966.

72. Chen, H.; Wang, L.; Yeh, J.; Wu, X.; Cao, Z.; Wang, Y.A.; Zhang, M.; Yang, L.; Mao, H. Reducing non-specific binding and uptake of nanoparticles and improving cell targeting with an antifouling PEO-b-PgammaMPS copolymer coating. *Biomaterials* 2010, *31*, 5397–5407.

73. Pellegrino, T.; Parak, W.J.; Boudreau, R.; le Gros, M.A.; Gerion, D.; Alivisatos, A.P.; Larabell, C.A. Quantum dot-based cell motility assay. *Differentiation* 2003, *71*, 542–548.

74. Parak, W.J.; Boudreau, R.; le Gros, M.; Gerion, D.; Zanchet, D.; Micheel, C.M.; Williams, S.C.; Alivisatos, A.P.; Larabell, C. Cell motility and metastatic potential studies based on quantum dot imaging of phagokinetic tracks. *Adv. Mater* 2002, *14*, 882–885.

75. Thein, M.; Cheng, A.; Khanna, P.; Zhang, C.; Park, E.J.; Ahmed, D.; Goodrich, C.J.; Asphahani, F.; Wu, F.; Smith, N.B.; Dong, C.; Jiang, X.; Zhang, M.; Xu, J. Site-specific sonoporation of human melanoma cells at the cellular level using high lateral-resolution ultrasonic micro-transducer arrays. *Biosens. Bioelectron* 2011, *27*, 25–33.

76. Susumu, K.; Uyeda, H.T.; Medintz, I.L.; Pons, T.; Delehanty, J.B.; Mattoussi, H. Design of biotin-functionalized luminescent quantum dots. *J. Am. Chem. Soc* 2007, *129*, 13987–13996.

77. Delehanty, J.B.; Medintz, I.L.; Pons, T.; Brunel, F.M.; Dawson, P.E.; Mattoussi, H. Self-assembled quantum dot-peptide bioconjugates for selective intracellular delivery. *Bioconjug. Chem* 2006, *17*, 920–927.

78. Rozenzhak, S.M.; Kadakia, M.P.; Caserta, T.M.; Westbrook, T.R.; Stone, M.O.; Naik, R.R. Cellular internalization and targeting of semiconductor quantum dots. *Chem. Commun* 2005,*17*, 2217–2219.

79. Lieleg, O.; Lopez-Garcia, M.; Semmrich, C.; Auernheimer, J.; Kessler, H.; Bausch, A.R. Specific integrin labeling in living cells using functionalized nanocrystals. *Small* 2007, *3*, 1560–1565.

80. Smith, B.R.; Cheng, Z.; De, A.; Koh, A.L.; Sinclair, R.; Gambhir, S.S. Real-time intravital imaging of RGD-quantum dot binding to luminal endothelium in mouse tumor neovasculature. *Nano Lett* 2008, *8*, 2599–2606.

81. Pan, Y.L.; Cai, J.Y.; Qin, L.; Wang, H. Atomic force microscopy-based cell nanostructure for ligand-conjugated quantum dot endocytosis. *Acta Biochim. Biophys. Sin* 2006, *38*, 646–652.

82. Yong, K.T.; Qian, J.; Roy, I.; Lee, H.H.; Bergey, E.J.; Tramposch, K.M.; He, S.; Swihart, M.T.; Maitra, A.; Prasad, P.N. Quantum rod bioconjugates as targeted probes for confocal and two-photon fluorescence imaging of cancer cells. *Nano Lett* 2007, *7*, 761–765.

83. Zhang, H.; Sachdev, D.; Wang, C.; Hubel, A.; Gaillard-Kelly, M.; Yee, D. Detection and downregulation of type I IGF receptor expression by antibody-conjugated quantum dots in breast cancer cells. *Breast Cancer Res. Treat* 2009, *114*, 277–285.

84. Lidke, D.S.; Nagy, P.; Heintzmann, R.; Arndt-Jovin, D.J.; Post, J.N.; Grecco, H.E.; Jares-Erijman, E.A.; Jovin, T.M. Quantum dot ligands provide new insights into erbB/HER receptor-mediated signal transduction. *Nat. Biotechnol* 2004, *22*, 198–203.

85. Liu, W.; Howarth, M.; Greytak, A.B.; Zheng, Y.; Nocera, D.G.; Ting, A.Y.; Bawendi, M.G. Compact biocompatible

quantum dots functionalized for cellular imaging. *J. Am. Chem. Soc*2008, *130*, 1274–1284.

86. Diagaradjane, P.; Orenstein-Cardona, J.M.; Colón-Casasnovas, N.E.; Deorukhkar, A.; Shentu, S.; Kuno, N.; Schwartz, D.L.; Gelovani, J.G.; Krishnan, S. Imaging epidermal growth factor receptor expression *in vivo*: Pharmacokinetic and biodistribution characterization of a bioconjugated quantum dot nanoprobe. *Clin. Cancer Res* 2008, *14*, 731–741.

87. Dudu, V.; Rotari, V.; Vazquez, M. Targeted extracellular nanoparticles enable intracellular detection of activated epidermal growth factor receptor in living brain cancer cells.*Nanomedicine* 2011, *7*, 896–903.

88. Chakraborty, S.K.; Fitzpatrick, J.A.; Phillippi, J.A.; Andreko, S.; Waggoner, A.S.; Bruchez, M.P.; Ballou, B. Cholera toxin B conjugated quantum dots for live cell labeling. *Nano Lett*2007, *7*, 2618–2626.

89. Jaiswal, J.K.; Goldman, E.R.; Mattoussi, H.; Simon, S.M. Use of quantum dots for live cell imaging. *Nat. Methods* 2004, *1*, 73–78.

90. Rajan, S.S.; Liu, H.Y.; Vu, T.Q. Ligand-bound quantum dot probes for studying the molecular scale dynamics of receptor endocytic trafficking in live cells. *ACS Nano* 2008, *2*, 1153–1166.

91. Rajan, S.S.; Vu, T.Q. Quantum dots monitor TrkA receptor dynamics in the interior of neural PC12 cells. *Nano Lett* 2006, *6*, 2049–2059.

92. Schroeder, J.E.; Shweky, I.; Shmeeda, H.; Banin, U.; Gabizon, A. Folate-mediated tumor cell uptake of quantum dots entrapped in lipid nanoparticles. *J. Control. Release* 2007, *124*, 28–34.

93. Dudu, V.; Ramcharan, M.; Gilchrist, M.L.; Holland, E.C.; Vazquez, M. Liposome delivery of quantum dots to the cytosol of live cells. *J. Nanosci. Nanotechnol* 2008, *8*, 2293–2300.

94. Voura, E.B.; Jaiswal, J.K.; Mattoussi, H.; Simon, S.M. Tracking metastatic tumor cell extravasation with quantum dot nanocrystals and fluorescence emission-scanning microscopy. *Nat. Med* 2004, *10*, 993–998.

95. Duan, H.; Nie, S. Cell-penetrating quantum dots based on multivalent and endosome-disrupting surface coatings. *J. Am. Chem. Soc* 2007, *129*, 3333–3338.

96. Fuente, J.M.; Fandel, M.; Berry, C.C.; Riehle, M.; Cronin, L.; Aitchison, G.; Curtis, A.S. Quantum dots protected with tiopronin: A new fluorescence system for cell-biology studies.*ChemBioChem* 2005, *6*, 989–991.

97. Farias, PM.; Santos, B.S.; Menezes, F.D. Core-shell CdS/$Cd(OH)_2$ quantum dots: Synthesis and bioconjugation to target red cells antigens. *J. Microsc* 2005, *219*, 103–108.

98. Coulon, J.; Thouvenin, I.; Aldeek, F.; Balan, L.; Schneider, R. Glycosylated quantum dots for the selective labelling of Kluyveromyces bulgaricus and Saccharomyces cerevisiae yeast strains. *J. Fluoresc* 2010, *20*, 591–597.

99. Bharali, D.J.; Lucey, D.W.; Jayakumar, H.; Pudavar, H.E.; Prasad, P.N. Folate-receptor-mediated delivery of InP quantum dots for bioimaging using confocal and two-photon microscopy. *J. Am. Chem. Soc* 2005, *127*, 11364–11371.

100. Clarke, S.; Nadeau, J.; Bahcheli, D.; Zhang, Z.; Hollmann, C. Quantum dots as phototoxic drugs and sensors of specific metabolic processes in living cells. Proceedings of the 27th Annual International Conference of the Engineering in Medicine and Biology Society, Shanghai, China, 17–18 January 2005; Volume 1. pp. 504–507.

101. Xue, X.; Pan, J.; Xie, H.; Wang, J.; Zhang, S. Fluorescence detection of total count of*Escherichia coli* and *Staphylococcus aureus* on water-soluble CdSe quantum dots coupled with bacteria. *Talanta* 2009, *77*, 1808–1813.

102. Clarke, S.J.; Hollmann, C.A.; Zhang, Z.J.; Suffern, D.; Bradforth, S.E.; Dimitrijevic, N.M.; Minarik, W.G.; Nadeau, J.L. Photophysics of dopamine-modified quantum dots and effects on biological systems. *Nat. Mater* 2006, *5*, 409–417.

103. Dubertret, B.; Skourides, P.; Norris, D.J.; Noireaux, V.; Brivanlou, A.H.; Libchaber, A. *In vivo*imaging of quantum dots encapsulated in phospholipid micelles. *Science* 2002, *298*, 1759–1762.

104. Medintz, I.L.; Pons, T.; Delehanty, J.B.; Susumu, K.; Brunel, F.M.; Dawson, P.E.; Mattoussi, H. Intracellular delivery of quantum dot-protein cargos mediated by cell penetrating peptides.*Bioconjug. Chem* 2008, *19*, 1785–1795.

105. Cai, W.B.; Shin, D.W.; Chen, K.; Gheysens, O.; Cao, Q.Z.; Wang, S.X. Peptide-labeled nearinfrared quantum dots for imaging tumor vasculature in living subjects. *Nano Lett* 2006,*6*, 669–676.

106. Yu, X.; Chen, L.; Li, K.; Li, Y.; Xiao, S.; Luo, X. Immunofluorescence detection with quantum dot bioconjugates for hepatoma *in vivo*. *J. Biomed. Opt* 2007, *12*, 1–5.

107. Weng, K.C.; Noble, C.O.; Papahadjopoulos-Sterberg, B.; Chen, F.F.; Drummond, D.C.; Kirpotin, D.B. Targeted tumor cell internalization and imaging of multifunctional quantum dot-conjugated immunoliposomes *in vitro and in vivo*. *Nano Lett* 2008, *8*, 2851–2857.

108. Janib, S.M.; Moses, A.S.; MacKay, J.A. Imaging and drug delivery using theranostic nanoparticles. *Adv. Drug Deliv. Rev* 2010, *62*, 1052–1063.

109. Matsuno, A.; Mizutani, A.; Okinaga, H.; Takano, K.; Yamada, S.; Yamada, S.M.; Nakaguchi, H.; Hoya, K.; Murakami, M.; Takeuchi, M.; Sugaya, M.; Itoh, J.; Takekoshi, S.; Osamura, R.Y. Molecular morphology of pituitary cells, from conventional immunohistochemistry to fluorescein imaging. *Molecules* 2011, *16*, 3618–3635.

110. Byers, R.J.; Hitchman, E.R. Quantum dots brighten biological imaging. *Prog. Histochem. Cytochem* 2011, *45*, 201–237.

111. Isse, K.; Grama, K.; Abbott, I.M.; Lesniak, A.; Lunz, J.G.; Lee, W.M.; Specht, S.; Corbitt, N.; Mizuguchi, Y.; Roysam, B.; Demetris, A.J. Adding value to liver (and allograft) biopsy evaluation using a combination of multiplex quantum dot immunostaining, high-resolution

whole-slide digital imaging, and automated image analysis. *Clin. Liver Dis* 2010, *14*, 669–685.

112. Tholouli, E.; Sweeney, E.; Barrow, E.; Clay, V.; Hoyland, J.A.; Byers, R.J. Quantum dots light up pathology. *J. Pathol* 2008, *216*, 275–285.

113. Smith, A.M.; Nie, S. Chemical analysis and cellular imaging with quantum dots. *Analyst*2004, *129*, 672–677.

114. Yaghini, E.; Seifalian, A.M.; MacRobert, A.J. Quantum dots and their potential biomedical applications in photosensitization for photodynamic therapy. *Nanomedicine* 2009, *4*, 353–363.

115. Rakovich, A.; Savateeva, D.; Rakovich, T.; Donegan, J.F.; Rakovich, Y.P.; Kelly, V.; Lesnyak, V.; Eychmüller, A. CdTe quantum dot/dye hybrid system as photosensitizer for photodynamic therapy. *Nanoscale Res. Lett* 2010, *5*, 753–760.

CHAPTER 5

QUANTUM DOTS: SYNTHESIS, BIOAPPLICATIONS, AND TOXICITY

Alireza Valizadeh, Haleh Mikaeili, Mohammad Samiei, Samad Mussa Farkhani, Nosratalah Zarghami, Mohammad kouhi, Abolfazl Akbarzadeh and Soodabeh Davaran

Keywords: QD delivery systems, Toxicity, Emission spectra, Luminescence characteristics

INTRODUCTION

In the past years, a new class of fluorescent particles emerged as a good candidate for single molecule and single particle tracking (SPT) in living cells and organisms, the semiconductor quantum dots [1]. Quantum dots (QDs), often described as 'artificial atoms,' exhibit discrete energy levels, and their bandgap can be precisely modulated by varying the size [2]. QDs are nanometer-scale semiconductor crystals composed of groups II to VI or III to V elements and are defined as particles with physical dimensions smaller than the exciton Bohr radius [3]. QDs exhibit unique luminescence characteristics and electronic properties such as wide and continuous absorption spectra, narrow emission spectra, and high light stability [4]. They absorb white light and then re-emit a specific color a few nanoseconds later depending on the bandgap of the material [5–7]. QDs are one of the first nanotechnologies to be integrated with the biological sciences [4, 8] and are widely anticipated to eventually find application in a number of commercial consumer and clinical products [9]. For example, CdSe/ZnS quantum dots are presently the most common commercially available product as secondary antibody conjugates that are composed of a core of cadmium selenide ranging from about 10 to 50 atoms in diameter and about 100 to 100,000 atoms in total [10]. QD range is typically between 2 and 10 nm in diameter. QDs consist of a semiconductor core, overcoated by a shell (e.g., ZnS) to improve optical properties, and a cap enabling improved solubility in aqueous buffers [11]. The application of QDs, as a new technology for biosystems, has been mostly studied on mammalian cells. There is an increasing tendency to apply QDs as markers in plant science [12–16]. The application of QDs as markers of the cells or their cell walls for plant bioimaging would be advantageous because of their small size, brightness, independence of emission on the excitation wavelength, and stability under relatively harsh environments. They also have excellent photostability [17] and overcome the limitations associated with photobleaching. Due to the small structures of QDs, some physical properties such as optical and electron transport characteristics are quite different from those of the bulk materials [18]. The study of the impurity states in these low dimensional structures is an important aspect to which many theoretical and experimental works based [16, 19–21]. This review introduces QDs and explores their properties, synthesis, applications, delivery systems in biology, and their toxicity.

SYNTHESIS

Several routes have been used to synthesize QDs [22] but, generally, techniques for QD synthesis used top-down processing methods and bottom-up approach. Top-down processing methods include molecular beam epitaxy (MBE), ion implantation, e-beam lithography, and X-ray lithography. Using the alternative bottom-up approach, colloidal QDs are prepared by self-assemblyin the solution following a chemical reduction [23–26].

In the approaches of top-down, for making the QDs, a bulk semiconductor is thinned. For the achieve QDs of diameter approximately 30 nm, electron beam lithography, reactive-ion etching, and/or wet chemical etching are commonly used. For systematic experiments on quantum confinement effect, controlled shapes and sizes are achievable with the desired

Citation: Alireza Valizadeh, Haleh Mikaeili, Mohammad Samiei, Samad Mussa Farkhani, Nosratalah Zarghami, Mohammad kouhi, Abolfazl Akbarzadeh and Soodabeh Davaran, Quantum dots: synthesis, bioapplications, and toxicity, doi:10.1186/1556-276X-7-480.

packing geometries. Alternatively, focused ion or laser beams have also been used to fabricate arrays of zero-dimension dots. Incorporation of impurities into the QDs and structural imperfections by patterning are major disadvantages with these processes [22].

A number of different self-assembly techniques (bottom-up) have been used to synthesize the QDs, and they may be broadly subdivided into wet-chemical and vapor-phase methods [22]: (a) wet-chemical methods mainly follow the conventional precipitation methods with careful control of parameters for a single solution or mixture of solutions. The precipitation process invariably involves both nucleation and limited growth of nanoparticles. Nucleation may be categorized as homogeneous, heterogeneous, or secondary nucleation [27]. Homogeneous nucleation occurs when solute atoms or molecules combine and reach a critical size without the assistance of a pre-existing solid interface. Wet-chemical methods are generally microemulsion, sol–gel [28–30], competitive reaction chemistry, hot-solution decomposition [31–33], sonic waves or microwaves [34], and electrochemistry. (b) Vapor-phase methods for producing QDs begin with processes in which layers are grown in an atom-by-atom process. Consequently, self-assembly of QDs occurs on a substrate without any patterning [35–38]. Self-assembly of nanostructures in material grown by MBE, sputtering, liquid metal ion sources, or aggregation of gaseous monomers are generally categorized under vapor-phase methods [22]. MBE has been mainly used to self-assemble QDs from III-V semiconductors and II-VI semiconductors using the large lattice mismatch, e.g., InAs on GaAs has a 7% mismatch and leads to SK growth [35].

APPLICATIONS

In this review, we evaluate few experiments that show the high potential of QDs in biological application, including tracking different macromolecules in the cell, tracking various cells in the tissue, labeling organelles and cells, clinical applications, and other applications [39–43].

QDs for Labeling Cells

Because QDs have constant and unique optical properties, they are the best candidate for cell labeling, as compared with organic dyes.

Use in Plant Bioimaging: There is an increasing application of QD as markers for the cells or cell walls (CWs) in plant science. A first target location for external agents in a plant cell is the CW [44]. Djikanović et al. demonstrated that CdSe QDs bind typically to cellulose and lignin in the cell wall of *Picea omorika* branch. Respectively, binding to lignin and cellulose are achieved by interaction with the chains of C=C and C-C alternating bonds and interaction with the OH groups [44]. Data showed that QDs are suitable for homogenous marking of the whole cell wall. This is a consequence of the structural arrangement of the cell wall polymers in the whole cell wall network as well as the extremely small size of the QDs. These characteristics enable a feasible penetration of the nanoparticles inside the polymer structures in the CW composite [44].

Use in Animal Bioimaging: Goldman et al. used biotinylated CTxB in conjunction with QD-avidin conjugates [45] for labeling of the live HeLa cells which Figure 1 shows an image of the lateral membrane staining for GM1 ganglioside using QDs (in red) and nuclear staining using Hoechst (in blue). Punctuate labeling of the cell surface by QD bioconjugate is typical for molecules such as GM1 that is present in membrane rafts [46].

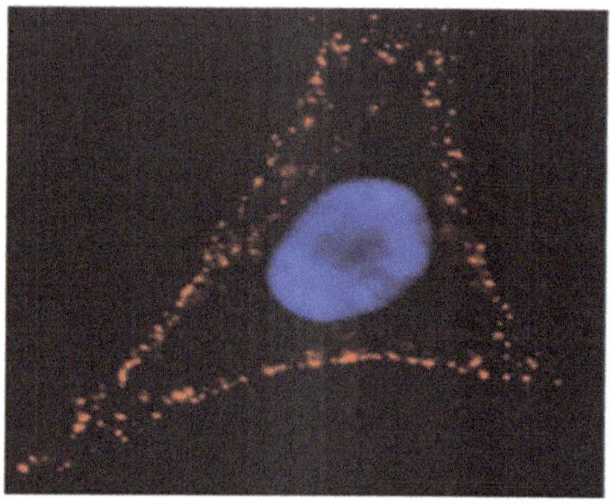

Figure 1 Live HeLa cells growing on a glass coverslip. Labeled with QD-avidin for GM1 (in red) and Hoechst 3342 for nuclear staining (in blue) [46].

In another study, they labeled live HeLa cells which were biotinylated using sulfo-NHS-SS biotinylating reagent and then incubated with the avidin-conjugated yellow-emitting QDs. It is shown in Figure 2[47].

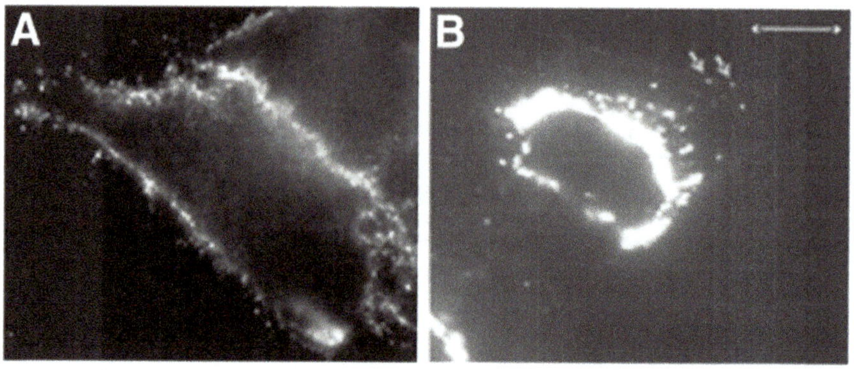

Figure 2 HeLa cells labeled with the avidin-conjugated yellow-emitting QDs.[47]. (A) Image of cells immediately after the unbound QDs were removed in which labeling is restricted to the cell surface. (B) Image of a cell that was allowed to grow for 2 h after washing out of unbound QDs.

For long-term live cell imaging, Hasegawa et al. used the $CHPNH_2$-QD complexes which were uniformly internalized into the cells without being aggregated. Therefore, $CHPNH_2$ nanogel has high potential for use in long-term live cell imaging. The interaction of QDs with cells was successfully controlled by the amino group content of the $CHPNH_2$ nanogel [48].

Use in prokaryote bioimaging Sensitive and selective staining of bacterial mutants using QD labels was demonstrated by Smith's group. This principle of detection is based on selective targeting affinity of Zn(II)-dipicolylamine coordination complex to phospholipids on the bacterial cell surface of specific strain as shown in Figure 3[49, 50].

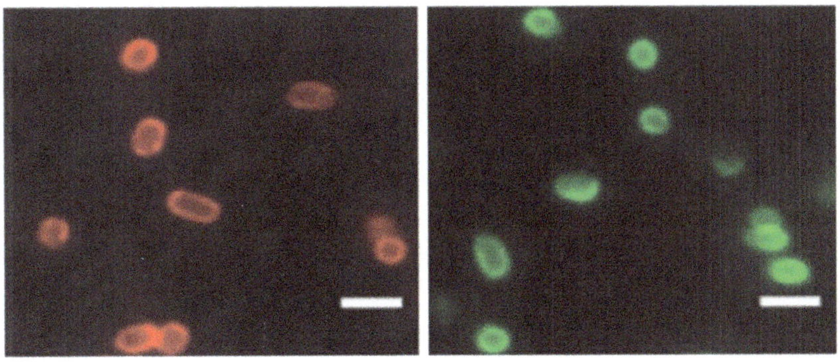

Figure 3 Imaging of rough *Escherichia coli* JM83 cells. Left imaging is red QD, and right imaging is green QD. Scale bar is 2 μM [49].

In another study, authors demonstrated the use of magnetic beads coated with anti-*E.coli O157*antibodies and streptavidin-coated QDs for measuring the bacterial cell concentration [51]. Yang and Li, using QDs with different emission wavelengths (525 nm and 705 nm), reported the simultaneous detection of *E. coli O157:H7* and *Salmonella typhimurium* [52].

Tracking Different Particles

With the application of new imaging methods and the use of brighter and more stable probes, such as QDs, single particle tracking has the potential to enter into a new era of high resolution and long timescale imaging [53–55]. SPT techniques allow scientists to follow single molecules in real time and visualize the actual molecular dynamics in their habitant environment.

For Extracellular Study: Because QDs do not require intracellular delivery through the impermeable plasma membrane, membrane receptors or membrane-associated proteins are intuitive targets for QD imaging [53]. Howarth et al. demonstrated a method to track endogenous cell-surface proteins without cross-linking by purifying monovalent antibody-QD conjugates. They approach to make monovalent tight-binding QDs, using mSA, which could be applied to other nanoparticles that show sufficient electrophoretic mobility. They applied sQD-mSA1 to study the mobility of a mutant of low-density lipoprotein (LDL) receptor with a truncated cytosolic tail, originally found from an individual with familial hypercholesterolemia. This mutant phenotype has been extensively investigated by following LDL, but Howarth and co-workers analyzed the behavior of the receptor itself (supplementary methods). They imaged single monovalent sQDs bound to the biotinylated AP-LDL receptor, as indicated by QD fluorescence intensity and blinking. The mobility of mutant receptors labeled with sQD-mSA1 was significantly greater than that of labeled wild-type LDL receptor ($P = 1.6 \times 10^{-14}$) [56].

In similar studies, recently, QDs used to target membrane proteins and investigate the mobility and entry-exit kinetics in several systems: (1) various transmembrane proteins, for example, integrins [57], channels [58], and aquaporines [59]; (2) receptors GABA [60], glycine [61], interferon [62], and HER [63, 64]; and (3) neurological synapse [65, 66].

For Intracellular Study: In one of the study, the advantages of the broad, continuous excitation spectrum were demonstrated in a dual-emission, single-excitation labeling experiment on mouse fibroblasts. These nanocrystal probes are, thus, complementary and, in some cases, may be superior to existing fluorophores [4]. Nonspecific labeling of the nucleus by both the red and the green probes resulted in a yellow color. The red actin filaments were specifically stained. Also, the green probes penetrate into the nucleus. Both are shown in Figure 4[4].

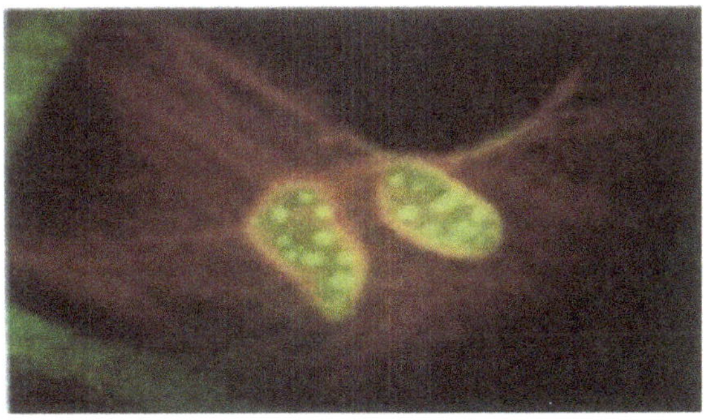

Figure 4 Image was obtained with 363-nm excitation and × 40 oil 1.3 numerical aperture objective.

This is shown as green color for nucleus and red color for actin filaments. Nonspecific labeling of the nucleus by both the red and the green probes resulted in a yellow color [4].

Superior stability of QD fluorophores gives the possibility to improve quantitation of FISH analysis of human chromosomal changes. Xiao and Barker have investigated coated (CdSe) ZnS QDs as fluorescence labels for FISH of biotinylated DNA to human lymphocyte metaphase chromosomes under conditions that approximate those commonly found in clinical cytogenetics laboratories [67]. They have also demonstrated the application of QDs to FISH detection of the clinically relevant HER2 locus in breast cancer cells (Figure 5).

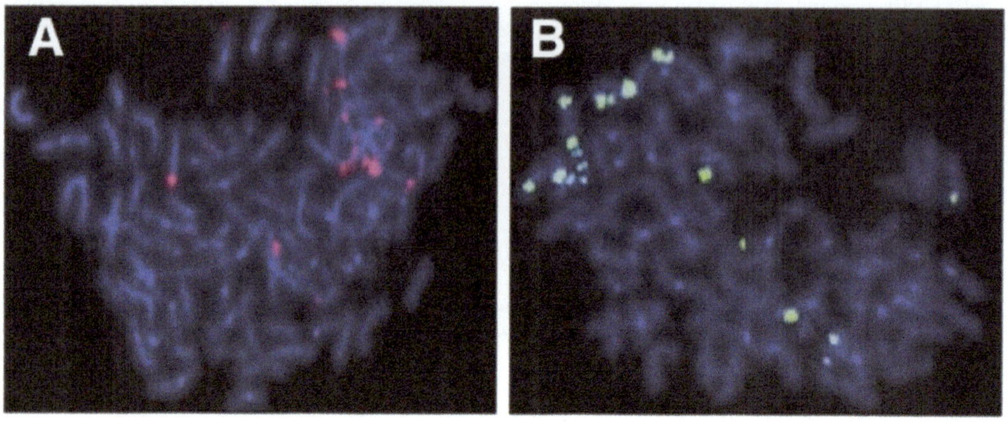

Figure 5 Qualitative FISH detection of HER2 gene-amplified SK-BR-3 breast cancer cells. With (A) streptavidin-conjugated Qdot605 and (B) FITC, respectively [67].

Pierobon et al. [68] and Nelson et al. [69] tagged myosin V molecules with QDS toestablish a link between *in vitro* and in-cell measurements of myosin V motors. Then, the complex myosin V/QD (MyoV::QD), using the pinocytic influx, was introduced into the cells.

Yoo et al. [70] and Courty et al. [71] characterized the dynamics of other major actors of intracellular transport: the kinesin-1, the actin filaments, and the microtubules [65].

Imaging in situ

Imaging of the satellite cells in rat intact and injured soleus muscles using quantum dots The employment of satellite cells, which are located between the basement membrane and the plasma membrane in myofibers, is required for myofiber repair after muscle injury or disease. Using QDs conjugated to anti-M-cadherin antibody, Ishido and Kasuga attempted the visualization of satellite cells in both intact and injured skeletal muscle of rat *in situ*. They demonstrated *in situ* real-time imaging of satellite cells localized within the skeletal muscle (Figure 6) [72].

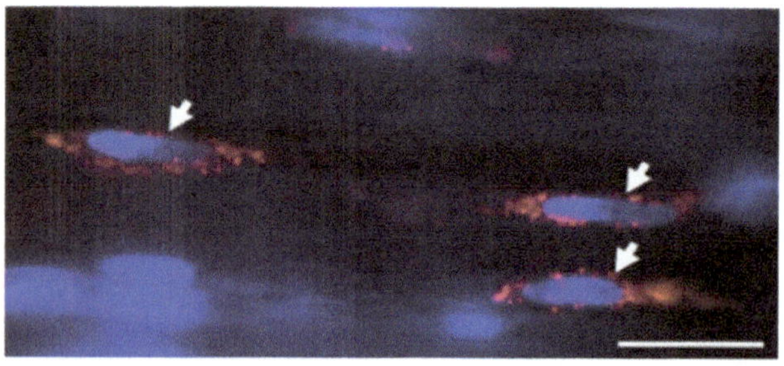

Figure 6 Double fluorescence staining to visualize the localization of M-cadherin (in red) and nuclei (in blue). Arrows indicate that M-cadherin-positive satellite cells were located within the intact soleus muscle *in situ* [72].

Imaging Morphogenesis in Xenopus with Quantum Dot Nanocrystals: Stylianou and Skourides are the first to report the use of near-infrared QDs to image mesoderm migration *in vivo* with single cell resolution and provide quantitative *in vivo* data regarding migration rates [73].

Navarro et al. experiments revealed that *Arabidopsis* exposed to QDs that are dispersed in Hoagland's solution for 1 to 7 days did not internalize intact QDs. Fluorescence microscopy showed strong evidence that the QDs were generally on the outside surfaces of the roots (Figure 7). The amount of QDs adsorbed is dependent on the stability of the QDs in suspension [74].

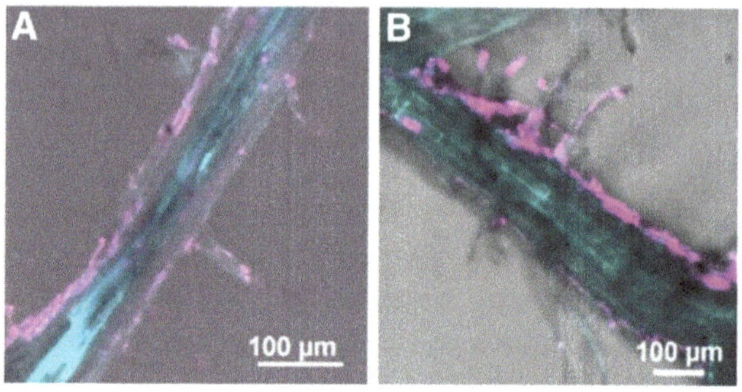

Figure 7 Imaging of roots from plants exposed to QD suspension in HS + HA. For (A) 1 and (B) 7 days [74].HS, Hoagland's solution; HA, one of the important groups of organic acids.

Using QDs in Clinical Applications

The development of multifunctional nanomaterials combining diagnostic and therapeutic purpose has recently attracted intensive interests [75–81]. In this paper, we have reviewed the clinical applications of QDs in the three categories that include: (1) biomarker detection in various cancers, (2) imaging and sensing of infectious diseases, and (3) other clinical therapeutic applications.

Biomarker Detection in Various Cancers Using QDs: The detection of cancer biomarkers is important for diagnosis, disease stage forecasting, and clinical management [82]. QDs with intense and stable fluorescent properties could enable the detection of tens to hundreds of cancer biomarkers in blood assays, on cancer tissue biopsies, or as contrast agents for medical imaging. Clinical outcome of cancer diagnosis is highly dependent on the stage at which the malignancy is detected, and therefore, early screening has become extremely important in any type of cancer [83].

- Multicolor and multiplexing potentialities of QDs are used for the detection of four protein biomarkers CD15, CD30, CD45, and Pax5 of Hodgkin's lymphoma from lymphoma tissues. Simultaneous visualization using multiplexed QD staining was advantageous for the selective identification of rare Hodgkin (Reed-Sternberg) cells, a primary diagnostic target for Hodgkin's disease, which was not achievable using traditional immunohistochemistry assays [84, 85].

- Yu et al. reported the use of GSH-TGA-QDs-ND-1 probes to label colorectal cancer cells CCL187. They prepared QDs, which were conjugated with monoclonal antibody ND-1 for specific reaction with antigen LEA [86].
- In the United States, pancreatic cancer is the fourth leading cause of cancer death (about 18,770 men and 18,030 women (36,800 people) in 2010) [87]. Using semiconductor QD-antibody conjugates, Lee et al. demonstrated quantitative profiling of biomarkers for pancreatic cancer at the single-cell level. Their results show the possibility of this method for staging and forecasting, such as prostate stem cell antigen claudin-4, and mesothelin, which are expressed in different stages of progression of pancreatic cancer [82]. Anyway, realizing quantitative profiling requires stable quantum yield, monodisperse QD-Ab conjugates, and well-defined surface chemistry [88].

There are evidences showing the application of QDs in micro- and nanoarrays for the detection of cancer biomarkers [83].

Imaging and Sensing of Infectious Diseases by QDs: QDs have become one of the most hopeful and interesting materials for diagnostic applications of bioimaging, labeling, and sensing for infectious diseases such as respiratory syncytial virus (RSV)that isone of the families of Paramyxoviridae [50]. In Table 1, some of the infectious diseases and QDs used to distinguish them are shown.

Table 1 Some of the infectious diseases and QDs used to distinguish them

Authors	**Type of infectious diseases**	***In vitro / in vivo***	**Type of modified QDs**
Tripp et al. [89]	RSV	*In vitro/ in vivo*	Antibody anti-F protein conjugated to QDs(CdTe)
Agrawal et al. [90]	Individual molecules of genes, proteins, and virus particles	*In vivo*	QD-antibody color-coded NP probes and two-color co-localization imaging
Bentzen et al. [91]	RSV	*In vivo*	Streptavidin-coated QDs conjugated to antibody anti-F and antibody anti-G
Dwarakanath et al. [92]	*S. typhimurium*	*In vivo*	Antibody-QD and DNA aptamer-QD
	E. coli		
	B. subtilis spores		
Goldman et al. [93]	Choleratoxin, ricin, shinga-like toxin1 and staphylococcal enterotoxin B	*In vitro*	Antibody-QD (CdSe/Zns)
Zhao et al. [94]	Food-borne pathogenic *E. coli* O157:H7,*S. typhimurium* and *S. flexneri*		QDanti-*S. flexneri*antibody,anti-*E. coli*antibody, anti-*S. typhimurium* antibody
Hahn et al. [95]	Single cells of *E.coli* O157:H7		Streptavidin-coated QDs conjugated to antibody
Mukhopadhyay et al. [96]	Detect *E. coli* at levels as low as 10^4bacteria/ml of sample		Mannose-conjugated QDs
Edgar et al. [97]	*Mycobacterium*	*In vivo*	Streptavidin-coated QDs conjugated to phage
	B. anthracis		
Zhu et al. [98]	*C. parvum*		QD-conjugated antibodies
	G. lamblia		
Klostranec et al. [99]	Biomarkers of the most globally prevalent blood-borne infectious diseases (i.e., hepatitis B, hepatitis C, and HIV) with low sample volume		QD-antibody
Gazouli et al. [100]	*Mycobacterium* genus	*In vivo*	Specific DNA sequences combining QDs with magnetic beads [101, 102]

Hahn et al. [95]	Individual pathogenic *E. coli O157:H7* in phosphate buffer saline solution		Streptavidin-coated Qdots labeled by antibody selectively targeted pathogenic *E. coli* O157:H7
Su and Li [51]	*E. coli O157*	*In vivo*	Streptavidin-coated QDs conjugated to anti-*E. coli* O157 antibody
Yang and Li [52]	*E. coli* O157:H7	*In vivo*	QDs with different sizes conjugated to anti-*E. coli* O157 and anti-*Salmonella*antibodies
	S. typhimurium		
			The bead-cell complexes reacted with QD-antibody conjugates to form bead-cell-QD complexes

Other clinical therapeutic applications.

- *QDs for assessing axon growth*: A major health problem with injuries to the spinal cord and brain is traumatic central nervous system injury reporting of approximately 265,000 and 1.5 million new injuries each year [103–105]. QDs represent a new device of significant potential in neuroscience research, and they are useful for experiments that are limited by the restricted anatomy of neuronal and glial interactions [106]. One of the problems in treatment is estimating its effectiveness. They allow the ability to visualize and track dynamic molecular processes over long times (Figure 8) [106]. Application of surface-engineered QDs is an area of nanotechnology probing the details of cellular and molecular processes in neuronal cells [4, 107–109]. QD bioconjugates based on surface chemistry can be broadly classified as follows: (1) QDs' surface modified by bioactive molecules and (2) QD-polymer nanocomposites [103]. This advance might be significantly important to assess axon growth pending the regeneration process [103]. Previous investigations were demonstrated in Table 2.
- *QD used as a probe in an anti-malarial drug-screening assay*: Malaria is a major global health problem, threatening over 300 million people and causing nearly one million deaths annually [114, 115]. Tokumasu et al. used QD-Ab to demonstrate the distinct pattern of distribution of protein and to observe erythrocyte membrane deformation occurring duringthe invasion of erythrocytes by *Plasmodium falciparum* [116]. Ku et al. showed a simple and efficient method to label *P. falciparum*-infected RBC using a QD-based probe and its applicability as an efficient probe for anti-malarial drug screening [115].

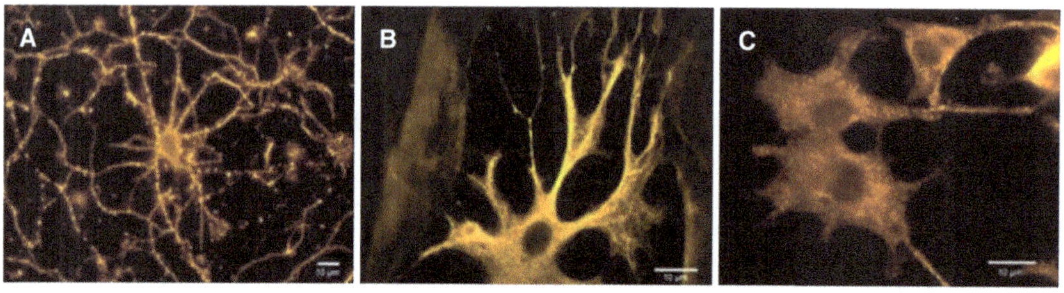

Figure 8 Using QD conjugated with antibody for labeling of neurons and glia. (A) Labeled β-tubulin in primary cortical neurons. (B) Labeled glial fibrillary acidic protein in primary cortical astrocytes. (C) Labeled for β-tubulin in PC12 cells [106].

Table 2 Applications of QDs in labeling neurons and glia cells

Authors	Type of QD used	Application of QD
Dahan et al. [61]	QD-GlyR	Target neurons to investigate a specific neurophysiological process(QDs to track individual glycine receptors and analyze their lateral dynamics in the neuronal membrane)
Pathak et al. [106]	Antibody-conjugated quantum dots	Performed the specific labeling of neurons and glia cells
Vu et al. [110]	Tagged nerve growth factor (βNGF) to QDs	Investigate the QD nanostructure's potential to assess the neurite outgrowth
Sundara Rajan et al. [111]	QD-anti-TrkA-TrkA receptor with transport by GFP	Immobilized QDs were conjugated with NGF, activate Trk receptors, and initiate neuronal differentiation in PC12 cells.

Howarth et al. [112]	Tagged cell surface proteins with a specific peptide (acceptor protein) that can be directly biotinylated as a target for streptavidin-conjugated quantum dots	Specifically label and track AMPA receptors on cultured hippocampal neurons
Prasad et al. [113]	Thioglycolic acid (TGA)-stabilized CdTe QDs	Performed imaging of PC12 cells

Other Applications

QDs as pH probes for the study of enzyme reaction kinetics: [117] Lately, worth advancement has been achieved in water-soluble QDs as ionic probe. Jin et al. reported the use of modified CdSe QDs for the sensitive determination of cyanide ions $[C{\equiv}N]^-$[117, 118]. Xie et al. reported the determination of Cu^{2+} by using CdSe/ZnS QDs modified with bovine serum albumin [119]. QDs also have been reported to be sensitive to pH [120–125]. The sensitivity of QDs' photoluminescence to pH, improve stability, and a monitoring range for the determination of proton concentration, which is maybe due to a function of surface modifications and effects on exciton trap sites, leads to applications utilizing QDs as pH probes [126]. Water-soluble QDs, ZnS, modified with mercaptoacetic acid (MAA) were sensitive to environmental factors and found to be a satisfactory pH probes that could have potential applications in chemical and biochemical sensing. Using the modified QD surface, they were applied as pH probes in monitoring the hydrolysis of glycidyl butyrate which is catalyzed by porcine pancreatic lipase (PPL) [117].

QDs use for protein micro- and nanoarrays to the detection of cancer biomarkers: Protein microarrays are useful device as highthroughput screening tools in proteomics [127–129], for biosensing purpose [130], new drug discovery [131], and enabling a quick parallel screening method for the detection of protein-protein interactions in case of large protein populations. There are various reports in which QDs have been used in microarray fabrication such as sandwich-based immunoassay type, RP protein microarray type, etc. [132–135]. Here, IgG detection was done on a glass chip using a QD-labeled secondary Abs as sandwich assay approach. In RP protein microarrays, Geho et al. used pegylated QDs conjugated with streptavidin as detection elements. In another study, Zajac et al. investigated the ability of the platform to detect different cytokines TNF-α, IL-8, IL-6, MIP-1β, IL-13, and IL-1β using two different models of quantum dot probes. Their results demonstrated high sensitivity of the investigated detection system with less than picomolar concentration [136]. Kerman et al. reported the use of QDs for detection cell lysates spiked with DNA-PK proteins with the help of mAb, in an RP protein microarray format. Kerman et al. make immunosensor based on QD for the detection of prostate specific antigen (PSA) in a sandwich assay approach for chip fabrication [134]. Gokarna et al. used pegylated QD-conjugated PSA Abs to demonstrate the fabrication of a cancer protein biochip for the detection of PSA, which is a biomarker for prostate cancer. The QD nonspecificity can show to be quite detrimental to some extent in case of multiplexed assay systems where multiple proteins are to be detected simultaneously [83].

QD delivery Due to the unique properties of QDs, they are best tools for intracellular studies such as visualizing the cellular structure, studying the dynamic cellular processes, and tracking single molecules in the cell [137, 138]. To achieve this goal, translocation of functionalized QDs into the cell for labeling organelles and tracking single molecules is important. QDs have hydrophobic surface and have a little toxicity, therefore cannot be applied *in vivo* unless their surface is modified. Thus, by surface modification, their hydrophilicity will increase but their toxicity will decrease.

Hasegawa et al. used nanogel-QD hybrid nanoparticles for live cell imaging [48]. They also confirmed the cellular uptake of $CHPNH_2$(15)-QD nanoparticles using other normal cells (TIG-3 and MRC-5) and cancer cells (T24, Saos-2, T98G, A549, MCF-7, and YKG-1) (Figure 9) [48].

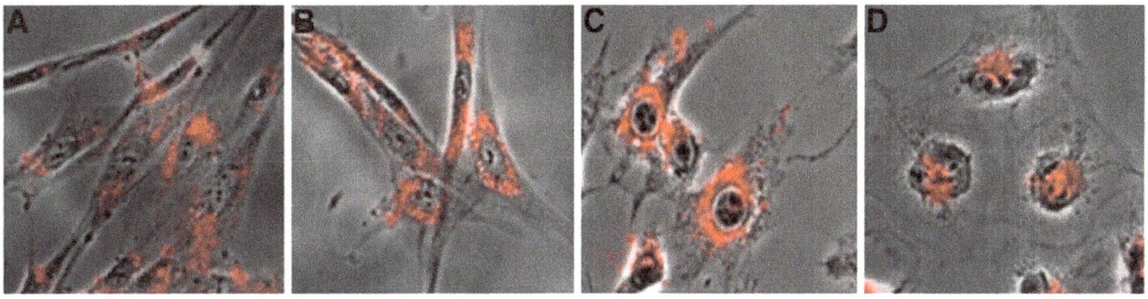

Figure 9 Confocal laser scanning fluorescence microscopyimages of cells labeled with $CHPNH_2$(15)-QD nanoparticle. (A) TIG-3 cells, (B) MRC-5 cells, (C) MCF-7 cells, and (D) YKG-1 cells [48].

In recent years, functional peptides that transmit biomaterials into cells have been developed in biomaterial research. Because of lysosomal trapping, QD delivery into cells with conjugated cell-penetrating peptides by the endocytic pathway was challenging in biomedical applications [139]. In another study, engineered peptides for producing QDs tagging protein ligands and biosensors to their surfaces, by appropriate cysteines or histidines, have served as ligands

[140]. Encapsulation of QDs in viral capsids provides a new tool which allows the design of intracellular microscopic probes and vectors [141]. More samples of QD delivery systems are shown in Table 3.

Table 3 QD delivery systems

Authors	Delivery system of QD	Use
Jia et al. [142]	Multiwalled carbon nanotube (MWNT) delivery system	MWNTs are containing antisense oligodeoxynucleotides and CdTe QDs via electrostatically layer-by-layer assembling.
Chen et al. [143], Xue et al. [144], Delehanty et al. [145], Ruan et al. [146], Wei et al.[147]	Tat peptide-mediated delivery system	QDs conjugated to the cell-penetrating peptide derived from the human immunodeficiency virus-1 transactivator protein
Lagerholm et al. [148]	Peptide delivery system	Nine residue biotinylated l-arginine peptide is used to enhance delivery of streptavidin-conjugated QDs into mammalian cells.
Bagalkot et al. [149]	A10 RNA aptamer	Functionalizes the surface of QD with the A10 RNA aptamer, which recognizes the extracellular domain of the prostate specific membrane antigen
Bakalova et al. [150]	Silica-shelled quantum dots	Based on silica-shelled single QD micelles with incorporated paramagnetic substances [tris(2,2,6,6-tetramethyl-3,5-heptanedionate)/gadolinium] into the micelle and/or silica coat
Yum et al. [151]	Nanoscale mechanochemical method	Using a membrane-penetrating nanoneedle
Yuan et al. [152]	Chitosan (N-acetylglucosamine) tumor-targeted drug delivery	QDs encapsulated with chitosan
Hasegawa et al. [48]	Nanogel-QD hybrid	Nanogels of CHPNH 2 with 15 amino groups per 100 glucose units and QDs that were conjugated with protein A molecules were mixed.
Dixit et al. [141]	Viral vectors	QDs encapsulation in viral capsids
Zhang and Liu [153]	Nonviral vectors	Cappingthe surface of ZnO QD with poly(2-(dimethylamino) ethyl methacrylate)
Jablonski et al. [154]	Cationic peptide and a hydrophobic counterion	Quantum dots have been delivered to the cytosol of living cells using a combination of a cationic peptide, polyarginine, and a hydrophobic counterion, pyrenebutyrate.
Qi and Gau [155]	QD-amphipol nanocomplex	Advantages include cytoplasm delivery and endosome escape.
Gao et al. [109]	Polymeric delivery system	The structural design involves encapsulating QDs with an ABC triblock copolymer and linking this amphiphilic polymer.
Duan and Nie [77]	Polymeric delivery system	QDs were encapsulated by PEI-g-PEG.

TOXICITY OF QDS

There are different opinions about the toxicity of QDs; therefore, we investigated their toxicity in amoeba as primary eukaryotes, in plant, and in animal.

In Amoeba

It has been determined that QD labeling had no detectable effect on cell growth and had no deleterious effects on cellular signaling and motility during development of the *Dictyostelium discoideum* cells [47].

In Plant

The ratio of reduced glutathione levels (GSH) relative to the oxidized glutathione (GSSG) in plants suggests that QDs caused oxidative stress on the plant at this condition [74].

In Animal

Yan et al. investigated the potential vascular endothelial toxicity of mercaptosuccinic acid (2-sulfanylbutanedioic acid)-capped QDs *in vitro*. Their results suggested that QDs could not only impair mitochondria but also exert endothelial toxicity through activation of mitochondrial death pathway and induction of endothelial apoptosis [156].

More recently, Chen et al. have studied the cytotoxicity of CdTe/CdS (core-shell) structured and also CdTe/CdS/ZnS (core-shell-shell) structured aqueous synthesized QDs, and their results suggest that the cytotoxicity of CdTe QDs not only comes from the release of Cd^{2+} ions but also intracellular distribution of QDs in cells and the associated nanoscale effects [157]. Table 4demonstrated more results for toxicity of QDs [158–162].

Table 4 More details for toxicity of QDs (modified from [163])

QD	Model	Administration	QD concentration	Exposure duration	Toxicity	Study
CdSe/ZnS-SSA	EL-4 cells	1×10^6 cells/well	0.1 to 0.4 mg/mL	0 to 24 h	Cytotoxic: 0.1 mg/mL altered cell growth; most cells nonviable at 0.4 mg/mL	Hoshino et al. 2004a
CdSe/ZnS-SSA	EL-4 cells	200-μL cell suspension injected (iv) into the mice	0.1 mg/mL QDs per 5×10^7 cells	2 h to 7 days	No toxicity in mice *in vivo*	Hoshino et al. 2004a (*in vivo*)
CdSe/ZnS conjugates: NH_2, OH, OH/COOH, H_2/OH, MUA, COOH	WTK1 cells	5×10^4 cells/mL	1 to 2 μM	12 h	2-μM QD-COOH-induced DNA damage at 2 h	Hoshino et al. 2004b
CdSe/ZnS-MUA	Vero, HeLa, and primary human hepatocytes	100-μL QDs/3×10^4 cells	0 to 0.4 mg/mL	24 h	Cytotoxic: 0.2 mg/mL, Vero; 0.1 mg/mL, HeLa; 0.1 mg/mL, hepatocytes	Shiohara et al. 2004
CdTe	Rat pheochromocytoma cells, murine, microglial cells	1×10^5 cells/cm^2	0.01 to 100 μg/mL	2 to 24 h	10 μg/mL cytotoxic	Lovric et al. 2005
CdSe-MAA, TOPO QDs	Primary rat hepatocytes		62.5-1,000 μg/mL	1 to 8 h	Cytotoxic: 62.5 μg/mL cytotoxic under oxidative/photolytic conditions No toxicity on addition of ZnS cap	Derfus 2004
QD micelles: CdSe/ZnS QDs in (PEG-PE) and phosphatydilcholine	*Xenopus* blastomeres	5×10^9 QDs/cell (approximately 0.23 pmol/cell)	1.5 to 3 nL of 2.3-μM QDs injected, approximately 2.1×10^9 to 4.2×10^9 injected QDs/cell	Days	5×10^9 QDs/cell: cell abnormalities, altered viability and motility No toxicity at 2×10^9 QDs/cell	Dubertret et al. 2002
CdSe/ZnS amp-QDs and mPEG QDs [158]	Mice	200-μL tail vein injection	Injections, approximately 180-nM QD, approximately 20-pmol QD/g animal weight	15-min cell incubations, 1 to 133 days *in vivo*	No signs of localized necrosis at the sites of deposition	Ballou et al. 2004
CdSe/ZnS-DHLA	*Dictyostelium discoideum* and HeLa cells		400 to 600 nM	45 to 60 min	No effects on cell growth	Jaiswal et al. 2003
Avidin-conjugated CdSe/ZnS QDs	HeLa cells		0.5 to 1.0 μM	15 min	No effect on cell growth and development	Jaiswal et al. 2003
CdSe/ZnS-amphiphilic micelle	Mice	Tail vein injection	60-μM QD/g animal weight, 1-μM and 20-nM final QD concentration	Not given	Mice showed no noticeable ill effects after imaging	Larson et al. 2003
CdSe/ZnS-DHLA QDs	Mice, B16F10 cells	5×10^4 B16F10 cells with 10-μL QDs (approximately 10 pmol), tail vein (iv) injection	100 μL of B16F10 cells used for tail vein injection, approximately 2×10^5 to 4×10^5 cells injected	4- to 6-h cell incubation, mice sacrificed at 1 to 6 h	No toxicity observed in cells or mice	Voura et al. 2004
CdSe/ZnS-MUA QDs, QD-SSA complexes [162]	Vero cells	0.4 mg/mL	0.24 mg/mL	2 h	0.4-mg/mL MUA/SSA-QD complexes did not affect viability of Vero cells	Hanaki et al. 2003
CdSe/ZnS	HeLa cells	1×10^6 cells	10-pmol QDs/1×10^5 cells (approximately 10 nM)	10 days (cell culture)	10-nM QD had minimal impact on cell survival	Chen and Gerion 2004
CdTe aqQDs	HEK293 cells	1×10^5 cells	300 or 600 nM	3 days	Nearly completely inhibited cell growth even from the very beginning	Nan Chen et al. 2012
CdTe-gelatinized/ nongelatinized	PC12 cells	1×10^5 cells/cm^2	1 to 100 nM	72 h	At 1 nM,did not initiate any detrimental effects; at 100 nM, resulted in the death of all cells	Babu R Prasad et al. 2010
CdTe, CdTe/CdS, CdTe/CdS/ZnS	K562 and HEK293T human cell lines	1×10^5 cells	0.2 to 3.0 μM	0 to 48 h	Cells treated with CdTe and CdTe/CdS QDs were mostly nonviable by 48 h (for all concentrations tested)	Su et al. 2009
CdSe/ZnS-PEG (EviTag T1 490 QD)	Caco-2 (human colon carcinoma) cell line	10^6 cells/ml, 0.2 ml/well	0.84 to 105 μM	0 to 24 h	Commercially available QD demonstrated low cytotoxicity but induced cell detachment	Wang et al. 2008
CdSe	Primary rat hippocampal neuron cells in culture	$10^4$to 10^5 cells/ml	1, 10, and 20nM	24 h	1-nM QD for 24 h showed no decrease in cell viability; in contrast, cells treated with 10- and 20-nM QD for 24 h showed decreases in cell viability on the order of 20 and 30%	Tang et al., 2008

CONCLUSIONS

In this review, we summarize few experiments that illustrate the high potential of QDs used for/as:

- labeling biomolecules and cells;
- tracer to follow the intracellular/extracellular dynamic of a single biomolecule/cell;
- localization of biomolecules *in vitro*/*in vivo*;
- imaging of biomolecules or cells *in vitro*/*in vivo*;
- assessing cell growth in damaged tissue;
- pH probes for the study of enzyme reaction kinetics;
- biomarker detection in various cancers;
- imaging and sensing of infectious diseases; and
- protein micro- and nanoarrays to the detection of cancer biomarkers.

These studies have been generated using QDs because of their small size, brightness, independence of emission on the excitation wavelength, and stability under relatively harsh environments which would be advantageous. In contrast, there are different opinions about the toxicity and fate of QDs *in vivo*. Therefore, more experiments should be done, and much more data should be available, to be sure to do clinical trials on humans.

FUTURE PROSPECTS

In the future, QDs will be used for identifying various categories of cancer cells, the molecular mechanisms of disease, and new drug action mechanisms, applying them in the intracellular/extracellular studies, and making new methods for biochemical assaying.

ACKNOWLEDGMENTS

The authors are grateful to the financial support from the Department of Medical Nanotechnology, Faculty of Advanced Medical Science, Iran, National Science Foundation Tabriz, Iran, and the Drug Applied Research Center Tabriz University of Medical Sciences.

Authors' contributions

SD conceived of the study and participated in its design and coordination. AA participated in the sequence alignment and drafted the manuscript. All authors read and approved the final manuscript.

REFERENCES

1. Pierobon P, Cappello G: Quantum dots to tail single biomolecules inside living cells. Adv Drug Deliv Rev 2012, 64(2):167–178.
2. Klimov VI: Spectral and dynamical properties of multiexcitons in semiconductor nanocrystals. Annu Rev Phys Chem 2007, 58:635–673.
3. Chan WC, Maxwell DJ, Gao X, Bailey RE, Han M: Luminescent quantum dots for multiplexed biological detection and imaging. Curr Opin Biotechnol 2002, 13(1):40–46.
4. Bruchez M Jr, Moronne M, Gin P, Weiss S, Alivisatos AP: Semiconductor nanocrystals as fluorescent biological labels. Science 1998, 281 (5385):2013–2016.
5. Deb P, Bhattacharyya A, Ghosh SK, Ray R, Lahiri A: Excellent biocompatibility of semiconductor quantum dots encased in multifunctional poly(N-isopropylacrylamide) nanoreservoirs and nuclear specific labeling of growing neurons. Appl Phys Lett 2011, 98(10):103702–103703.
6. Dabbousi BO, Rodriguez-Viejo J, Mikulec FV, Heine JR, Mattoussi H, Ober R: (CdSe)ZnScore – shell quantum dots: synthesis and characterization of a size series of highly luminescent nanocrystallites. J Phys Chem B 1997, 101(46):9463–9475.
7. Bakalova R, Ohba H, Zhelev Z: Quantum dots as photosensitizers? Nat Biotechnol 2004, 22(11):1360–1361.
8. Chan WC, Nie S: Quantum dot bioconjugates for ultrasensitive nonisotopic detection. Science 1998, 281(5385):2016–2018.
9. Azzazy HM, Mansour MM, Kazmierczak SC: From diagnostics to therapy: prospects of quantum dots. Clin Biochem 2007, 40(13–14):917–927.
10. Deerinck TJ: The application of fluorescent quantum dots to confocal, multiphoton, and electron microscopic imaging. Toxicol Pathol 2008, 36(1):112–116.
11. Ghasemi Y, Peymani P, Afifi S: Quantum dot: magic nanoparticle for imaging, detection and targeting. Acta Biomed 2009, 80(2):156–165.
12. Corredor E, Testillano PS, Coronado MJ, González-Melendi P, FernándezPacheco R, Marquina C: Nanoparticle penetration and transport in living pumpkin plants: in situ subcellular identification. BMC Plant Biol 2009, 9:45.
13. Lin S, Meyer DE, Curran MA: Uptake, translocation, and transmission of carbon nanomaterials in rice plants. Small 2009, 5(10):1128–1132.
14. Muller F, Houben A, Barker PE, Xiao Y, Käs JA, Melzer M: Quantum dots – a versatile tool in plant science? J Nanobiotechnology 2006, 4:5.
15. Santos AR, Miguel AS, Tomaz L, Rui Malhó, Christopher: The impact of CdSe/ZnS quantum dots in cells of Medicago sativa in suspension culture. J Nanobiotechnology 2010, 8:24.
16. Wu YL, Lim CS, Fu S, Tok AIY, Lau HM, Boey FYC: Surface modifications of ZnO quantum dots for bio-imaging. Nanotechnology 2007, 18(21):215604.
17. Wu YL, Lim CS, Fu S, Tok AIY, Lau HM, Boey FYC: Water-soluble quantum dots for biomedical applications. Biochem Biophys Res Commun 2006, 348 (3):781–786.
18. Arnot HEG, Watt M, Sotomayor-Torres CM, Glew R, Cusco R, Bates J, Beaumont SP: Photoluminescence of overgrown GaAs-GaAlAs quantum dots. Superlattices and Microstructures 1989, 5(3):459–463.
19. Li SS, Xia JB: Electronic structure and binding energy of a hydrogenic impurity in a hierarchically self-assembled GaAs/AlxGa1 - xAs quantum dot. J Appl Phys 2006, 100(8):083714.
20. Li S-S, Xia J-B: Electronic states of a hydrogenic donor impurity in semiconductor nano-structures. Physics Letters A 2007, 366(1–2):120–123.
21. Li SS, Kong XJ: Hydrogenic impurities in GaAs-Ga1-xAlxAs superlattices in an axial magnetic field. J Phys Condens Matter 1992, 4(20):4815.
22. Bera D, Qian L, Tseng T-K, Holloway PH: Quantum dots and their multimodal applications: a review. Materials 2010, 3(4):2260–2345.
23. Mattoussi H, Palui G, Na HB: Luminescent quantum dots as platforms for probing in vitro and in vivo biological processes. Adv Drug Deliv Rev 2012, 64(2):138–166.
24. Birudavolu S, Nuntawong N, Balakrishnan G, Xin YC, Huang S, Lee SC, Brueck SRJ, CP: Selective area growth of InAs quantum dots formed on a patterned GaAs substrate. Appl Phys Lett 2004, 85(12):2337–2339.

25. Nakata Y, Mori T, Seki H: Molecular beam epitaxial growth of InAs selfassembled quantum dots with light-emission at 1.3 μm. Journal of Crystal Growth 2000, 208(1–4):93–99.
26. Yamilov A, Herrera MR, Bertino MF: Quantum dots by ultraviolet and x-ray lithography. Nanotechnology 2007, 18(31):315603.
27. Burda C, Chen X, Narayanan R, El-Sayed MA: Chemistry and properties of nanocrystals of different shapes. Chem Rev 2005, 105(4):1025–1102.
28. Bang J, Fau Yang H, Holloway PH: Enhanced and stable green emission of ZnO nanoparticles by surface segregation of Mg. Nanotechnology 2006, 17(14):973.
29. Spanhel L, Anderson MA: Semiconductor clusters in the sol–gel process: quantized aggregation, gelation, and crystal growth in concentrated zinc oxide colloids. J Am Chem Soc 1991, 113(8):2826–2833.
30. Bera D, Qian L, Sabui S, Santra S: Photoluminescence of ZnO quantum dots produced by a sol–gel process. Opt Mater 2008, 30(8):1233–1239.
31. Qu L, Peng X: Control of photoluminescence properties of CdSe nanocrystals in growth. J Am Chem Soc 2002, 124(9):2049–2055.
32. Murray CB, Norris DJ, Bawendi MG: Synthesis and characterization of nearly monodisperse CdE (E = sulfur, selenium, tellurium) semiconductor nanocrystallites. J Am Chem Soc 1993, 115(19):8706–8715.
33. Qu L, Peng ZA, Peng X: Alternative routes toward high quality CdSe nanocrystals. Nano Lett 2001, 1(6):333–337.
34. Li L, Qian H, Ren J: Rapid synthesis of highly luminescent CdTe nanocrystals in the aqueous phase by microwave irradiation with controllable temperature. Chem Commun (Camb) 2005, doi:10.1039/B412686F.
35. Xin SH, Yin A, Kim C, Dobrowolska M, Merz JL: Formation of selfassembling CdSe quantum dots on ZnSe by molecular beam epitaxy. Appl Phys Lett 1996, 69(25):3884–3886.
36. Leonardi K, Selke H, Heinke H, Ohkawa K, Hommel D, Gindele F, Woggon U: Formation of self-assembling II–VI semiconductor nanostructures during migration enhanced epitaxy. J Crystal Growth 1998, 184–85:259–263.
37. Kurtz E, Shen J, Schmidt M, Grun M, Hong SK, Litvinov D: Formation and properties of self-organized II–VI quantum islands. Thin Solid Films 2000, 367(1–2):68–74.
38. Swihart MT: Vapor-phase synthesis of nanoparticles. Curr Opin Colloid Interface Sci 2003, 8(1):127–133.
39. Zhang ZY, Oehler AEH, Resan B, Kurmulis S, Zhou KJ, Wang Q, Mangold M, Süedmeyer T, Keller U, Weingarten KJ, Hogg RA: 1.55 μm InAs/GaAs quantum dots and high repetition rate quantum dot SESAM modelocked laser. Nano Lett 2010, 10:1512–1516.
40. Yang H, Luan W, Shan-tung T, Wang ZM: Synthesis of nanocrystals via microreaction with temperature gradient: towards separation of nucleation and growth. Lab Chip 2008, 8:451–455.
41. Jiang W, Wang ZM, Dorogan VG, Mazur YI, Shibin L: Gregory, Insight into optical properties of strain-free quantum dot pairs. Journal of Nanoparticle Research 2011, 13:947–952.
42. Yuechao J, Xiaoyong G, Jingxiao L, Yongsheng C, Jianpeng Z, Xinli L: A novel method for PbS quantum dot synthesis. Mater Lett 2012, 72:116–118.
43. Zhiming M, Wang D, Liang B, Sablon KA, Jihoon L, Yuriy I, Mazur D, Strom NW, Gregory J, Salamo D: Self-organization of InAs quantum-dot clusters directed by droplet homoepitaxy. Small 2007, 3:235–238.
44. Kalauzi A, Mutavdžić D, Djikanović D, Radotić K, Jeremić M: Interaction of the CdSe quantum dots with plant cell walls. Colloids Surf B Biointerfaces 2012, 91:41–47.
45. Goldman ER, Balighian ED, Mattoussi H, Kuno MK, Mauro JM, Tran PT, Anderson GP: Avidin: a natural bridge for quantum dot-antibody conjugates. J Am Chem Soc 2002, 124(22):6378–6382.
46. Jaiswal JK, Mattoussi H, Mauro JM, Simon SM: Use of quantum dots for live cell imaging. Nat Meth 2004, 1(1):6.
47. Jaiswal JK, Goldman ER, Mattoussi H, Simon SM: Long-term multiple color imaging of live cells using quantum dot bioconjugates. Nat Biotech 2003, 21(1):47–51.
48. Hasegawa U, Nomura SM, Kaul SC: Nanogel-quantum dot hybrid nanoparticles for live cell imaging. Biochem Biophys Res Commun 2005, 331(4):917–921.
49. Leevy WM, Lambert TN, Johnson JR, Morris J, Smith BD: Quantum dot probes for bacteria distinguish Escherichia coli mutants and permit in vivo imaging. Chem Commun 2008, 20:2331–2333.
50. Tallury P, Malhotra A, Byrne LM, Santra S: Nanobioimaging and sensing of infectious diseases. Adv Drug Deliv Rev 2010, 62(4–5):424–437.
51. Su X-L, Li Y: Quantum dot biolabeling coupled with immunomagnetic separation for detection of Escherichia coli O157:H7. Anal Chem 2004, 76(16):4806–4810.
52. Yang L, Li Y: Simultaneous detection of Escherichia coli O157[ratio]H7 and Salmonella Typhimurium using quantum dots as fluorescence labels. Analyst 2006, 131(3):394–401.
53. Chang Y-P, Pinaud F, Antelman J, Weiss S: Tracking bio-molecules in live cells using quantum dots. J Biophotonics 2008, 1(4):287–298.
54. Cherry RJ: Keeping track of cell surface receptor. Trends Cell Biol 1992, 2(8):242–244.
55. Saxton MJ, Jacobson K: Single-particle tracking: applications to membrane dynamics. Annu Rev Biophys Biomol Struct 1997, 26:373–399.
56. Howarth M: Monovalent, reduced-size quantum dots for imaging receptors on living cells. Nat Methods 2008, 5(5):397–399.
57. Chen H: Altered membrane dynamics of quantum dot-conjugated integrins during osteogenic differentiation of human bone marrow derived progenitor cells. Biophys J 2007, 92(4):1399–1408.

58. Haggie PM: Tracking of quantum dot-labeled CFTR shows near immobilization by C-terminal PDZ interactions. Mol Biol Cell 2006, 17(12):4937–4945.

59. Crane JM, Verkman AS: Long-range nonanomalous diffusion of quantum dot-labeled aquaporin-1 water channels in the cell plasma membrane. Biophys J 2008, 94(2):702–713.

60. Bouzigues C: Asymmetric redistribution of GABA receptors during GABA gradient sensing by nerve growth cones analyzed by single quantum dot imaging. Proc Natl Acad Sci 2007, 104(27):11251–11256.

61. Dahan M: Diffusion dynamics of glycine receptors revealed by singlequantum dot tracking. Science 2003, 302(5644):442–445.

62. Roullier V: High-affinity labeling and tracking of individual histidinetagged proteins in live cells using Ni2+ tris-nitrilotriacetic acid quantum dot conjugates. Nano Lett 2009, 9(3):1228–1234.

63. Lidke DS: Quantum dot ligands provide new insights into erbB/HER receptor-mediated signal transduction. Nat Biotechnol 2004, 22(2):198–203.

64. Watanabe TM, Higuchi H: Stepwise movements in vesicle transport of HER2 by motor proteins in living cells. Biophys J 2007, 92(11):4109–4120.

65. Groc L: Differential activity-dependent regulation of the lateral mobilities of AMPA and NMDA receptors. Nat Neurosci 2004, 7(7):695–696.

66. Choquet D, Triller A: The role of receptor diffusion in the organization of the postsynaptic membrane. Nat Rev Neurosci 2003, 4(4):251–265.

67. Xiao Y, Barker PE: Semiconductor nanocrystal probes for human metaphase chromosomes. Nucleic Acids Res 2004, 32(3):e28–e28.

68. Pierobon P: Velocity, processivity, and individual steps of single myosin V molecules in live cells. Biophys J 2009, 96(10):4268–4275.

69. Nelson SR: Random walk of processive, quantum dot-labeled myosin Va molecules within the actin cortex of COS-7 cells. Biophys J 2009, 97(2):509–518.

70. Yoo J: Intracellular imaging of targeted proteins labeled with quantum dots. Exp Cell Res 2008, 314(19):3563–3569.

71. Courty S: Tracking individual kinesin motors in living cells using single quantum-dot imaging. Nano Lett 2006, 6(7):1491–1495.

72. Ishido M, Kasuga N: In situ real-time imaging of the satellite cells in rat intact and injured soleus muscles using quantum dots. Histochem Cell Biol 2011, 135(1):21–26.

73. Stylianou P, Skourides PA: Imaging morphogenesis, in Xenopus with quantum dot nanocrystals. Mech Dev 2009, 126(10):828–841.

74. Navarro DA, Bisson MA, Aga DS: Investigating uptake of water-dispersible CdSe/ZnS quantum dot nanoparticles by Arabidopsis thaliana plants. J Hazard Mater 2012, 211–212:427–435.

75. Gautrot JE, Zhu XX: Macrocyclic bile acids: from molecular recognition to degradable biomaterial building blocks. J Mater Chem 2009, 19(32):5705–5716.

76. Lim IIS: Gold and magnetic oxide/gold core/shell nanoparticles as biofunctional nanoprobes. Nanotechnology 2008, 19(30):305102.

77. Duan H, Nie S: Cell-penetrating quantum dots based on multivalent and endosome-disrupting surface coatings. J Am Chem Soc 2007, 129(11):3333–3338.

78. Smith AM: Bioconjugated quantum dots for in vivo molecular and cellular imaging. Adv Drug Deliv Rev 2008, 60(11):1226–1240.

79. Wang L: Core@shell nanomaterials: gold-coated magnetic oxide nanoparticles. J Mater Chem 2008, 18(23):2629–2635.

80. Park K: New generation of multifunctional nanoparticles for cancer imaging and therapy. Adv Funct Mater 2009, 19(10):1553–1566.

81. Wu W: In-situ immobilization of quantum dots in polysaccharide-based nanogels for integration of optical pH-sensing, tumor cell imaging, and drug delivery. Biomaterials 2010, 31(11):3023–3031.

82. Lee KH: Quantitative molecular profiling of biomarkers for pancreatic cancer with functionalized quantum dots. Nanomedicine, . in press.

83. Gokarna A: Quantum dot-based protein micro- and nanoarrays for detection of prostate cancer biomarkers. Proteomics 2008, 8(9):1809–1818.

84. Liu J: Multiplexed detection and characterization of rare tumor cells in Hodgkin›s lymphoma with multicolor quantum dots. Anal Chem 2010, 82(14):6237–6243.

85. Ray S: Emerging nanoproteomics approaches for disease biomarker detection: A current perspective. J Proteomics 2011, 74(12):2660–2681.

86. Yu Y: Hydrothermal synthesis of GSH–TGA co-capped CdTe quantum dots and their application in labeling colorectal cancer cells. Colloids Surf B Biointerfaces 2012, 95:247–253.

87. American Cancer Society: Pancreatic cancer. http://www.cancer.org/cancer/ pancreaticcancer/detailedguide/pancreatic-cancer-key-statistics.

88. Resch-Genger U: Quantum dots versus organic dyes as fluorescent labels. Nat Meth 2008, 5(9):763–775.

89. Tripp RA: Bioconjugated nanoparticle detection of respiratory syncytial virus infection. Int J Nanomedicine 2007, 2(1):117–124.

90. Agrawal A: Counting single native biomolecules and intact viruses with color-coded nanoparticles. Anal Chem 2006, 78(4):1061–1070.

91. Bentzen EL: Progression of respiratory syncytial virus infection monitored by fluorescent quantum dot probes. Nano Lett 2005, 5(4):591–595.

92. Dwarakanath S: Quantum dot-antibody and aptamer conjugates shift fluorescence upon binding bacteria. Biochem Biophys Res Commun 2004, 325(3):739–743.

93. Goldman ER: Multiplexed toxin analysis using four colors of quantum dot fluororeagents. Anal Chem 2003, 76(3):684–688.

94. Zhao Y: Simultaneous detection of multifood-borne pathogenic bacteria based on functionalized quantum

dots coupled with immunomagnetic separation in food samples. J Agric Food Chem 2008, 57(2):517–524.

95. Hahn MA, Tabb JS, Krauss TD: Detection of single bacterial pathogens with semiconductor quantum dots. Anal Chem 2005, 77(15):4861–4869.

96. Mukhopadhyay B: Bacterial detection using carbohydrate-functionalised CdS quantum dots: a model study exploiting E. coli recognition of mannosides. Tetrahedron Lett 2009, 50(8):886–889.

97. Edgar R: High-sensitivity bacterial detection using biotin-tagged phage and quantum-dot nanocomplexes. Proc Natl Acad Sci U S A 2006, 103(13):4841–4845.

98. Zhu L, Ang S, Liu W-T: Quantum dots as a novel immunofluorescent detection system for Cryptosporidium parvum and Giardia lamblia. Appl Environ Microbiol 2004, 70(1):597–598.

99. Klostranec JM, Xiang Q, Farcas GA: Convergence of quantum dot barcodes with microfluidics and signal processing for multiplexed high-throughput infectious disease diagnostics. Nano Lett 2007, 7(9):2812–2818.

100. Gazouli M, Liandris E, Andreadou M, Sechi LA, Masala S, Paccagnini D, Ikonomopoulos J: Specific detection of unamplified mycobacterial DNA by use of fluorescent semiconductor quantum dots and magnetic beads. J Clin Microbiol 2010, 48(8):2830–2835.

101. Griffith DE: An official ATS/IDSA statement: diagnosis, treatment, and prevention of nontuberculous mycobacterial diseases. Am J Respir Crit Care Med 2007, 175(4):367–416.

102. Veigas B, Doria G, Baptista PV: Nanodiagnostics for tuberculosis. In Understanding Tuberculosis - Global Experiences and Innovative Approaches to the Diagnosis. Edited by Cardona PJ. Rijeka: InTech; 2012:20.

103. GhoshMitra S: Role of engineered nanocarriers for axon regeneration and guidance: current status and future trends. Adv Drug Deliv Rev 2012, 64(1):110–125.

104. Langlois JA, Rurland-Brown W, Thomas KE: Traumatic brain injury in the United States: emergency department visits, hospitalizations, and deaths. Atlanta, GA: Dept. of Health and Human Services (US), Centers for Disease Control and Prevention, National Center for Injury Prevention and Control; 2004.

105. National Spinal Cord Injury Statistical Center: Facts and figures at a glance. http://www.nscisc.uab.edu/public_content/pdf/Facts%202011%20Feb% 20Final.pdf 2011.

106. Pathak S: Quantum dot applications to neuroscience: new tools for probing neurons and glia. J Neurosci 2006, 26(7):1893–1895.

107. Alivisatos P: The use of nanocrystals in biological detection. Nat Biotech 2004, 22(1):47–52.

108. Smith A: Engineering luminescent quantum dots for in vivo molecular and cellular imaging. Ann Biomed Eng 2006, 34(1):3–14.

109. Gao X: In vivo cancer targeting and imaging with semiconductor quantum dots. Nat Biotech 2004, 22(8):969–976.

110. Vu TQ: Peptide-conjugated quantum dots activate neuronal receptors and initiate downstream signaling of neurite growth. Nano Lett 2005, 5 (4):603–607.

111. Sundara Rajan S, Vu TQ: Quantum dots monitor TrkA receptor dynamics in the interior of neural PC12 cells. Nano Lett 2006, 6(9):2049–2059.

112. Howarth M: Targeting quantum dots to surface proteins in living cells with biotin ligase. Proc Natl Acad Sci USA 2005, 102(21):7583–7588.

113. Prasad B: Long-term exposure of CdTe quantum dots on PC12 cellular activity and the determination of optimum non-toxic concentrations for biological use. J Nanobiotechnology 2010, 8(1):7.

114. WHO: World Malaria Report 2010. Geneva: WHO; 2010.

115. Ku MJ: Quantum dots: a new tool for anti-malarial drug assays. Malar J 2011, 10:118.

116. Tokumasu F: Band 3 modifications in Plasmodium falciparum-infected AA and CC erythrocytes assayed by autocorrelation analysis using quantum dots. J Cell Sci 2005, 118(Pt 5):1091–1098.

117. Wu D, Chen Z: ZnS quantum dots as pH probes for study of enzyme reaction kinetics. Enzyme Microb Technol 2012, 51(1):47–52.

118. Jin WJ: Photoactivated luminescent CdSe quantum dots as sensitive cyanide probes in aqueous solutions. Chem Commun (Camb) 2005, doi:10.1039/B414858D.

119. Xie H-Y: Luminescent CdSe-ZnS quantum dots as selective Cu2+ probe. Spectrochim Acta A Mol Biomol Spectrosc 2004, 60(11):2527–2530.

120. Tomasulo M, Yildiz I, Raymo FM: pH-sensitive quantum dots. J Phys Chem B 2006, 110(9):3853–3855.

121. Sun YH: Photostability and pH sensitivity of CdSe/ZnSe/ZnS quantum dots in living cells. Nanotechnology 2006, 17(17):4469.

122. Wang Y-Q: Cadmium telluride quantum dots as pH-sensitive probes for tiopronin determination. Anal Chim Acta 2008, 610(1):50–56.

123. Liu Y-S: pH-Sensitive photoluminescence of CdSe/ZnSe/ZnS quantum dots in human ovarian cancer cells. J Phys Chem C 2007, 111(7):2872–2878.

124. Yun Z: Using cadmium telluride quantum dots as a proton flux sensor and applying to detect H9 avian influenza virus. Anal Biochem 2007, 364 (2):122–127.

125. Deng Z: Green and orange CdTe quantum dots as effective ph-sensitive fluorescent probes for dual simultaneous and independent detection of viruses. J Phys Chem B 2007, 111(41):12024–12031.

126. Gao M, Kirstein S: Mo1hwald H: Strongly photoluminescent CdTe nanocrystals by proper surface modification. J Phys Chem B 1998, 102 (43):8360–8363.

127. Lueking A: Protein microarrays for gene expression and antibody screening. Anal Biochem 1999, 270(1):103–111.

128. MacBeath G, Schreiber SL: Printing proteins as microarrays for highthroughput function determination. Science 2000, 289(5485):1760–1763.

129. Hergenrother PJ, Depew KM, Schreiber SL: Small-molecule microarrays: covalent attachment and screening

of alcohol-containing small molecules on glass slides. J Am Chem Soc 2000, 122(32):7849–7850.

130. Nelson RW, Nedelkov D, Tubbs KA: Biosensor chip mass spectrometry: a chip-based proteomics approach. Electrophoresis 2000, 21(6):1155–1163.
131. Schweitzer B, Kingsmore SF: Measuring proteins on microarrays. Curr Opin Biotechnol 2002, 13(1):14–19.
132. Sun B: Microminiaturized immunoassays using quantum dots as fluorescent label by laser confocal scanning fluorescence detection. J Immunol Methods 2001, 249(1–2):85–89.
133. Shingyoji M: Quantum dots-based reverse phase protein microarray. Talanta 2005, 67(3):472–478.
134. Kerman K: Quantum dot-based immunosensor for the detection of prostate-specific antigen using fluorescence microscopy. Talanta 2007, 71 (4):1494–1499
135. Geho D: Pegylated, steptavidin-conjugated quantum dots are effective detection elements for reverse-phase protein microarrays. Bioconjug Chem 2005, 16(3):559–566.
136. Zajac A: Protein microarrays and quantum dot probes for early cancer detection. Colloids Surf B Biointerfaces 2007, 58(2):309–314.
137. Stephens DJ, Allan VJ: Light microscopy techniques for live cell imaging. Science 2003, 300(5616):82–86.
138. Weijer CJ: Visualizing signals moving in cells. Science 2003, 300(5616):96–100.
139. Liu BR: Intracellular delivery of quantum dots mediated by a histidineand arginine-rich HR9 cell-penetrating peptide through the direct membrane translocation mechanism. Biomaterials 2011, 32(13):3520–3537.
140. Zhou M, Ghosh I: Quantum dots and peptides: a bright future together. Pept Sci 2007, 88(3):325–339.
141. Dixit SK: Quantum dot encapsulation in viral capsids. Nano Lett 2006, 6(9):1993–1999.
142. Jia N: Intracellular delivery of quantum dots tagged antisense oligodeoxynucleotides by functionalized multiwalled carbon nanotubes. Nano Lett 2007, 7(10):2976–2980.
143. Chen B: Transmembrane delivery of the cell-penetrating peptide conjugated semiconductor quantum dots. Langmuir 2008, 24(20):11866– 11871.
144. Xue F: Enhancement of intracellular delivery of cdte quantum dots (qds) to living cells by tat conjugation. J Fluoresc 2007, 17(2):149–154.
145. Delehanty JB: Self-assembled quantum dot–peptide bioconjugates for selective intracellular delivery. Bioconjug Chem 2006, 17(4):920–927.
146. Ruan G: Imaging and tracking of Tat peptide-conjugated quantum dots in living cells: new insights into nanoparticle uptake, intracellular transport, and vesicle shedding. J Am Chem Soc 2007, 129(47):14759–14766.
147. Wei Y: Surface coating directed cellular delivery of TAT-functionalized quantum dots. Bioconjug Chem 2009, 20(9):1752–1758.
148. Lagerholm BC: Multicolor coding of cells with cationic peptide coated quantum dots. Nano Lett 2004, 4(10):2019–2022.
149. Bagalkot V: Quantum dot–aptamer conjugates for synchronous cancer imaging, therapy, and sensing of drug delivery based on bi-fluorescence resonance energy transfer. Nano Lett 2007, 7(10):3065–3070.
150. Bakalova R: Multimodal silica-shelled quantum dots: direct intracellular delivery, photosensitization, toxic, and microcirculation effects. Bioconjug Chem 2008, 19(6):1135–1142.
151. Yum K: Mechanochemical delivery and dynamic tracking of fluorescent quantum dots in the cytoplasm and nucleus of living cells. Nano Lett 2009, 9(5):2193–2198.
152. Yuan Q, Hein S, Misra RDK: New generation of chitosan-encapsulated ZnO quantum dots loaded with drug: synthesis, characterization and in vitro drug delivery response. Acta Biomater 2010, 6(7):2732–2739.
153. Zhang P, Liu W: ZnO QD@PMAA-co-PDMAEMA nonviral vector for plasmid DNA delivery and bioimaging. Biomaterials 2010, 31(11):3087–3094.
154. Jablonski AE, Humphries WH, Payne CK: Pyrenebutyrate-mediated delivery of quantum dots across the plasma membrane of living cells. J Phys Chem B 2008, 113(2):405–408.
155. Qi L, Gao X: Quantum dot–amphipol nanocomplex for intracellular delivery and real-time imaging of siRNA. ACS Nano 2008, 2(7):1403–1410.
156. Yan M: An in vitro study of vascular endothelial toxicity of CdTe quantum dots. Toxicology 2011, 282(3):94–103.
157. Chen N: The cytotoxicity of cadmium-based quantum dots. Biomaterials 2012, 33(5):1238–1244.
158. Akbarzadeh A, Asgari D, Zarghami N, Mohammad R, Davaran S: Preparation and in vitro evaluation of doxorubicin-loaded Fe3O4magnetic nanoparticles modified with biocompatible co-polymers. Int J Nanomedicine 2012, 7:511–526.
159. Akbarzadeh A, Zarghami N, Mikaeili H, Asgari D, Goganian AM, Khiabani HK, Samiei M, Davaran S: Synthesis, characterization, and in vitro evaluation of novel polymer-coated magnetic nanoparticles for controlled delivery of doxorubicin. Nanotechnol Sci Appl 2012, 5:13–25.
160. Akbarzadeh A, Samiei M, Davaran S: Magnetic nanoparticles: preparation, physical properties, and applications in biomedicine. Nanoscale Res Lett 2012, 7:144.
161. Wang ZM, Kunets VP, Xie YZ, Schmidbauer M, Dorogan VG, Mazur YI, Salamo GJ: Multilayer self-organization of InGaAs quantum wires on GaAs surfaces. Phys. Lett. A 2010, 375:170–173.
162. Passmore BS, Wu J, Manasreh MO, Kunets VP, Lytvyn PM, Salamo GJ: Room temperature near-infrared photoresponse based on interband transition in In 0.35 Ga 0.35 As multiple quantum dot photodetector. IEE Electron Device Letters 2008, 29:224–227.
163. Hardman R: A toxicologic review of quantum dots: toxicity depends on physicochemical and environmental factors. Environ Health Perspect 2006, 114(2):165–172.

CHAPTER 6

CURRENT STATUS AND FUTURE SCOPE FOR NANOMATERIALS IN DRUG DELIVERY

Biswajit Mukherjee, Niladri Shekhar Dey, Ruma Maji, Priyanka Bhowmik, Pranab Jyoti Das and Paramita Paul

INTRODUCTION

Nanotechnology is a revolutionary field of micro manufacturing involving physical and chemical changes to produce nano-sized materials. The word "nano" is a Latin word meaning "dwarf". Mathematically a nanometer is equal to one thousand millionth of a meter [1]. A nanomaterial consists of aggregated as well as unbound particles. Nanotechnology in scientific terms is defined as the science which deals with processes that occur at molecular and atomic level or at nanolength size. It involves designing, synthesis and characterization of material structure by controlling the shapes and sizes at nano scale. The conversion of a particle to nano scale size changes the properties of the material such as increase in surface area, dominance of quantum effects often associated with minute sizes, higher surface area to volume ratio etc. and varies material's magnetic, thermal and electrical property. For example, copper which is opaque at macro scale becomes transparent at nano scale. Similarly the properties of gold at nanoscale causes change in melting point from 200°C to 1068°C and colour changes from yellow to blue to violet along with the change in its catalytic property [2]. Nanoparticles are persistent in nature as well. Functional proteins may be classified as nanoparticles. Some biological system consists of nanoparticles which are devoted to locomotory function. The colours on butterfly's wings are due to light being bounced off nanoscale layers in the structure of the wings. The red and yellow colours seen at sunset are also due to nanoparticles [3]. Super paramagnetic iron oxide less than 50 μm are used for imaging of organs. They can be even used for treating complicated brain disorder bio-imaging at nano scale size [9].

Indian craftsman and artisan used nanotechnology for designing weapons in early times. The first observation and size measurement of nanoparticle were carried out using an ultra microscope by Richard Zsigmondy in 1902. The term nanotechnology was first time used by a researcher named Norio Taniguchi in University of Tokyo in 1974. In 1980 the inventions of two atoms further advanced the field of Nanotechnology. In 1985 fullerene C_{60} was discovered by Kroto's and Smalley's research team. In 1991 carbon tubes were discovered by Saumio lijima and by 2000 National Nanotechnology Initiative (NNI), The United States was launched which paved the way for future development in nanotechnology [2].

Nanotechnology may be considered as one of the main propellants for technological, economical change as industrial competeition. Nanotechnology has integrated various disciplines including biomedicine, engineering and technology. Nanotechnology is being used for improving the existing products and to create new products. The strength can be varied accordingly with the requirements of engineering. It can be used to make the water cleaner by remediation to remove its pollutant. It has helped to clean the environment by removing pollutants and has generated cleaner and cheaper energy. It has improved the healthcare system by introducing new devices for diagnosis, monitoring, treatment of diseases and drug-delivery [1].

Nanomaterials have wide applications in pharmaceutical sciences and technology. Few other predominant areas of use of nanotechnology are in drug delivery, and as diagnostic imaging and biosensor. These devices of nanoscale size are

Citation: Biswajit Mukherjee, Niladri Shekhar Dey, Ruma Maji, Priyanka Bhowmik, Pranab Jyoti Das and Paramita Paul (2014). Current Status and Future Scope for Nanomaterials in Drug Delivery, Application of Nanotechnology in Drug Delivery, PhD. Ali Demir Sezer (Ed.), InTech, DOI: 10.5772/58450.

popularly known as nanomedicine. Thus nanomedicines are sub-micron size materials (<1 μm) which are used for treatment, monitoring and diagnostic purposses. In the present chapter we will discuss on the current status and future strategies of nanosize drug delivery systems.

SIGNIFICANCE OF NANOMATERIALS IN DRUG DELIVERY

There are many reasons for which nanoscale size drug delivery systems are attractive to formulation scientists. The most important reason is that number of surface atoms or molecules to the total number of atoms or molecules increases in drug delivery systems. Thus the surface area increases. This helps to bind, adsorb and carry with other compounds such as drug, probes and proteins. The drug particles itself can be engineered to form nanoscale size materials too [4]. The nanosize device systems, sizes smaller than eukaryotic or prokaryotic cells, can eventually much more in amount reach in generally inaccessible areas such as cancer cells, inflamed tissues etc. due to their enhanced permeability and retention effect (EPR) and can impair lymphatic drainage thus that can be used for administration of genes, proteins through the peroral route of administration [5]. They can be used to target the reticuloendothelial cells, thereby facilitating passive targeting of drug to the macrophages of liver and spleen and thus enabling a natural system for treating intracellular infections [6]. The nanomaterials used for the purpose should be soluble, safe and biocompatible as well as bioavailable. They should not occlude blood vessel and less invasive and the toxicity associated with the nanomaterials for drug delivery should be very low so that they can be used to target the specific diseased tissue in a safe concentration [7]. They need protecting drug from enzymatic and hydrolytic degradation in the gastrointestinal tract and help in bypassing the "first-pass" metabolism in the liver. They generally remain in the circulation for longer time especially those coated with hydrophilic polymers and hence suitable for enhancing the efficacy of drugs with short half-lives and can be used to monitor drug as sustained release formulation as well as for delivering DNA [8]. The dissolution rate of drug is enhanced, onset of therapeutic action is increased, and the dose is reduced. The premature loss of drug through rapid clearance and metabolism can also be prevented [6]. They also increase retention due to bio-adhesion.

Nanoscale drug delivery systems such as nanoparticles, nanoliposomes, dendrimers, fullerence, nanopores, nanotubes, nanoshells, quantum dots, nanocapsule, nanosphere, nanovaccines, nanocrystals etc. are believed to have potentials to revolutionize drug delivery systems. Further nanomaterials on chips, nano robotics, and magnetic nanoparticles attached to specific antibody, nanosize empty virus capsids and magnetic immunoassay are new dimensions of their use in drug delivery. Thus nanomaterials can be used for strategic development of new drug delivery systems and reformulating existing drugs to enhance the effectiveness, patent protection, patient-compliance, safety of drugs and decreasing the cost of health care [9].

VARIOUS NANOSCALE DRUG DELIVERY SYSTEMS

Nanoparticles

Nanoparticles are submicron-sized polymeric colloidal particles with therapeutic agents of interest encapsulated or dispersed within their polymeric matrix or adsorbed or conjugated onto the surface. Commonly used synthetic polymers to prepare nanoparticles for drug delivery are generally biodegradable [10]. Nanoparticles may also be composed of or transport a variety of substances such as silica, gold or other heavy metals, medicaments, quantum dots, nanocrystals, quantum rods and various contrast agents [11]. Nanoparticle systems offer major improvements in therapeutics through site specificity, their ability to escape from multi-drug resistance and the efficient delivery of an agent. They can be used for active drug targeting attaching ligand such as antibody on their surface (Figure 1).

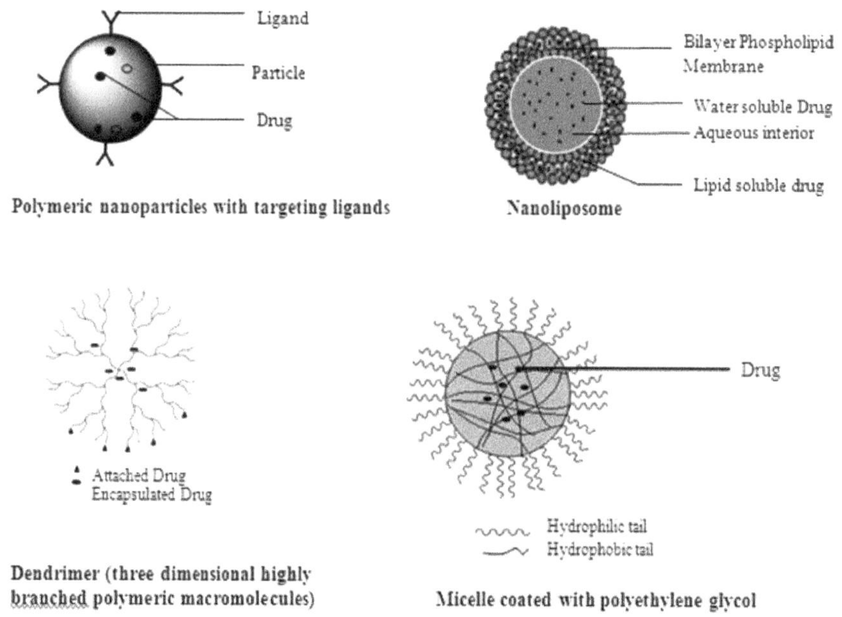

Figure 1. Different Types of Nanocarriers for drug delivery.

Solid lipid nanoparticles (SLNs) refer to as lipospheres or solid lipid nanospheres, or particles and are generally solid at human physiological temperature (37°C) and have a diameter less than 1000 nm [12]. They can be formed from a range of lipids, including mono-, di- and triglycerides, fatty acids, waxes and combinations there of. SLNs must be stabilized by surfactants to form administrable emulsions. SLNs form a strongly lipophilic matrix into which drugs can be loaded for subsequent release. SLNs have been investigated for the delivery of various cancer treatments like colon cancer, breast cancer [13].

Polymer-based nanoparticles have been extensively investigated as drug nanocarriers. The most widely researched synthetic polymers include polylactide (PLA), poly (D,L-lactide-co-glycolide) (PLGA) and poly ethylene glycol (PEG). All three polymers are hydrolized *in vivo* and are biodegradable. Other polymers based on biological polysaccharides have been extensively investigated, including chitosan, Clycodextrin and dextrans [14].

Gold nanoparticles (NPs) consist of a core of gold atoms that can be functionalized by addition of a monolayer of moieties containing a thiol (SH) group. Gold NPs can be synthesized using $NaBH_4$ to reduce $AuCL_4^-$salts in the presence of thiol containing moieties that subsequently form a monolayer around the core gold atom, depending on the stoichiometric gold/ thiol ratio [15]. Drug delivery using gold NPs has been made in DNA delivery for gene therapy and imaging [16]. PEG coated micelles containing drug are also used to deliver drug as new delivery system (Figure 1). Many other nanoparticulate synthetic, semisynthetic, natural and metals are under investigation to know their potentials as drug delivery materials.

Polymeric nanoparticles may adhere to the cell surface and release drug molecules by diffusion which may enter inside the cell to work. However the entire polymeric nanoparticles can also enter the cell by endocytosis. They bind with the cell surface receptor and formation of endosome takes place. Endosome may be lysed with the help of lysosomal enzymes and the nanoparticles release in the cytoplasm (Figure 2).

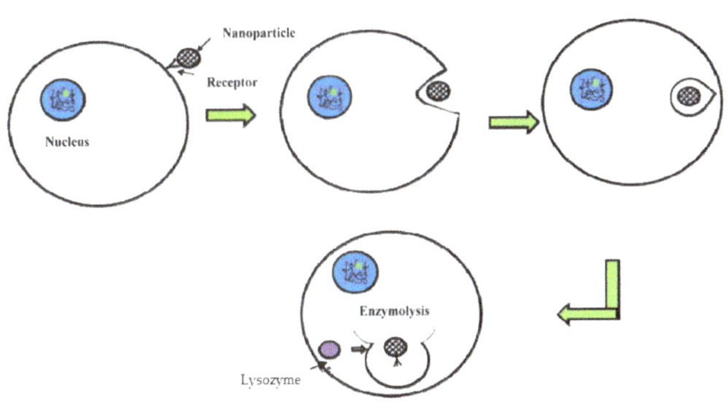

Figure 2. Endocytosis mediated cellular internalization of drug nano-carriers.

Nanoliposomes

Nanoliposomes are the nanosize vesicles made of bilayered phospholipid membranes generally unilamellar with an aqueous interior (Figure 1) [17]. They can be used for the delivery of low molecular weight drugs, imaging agents, peptides, proteins, and nucleic acids. Different anticancer, antiviral drugs are incorporated within the liposomes [18]. Nanoliposomes can also provide slow release of an encapsulated drug, resulting in sustained exposure to the site of action and enhanced efficacy. Usually hydrophilic drugs can be loaded in aqueous compartment and lipophilic drugs are incorporated in the phospholipid layer [19]. However unlike liposome nanoliposome does not undergo rapid degradation and clearance by liver macrophages. As for the targeted drug delivery, nanoliposome plays an important role. It can be used for passive targeting or active targeting [20]. Due to the leaky vascular structure of the tumor tissue nanoliposomes get predominantly accumulated in the tumor and release the drug for a prolonged period of time in passive targeting. Active targeting is achieved by incorporating antibody, ligands etc. on the nanoliposomal surface. By active targeting liposomes directly go to the targeted organs or tissues, and release drug for a prolonged period of time, so that the normal cells are not affected and only the diseased cells are affected [21]. Targeted nanoliposomal drug delivery is more efficacious than the non-targeted drug delivery systems. C6-ceremide ligand induced nanoliposome used to treat the blood cancer directly targets the over expressed lukemic cells and decreases the high epxpression of survivin protein in leukemic cells [22]. The concept of long-circulating or sterically stabilized nanoliposomes is derived for novelibility of delivery systems which can circulate in the blood for a long period of time. Nanoliposomal formulations containing polyethylene glycol (PEG) alter the pharmacokinetic properties of various drug molecules leading to long elimination half-life [23]. Nanoliposomes are expected to bring lots of change in drug delivery in near future.

Dendrimers

Dendrimers are branched polymers, resembling the structure of a tree (Figure 1). Dendrimers represent three dimensional highly branched polymeric macromolecules with the diameter varying from 2.5 to 10 nm. It can be synthesized from both synthetic and natural monomers e.g. aminoacids, monosaccharides and nucleotides. Two classes of dendrimers commonly used for biomedical applications are polyamidoamines and polypropyleneimines [24]. A dendrimer is typically symmetric around the core, and when sufficiently extended it often adopts a spheroidal three-dimensional morphology in water. A central core can be recognized in their structure with at least two identical chemical functionalities. Starting from those groups, repeated units of other molecules can originate with at least one junction of branching. The repetitions of chains and branching result in a series of radially concentric layers with increased crowding [25].

The overall shapes of dendrimers range from spheres to flattened spheroids (disks) to amoeba-like structures, especially in cases where surface charges exist and give the macromolecule a "starfish"-like shape. Branching of dendrimers depends on the synthesis processes. Low molecular weight drugs can be placed into the cavities within the dendrimer molecules and are temporarily immobilized there with hydrophobic forces, hydrogen and covalent bonds [26]. The two processes for the synthesis of dendrimers are divergent and convergent methods. In the divergent method dendrimer grows outwords from a multifunctional core molecule. The core molecule reacts with monomer molecules containing one reactive and two dormant groups giving the first generation dendrimer. The convergent method is developed as a response to the weakness of the divergent synthesis. In the convergent approach, the dendrimer is constructed stepwise, starting from the end groups and progressing inwards. When the growing branched polymeric arms, called dendrons, are large enough, they are attached to a multifunctional core molecule. The convergent method is relatively easy to purify the desired product and the occurrence of defects in the final structure is minimised [27]. Due to classical polymerization dendrimers have a negligible degree of polydispersity. They are random in nature and produce molecules of various sizes. The size of dendrimers can be carefully controlled during the process of synthesis of dendrimers. Scientists are focusing on newer approaches for speeding up the synthesis process by preassembly of oligomeric branches which can be linked together to reduce the number of synthesis steps involved and also increase the dendrimer yield [28].

Dendrimers are popularly used for transfer of genetic materials in cancer therapy or other viral diseases in different organs because of their monodisperisity, high density of functional groups, well-defined shape and multivalency. In gene delivery polyamidoamines (PAMAM) dendrimer is widely used. Some other types of dendrimers are peptide dendrimers, glycodendrimers, polypropilimine dendrimers, Polyethyleneimine (PEI) dendrimers etc.

Nanoshells

Nanoshells (100-200 nm) may be used for drug carrier of both imaging and therapy. Nanoshells consist of nanoparticles with a core of silica and a coating of thin metallic shell [29]. They can be targeted to a tissue by using immunological methods. Nanoshells can also be embedded in a hydrogel polymer [30]. Nanoshells are currently being investigated for prevention of micrometastasis of tumors and also for the treatment of diabetes. Nanoshells are useful for diagnostic purposes in whole blood immunoassays [31].

Fullerenes and Nanotubes

Fullerenes composed of carbon in the form of a hollow sphere or ellipsoid tube. These are also known as 'bucky balls' because of their resemblance to the geodesic dome design of Buck minster Fuller. Fullerenes are being investigated for drug transport of antiviral drugs, antibiotics and anticancer agents [32]. Fullerenes have the potential to stimulate host immune response and productions of fullerene specific antibodies. Soluble derivatives of fullerenes such as C60 have shown great utility as pharmaceutical agents.

Nanotubes are nanometer scale tube like structure and they are of different types like carbon nanotube, inorganic nanotube, DNA nanotube, membrane nanotube etc. [33]. Carbon nanotubes can be made more soluble by incorporation of carboxylic or ammonium groups to their structures and can be used for the transport of peptides, nucleic acids and other drug molecules. The ability of nanotubes to transport DNA across cell membrane is used in studies involving gene therapy. DNA can be attached to the tips of nanotubes or can be incorporated within the tubes [34].

Nanopores

Nanopores (20 nm in diameter) consist of wafers with high density of pores which allow entry of oxygen, glucose and other chemicals such as insulin to pass through. Nanopores can be used as devices to protect transplanted tissues from the host immune system, at the same time, utilizing the benefit of transplantation [35]. β-Cells of pancreas can be enclosed within the nanopore device and implanted in the recipient's body. Nanopores can also be employed in DNA sequencing. Nanopores are also being developed with an ability to differentiate purines from pyrimidines [36].

Quantum Dots

Quantum dots (QD) are tiny semiconductor nanocrystals type of particles generally no larger than 10 nanometers that can be made to fluoresce in different colours when stimulated by light. The biomolecule conjugation of the QD can be modulated to target various biomarkers [37]. They can be tagged with biomolecules and used as highly sensitive probes. QD can also be used for imaging of sentinel node in cancer patients for tumour staging and planning of therapy. This technology also outlines some early success in the detection and treatment of breast cancer [38]. QD may provide new insights into understanding the pathophysiology of cancer and real time imaging and screening of tumors.

Bioconjugated QD are collections of variable sizes of nanoparticles embedded in tiny beads made of polymer material. In a process called "multiplexing," they can be finely tuned to a myriad of luminescent colors that can tag a multitude of different protein biomarkers or genetic sequences in cells or tissues [39]. The new class of quantum dot conjugate contains an amphiphilic triblock copolymer layer for *in vivo* protection and multiple PEG molecules for improved biocompatibility and circulation, making it highly stable and able to produce bright signals. Another advantage is that quantum dot probes emitting at different wavelengths can be used together for imaging and tracking multiple tumor markers simultaneously, potentially increasing the specificity and sensitivity of cancer detection [40]. Recent progress in the surface chemistry of QD has expanded their use in biological applications, reduced their cytotoxicity and rendered quantum dots a powerful tool for the investigation of dinstinct cellular processes, like uptake, receptor trafficking and intracellular delivery. Another application of QD is for viral diagnosis. Rapid and sensitive diagnosis of Respiratory Syncytial Virus (RSV) is important for infection control and development of antiviral drugs. Antibody-conjugated nanoparticles rapidly and sensitively detect RSV and estimate relative levels of surface protein expression. A major development is the use of dual-colour QD or fluorescence energy transfer nanobeads that can be simultaneously excited with a single light source [41]. QD linked to biological molecules, such as antibodies, have shown promise as a new tool for detecting and quantifying a wide variety of cancer-associated molecules. In the field of nanomedicine, QD can make a worthy contribution to the development of new diagnostic and delivery systems as they offer unique optical properties for highly sensitive detection and they are well defined in size and shape and can be modified with various targeting principles.

APPLICATIONS OF NANOSCALE DRUG DELIVERY SYSTEMS

Nanotechnology for Brain Drug Delivery

The blood brain barrier (BBB) is a structure formed by a complex system of endothelial cells, astroglia, pericytes, and perivascular mast cells, preventing the passage of most circulating cells and molecules [42]. The tightness of the BBB is attributed mainly to the vascular layer of brain capillary endothelial cells which are interconnected side-by-side by tight and adherens junctions. Among the different nanodevices, nanosize drug delivery systems between 1 and 100 nm work as a whole unit in terms of transport to cross BBB [43]. Nanosize brain drug delivery systems may promote the targeting ability of drug in brain and at the same time enhance the permeability of molecules through BBB. However crossing of BBB by the nano drug carriers will depend completely on the physicochemical and biomimetic features and does not depend on the chemical structure of drug, inside the nanoparticles [44]. Nanosize drug carriers which do not cross BBB generally can be made "stealth" coated with some polymeric materials or other chemicals to avoid the reticuloëndothelial system, to display long circulation time and stability in blood, and may be functionalized to successfully cross the BBB and target brain [45].

Nanosize Drug Carriers in Ocular Drug Delivery

Drug loaded nanoparticles with favourable biological properties include prolonging the residence time, decreasing toxicity and high ability of drug penetration into the deeper layers of the ocular structure and minimizing precorneal drug loss by the rapid tear fluid turnover [46]. Nanoparticles could target at cornea, retina and choroid by surficial applications and intravitreal injection. Nanocarrier based drug delivery is suitable in the case of the retina, as it has no lymph system, hence retinal neovascularisation and choroidal neovascularization have similar environments to that of solid tumors, and the EPR effect as available for solid nanoparticles in case of solid tumor may be also available for drug delivery targeted to eyes by nanoparticles [47]. Nanoparticles can deliver ocular drugs to the target sites for the treatment of various diseases such as glaucoma, corneal diseases, diabetic retinopathy etc. The uses of nanotechnology based drug delivery systems like nanosuspensions, SLNs and nanoliposomes have greater effect for ocular therapeutic efficacy [48]. Nanotechnology-based drug delivery is also very efficient in crossing membrane barriers, such as the blood retinal barrier in the eye.

Nanoparticle Loaded Contact Lenses

Contact lenses loaded with nanoparticles can be effective for topical administration of ophthalmic drugs. Drug loaded contact lenses can also provide continuous drug release because of slow diffusion of the drug molecules through the lens matrix. The soaked contact lenses also delivered drugs only for a period of few hours for some typical drugs [49]. The duration of drug delivery from contact lenses can be significantly increased if the drug is first entrapped in nanoformulations, such as nanoliposomes, nanoparticles, or microemulsions. Such drug nanocarriers can then be dispersed throughout the contact lens material. The entrapment of drug in nanocarriers also prevents the interaction of drug with the polymerization mixture. This provides additional resistance to drug release, as the drug must first diffuse through the nanocarriers and penetrate the drug carrier surface to reach the contact lens matrix [50].

Biodistribution of Nanoparticles in the Retina

The ocular biodistribution of nanoparticles can provide insights into the bioavailability, cellular uptake, duration of drug action and toxicity. Factors such as particle size, composition, surface charge and mode of administration influence the biodistribution in the retinal structures and also their drainage from the ocular tissues [51]. Larger particles (2 μm) were found to remain in vitreous cavity near the trabecular meshwork from which they are discharged out from the ocular tissue within 6 days, whereas the particles 200 nm were found evenly distributed in the vitreous cavity, and the inner limiting membrane. The smaller particles ~50 nm crossed the retinal barriers, and was detected in the retina even after 2 months post injection [52]. The surface chemistry can also affect nanoparticle distribution. Positively charged nanoparticles can adhere to the anionic vitreous network components and aggregate within the vitreous network. The surface chemistry can also affect nanoparticle distribution. Posively charged nanoparticles can adhere to the anionic vitreous network components and aggregate within the vitreous humor [53]. Anionic nanoparticles were found to diffuse through the vitreous humor and could even penetrate the retinal layers to be taken up by Muller Cells [54]. Vitreous humor is regarded as the barrier for non-viral ocular gene therapy because of the strong interaction of conventional cationic nature of non-viral gene vectors with the anionic vitreous humor [53]. The cationic PEI nanoparticles aggregated within vitreous humor and were prevented from distributing to the retina by the vitreal barrier. In contrast, cationic glycol chitosan (GC) nanoparticles and GC/PEI blended nanoparticles could penetrate the vitreal barrier and even reach at the inner limiting membrane because of the existence of glycol groups on nanoparticles [55].

Nanoparticles in Cancer

Cancer cells are more vulnerable than normal cells to the effect of chemotherapeutic agents and the most of the anticancer drugs can cause injury to the normal cells. Optimum dose and frequency are both important factors in the persistence of cancer cells during cancer chemotherapy [56]. Now attempts are focused on efforts to kill cancer cells by more specific targeting while sparing the normal cells.

Nanoparticulate delivery systems in cancer therapies provide better penetration of therapeutic and diagnostic substances within the cancerous tissue in comparison to conventional cancer therapies [57]. Nanoparticles are constructed to take advantages of fundamental cancer morphology and modes of development such as rapid proliferation of cells, antigen expression, and leaky tumor vasculature. Nanoparticulate drug delivery systems are being developed to deliver smaller doses of chemotherapeutic agents in an effective form and control drug distribution within the body [58]. Nanocarriers can offer many advantages over free drugs in cancer chemotherapy such as they protect the drug from premature degradation, prevent drugs from prematurely interacting with the biological environment, enhance absorption of the drugs into a selected tissue (solid tumour), control the pharmacokinetic and drug tissue distribution profile and improve intracellular penetration [59].

Nanoparticulate delivery systems utilize specific targeting agents for cancer cells minimizing the uptake of the anticancer agent by normal cells and enhance the entry and retention of the agent in tumor cells (Figure 3) [60]. Nanocarriers may actively bind to the specific cancer cells by attaching targeting agents with the help of ligand molecules to the surface of the nanocarriers that bind to specific receptor antigens on the cell surface. Nanocarriers will recognize and bind to target cells through ligand receptor interactions. It is even possible to increase the drug targeting efficacy with the help of antibodies by conjugating a therapeutic agent directly to it for targeted delivery [61].

Like receptor targeting, targeting of angiogenic factors also takes advantage of properties unique to cancer cells. Anti-angiogenic treatment is the use of drugs or other substances to stop tumors from developing new blood vessels. In a study nanoparticles were formulated comprising a water-based core of Vickers microhardness sodium alginate, cellulose sulphate, and anti-angiogenic factors such as thrombospondin (TSP)-1 or TSP-517, crosslinked with dextran polyaldehyde with calcium chloride or conjugated to heparin sulphate with sodium chloride. In addition bioluminescent agent, luciferase, or contrast agent, polymeric gadolinium was located within the polyanionic core [62] for drug targeting and detection. Similarly, many efforts are on for cancer cell targeting specifically with drug nanocarriers.. Thus the drug nanocarriers are of great hope for future cancer therapy.

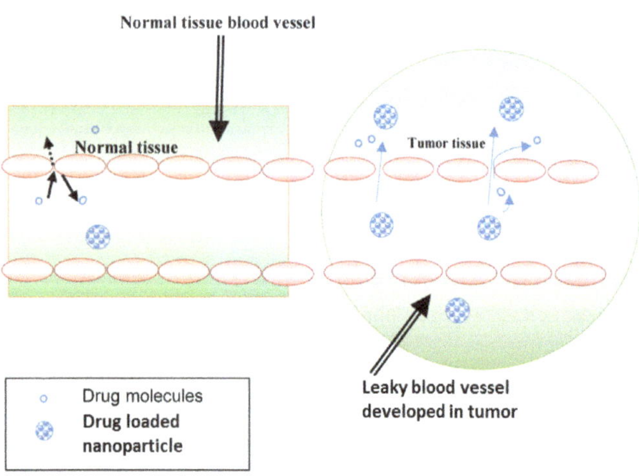

Figure 3. Schematic diagram of nanoparticle permeation and retention effect in normal and tumour tissues. Normal tissue vasculatures are lined by tight endothelial cells, hereby preventing nanoparticulate drug delivery system from escaping, whereas tumor tissue vasculatures are leaky and hyperpermeable allowing preferential accumulation of nanoparticles or nanoliposomes in the tumor interstial space by passive targeting.

Gene Delivery

Transfer of genetic material in nanocarriers may be an approach for the treatment of various genetic disorders such as diabetes mellitus, cystic fibrosis, alpha 1 antitrypsin deficiency and may more. A number of systemic diseases are caused by lack of enzymes factors that are due to missing or defective genes [63]. Previously gene therapy which was used to treat genetic disorders nowadays being contemplated as carrier systems which could be implanted for combating diseases other than genetic disorder like malignant form of cancer, heart diseases and nervous diseases [64]. Nanoliposomes can be used to deliver genetic materials into cells. Nanoliposomes incorporated with PEG and galactose target liver cells effectively due to their rapid uptake by liver Kupffer cells. Gene therapy may be tried with liposomal nanocarriers for liver disorders such as Wilson's and hereditary hemochromatosis. Cationic nanoliposomes have been considered as potential non-viral human gene delivery system [65]. Another effective method for administering nanoliposomes is by using ligand receptor complex using EGF-EGFR system for targeting purpose by nanoliposomes where EGF is a small protein which binds with receptor EGFR. Also mixing cationic lipids with plasmid DNA leads to the formation of lipoplexes where the process is driven by electrostatic interactions [66]. The negatively charged genetic material (e.g. plasmid) is not encapsulated in nanoliposomes but complexed with cationic lipids by electrostatic interactions. Plasmid liposome complexes can enter the disease cells by infusion with the plasma or endosome membrane. Allovectin-7 (gene transfer product) is composed of a plasmid containing the gene for the major histocompatibility complex antigene HLA-B7 with B2 microglobulin formulated with the cytofectin [67]. The nature of a composed lipid decides the unloading of the gene from nanoliposomes which enables control over the mode of release, doping of nanoliposomes with neutral lipids such as 1,2-Dioleoyl-sn-glycero-3-phosphoethanolamine (DOPE) which helps in endosomal membrane fusion by recognizing and destabilizing the phospholipids in a flip flop manner which paves way for the liposomes to integrate in the membrane with the dissociation of nucleic acid into the cytoplasm [64].

Viral system based gene carrier had the ability to overcome the biological barriers in the body and then access to the host nucleus replicative machinery which resulted in the exploitations of the system for drug delivery using nanotechnology [64]. The development of a non-viral method for *in vivo* gene transfer was designed where the vector was packed into compact nanoparticles by successive additions of oppositely charged polyelectrolytes including an incorporation of ligands into the DNA-polyelectrolyte shells which were mixed with Pluronic F127 gel serving as a biodegradable adhesive to keep shells in contact with the targeted vessel [68].

A novel method of gene delivery is with viruses such as adeno associated virus (AAV) which have their virulent genes removed with lentiviruses, clearly showing their efficiency [64].

DRUG DELIVERY WITH THE HELP OF EMPTY VIRUS CAPSID

The viral nanoparticles (VNPs) consist of protein core which ranges in complexity from small capsid-protein homomers to larger protein-based heteromers capable of internalizing oligonucleotides and being enveloped by lipids. Chemical modification process and genetic mutation provide the viral coat proteins with receptor binding domain that helps in cell specific targeting of VNPs [69]. Even fusion of terminal

/ internal proteins on the surface or inside the VNPs can be utilized for introduction of heterologous peptides, and in some cases entire proteins. VNPs can be genetically engineered by inserting amino acids for bioconjugation, peptide based affinity tags and peptides as targeting ligands for stimulation of immune response. [70].

High sequence variability due to the influence of the immune system in viral life-cycles is often seen on the surface loops of viral capsid proteins. This variability makes the loops highly susceptible to insertion of foreign sequences. VP1, the major coat protein of viruses of Polyomaviridae family, when expressed in insect cells, yeast and *Escherichia coli* self-assembles as protein cages and shows natural affinity for a cell surface glycoprotein with a terminal a 2,3-linked N-acetylneuraminic acid and attaches to a4h1-integrin receptors [71]. Virus like particle (VLPs) constructed from the virus are used to deliver therapeutic genes to human fetal glial cells. Another technique Cell-docking involves attachment of antibodies to the surface of brain natriuretic peptide (BNPs). Coupling reaction between murine polyomavirus and antitumor antibody B3 yielded polyoma VLPs with 30 to 40 antibody fragments bound to the surface, allowing the modified VLPs to bind to the breast carcinoma cells with high efficiencies [72].

A GLIMPSE TO FUTURE OF NANOSIZE DRUG DELIVERY SYSTEMS

Advancement of nanosize drug delivery systems establishes a new paradigm in pharmaceutical field. Convergence of science and engineering leads a new era of hope where medicines will act with increase efficacy, high bioavailability and less toxicity. Several nanoscale drug delivery systems are currently in clinical trials and few of them are already commercially available. Examples of such products are Abeicet (for fungal infection), Doxil (antineoplastic), Abraxane (metastatic breast cancer), Emend (antiemetic) etc. Despite the impressive progress in the field, very few nanoformulations have been approved by US-FDA (United States Food and Drug Administration) and even reached market in recent years. Although nanocarriers have lots of advantages because of the unique properties they have, there are many clinical, toxicological and regulatory aspects which are the matters of concern too. The biocompatibility of nanomaterials is of atmost importance because of the effect of the nanomaterials in the body ranging from cytotoxicity to hypersensitivity [8]. With the advancement of nanotechnology, the biological phenomenon such as host response to a specific nanomaterial should also be clinically transparent [9]. Therefore it is quite essential to introduce cost effective, better and safer nanobiomaterials which will provide efficient drug loading and controlled drug release of some challenging drug moieties for which there is no other suitable delivery available yet.

Nanoliposomes are well developed and presently possess the highest amount of clinical trials among other nanomaterials with some formulations currently in the market. This may be due to the fact that other materials have not been investigated for the same duration and are relatively newer in comparison. However polymer based nanomaterial, carbon nanotubes, gold nanoparticles etc. should not be overlooked because of less number of clinical trials [7].

Genexol-PM is an example which was undergone recent clinical trial. This is an amphiphilic diblock co-polymer (PEG-D, L-Lactic acid) that delivers paclitaxel. Clinical trial currently is in phase IV using Genexol-PM for recurrent breast cancer and phase III for breast cancer. Fungal infections associated with acute leukemia and for central line fungal infections, amphotericin B containing nanoliposomes are in phase IV clinical trial. ThermoDox (Doxorubicin loaded nanoliposome) is currently in phase III trials for hepatocellular carcinoma. Similarly Caelyx, a doxorubicin HCl loaded nanoliposome that is pegylated, is currently in phase IV trials for ovarian neoplasms [7]. Some recent clinical trials are shown in Table 1.

Ligand or antibody conjugated nanoformulation, bifunctional and multifunctional nanoparticles are the newer research approaches through which detection and treatment of cancerous cells can be achieved. Nanomachines are also largely in the research-and-development phase, but some primitive molecular machines have been tested. An example is nanorobot which is capable of penetrating the various biological barriers of human body to identify the cancer cells. Thus, nanodrug delivery systems have a leading role to play in nanomedicine in near future.

Table 1. Recent Nanodrug Carriers in Clinical Trials (Source: Clinicaltrials.gov)

Product name	**Delivery material**	**Phase**	**Condition**	**Therapeutic Delivered**	**Sponsor**	**Clinicaltrials. gov Identifier**
Genexol-PM	Amphilic diblock Copolymer forming micelle	I	Non small Cell lung cancer	Paclitaxel	Samyang Biopharmaceutical Corp	NCT01023347
Docetaxel-PNP	Polymeric nanoparticles	I	Advanced solid malignancies	Docetaxel	Samyang Biopharmaceutical Corp	NCT01103791
CYT-6091	AuNP	I	Unspecified adult solid tumor	TNF	NCI	NCT00356980
Paclitaxel poliglumex	Drug Polymer Conjugate	II	Prostate cancer	Paclitaxel	OHSU Knight Cancer Institute	NCT00459810

Kogenate FS	PEG-liposome	I	Hemophilia A	Recombinant factor VIII	Bayer	NCT00629837
Long-circulating liposomal prednisolone disodium phosphate	Liposome	II	Rheumatoid arthritis	Prednisolone	Radboud University	NCT00241982
LE-DT	Liposome	II	Pancreatic cancer	Doxetaxel	Insys Therapeutics Inc	NCT01186731
Cisplatin and Liposomal Doxorubicin	Liposome	I	Advanced cancer	Cisplatin and doxorubicin	M.D. Anderson Cancer Center	NCT00507962
Liposomal doxorubicin and bevacizumab	Liposome	II	Kaposi's sarcoma	Doxorubicin and bevacizumab	NCI	NCT00923936
AP5346	Drug polymer conjugate	Not stated	Head and neck cancer	AP5346 and oxaliplatin	University of California, San Diego	NCT00415298

[i] - Abbreviations: PEG-Polyethylene glycol, TNF-Tumor necrosis factor, NCI-National Cancer Institute, AuNP-Gold nanoparticles

Nanocarriers may lead to a solution to major unsolved medical problems which will aggressively enhance quality of life.

***Regulatory Aspect*:** One of the main areas related to the safety aspects of drug-nanocarrier systems is to encourage academic organizations, industry and regulatory governmental agencies to establish convincing testing procedures on the safety aspects of the nanomaterials. The global importance of trade for nanomaterials has established new international organizations, such as the International Council on Nanotechnology (ICON), the International Organization for Standardization (Geneva, Switzerland) etc. for sharing responsibilities in this field. In the year 1996 the NNI was established in the United States of America to coordinate governmental multi-agencies such as the Food and Drug Administration (FDA), the Department of Labor through the Occupational Safety and Health Administration (OSHA), the National Institute for Occupational Safety and Health (NIOSH), and the Environmental Protection Agency (EPA), for the development of nanoscience and technology.

CONCLUSION

Last few years several new technologies have been developed for the treatment of various diseases. The use of nanotechnology in developing nanocarriers for drug delivery is bringing lots of hope and enthusiasm in the field of drug delivery research. Nanoscale drug delivery devices present some advantages which show higher intracellular uptake than the other conventional form of drug delivery systems. Nanocarriers can be conjugated with a ligand such as antibody to favor a targeted therapeutic approach. The empty virus capsids are also being tried to use for delivering drugs as a new therapeutic strategy. Thus, nanoscale size drug delivery systems may revolutionize the entire drug therapy strategy and bring it to a new height in near future. However, toxicity concerns of the nanosize formulations should not be ignored. Full proof methods should be established to evaluate both the short-term and long-term toxicity analysis of the nanosize drug delivery systems.

REFERENCES

1. Ochekpe NA, Olorunfemi PO, Ngwuluka NC. Nanotechnology and drug delivery part 1: background and applications. Trop J Pharm Res. 2009; 8(3): 265-274.
2. Commission Staff Working Paper. Types and uses of nanomaterials including safety aspects accompanying the communication from the commission to the European Parliament, the council and the European and Economic Social Committee on the Second Regulatory Review on Nanomaterials. SWD 2012; 288 Final.
3. Svenson S, Prud'homme RK. Multifunctional nanoparticles for drug delivery applications. The nanotech revolution in drug delivery. Cientifica Ltd. 2007; 978-1-4614- 2304-1.
4. Hadjipanayis CG, Machaidze R, Kaluzova M, Wang L, Schuette AJ, Chen H, Wu X, Mao H. EGFRvIII antibody-conjugated iron oxide nanoparticles for magnetic resonance imaging- guided convection-enhanced delivery and targeted therapy of glioblastoma. Cancer Res. 2010; 70(15): 6303- 6312.
5. Jong WHD, Borm PJA. Drug delivery and nanoparticle applications and hazards. Int J Nanomedicine. 2008; 3(2): 133- 149.
6. Sahoo SK, Parveen S, Panda JJ. The present and future of nanotechnology in human healthcare. Nanomed Nanotech Biol Med. 2007; 3(1): 20-31.
7. Webster DM, Sundaram P, Byrne ME. Injectable nanomaterials for drug delivery: carriers, targeting moieties, and therapeutics. Eur J Pharm Biopham. 2013; 84(1): 1-20.

8. Chakroborty G, Seth N, Sharma V. Nanoparticles and nanotechnology: clinical, toxicological, social, regulatory and other aspects of nanotechnology. JDDT. 2013; 3(4): 138-141.
9. Couvreur P. Nanoparticles in drug delivery: past, present and future. Adv Drug Delivery Rev. 2013; 65(1): 21-23.
10. Panyam J, Labhasetwar V. Biodegradable nanoparticles for drug and gene delivery to cells and tissue. Adv Drug Delivery Rev. 2003; 55(3):329-347.
11. Praetorius NP, Mandal TK. Engineered nanoparticles in cancer therapy. Recent Pat Drug Deliv and Formul. 2007; 1(1): 37-51.
12. Pardeike J, Hommoss A, Müller RH. Lipid nanoparticles (SLN, NLC) in cosmetic and pharmaceutical dermal products. Int J Pharm. 2009; 366(1-2): 170-184.
13. Ogawara KI, Un K, Tanaka K, Higaki K, Kimura T. *In vivo* anti-tumor mal doxorubicin (DOX) in Dox-resistant tumor-bearing mice: involvement of cytotxic effect on vascular endothelial cells. J Cont Rel. 2009; 133(1): 4-10.
14. Chan JM, Zhang L, Yuet KP, Liao G, Rhee JW, Langer R, Farokhzad OC. PLGA-lecithin-PEG core-shell nanoparticles for controlled drug delivery. Biomaterials. 2009; 30(8): 1627-1634.
15. Huang X , Jain PK, El-Sayed IH, El-Sayed MA. Gold nanoparticles: interesting optical properties and recent applications in cancer diagnostics and therapy. Nanomedicine. 2007; 2: 681-693.
16. Wei XL, Mo ZH, Li B, Wei JM. Disruption of Hep G2 cell adhesion by gold nanoparticle and paclitaxel disclosed by *in situ* QCM measurement. Colloids Surf B Biointerfaces. 2007; 59: 100-104.
17. Yousefi A, Esmaeili F, Rahimian S, Atyabi F, Dinarvand R. Preparation and *in vitro* evaluation of a pegylated nano-liposomal formulation containing docetaxel. Sci Pharm. 2009; 77: 453-464.
18. Patel S, Bhirde AA, Rusling JF, Chen X, Gutkind JS, Patel V. Nano delivers big: designing molecular missiles for cancer therapeutics. Pharmaceutics. 2011; 3(1): 34-52.
19. Patel RP, Patel H, Baria AH. Formulation and evaluation of liposomes of ketokonazole. Int J Drug Deliv Technol. 2009; 1(1): 16-23.
20. Kumar A, Badde S, Kamble R, Pokharkar VB. Development and characterization of liposomal drug delivery system for nimesulide. Int J Pharm Pharm Sci. 2010; 2(4): 87-89.
21. Arab Tehrany E, Kahn CJ, Baravian C, Maherani B, Belhaj N, Wang X, Linder M. Elaboration and characterization of nanoliposome made of soya, rapeseed and salmon lecithins: application to cell culture. Colloids Surf B Biointerfaces. 2012; 95: 75-81.
22. Kumar KPS, Bhowmik D, Deb L. Recent trends in liposomes used as novel drug delivery system. J Pharma Innovation. 2012; 1(1): 26-34.
23. Dadashzadeh S, Vali AM, Rezaie M. The effect of PEG coating on *in vitro* cytotoxicity and *in vivo* disposition of topotecan loaded liposomes in rats. Pharmaceutical Nanotechnology. 2008; 353: 251-259.
24. Dhanikula RS, Hammady T, Hildgen P. On the mechanism and dynamics of uptake and permeation of polyether copolyester dendrimers across an*in vitro* blood-brain barrier model. J Pharm Sci. 2009; 98(10): 3748-3760.
25. Albertazzi L, Gherardini L, Brondi M, Sulis Sato S, Bifone A, Pizzorusso T, Ratto GM, Bardi G. *In vivo* distribution and toxicity of PAMAM dendrimers in the central nervous system depend on their surface chemistry. Mol Pharmacol. 2013; 10(1): 249-260.
26. Kannan S, Dai H, Navath RS, Balakrishnan B, Jyoti A, Janisse J, Romero R, Kannan RM. Dendrimer-based postnatal therapy for neuroinflammation and cerebral palsy in a rabbit model. Sci Transl Med. 2012; 4(130): 130ra46.
27. Klajnert B, Bryszewska M. Dendrimers: properties and applications. Acta Bioch Pol. 2001; 48(1): 199-208.
28. Huang RQ, Qu YH, Ke WL, Zhu JH, Pei YY, Jiang C. Efficient gene delivery targeted to the brain using a transferrin-conjugated polyethylene glycol-modified polyamidoamine dendrimer. Drug Delivery Systems: Advanced Technologies Potentially Applicable in Personalised Treatment. 2007; 21(4): 1117-1125.
29. Kherlopian AR, Song T, Duan Q, Neimark MA, Po MJ, Gohagan JK, Laine AF. A review of imaging techniques for systems biology. BMC Syst Biol. 2008; 2: 74.
30. Lowery AR, Gobin AM, Day ES, Halas NJ, West JL. Immunonano shells for targeted photothermal ablation of tumor cells. Int J Nanomed. 2006; 1(2): 149-154.
31. Hirsch LR, Stafford RJ, Bankson JA, Sershen SR, Rivera B, Price RE. Nanosell-mediated near-infrared thermal therapy of tumors under magnetic resonance guidance. Prof Natl Acad Sci USA. 2003; 100: 13549-13554.
32. Reilly RM. Carbon nanotubes: potential benefits and risks of nanotechnology in nuclear nuclear medicine. J Nucl Med. 2007; 48(7): 1039-1042.
33. Prato M, Kostarelos K, Bianco A. Functionalized carbon nanotubes in drug design and discovery. Acc Chem Res. 2008; 41(1): 60-68.
34. Mc Devitt MR, Chattopadhyay D, Kappel BJ, Jaggi JS, Schiffman SR, Antczak C, Njardarson JT, Brentjens R, Scheinberg DA. Tumor targeting with antibody functionalized, radiolabelled carbon nanotubes. J Nucl Med. 2007; 48: 1180-1189.
35. Freitas RA. Current status of nanomedicine and medical nanorobotics. J Comput Theor Nanosci. 2005; 2: 1-25.
36. Yadav A, Ghune M, Jain DK. Nanomedicine based drug delivery system. JAPER. 2011; 1(4): 201- 213.
37. Gangrade SM. Nanocrystals-a way for carrier free drug delivery. Pharma Buzz. 2011; 6: 26-31.
38. Gao X, Dave SR. Quantum dots for cancer molecular imaging. Adv Exp Med Boil. 2007; 620: 57-73.
39. Mulder WJM, Koole R, Brandwijk RJ, Storm G, Chin PTK, Strijkers GJ, Donega CM, Nicolay K, Griffioen AW. Quantum dots with a paramagnetic coating as a

bimodal molecular imaging probe. Nano letter. 2006; 6: 87-90.

40. Jain S, Shukla K, Jain V, Saraf S. Nanoparticles: emerging carriers for delivery of bioactive agents. Pharma Times. 2007; 39: 30-35.

41. Amiot CL, Xu S, Liang S, Pan L, Zhao JX. Near-infrared fluorescent materials for sensing of biological targets. Sensors. 2008; 8(5): 3082-3105.

42. Petty MA, Lo EH. Junctional complexes of the blood brain barrier: permeability changes in neuroinflammation. Prog Neurobiol. 2002; 68(5): 311-323.

43. Rabanel JM, Aoun V, Elkin I, Mokhtar M, Hildgen P. Drug loaded nanocarriers: passive targeting and crossing of biological barriers. Curr Med Chem. 2012; 19(19): 3070-3102.

44. Youns M, Hoheisel JD, Efferth T. Therapeutic and diagnostic applications of nanoparticles. Curr Drug Targets. 2011; 12(3): 357-365.

45. Gabathuler R. Approaches to transport therapeutic drugs across the blood- brain barrier to treat brain diseases. Neuro biology of Disease. 2010; 37(1): 48-57.

46. Kesavan K, Balasubramaniam J, Kant S, Singh PN, Pandit JK. Newer approaches for optimal bioavailability of ocularly delivered drugs: review. Curr Drug Deliv. 2011; 8(2): 172-193.

47. Yasukawa T. Drug delivery systems for vitreoretinal diseases. Prog Retin Eye Res. 2004; 23(3): 253-281.

48. Behar-Cohen F. Drug delivery to target the posterior segment of the eye. Med Sci (Paris). 2004; 20(6-7): 701-706.

49. Gulsen D, Chauhan A. Ophthalmic drug delivery through contact lenses. Invest Ophthalmol Vis Sci. 2004; 45(7): 2342-2347.

50. Gulsen D, Chauhan A. Dispersion of microemulsion drops in HEMA hydrogel: a potential ophthalmic drug delivery vehicle. Int J Pharm. 2005; 292(1-2): 95-117.

51. Amrite AC, Edelhauser HF, Singh SR, Kompella UB. Effect of circulation on the and ocular tissue distribution of 20 nm nanoparticles after periocular administration. Mol Vis. 2008; 14: 150-160.

52. Kim JH, Kim JH, Kim KW, Kim MH, Yu YS. Intravenously administered gold nanoparticles passing through the blood retinal barrier depending on the particle size, and induce no retinal toxicity. Nanotechnology. 2009; 20(50): 505101.

53. Koo H, Moon H, Han H, Na JH, Huh MS, Park JH, Kwon IC, Kim K, Kim H. The movement of self- assembled amphiphilic polymeric nanoparticles in the vitreous and retina after intravitreal injection. Biomaterials. 2012; 33(12): 3485-3493.

54. Peeters L, Sanders NN, Braeckmans K, Boussery K, Van de Voorde J, De Smedt SC, Demeester J. Vitreous: a barrier to nonviral ocular gene therapy. Invest Ophthalmol Vis Sci. 2005; 46(10): 3553-3561.

55. Kim H, Robinson SB, Csaky KG. Investigating the movement of intravitreal human serum albumin nanoparticles in the vitreous and retina. Pharm Res. 2009; 26(2): 329-337.

56. Surendiran A, Sandhiya S, Pradhan SC, Adithan C. Novel applications of nanotechnology in medicine. Indian J Med Res. 2009; 130: 689-701.

57. Malam Y, Loizidou M, Seifalian AM. Liposomes and nanoparticles: nanosized vehicles for drug delivery in cancer. Trends Pharmacol Sci. 2009; 30(11): 592-599.

58. Byrne JD, Betancourt T, Brannon-Peppas L. Active targeting schemes for nanoparticle systems in cancer therapeutics. Adv Drug Delivery Rev. 2008; 60: 1615-1626.

59. Ruoslahti E, Bhatia SN, Sailor MJ. Targeting of drugs and nanoparticles to tumors. J Cell Biol. 2010; 188 (6): 759-768.

60. Praetorius NP, Mandal TK. Engineered nanoparticles in cancer therapy. Recent Pat Drug Deliv Formul. 2007; 1(1): 37-51.

61. Peer D, Karp JM, Hong S, Farokhzad OC, Margalit R, Langer R. Nanocarriers as an emerging platform for cancer therapy. Nature Nanotech. 2007; 2: 751-760.

62. Lammers T, Hennink WE, Storm G. Tumour-targeted nanomedicines: principles and practice. British Journal of Cancer. 2008; 99: 392-397.

63. Davis PB, Cooper MJ. Vectors for airway gene delivery. AAPS J. 2007; 9: E11-E17.

64. Alex SM, Sharma CP. Nanomedicine for gene therapy. Drug Deliv Transl Res. 2013; 3: 437-445.

65. Pathak A, Vyas SP, Gupta KC. Nano-vectors for efficient liver specific gene transfer. Int J Nanomed. 2008; 3: 31-49.

66. Bunuales M, Duzgunes N, Zalba S, Garrido MJ, Ilarduya CT. Efficient gene delivery by EGF lipoplexes *in vitro* and *in vivo*. Nanomedicine. 2011; 6(1): 89-98.

67. Boettger M, Zaitsev S, Cartier R, Haberland A, Sukhorukov G, Moehwald H, Zastrow H, Schneider M. WO03087384A1 (2003).

68. Niu L, Xu YC, Dai Z, Tang HQ. Gene therapy for type 1 diabetes mellitus in rats by gastrointestinal administration of chitosan nanoparticles containing human insulin gene. World J Gastroenterol. 2008; 14: 4209-4215.

69. Caruso M, Belloni L, Sthandier O, Amati P, Garcia MI. Alpha4beta1 integrin acts as a cell receptor for murine polyomavirus at the level. J Virol. 2003; 77: 3913- 3921.

70. Young M, Willits D, Uchida M, Douglas T. Plant viruses as biotemplates for materials and their use in nanotechnology. Annu Rev Phytopathol. 2008; 46: 361-384.

71. Gleiter S, Lilie H. Cell-type specific targeting and gene expression using a polyoma VP1 virus-like particles. Biol Chem. 2003; 384(2): 247- 255.

72. Wang Q, Chan TR, Hilgraf R, Fokin VV, Sharpless KB, Finn MG. Bioconjugation by copper (I)-catalyzed azide-alkyne [3+2] cycloaddition. J Am Chem Soc. 2003; 125(11): 3192 - 3193.

CHAPTER 7

SINGLE MOLECULE DNA DETECTION WITH AN ATOMIC VAPOR NOTCH FILTER

Denis Uhland, Torsten Rendler, Matthias Widmann, Sang-Yun Lee, Jörg Wrachtrup and Ilja Gerhardt

Keywords: DNA detection, fluorescence microscopy, single molecules, atomic filtering, sodium spectroscopy

INTRODUCTION

The optical detection of single molecules [1–3] has facilitated important progress in various fields of research. Especially in microbiology the localization and tracking of labeled biomolecules [4–6] or other relevant structures like DNA molecules [7–10] reveals underlying compositions and functionalities of living cells. Normally the red-shifted fluorescence of single molecule labels is detected. Besides probe specific properties like the (non-)radiative decay rate of the involved electronic transitions also experimental parameters like sample absorption, collection and detection efficiency of the given setup determine the number of detectable photons. Under ambient conditions the total number of emitted photons of an organic fluorophore is limited by photo-bleaching [11]. This poses a serious problem, since parameters such as the localization accuracy are limited by the finite number of detected photons. In single molecule studies the precision is usually shot-noise limited, scaling as $1/N$—$\sqrt{1/N}$, where N is the total number of detected photons. Once a fluorescent dye is chosen there remain two key parameters to optimize: The collection and detection efficiency of the microscope.

In the last decade there have been many efforts to optimize all involved experimental parameters. For example photo-stable alternatives to dyes like defect centers in diamond [12,13] and semiconductor nano-crystals [14] were investigated. Non-fluorescent structures can be detected with a high signal-to-noise ratio [15], or novel techniques extract a higher amount of information from an acquired image [16]. Other attempts have been made to increase the extraction efficiency from the structure under study [17, 18]. Additionally, the detectors have been continuously improved, such that nowadays silicon avalanche photon diodes (APDs) exhibit more than 70% quantum efficiency in the range of 650-750 nm and sensitive CCD-cameras more than 95% from 500-650 nm. Another way to enhance the overall detection efficiency is an optimized filtering scheme. An ideal filter solely blocks the excitation light which is scattered by the sample and transmits all photons originated by fluorescence. The edge-steepness should be ideally a step-function as sharp as the spectra laser line, but is often limited by material properties and technical imperfections to some nanometers which are common in dichroic mirrors and filters.

Hot atomic vapors can also allow for optical filtering. These are generally easy to handle in evacuated glass reference cells. They are typically composed of an evacuated glass cylinder with optical windows, in which a small amount of alkali metals (hundreds of milligram) is present. Such cells can exhibit a large optical depth and simultaneously ensure a few GHz spectral width. The optical rejection obeys Beer-Lambert's law and is a function of the vapor density, which rises approximately exponentially with temperature. The spectral width is to a first approximation given by the Doppler broadening of an atomic vapor in the range of a few GHz at ambient conditions up to a few hundred degree centigrade [19]. In atom optics experiments such filtering schemes are common and have been characterized for different alkali metals like rubidium [20, 21]. One common use is to filter the emission of one isotope of rubidium with another, an application which is commonly implemented in atomic clocks [22]. These filters are also suitable for other applications like Raman spectroscopy [23–26] and are intrinsically matched to atomic transitions.

Citation: Denis Uhland, Torsten Rendler, Matthias Widmann, Sang-Yun Lee, Jörg Wrachtrup and Ilja Gerhardt, Single molecule DNA detection with an atomic vapor notch filter, doi:10.1140/epjqt/s40507-015-0033-1

Here we compare the filtering performance of a high-end commercial filter with a hot atomic sodium vapor cell in context of detection of single fluorescing molecules in a confocal [27] and a wide-field microscope [28]. Unlike many other atomic filters, sodium matches the visible range of many common dye systems which are used for biological labeling. For this demonstration we study single stranded DNA molecules, labeled with a commonly used fluorescent dye (Atto 590, ATTO-TEC). Both microscopic schemes, confocal and wide-field, have their specific advantages and disadvantages. We show that filtering with atomic vapor is able to facilitate an enhanced detection of the fluorescence which originates from a single molecule. The first steps of this study were presented for single molecules under cryogenic conditions [29], and in a micro-fluidic configuration [30].

EXPERIMENTAL CONFIGURATION

The experimental configuration consists of a combined confocal and wide-field microscope (Figure 1(b)). Both experimental configurations are described below. Filtering is performed with a commercial filter and an atomic vapor cell.

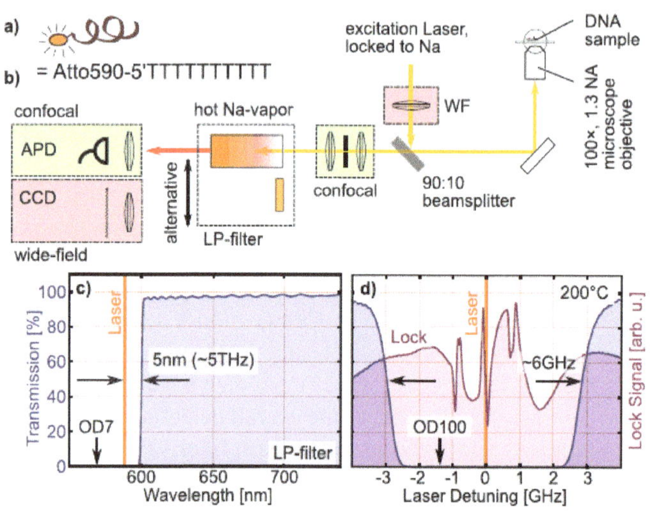

Figure 1 Experimental configuration. (a) Single stranded DNA under study, labeled at the 5'-end with the organic dye 'Atto590'. (b) Confocal and wide-field microscope. The filter configuration can be changed in approx. 1 s. WF = wide-field configuration; APD = avalanche photo diode; LP-filter = long-pass filter; CCD = charged coupled device, camera; (c) Laser emission and filter function of the commercial long pass filter. OD = optical density; (d) Spectrum of the sodium reference cell scanning over the D_2resonance frequency. The purple line shows the Doppler-free difference laser lock signal. The blue line shows the measured absorption spectrum of the atomic vapor cell at 200°C. The block band is approx. 6 GHz broad and shows an optical rejection of more than 6 orders of magnitude (measured). The calculated optical density is much higher (OD 100).

Excitation Laser

The excitation laser is a dye ring laser (699-21, Coherent), which can be locked to a sodium transition. The lock-signal is provided by Doppler-free dichroic atomic vapor laser lock (DAVLL, see e.g. [31]) with a 140°C hot sodium vapor cell. For optimal optical rejection the laser is locked to the cross-over resonance, midway between the two F=1F=1 and F=2F=2 ground states of the D_2-line (see Figure 1(d)). This defines for further filtering purposes a preferred point, since both optical transitions add up due to their Doppler broadening. Note, that this zero-point does not represent the 'center of gravity' of the unshifted sodium transition. The dispersive lock signal depicted in Figure 1(d) is robust against external influences such as mechanical noise on the laser table. For the rare case the laser jumps out of lock, the signal is monitored with an oscilloscope (LeCroy, WavePro 7K), where the pass-fail output controls an optical shutter and blocks the laser if required to prevent damage to the used single photon detector. A 20 m long optical single mode glass fiber guides the laser beam to the setup. The light is then filtered with a narrow band-pass filter (589 ± 0.2 nm, Omega Optical), since our analysis showed that we can further suppress the background level (data not shown). The microscope is built around a commercial inverted microscope (Olympus IX71). In both configurations the sample is excited and detected via a 100×, 1.3 NA microscope objective (UPLANFL, Olympus).

Confocal Microscope

For the confocal configuration, the collimated beam is reflected into the microscope via a quartz wedge (10% reflectivity). This is not realized by a dichroic beam-splitter to avoid additional spectral cut-off of the detected fluorescence signal. The light is focused onto and collected from the sample with the microscope objective. Spatial discrimination is performed by focusing the resulting beam onto a 50 μm sized pinhole. From there, it is again 1:1 collimated and passes through the laser suppression filter under study. The filtered light is then focused onto a single photon counting module (SPCM-AQR-14, Excelitas). Addressing different locations on the sample and focusing is realized by scanning the sample with a 3D-piezo actuator (P527.3CL, Physik Instrumente). A pixel size of 100 nm was used with a typical integration time of 10 ms per pixel. The entire detection scheme is carefully optically shielded from the environment.

To acquire single molecule spectra, a flip-mirror is introduced into the confocal configuration to divert the light from the single photon detector to a Peltier cooled CCD spectrometer (Princeton Instruments, Acton, 300 mm, camera: 'Pixis'). An acquisition time of 30 s is used.

Wide-field Microscope

For the wide-field microscope experiments, the incident laser light is focused by a pair of achromatic lenses (100 mm and 50 mm focal length) into the back-focal plane of the microscope objective. The resulting illuminated area in the objective plane is 25 μm in diameter. For imaging, a Peltier cooled CCD camera is used (Photometrics Cascade 512B) with a pixel size of 16 × 16 μm² and 512 × 512 pixels. The objective plane is projected via a 250 mm achromatic lens onto the CCD chip. As a consequence one pixel represents a 115 × × 115 nm² sized area on the sample. A typical acquisition was performed with an exposure time of 10 s and no internal gain. For comparison between the two microscope configurations we convert the output by the camera, which is give in analog-digital units (ADUs) to an incident photon equivalent onto the CCD chip. In the used configuration 4 detected photons correspond to 1 ADU.

The Sample

The single stranded deoxyribonucleic acid (DNA) under study consists of 10 bases (5′-TTTTTTTTTT, see also Figure 1(a)) and is labeled on the 5′-end with Atto 590 (Thermo Scientific). This dye is a common dye label in micro-biology, which spectral matches to atomic sodium vapor. The sample is produced by first dissolving and diluting the dried DNA in sterile water. Then, an aqueous polyvinyl alcohol (PVA) solution (3 mg/ml) is stepwise mixed with the DNA solution until a relative DNA concentration of $1{:}10^{14}$-$1{:}10^{15}$ is reached. Finally the mixture is spin-coated onto a nitrogen-plasma cleaned cover slides. For most of the experiments a concentration of $1{:}10^{15}$ was used because it results in single molecules separated by several μm, whereas the $1{:}10^{14}$ diluted sample shows a too dense concentration for our automated peak-finding routine. The thickness of the sample has been calculated to be below 1 μm.

To verify that the experiments are performed on the single molecule level, the typical blinking behavior of single molecules is confirmed in the wide-field configuration with millisecond integration times. In the confocal configuration single step bleaching around t=500 st=500 s can be seen as shown in Figure 2(b), which is used for the verification of single molecules. Note, that no telegraph function or triplet blinking is observed during this acquisition.

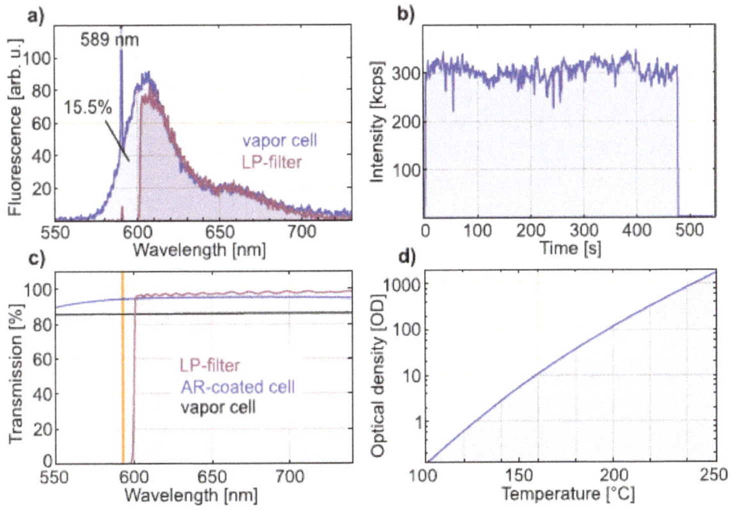

Figure 2 Relevant spectra. (a) Emission spectra of the dye Atto 590 under illumination with laser light at 589 nm, filtered solely with a hot (200°C) atomic vapor cell. Purple: Same spectrum, filtered with the commercial long-pass filter. The integral enhancement between atomic and the commercial filter is 15.5%. LP-filter = long-pass filter. (b) Single step bleaching of individual molecules proves the single emitter nature. (c) Transmission curves of the filter and different atomic vapor cells measured with a UV-VIS spectrometer (Lambda 16, PerkinElmer). (d) Calculated optical density of the filter (100 mm optical length) against temperature.

Filtering the Excitation Light

The optical filtering from the microscope is performed with two possible filter configurations: a commercial filter and an atomic sodium vapor cell.

The commercial filter (Semrock, FF01-593/LP-25) efficiently rejects the excitation light and exhibits around 97% transmission in its pass-band. A spectrum of the filter was recorded in a commercial absorption spectrometer (PerkinElmer, Lambda 16) and is depicted in Figures 1(c) and 2(c). Please note, that this is a representation in a linear scale and the 50% point is observed around 600 nm. Even if the filter is designed to have a six to seven orders of magnitude rejection from the blue region of the spectrum to 593 nm, it is probably possible to slightly tilt the filter to blue shift its transmission closer to the excitation laser. This may result in altering the performance (transmission and rejection) of the filter and was therefore not pursued.

Alternatively, the excitation light, scattered and reflected from the sample, is blocked solely by a hot atomic vapor cell. A simple calculation [19] of the six D_2 transitions (32S1/2→32P3/232S1/2→32P3/2, three from each ground state) allows to estimate the optical density for the 100 mm long cell (30 mm diameter) as a function of the temperature (Figure 2(d)). This is an idealized picture, since we do not account for any forward scattering, non-linear or saturation effects. Furthermore, the laser is represented by a delta-peak and locked to the cross-over resonance of the sodium D_2-line. The D_2-line is generally preferred for filtering, due to its higher oscillator strength by a factor of two against the D_1-line. Nevertheless, we measure a suppression of better than 6 orders of magnitude with our filter at a temperature of 200°C, which is chosen as the operation temperature for all further experiments. The measured filter function of the sodium vapor is depicted in Figure 1(d). The optical density of about 100 represents the calculated value at this temperature. The vapor pressure at 200°C is calculated to be ~10−4 mbar~10−4 mbar.

Atomic sodium tends to diffuse into the usual boro-silicate glass cell and darken the windows. Therefore, the cells were made out of quartz glass. We note that special glasses and coatings also allow a suppression of this darkening effect [32, 33]. To enhance the transmission on the cell windows, we used commercial anti-reflection coated sodium cells (Triad technologies, Longmont, Colorado). Although unspecified, the supplied coating works well up to 200°C. In addition another batch of cells was produced in house, but without an anti-reflection coating. The transmission spectra of these cells and the commercial filter is shown in Figure 2(c). As one can see, the transmission of the anti-reflection coated cell is 7-9% higher than for the uncoated one. With the resolution of the absorption spectrometer, the GHz-wide notch filtering for sodium light is not observed.

To directly compare the two filter configurations in the microscope, both filters are mounted onto a small optical bread board, which slides on rails on the optical table. The filter configuration can therefore be switched back and forth within less than a second. The area between the filters was filled with an opaque material, such that the filter configuration can be changed during a running experiment. Furthermore, the detection path was optically shielded using a black cardboard box. To ensure a well defined convection of air inside the box, it was designed such that a laminar flow was realized from bottom to top by small holes at the bottom.

RESULTS

Acquired Spectra

We first compare the different spectra, acquired from a single molecule in both filter configurations. When a single molecule was identified in the confocal microscope, the spectrometer was introduced and a spectrum was acquired. Figure 2(a) shows the single molecule spectra in both filter configurations. The filtering with the atomic vapor shows the entire spectrum of the single molecule as it would be excited with a more blue wavelength. On first sight, this violates energy conservation, but most likely the more blue components are introduced by anti-Stokes processes. On the other hand, we find the spectrum acquired with the commercial 593 nm long-pass filter to be cutoff until 600 nm. Furthermore, compared to the spectrum provided by the dye producer, we find the single molecule fluorescence to be spectral shifted to the blue by about 5 nm. This is not untypical for organic dyes that their fluorescent properties critically depend on their chemical environment [34, 35].

Assuming the quantum efficiency of the single photon detector is a fixed value in the range of 570-700 nm, we can directly compare the integral contribution of the dye spectra as proportional to the detected signal on the photo detector. The result is a 15.5% higher signal with atomic filtering than with a commercial long-pass filter.

Unfortunately, although the sodium vapor filter increases the overall signal from the molecule studied, it also introduced a higher laser background. In particular we find around 20 times increased laser background even though the optical rejection is calculated to be much higher than with the commercial filter. There are two explanations for this finding: (a) the filter does not have such a high rejection due to non-linearities or saturation effects in the vapor. Or, (b), the atomic filter or some components of our setup (e.g.: laser, optical fiber, sample) spectral shifts the scattered light far enough away from the laser wavelength that it cannot be blocked. To investigate this problem, we performed simple measurements of the atomic filter with laser light via bypassing the microscope with a mirror. This resulted in a measured optical rejection of more than 6 orders of magnitude. It was necessary to perform the measurements above the saturation intensity (9.4 mW/cm^2) to determine the optical rejection of the filter due to the weak signal. Generally, a weak laser background contribution is not relevant for the acquisition of single molecule signals, since this simply adds a constant background, as long as laser power fluctuations are not too large. Following this assumption the background can be simply subtracted.

Data Analysis

In the following paragraph we introduce the protocol for data analysis. For both, the confocal and the wide-field imaging, data processing was performed as follows: Since the main goal was to compare the two filter configurations, an image was acquired with each filter at the exact same settings (excitation intensity, acquisition time, etc.). Both images were acquired with minimal time delay and usually an acquisition sequence alternating several times between the commercial filter and the atomic filter was used. To avoid additional systematic errors, for example by photobleaching or mechanical drift out of the objective plane, also the filter which was used first had been alternated from sequence to sequence. Then we compared within a sequence the directly following images. The two corresponding images were processed by an automated peak-find routine to identify the molecules. Then molecule pairs were identified by direct correlation of all emitter positions in a corresponding set of images (with the commercial and the atomic filter). The algorithm checks for the nearest neighbor in both images and saves the result with a unique pair ID. Several emitters too close to each other in one image (±3 pixel) were not considered for further evaluations. In addition, all results (emitter pair correlations) had to be checked and confirmed manually. Each identified emitter, which belongs to a former defined pair, was then fitted by using the least-square method to a 2D-Gaussian (symmetric in *x*and *y*). The extracted fit parameters are used for further analysis. Note, that we only consider the changes we find for the directly correlated pairs. Since the molecules tend to bleach and blink at ambient conditions, this required a statistical analysis. To determine if molecules are brighter or dimmer after change the filter, we define the quantity relative enhancement ΔI as shown in equation (1):

$$\Delta I = \frac{I_{\text{vapor}} - I_{\text{comm}}}{1/2 \cdot (I_{\text{vapor}} + I_{\text{comm}})}. \tag{1}$$

In case of the confocal configuration I_{vapor} and I_{comm} are the maximum of the fitted 2D-Gaussian for the vapor cell and the commercial filter, respectively. In case of the wide-field configuration I_{vapor} and I_{comm} are the integrals of the fitted 2D-Gaussian. This is attributed to the information content of the images: In the confocal case one obtains the integrated signal from the maximum of the emitter position; whereas in the case of the wide-field one acquires a photon distribution which obeys the point spread function of the system. Subsequently, an integration of the signal is required before the total photon flux is accessible.

To estimate the background and its fluctuation, all images were analyzed in an area where no molecules were present. Due to the fact that we can analyze many more molecules than images, the statistical fluctuation is higher. With this data, a histogram of signal to background (SBR) and signal to noise (SNR) ratios can be determined. Any fluctuation in the molecules emission rate was not accounted for. Assuming a non-fluctuating background contribution, this can be simply subtracted from the original data. The analysis presented below is always a statistical comparison between both filter configurations and many co-localized single molecules.

Confocal Imaging

Now we turn to the confocal configuration of the microscope. Initially, the experimental configuration was set up differently than shown in Figure 1(b): The filters were placed between the microscope and the pinhole. However, severe power fluctuations were found in the single molecule signals acquired with the atomic filter on the order of one magnitude. Due to hot air convection surrounding the vapor cell, the collimated beam wanders on the pinhole. Therefore, the confocal configuration was changed and the atomic filter was placed between the pinhole and the single photon detector. This configuration is also described in [26]. To estimate the fluctuation of the wandering beam on the detector, we placed a camera at the location of the avalanche photo diode and monitor the fluorescence of a dense labeled DNA droplet. At high speed (10 ms), a 71 μm spot size ($1/e^2$) is observed, which is the same as if no thermal fluctuations are present. With an integration over 60 s, a spot size of 82 μm is observed. The used avalanche photo diode has an active detector size of around 170 μm, so the configuration of the vapor cell behind the pinhole does not alter the performance of our setup. We experience no reduction in the spatial resolution compared to the commercial filter. For the vapor cell we find the point spread function of individual emitters to have an average full width half maximum (FWHM) of 291 nm compared to 294 nm with the commercial filter. For comparison the Airy disk should have a radius of around 277 nm for the given experimental parameters.

A raw confocal image of the single molecule sample in both filter configurations is shown in Figure 3(a) and (b). Single step blinking of a molecule is observed in the commercial filter configuration (x, y=4x, y=4, 8.5 μm). Visually, both images are comparable, but a higher background contribution around 15 kcps is observed in atomic filtering vs. very stable 2 kcps with the commercial filter. This leads to a relative increase of the background by a factor of 7.5 with the atomic filter. Subsequently, we also estimate the signal to noise ratio to be an order of magnitude larger with the commercial filter (SNR = 800!). This is different than in the experiments under cryogenic conditions [29]. The increase in excitation power at ambient conditions (μW instead of nW at cryogenic conditions) seem to have an influence on the signal to noise ratio. This could be a hint, that saturation effects in the vapor increases the background.

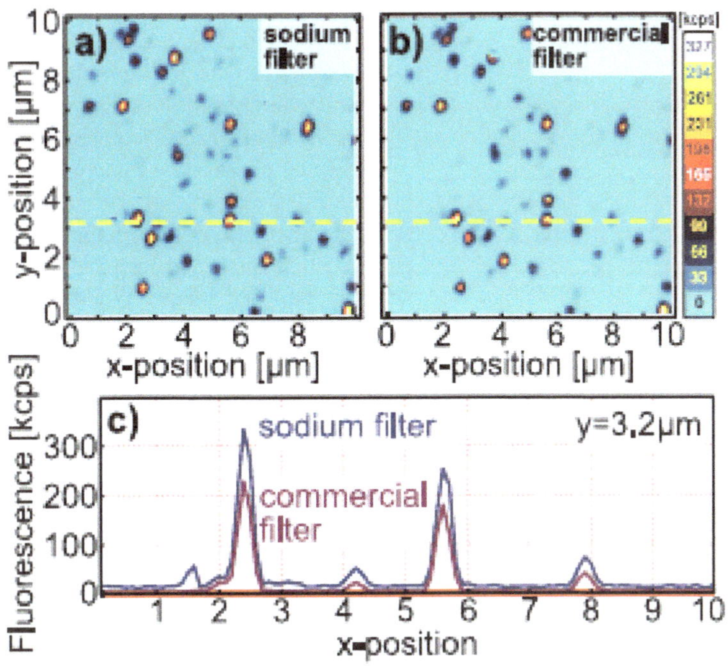

Figure 3 Confocal image of the sample. (a) Filtering with the atomic vapor cell (T = 200°C). The pixel size is 100 nm. The integration time per pixel is 10 ms. (b) Same area as before by using the commercial filter. (c) Line-cut of single molecules emission extracted from the above image (yellow dashed line).

The line cuts in Figure 3(c) illustrate the background level. The image background acquired with atomic filtering is higher by a factor of 7-8 based on a laser-power of 1 μW into the microscope. This is fully consistent with the measurements of the optical density: The commercial filter shows an optical density of about seven, whereas the atomic vapor cell was determined to show six orders of magnitude optical suppression.

In an statistical analysis of 963 molecules, when each molecule is compared in both filter configurations, we observe an enhancement in the overall detected counts per emitter of 15.4% (calculated by using equation (1)). Figure 5(a) shows a histogram for all recorded molecule pairs. The statistical error defined as $\sigma/N--\sqrt{\sigma}/N$ is 1.5%. The atomic filter therefore increases the number of collected photons, but the background suppression is one order of magnitude smaller for the vapor cell to the commercial filter.

Wide-field Imaging

In the wide-field experiment no pinhole is introduced. Therefore, clipping of a wandering beam is not critical. Instead, convection of hot air originating from the vapor cell leads to a shifted or blurred image. In fact, the image is found slightly blurred. The average FWHM of the fluorescence molecules for the atomic cell is 427 nm and for the commercial filter 365 nm. Following the Rayleigh criteria this would clearly result in a reduction of the resolution for the atomic vapor cell compared to the confocal configuration. For localization spectroscopic methods [5, 6] also the number of detected photons directly influences the localization precision: For the used filters one would find a reduction of 10% compared to 17% when only considering the Rayleigh criteria. Of course, this implies that one can treat the wandering of the beam on the CCD chip as a stochastic distribution within the measurement time.

The images looks as in previous case comparable between the two filter configurations (Figure 4(a) and (b)). The background contribution in the atomic filter case is increased, but not as significant as in the confocal configuration. In an analysis of all acquired images, the mean background contribution is increased by 30% from 70 to 100 cps. This increase is also visible in the line-cut shown in Figure 4(c).

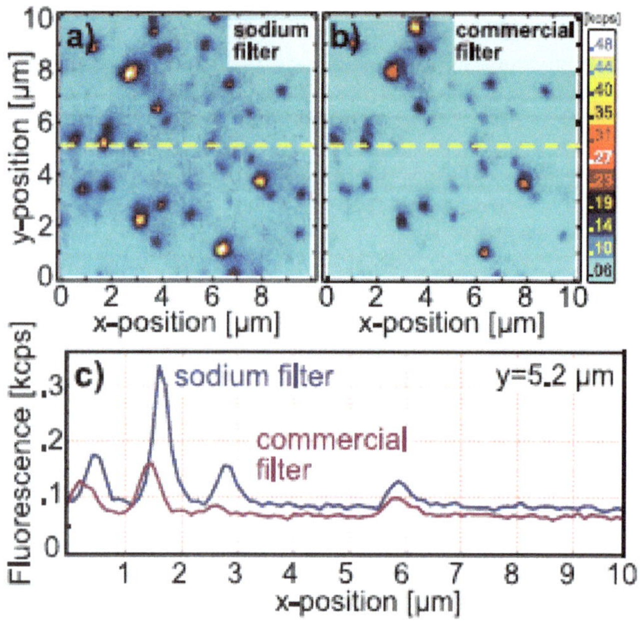

Figure 4 Wide-field image of the DNA sample. (a) Unter atomic filtering (T = 200°C). The acquisition time per frame is 10 s. (b) Same with filtering by the commercial filter. (c)Line-cut of a single molecule (yellow dashed line). The wide-field configuration is more sensitive to scintillations that causes temporal changes in the refractive index of the air. As a consequence the image is blurred out.

For the detection efficiency calculated by integrating over the emitters point spread function, an enhancement of 18.7% with a standard deviation of σ=28.4σ=28.4% is observed. Taking 2,337 measured molecules into account, this leads to a statistical uncertainty of 0.6%. The noise level of the commercial filter and sodium filter is determined to be 30 vs. 40 photons per seconds, based on a laser excitation of 50 µW into the microscope. The increased noise of the atomic filter tends to lower the SNR. However, the higher signal not only compensates this drawback, but gives an increase in the SNR. We find an SNR of 60 for atomic filtering vs. 31 with the commercial filter. The intrinsic noise level of the camera is at least one order of magnitude less and does not play a role here. In summary, we achieve an increased signal, as well as an increased signal to noise ratio for the atomic filter vs. the commercial filter in the wide-field imaging configuration.

Overall Enhancement

Figure 5 shows a histogram of the determined integrated count rates for the commercial and the atomic filtering schemes. Molecules tend to blink and bleach and it is possible that a molecule was fully bright in one image and much dimmer in the next image, acquired with the other filter. The higher fluctuation on the camera (wandering image) leads to an increased spread of the resulting count rate. Therefore, a much wider distribution than in the confocal case is observed. Another important factor is the signal to noise ratio (SNR) in the analyzed image. When e.g. the background noise in the image is higher, the resulting fit outcome for a single molecule shows higher fluctuations. Thereby, Figure 5 does not only represent the overall enhancement of the signal, but also represents the SNR.

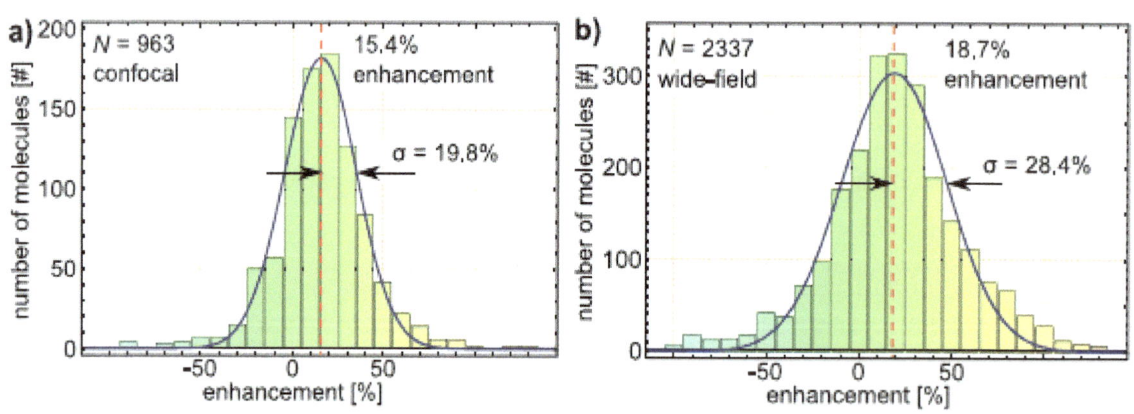

Figure 5 Direct comparison of the enhancement of multiple molecules in the different filter configurations. Molecules tend to blink and bleach, such that statistical data processing is necessary. (a) 963 molecules, detected in a confocal configuration. The signal for the atomic notch filter is improved by 15.4% for the atomic notch filter. N = number of molecules. (b) 2337 molecules, detected in a wide-field microscope. The enhancement is shown to be 18.7%. The statistical distribution of the enhancement is wider than in the confocal case due to the generally lower SNR than in the confocal configuration.

In summary, the confocal images show a lower SNR and SBR for an atomic vapor cell filter compared to a commercial filter. But the lateral spread of the molecules does not change and we find an enhancement in the total number detected photons of 15.4 ± 1.5%.

The wide-field images show a comparable SBR for both filters. For wide-field applications the vapor-cell filter exhibits a better detection efficiency than the commercial filter by 18.7 ± 0.6% for the total detected signal and a factor two increased SNR.

In both configurations, atomic filtering results in an enhancement of the number of detectable photons on the order of 15%. This is fully consistent with the single molecule spectra as shown in Figure 2(a), which also indicate an overall enhancement of approximately 15%.

CONCLUSION AND OUTLOOK

The atomic notch filter poses an alternative to commercial available filters. The improvement of about 15% is accounted to rejecting solely the laser, while allowing the entire fluorescence spectrum of the emitter to pass the filter. This also allows for studies on weakly spectral shifted signals against the excitation wavelength. This can be a crucial enhancement for microscopy and sensing. It allows for studies on low fluorescing samples down to the single molecule level. The increased background fluctuations in the confocal case need to be addressed, eventually with a widened beam to lower the intensity inside the vapor cell. Also a filter cavity should be used to clean up the laser light leaving the fiber. In Raman spectroscopy, atom-based narrow-band filters were explored in the near-infrared part of the spectrum [26], but such experiments were not performed with atomic sodium and yellow light. We underline that the introduced sodium filter matches well to many dyes used in (micro-)biological imaging. So far, other vapor cells, such as Cs, Rb, have been subject for study. Such filters can allow for the sensitive detection of weakly fluorescing defect centers, such as defects in silicon-carbide [36], which could be eventually combined with e.g. atomic rubidium or cesium.

The experimental challenges of using atomic vapor cells are not necessarily easy to solve in micro-biology labs. However, the introduced filtering option allows for enhancing the overall collection efficiency after everything else has been optimized. A complementary feature to this atomic notch filter is the use of a Faraday anomalous dispersion optical filter (FADOF), which represents a GHz-wide band-pass configuration [37, 38]. In the future, when convenient diode lasers operating at the sodium wavelength are available, the technique of filtering a single molecule with an atomic vapor notch may give a small but crucial enhancement of detection efficiency, which in turn would allow better sensing and localization accuracies in material-science and applications in micro-biology.

ACKNOWLEDGEMENTS

We thank Andrea Zappe for DNA preparation and Helmut Kammerlander from the glass shop for producing several sodium cells. We also thank Philip Hemmer and Nathan Chejanovsky for carefully proofreading the manuscript. Initial experiments were conducted by Guilherme Stein & Petr Siyushev. We acknowledge funding from the Max Planck Society (JW, Max Planck fellowship).

COMPETING INTERESTS

The authors declare no competing financial interests.

AUTHORS' CONTRIBUTIONS

IG envisioned and prepared the experiments. TR helped with the configuration and the design of the microscope. DU and TR performed the measurements. IG, DU, TR, and MW processed the data. JW, SL and IG supervised the team. IG, MW and TR wrote the manuscript.

REFERENCES

1. Hirschfeld T. Optical microscopic observation of single small molecules. Appl Opt. 1976;15(12):2965-6. doi:10.1364/AO.15.002965.
2. Moerner W, Kador L. Optical detection and spectroscopy of single molecules in a solid. Phys Rev Lett. 1989;62(21):2535-8. doi:10.1103/PhysRevLett.62.2535.
3. Orrit M, Bernard J. Single pentacene molecules detected by fluorescence excitation in a *p*-terphenyl crystal. Phys Rev Lett. 1990;65:2716-9. doi:10.1103/PhysRevLett.65.2716.
4. Hell SW, Stelzer EHK, Lindek S, Cremer C. Confocal microscopy with an increased detection aperture: type-B 4Pi confocal microscopy. Opt Lett. 1994;19:222-4. doi:10.1364/OL.19.000222.
5. Betzig E, Patterson GH, Sougrat R, Lindwasser OW, Olenych S, Bonifacino JS, Davidson MW, Lippincott-Schwartz J, Hess HF. Imaging intracellular fluorescent proteins at nanometer resolution. Science. 2006;313(5793):1642-5. doi:10.1126/science.1127344.
6. Rust MJ, Bates M, Zhuang X. Sub-diffraction-limit imaging by stochastic optical reconstruction microscopy (storm). Nat Methods. 2006;3(10):793-6. doi:10.1038/nmeth929.
7. Schafer D, Gelles J, Sheetz M, Landick R. Transcription by single molecules of RNA polymerase observed by light microscopy. Nature. 1991;352(6334):444-8. doi:10.1038/352444a0.
8. Nie S, Chiu D, Zare R. Probing individual molecules with confocal fluorescence microscopy. Science. 1994;266(5187):1018-21. doi:10.1126/science.7973650.
9. Nie S, Zare RN. Optical detection of single molecules. Annu Rev Biophys Biomol Struct. 1997;26(1):567-96. doi:10.1146/annurev.biophys.26.1.567.
10. Weiss S. Fluorescence spectroscopy of single biomolecules. Science. 1999;283(5408):1676-83. doi:10.1126/science.283.5408.1676.
11. Gordon MP, Ha T, Selvin PR. Single-molecule high-resolution imaging with photobleaching. Proc Natl Acad Sci USA. 2004;101(17):6462-5. doi:10.1073/pnas.0401638101.
12. Jelezko F, Tietz C, Gruber A, Popa I, Nizovtsev A, Kilin S, Wrachtrup J. Spectroscopy of single N-V centers in diamond. Single Mol. 2001;2:255-60. doi:10.1002/1438-5171(200112)2:4<255::AID-SIMO255>3.0.CO;2-D.
13. Vlasov II, Shiryaev AA, Rendler T, Steinert S, Lee S-Y, Antonov D, Voros M, Jelezko F, Fisenko AV, Semjonova

LF, Biskupek J, Kaiser U, Lebedev OI, Sildos I, Hemmer PR, Konov VI, Gali A, Wrachtrup J. Molecular-sized fluorescent nanodiamonds. Nat Nanotechnol. 2014;9(1):54-8. doi:10.1038/nnano.2013.255.

14. Biju V, Itoh T, Anas A, Sujith A, Ishikawa M. Semiconductor quantum dots and metal nanoparticles: syntheses, optical properties, and biological applications. Anal Bioanal Chem. 2008;391(7):2469-95. doi:10.1007/s00216-008-2185-7.

15. Sandoghdar V, Klotzsch E, Jacobsen V, Renn A, Hakanson U, Agio M, Gerhardt I, Seelig J, Wrigge G. Optical detection of very small nonfluorescent nanoparticles. Chimia. 2006;60:761-4. doi:10.2533/chimia.2006.761.

16. Gerhardt I, Wrigge G, Hwang J, Zumofen G, Sandoghdar V. Coherent nonlinear single molecule microscopy. Phys Rev A. 2010;82:063823. doi:10.1103/PhysRevA.82.063823.http://journals.aps.org/pra/abstract/10.1103/PhysRevA.82.063823.

17. Lee KG, Chen WX, Eghlidi H, Kukura P, Lettow R, Renn A, Sandoghdar V, Götzinger S. A planar dielectric antenna for directional single-photon emission and near-unity collection efficiency. Nat Photonics. 2011;5(3):166-9. doi:10.1038/nphoton.2010.312.

18. Jamali M, Gerhardt I, Rezai M, Frenner K, Fedder H, Wrachtrup J. Microscopic diamond solid-immersion-lenses fabricated around single defect centers by focused ion beam milling. Rev Sci Instrum. 2014;85(12). doi:10.1063/1.4902818.

19. Domenico GD, Weis A. Spectra of the D-lines of alkali vapors. Wolfram Demonstrations Project. 2011. http://demonstrations.wolfram.com/SpectraOfTheDLinesOfAlkaliVapors/.

20. Heifetz A, Agarwal A, Cardoso GC, Gopal V, Kumar P, Shahriar MS. Super efficient absorption filter for quantum memory using atomic ensembles in a vapor. Opt Commun. 2004;232(1-6):289-93. doi:10.1016/j.optcom.2004.01.006.

21. Weller L, Kleinbach KS, Zentile MA, Knappe S, Adams CS, Hughes IG. Absolute absorption and dispersion of a rubidium vapour in the hyperfine Paschen-Back regime. J Phys B, At Mol Opt Phys. 2012;45(21):215005. doi:10.1088/0953-4075/45/21/215005.

22. Camparo J. The rubidium atomic clock and basic research. Phys Today. 2007;60(11):33-9. doi:10.1063/1.2812121.

23. Pelletier MJ. Ultraviolet Raman spectroscopy using an atomic vapor filter and incoherent excitation. Appl Spectrosc. 1992;46(3):395-400. doi:10.1366/0003702924125320.

24. Chen H, White MA, Krueger DA, She CY. Daytime mesopause temperature measurements with a sodium-vapor dispersive Faraday filter in a lidar receiver. Opt Lett. 1996;21(15):1093-5. doi:10.1364/OL.21.001093.

25. Junxiong T, Qingji W, Yimin L, Liang Z, Jianhua G, Minghao D, Jiankun K, Lemin Z. Experimental study of a model digital space optical communication system with new quantum devices. Appl Opt. 1995;34(15):2619-22. doi:10.1364/AO.34.002619.

26. Lin J, Li Y-Q. Ultralow frequency Stokes and anti-Stokes Raman spectroscopy of single living cells and microparticles using a hot rubidium vapor filter. Opt Lett. 2014;39(1):108-10. doi:10.1364/OL.39.000108.

27. Minsky M. Microscopy apparatus. 1961.http://worldwide.espacenet.com/publicationDetails/biblio?CC=US&NR=3013467&KC=&FT=E&locale=en_EP.

28. Singer E. A microscope for observation of fluorescence in living tissues. Science. 1932;75(1941):289-91. doi:10.1126/science.75.1941.289-a.

29. Siyushev P, Stein G, Wrachtrup J, Gerhardt I. Molecular photons interfaced with alkali atoms. Nature. 2014;509(7498):66-70. doi:10.1038/nature13191.

30. Lee Y-H, Maus RG, Smith BW, Winefordner JD. Laser-induced fluorescence detection of a single molecule in a capillary. Anal Chem. 1994;66:4142-9. doi:10.1021/ac00095a005.

31. Petelski T, Fattori M, Lamporesi G, Stuhler J, Tino GM. Doppler-free spectroscopy using magnetically induced dichroism of atomic vapor: a new scheme for laser frequency locking. Eur Phys J, D, At Mol Opt Phys. 2003;22(2):279-83. doi:10.1140/epjd/e2002-00238-4. ADSMATH

32. Laux L, Schulz G. A sodium-resistant glass cell by coating with caf2. J Phys E, Sci Instrum. 1980;13(8):823.

33. Sakurai T, Tanaka K, Miyazaki H, Ichimoto K, Sakata A, Wada S. Construction of long-life magneto-optical filters for helioseismology observations. In: Osaki Y, Shibahashi H, editors. Progress of seismology of the sun and stars. Lecture notes in physics vol. 367. Berlin: Springer; 1990. p. 277-80. doi:10.1007/3-540-53091-6_92.

34. Kubin RF, Fletcher AN. Fluorescence quantum yields of some rhodamine dyes. J Lumin. 1983;27(4):455-62. doi:10.1016/0022-2313(82)90045-X.

35. Arbeloa FL, Aguirresacona IU, Arbeloa IL. Influence of the molecular structure and the nature of the solvent on the absorption and fluorescence characteristics of rhodamines. Chem Phys. 1989;130(1-3):371-8. doi:10.1016/0301-0104(89)87066-1.

36. Widmann M, Lee S-Y, Rendler T, Son NT, Fedder H, Paik S, Yang L-P, Zhao N, Yang S, Booker I, Denisenko A, Jamali M, Momenzadeh SA, Gerhardt I, Ohshima T, Gali A, Janzén E, Wrachtrup J. Coherent control of single spins in silicon carbide at room temperature. Nat Mater. 2015;14(2):164-8. doi:10.1038/nmat4145.

37. Dick DJ, Shay TM. Ultrahigh-noise rejection optical filter. Opt Lett. 1991;16(11):867-9. doi:10.1364/OL.16.000867.

38. Kiefer W, Löw R, Wrachtrup J, Gerhardt I. Na-Faraday rotation filtering: the optimal point. Sci Rep. 2014;4:6552. doi:10.1038/srep06552.

CHAPTER 8

LOW DOSE X-RAY SOURCES AND HIGH QUANTUM EFFICIENCY SENSORS

Arnav R. Mistry, Daniel Uzbelger Feldman, Jie Yang and Eric Ryterski

INTRODUCTION

With all other technical factors (e.g., kilovolts, distance, time, etc.) held constant, patient radiation dose is directly proportional to the milliamperes (mA). A 50% reduction in mA would result in a decrease in radiation dose by 50% [1]. Previously, the mA range has not been taken into consideration in any attempt to reduce radiation dose to which dental patients are being exposed [2–6]. The sensitivity of a digital sensor is measured at a constant wavelength in nanometers (nm) on the basis of the detective quantum efficiency (DQE). This value is used primarily to describe imaging detectors in optical imaging and medical radiography [7]. The quantum efficiency (QE) is the ratio of impinging photons on a pixel to the number of collected electrons. The QE of the pixel is equal to the QE of the complementary metal oxide semiconductor (CMOS) photodiode multiplied for the fill factor of the pixel [8]. Fluoroscopy is a dynamic X-ray or X-ray movie showing images of video frame rates produced by a low mA X-ray source and image intensification at the detector level [9]. An image intensifier unit is capable of multiplying 1,000 to 20,000 times, electron-by-electron, of the produced image, therefore increasing the system QE while allowing dose reduction [10, 11]. Unfortunately, image intensifier units and direct radiography large flat panel detectors have heretofore been too bulky to be used inside the mouth as well as being expensive for a dental setting [12–14]. Another disadvantage of intensifiers is image distortion [15]. With regard to visualization of a stent created from 50 microns (μm) diameter wires in flat panel X-ray fluoroscopy, for the idealized direct detector, the 100 μm pixel size resulted in maximum measured contrast sensitivity. For an idealized indirect detector, with a scintillating layer, the maximal measured contrast sensitivity was obtained at 200 μm pixel size [16].

The major challenge encountered to decrease the mA level in X-ray imaging systems is the quantum noise phenomena. In electronic imaging systems such as fluoroscopy or intraoral radiography digital sensors, we find three principal sources of noise. The first and most relevant arises from quantum statistics, in which the discrete nature of the radiographic signal (which often is photon-starved) introduces uncertainty into the image. The second is electronic noise which is generated in the detector or detector electronics. The third is quantization error that occurs in digital electronic imaging systems when the signal is digitized. For a quantum statistics noise limited system, if the number of photons used is quadrupled, the noise in the resultant image should be halved [17–20]. Not only X-ray systems but also digital cameras are noise limited and quantum limited. Randomly spaced speckles, called noise, can appear in digital images. Noise is similar to grain that appears in photos taken with traditional cameras using high International Organization of Standardization (ISO) films. Noise increases in photos taken with a digital camera using a high ISO number. The higher ISO number leads to more noise. When noise is present, image detail and clarity are reduced, sometimes significantly. The ISO level indicates the film and digital camera's sensitivity to light. According to the ANSI/ISO classification, a dental film with raw speed of ISO 29–56 would be classified as E speed, while one with speed of ISO 57–112 would be classified as F speed [21]. Photographic film typically has a QE of much less than 10% [22]. Current digital cameras have improved their ISO settings which can achieve

Citation: Arnav R. Mistry, Daniel Uzbelger Feldman, Jie Yang, and Eric Ryterski, "Low Dose X-Ray Sources and High Quantum Efficiency Sensors," Radiology Research and Practice, vol. 2014, Article ID 543524, 7 pages, 2014. doi:10.1155/2014/543524.

up to ISO 204,800 [23–25]. To improve the sensitivity or QE of front illuminated charged couple device (CCD) and CMOS image sensors without increasing their pixel size, digital cameras manufacturers apply a thin (0.5–1.0 mm thickness) and inexpensive microlens array to the sensors to reach high ISO levels [26]. The microlens principle was invented in the 17th century where Hooke and Van Leeuwenhoek developed techniques to make small glass lenses for use with their microscopes [27]. The microlens collects and focuses light that would have otherwise fallen onto the nonsensitive areas of the sensor chip, improving the QE significantly (Figure 1) [28].

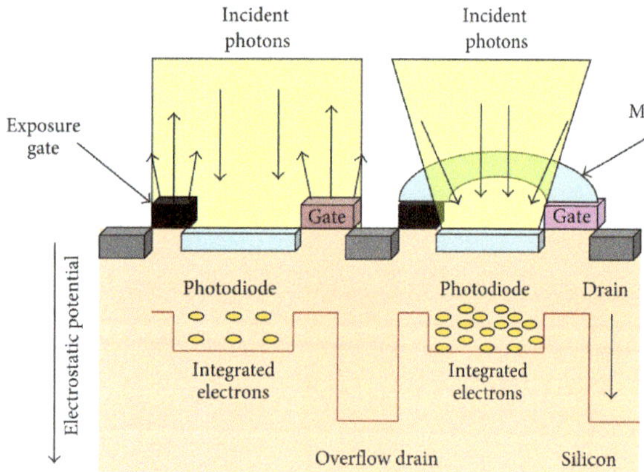

Figure 1: CMOS sensor and microlens (right side) architecture scheme.

Attempts for improving digital sensors QE without compromising the system's noise have been made through the introduction of back illuminated and electron multiplied CCD and CMOS image sensors. Their major drawbacks are complicated manufacturing processes and elevated cost [29, 30]. As a result, the most common used image sensor in dentistry is the front illuminated type [31].

Consequently, a lower milliamperes (mA) setting at the X-ray source and the use of front illuminated sensors with microlens or back illuminated sensors at the detector level for an increased QE that reduces the required radiation dose and sensor pixel size without impacting image quality should have a dramatic positive impact on dental radiology and oral diagnosis. The purpose of this investigation was to prove the concept of radiation exposure reduction and dynamic fluoroscopy feasibility in dentistry without compromising image quality by testing a low-dose X-ray imaging (LDXI) prototype comprising a low-dose X-ray source and a high QE front illuminated sensor with microlens and comparing it to standard of care in terms of dose exposure in milligrays (mGy) and image resolution in lines per millimeters (lp/mm) [6].

MATERIALS AND METHODS

The Temple University Environmental Health and Radiation Safety Institutional Department approved this study. A LDXI prototype (Real Time Imaging Technologies, LLC, Cleveland, OH) was used. The LDXI was comprised of a 35–80 kilovolt peak (kVp), 0.1–0.5 mA X-ray source (9.95′L × 5.27′W × 5.35′H) (SourceRay, Bohemia, NY), and a 8′ rectangular collimator (Margraf Dental, Jenkintown, PA) and an X-ray detector utilizing a CMOS front illuminated sensor (EOS 5D Mark III, Canon, Japan) having 36 × 24 mm effective area, 6.25 × 6.25 um pixel size, 22.3 megapixels resolution, and 49% [32] QE with microlens and capable of performing up to 30 frames per second (fps) for the dynamic video mode. The CMOS sensor modular transfer function (MTF) was >30% at the 18 lp/mm range. Upon low-pass filter and sensor removal, the scintillator/fiber optic plate (AppliedScintech, UK) was coupled at the sensor [33, 34] on top of the microlens [35] with optic glue (BEW Engineering, Ketsch, Germany). The scintillator/fiber optic plate measured 20.8×20.8 mm and was comprised of a 6 um columnar cesium iodide with thallium (CsI:TI) coating with a MTF curve showing and ultimate resolution >18 lp/mm at 60 kVp with 98% attenuation at 70 kVp. Camera software (EOS Utility Ink, Japan) and a kid's watch (Disney Store, Orlando, FL) were used in order to establish and calibrate sensor's ISO settings. Images obtained were raw positive still shots and video. Raw images (A) were converted to negative radiographic images (B) through a software and then cropped (Microsoft Digital Suite 2006 Editor, Microsoft, Redmond, WA) (Figure 2).

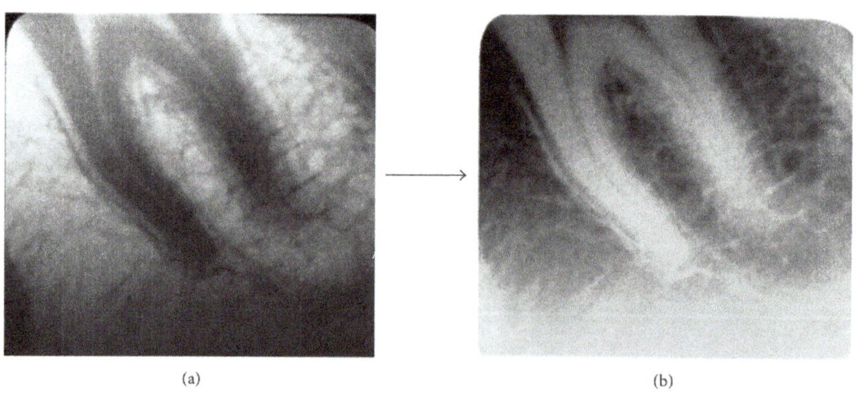

Figure 2: (a) Molar tooth raw positive image obtained at 0.2 mA and 1/30 camera shutter (0.03 seconds) with collimation and (b) negative image.

Ionization chambers Radcal 9010 (Radcal Corp., Monrovia, CA) and RaySafe Xi (RaySafe, Hopkinton, MA) were used to measure the dose exposure in mGy (obtained through rad formula conversion) received by a patient phantom (DXTTR, RINN, Elgin, IL) in all groups and sixteen dosimeters (Landauer, Inc., Glenwood, IL) were utilized for obtaining the dose equivalent in millisievert (mSv) (obtained through rem formula conversion) received by the operator simulated distance at 30 cm for LDXI and fluoroscopy (Figure 3).

(a)

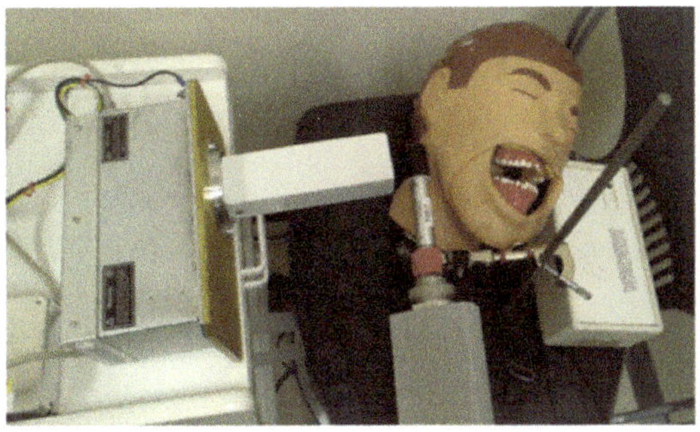

(b)

Figure 3: (a) Dental size sensor with microlens coupled with scintillator/FOP (front view) and (b) LDXI prototype testing on the DXTTR phantom at 0.2 mA and 80 kVp.

ISO settings were established at ISO 5,000 for group 1 and ISO 12,800 for group 2 and image acquisition was made at 1/30 shutter (0.03 seconds). Group 1 was exposed to the LDXI prototype (Real Time Imaging Technologies, LLC, Cleveland, OH) at 0.2 mA and 80 kVp in static mode during 10 intervals from 0 to 27 continuous seconds with a 6.25 × 6.25 um pixel size sensor and 49% QE. Group 2 was exposed to the LDXI prototype at 0.2 mA and 80 kVp in the dynamic video mode during 31 intervals from 0 to 300 seconds with a 6.25 × 6.25 um pixel size sensor and 49% QE. Group 3 was exposed to medical fluoroscopy (GE C-arm OEC 9800, Cleveland, OH) at 0.038 mA and 55 kVp in 31 intervals from 0 to 300 seconds with a 12.8 × 12.8 um pixel size sensor and a 9′ image intensifier with 65% QE at 550 nm [36]. Group 4 was exposed to digital intraoral radiography (Gendex GX-770, Hatfield, PA/ Planmeca Dixi 2 v3, Roselle, IL) at 7 mA and 70 kVp in 29 intervals from 0 to 1.65 seconds with a 19 × 19 um pixel size sensor. Group 5 was exposed to CBCT Scan (iCat Imaging Sciences International Inc., Hatfield, PA) at 5 mA and 120 kVp from 0 to 26.9 seconds in 28 intervals and several modalities with a 125 um voxel size sensor. Six thermoluminescent Luxel and ten optical stimulated luminescence technology Nanodots dosimeters (Landauer, Inc., Glenwood, IL) were used at the patient phantom, area monitor, and positive and negative controls for the LDXI. After research completion, all the dosimeters were mailed back to Landauer for analysis purposes. Image resolution for the LDXI was measured with a resolution test pattern (Fluke Corp., Cleveland, OH) in lp/ mm. An endodontic file size 10 (Dentsply, Maillefer, York, PA) was placed within phantom's tooth number 27 at the radiographic apex for intraoral imaging subjective resolution assessment purposes in which two endodontists and one oral and maxillofacial radiologist from the institution were asked to confirm tooth's working length [6]. Dose exposure measurements and image resolution were calculated for all groups and dosimeters were analyzed.

RESULTS

The Mann-Whitney test showed no statistical significance (($\alpha = 0.01$)) in dose exposure (mGy) between groups 1 and 3 and 1 and 4 and timing exposure (seconds) between groups 1 and 5 and 2 and 3 (Figure 4).

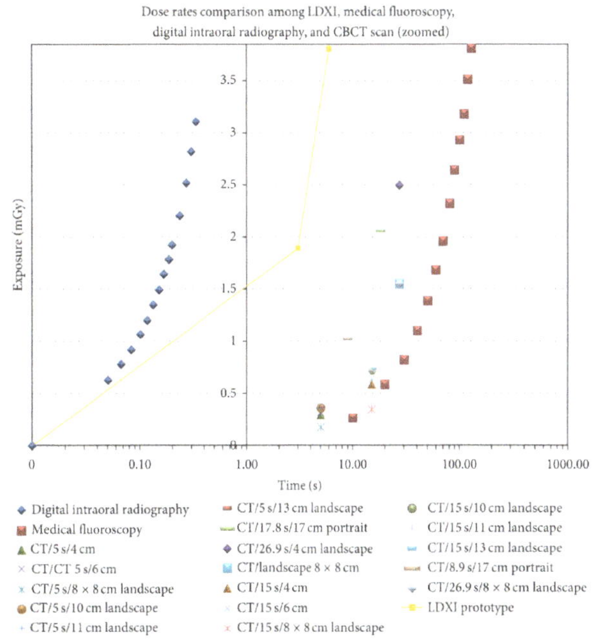

Figure 4: Radiation and timing exposures comparison amongst all groups.

Dosimeters for the LDXI operator simulated distance and controls did not register significant dose equivalent (mSv).

Image resolution test showed LDXI 1 > digital intraoral radiography > LDXI 2 > medical fluoroscopy > CBCT scan (Table 1).

Table 1: mA, kVp, pixel size settings, and resolution outcome comparison of different devices

Device type	mA	kVp	Pixel/voxel size (um)	Resolution (lp/mm)
LDXI 1 (static)	0.2	80	6.25 × 6.25	18
LDXI 2 (dynamic)	0.2	80	6.25 × 6.25	10
Medical fluoroscopy (image intensified)	0.038	55	12.8 × 12.8	1.75
Digital intraoral radiography	7	70	19 × 19	16
CBCT scan	5	120	125	1.6

File size 10 was observed at the working length as confirmed by two endodontists and one oral and maxillofacial radiologist (Figure 5).

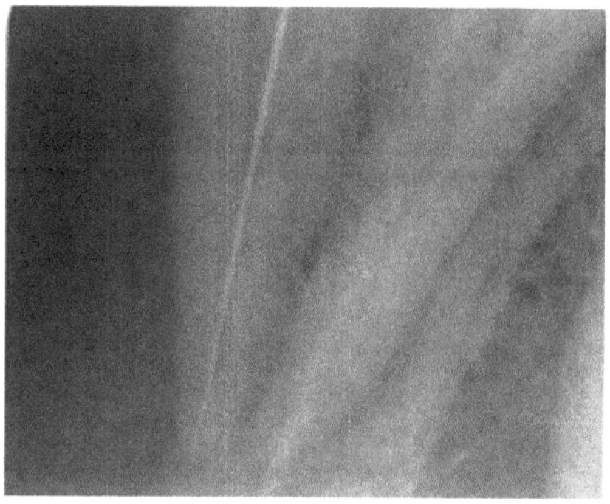

(a)

(b)

Figure 5: Group 1 LDXI static: (a) endodontic file number 10 at working length and (b) 18 lp/mm.

DISCUSSION

Technical advances enable cameras to better capture the drama of low-light photography/video. As a rule, bigger is indeed better when it comes to the overall dimensions of a sensor because a bigger sensor provides not only more pixels, but also bigger pixels, which more efficiently gather light. However, of all the improvements in the imaging field, perhaps the most notable are not the increases in sensor size but the innovations in increasing a sensor's ability to gather light in low-light situations and to record a wider range of light. Much improved sensors with microlens now allow one to obtain detailed low-light images by allowing shots at high ISO levels that previously generated unflattering digital "noise" or graininess in an image at smaller pixel sizes. From an image resolution dental perspective, studies have demonstrated that spatial resolution affects bone loss and caries diagnosis in dentistry. A research undertaken to determine the effect of X-ray beam alignment and spatial resolution on quantification of alveolar bone using radiometric techniques concluded that 50 um pixel spatial resolution is apparently superior to 200 um pixel images if radiometric data is to be evaluated [37]. Another study determined that spatial resolution affected radiometric analyses aimed at detecting progressive enamel loss. Cumulative percent histograms shifts associated with the smaller 59 um pixels accounted for 68% of the variation in weights caused by enamel reduction, whereas the shifts associated with the larger 200 um pixels accounted for 50%. The results indicated that pixel size does affect radiometric determinations of enamel reduction [38, 39]. In summary, the smaller the pixel size and the more pixels are arranged in a sensor, the better the quality of the image that is captured. In this study, microlens also allows decreasing the sensor's pixel size (6.25 × 6.25 um), therefore increasing image spatial resolution as compared to the larger pixel size of current sensors used in digital intraoral radiography (19 × 19 um), CBCT scans (125 um voxel), and medical fluoroscopy (12.5 × 12.5 um for image intensifiers and um for flat panel detectors). As compared to similar front illuminated intraoral sensors used in dentistry (DQE 3% CCD and DQE 18% CMOS) [40, 41] the LDXI prototype has used a large size (35 mm), full frame, and front illuminated CMOS sensor with a thin microlens array for improved DQE (49%) directly coupled to the scintillator/fiber optic plate, capable of acquiring low mA images at ISO 5,000 and ISO 12,800 as well as in dynamic video frame rates. From a radiation exposure perspective, we found that 3.05 and 3.95 seconds of LDXI are comparable to one single shot of 0.2 seconds of digital intraoral radiography and 26.9 seconds 4 cm landscape single area CBCT scan, respectively. As a result, the LDXI used for 0.2 seconds would reduce 93.4% dose exposure as compared to digital intraoral radiography used for 0.2 seconds. For a conventional root canal requiring a pre-op, working length, master cone, partial condensation, and final X-ray, the LDXI prototype could be used during 15 seconds without producing more dose exposure due to the addition of microlens for increased QE as compared to digital intraoral radiography. As a result, lower mA settings (up to 0.2 mA) as compared to digital intraoral radiography (7 mA) could be captured at 18 lp/mm and 10 lp/mm image resolution for the static and dynamic modalities, respectively [6]. In addition, an endodontic file size 10 was observed at the working length at the phantom as confirmed by two endodontists and one oral and maxillofacial radiologist from the institution (Figure 5). In addition to root canals, other clinical applications for low dose X-ray sources and high QE image sensors in the static mode would be full mouth X-rays, pre- and postradiographs, bitewings, and panoramic, cephalometric, and CBCT scans for oral diagnosis while the dynamic mode would allow the introduction of fluoroscopy in dentistry for dental implant placement, temporomandibular joint analysis, maxillofacial surgeries, and postplacement. Since the scintillator/fiber optic plate was coupled in the central area of the sensor, the effect of X-rays on the surrounding, residual, and uncovered area caused significant noise causing black spots at the image. In addition, despite the fact that sensor's housing was internally masked against light, light pollution was observed on one corner (Figure 6).

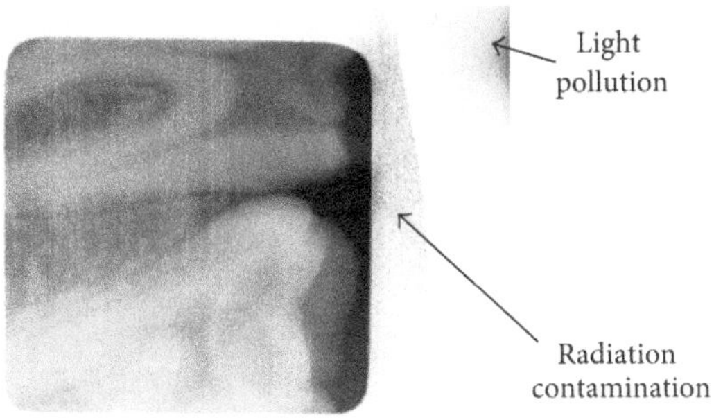

Figure 6: Preventable artifacts obtained in images.

These artifacts should be avoided in future studies by providing complete sensor shielding and a 100% light proof housing. In addition to further evaluations with higher QE (>65% at 550 nm) new generation front illuminated sensors with microlens, the authors recommend testing a back illuminated sensor due to its increased QE as compared to the front illuminated sensor used in this study [14, 35]. Finally, we propose testing the next generation LDXI using head and neck anthropomorphic phantoms following Sections C and D of the FDA guidance for Solid State X-Ray Imaging Devices (MTF, detective quantum efficiency, and signal-to-noise ratio) and compare it to digital intraoral radiography, medical fluoroscopy, and CBCT scan imaging devices.

CONCLUSIONS

The LDXI proved the concept for obtaining a high definition image resolution for static and dynamic radiography at lower or similar dose exposure and smaller pixel size, respectively, when compared to traditional imaging devices. Lower mA at the X-ray source and high QE at the detector level principles with microlens could be applied not only to digital intraoral radiography and dental fluoroscopy but also to panoramic, cephalometric, and CBCT scans devices to considerably reduce X-ray source dose exposure as well as sensor pixel size and more research is recommended to demonstrate this further.

DICLOSURE

The paper, or any part of it, has not been submitted or published and will not be submitted elsewhere for publication while being considered by the journal. The research has been shown as a poster presentation at Temple University Kornberg School of Dentistry Research Day and has been presented at the 64th AAOMR Annual Session as an abstract and as part of the 2013 William H. Rollins Award Lecture.

CONFLICT OF INTERESTS

Jie Yang is a consultant to and Daniel Uzbelger Feldman has equity interest in Real Time Imaging Technologies, LLC. Arnav R. Mistry and Eric Ryterski had no conflict of interests. At the time this study was conducted, Arnav R. Mistry was an endodontology postgraduate student at Temple University Kornberg School of Dentistry.

ACKNOWLEDGMENTS

This study has been possible thanks to the funds received by Real Time Imaging Technologies, LLC (Cleveland, OH), from the Great Lakes Innovation & Development Enterprise Grant ($25,000) and the Indus Entrepreneurs (TiE) Ohio Business Plan Competition Award ($2,500) as well as by the research team from Temple University Kornberg School of Dentistry Student Research Grant ($2,000).

REFERENCES

1. Q. B. Carroll, Fuchs's Radiographic Exposure and Quality Control, Charles C Thomas Publisher, 7th edition, 2003.
2. D. Uzbelger, "Comparison between medical fluoroscopy, digital dental imaging and intraoral radiography," Journal of Dental Research, vol. 97, p. 701, 2005.
3. D. Uzbelger-Feldman, C. Susin, and J. Yang, "The use of fluoroscopy in dentistry: a systematic review,"Oral Surgery, Oral Medicine, Oral Pathology, Oral Radiology, and Endodontology, vol. 105, article e61, 2008.
4. D. Uzbelger Feldman, J. Yang, and C. Susin, "A systematic review of the uses of fluoroscopy in dentistry,"The Chinese Journal of Dental Research, vol. 13, no. 1, pp. 23–29, 2010.
5. D. Uzbelger Feldman, P. M. Panchal, and J. Yang, "Resolution and dose rates comparison among fluoroscopy, digital/film-based intraoral radiography and CBCT scan," in Proceedings of the 61th Annual Meeting of the American Academy of Oral & Maxillofacial Radiology, San Diego, Calif, USA, 2010.
6. D. Uzbelger Feldman, A. Mistry, and J. Yang, "Radiation exposure and image resolution testing of a low-dose X-ray imaging," in Proceedings of the 64th Annual Meeting of American Academy of Oral & Maxillofacial Radiology, Beverly Hills, Calif, USA, October 2013.
7. A. R. Cowen, A. G. Davies, and M. U. Sivananthan, "The design and imaging characteristics of dynamic, solid-state, flat-panel x-ray image detectors for digital fluoroscopy and fluorography," Clinical Radiology, vol. 63, no. 10, pp. 1073–1085, 2008.
8. G. C. Holst and T. S. Lomheim, CMOS/CCD Sensors and Camera Systems, SPIE Press Book, Bellingham, Wash, USA, 2nd edition, 2011.
9. B. D. Giordano, S. Ryder, J. F. Baumhauer, and B. F. DiGiovanni, "Exposure to direct and scatter radiation with use of mini-c-arm fluoroscopy," Journal of Bone and Joint Surgery—Series A, vol. 89, no. 5, pp. 948–952, 2007.
10. J. T. Thirlwall, "The detective quantum efficiency of medical x-ray image intensifiers," Review of Scientific Instruments, vol. 69, no. 11, pp. 3953–3957, 1998.
11. S. Hasham, F. D. Burke, S. J. Evans, M. K. Arundell, and D. N. Quinton, "An audit of the safe use of the mini c-arm image intensifier in the out-patient setting," Journal of Hand Surgery, vol. 32, no. 5, pp. 563–568, 2007.
12. J. A. Seibert, "Flat-panel detectors: how much better are they?" Pediatric Radiology, vol. 36, no. 14, pp. 173–181, 2006.

13. D. Uzbelger Feldman, "Real Time Imaging Technologies, LLC, Assignee. Apparatus for Diagnosis and/or Treatment in the Field of Dentistry using Fluoroscopic and Conventional Radiography," US Patent 6543936, 2003.

14. D. Uzbelger Feldman, "Real Time Imaging Technologies, LLC, assignee. Dental Fluoroscopic Imaging System," US patent 8,430,563, April 2013.

15. D. P. Chakraborty, "Image intensifier distortion correction," Medical Physics, vol. 14, no. 2, pp. 249–252, 1987.

16. Y. Jiang and D. L. Wilson, "Optimization of detector pixel size for stent visualization in x-ray fluoroscopy," in Medical Imaging 2004: Image Perception, Observer Performance, and Technology Assessment, vol. 5372 of Proceedings of SPIE, p. 311, May 2004.

17. B. A. Arnold and P. O. Scheibe, "Noise analysis of a digital radiography system," American Journal of Roentgenology, vol. 142, no. 3, pp. 609–613, 1984

18. D. W. Mah, J. A. Rowlands, and J. A. Rawlinson, "Measurement of quantum noise in fluoroscopic systems for portal imaging," Medical Physics, vol. 23, no. 2, pp. 231–238, 1996.

19. M. Hensel, T. Pralow, and R. R. Grigat, "Modeling and real-time estimation of signal-dependent noise in quantum-limited imaging," in Proceedings of the 6th WSEAS International Conference on Signal Processing, Robotics and Automation (ISPRA '07), World Scientific and Engineering Academy and Society, Corfu Island, Greece, February 2007.

20. S. Suh, S. Itoh, S. Aoyama, and S. Kawahito, "Column-parallel correlated multiple sampling circuits for CMOS image sensors and their noise reduction effects," Sensors, vol. 10, no. 10, pp. 9139–9154, 2010.

21. American National Standards Institute, Photography—Direct-Exposing Medical and Dental Radiographic Film/Process Systems—Determination of ISO Speed and ISO Average Gradien, American National Standards Institute ISO, New York, NY, USA, 1991.

22. F. Träger, Handbook of Lasers and Optics, Springer, Berlin, Germany, 2012.

23. M. K. Khoury, I. Parker, and D. W. Aswad, "Acquisition of chemiluminescent signals from immunoblots with a digital single-lens reflex camera," Analytical Biochemistry, vol. 397, no. 1, pp. 129–131, 2010.

24. T. Ang, Digital Photography: An Introduction, DK Publishing, New York, NY, USA, 4th edition, 2012.

25. The British Journal of Photography Report. The 50 Best Photography Products of 2012. The British Journal of Photography, Apptitude Media Ltd., London, UK, 2012.

26. R. F. Stevens and N. Davies, "Lens arrays and photography," The Journal of Photographic Science, vol. 39, pp. 199–208, 1991.

27. R. Hooke, Micrographia: or some Physiological Descriptions of Minute Bodies made by Magnifying Glasses, The Royal Society of London, London, UK, 1st edition, 1665.

28. P. H. Nussbaum, R. Völkel, H. P. Herzig, M. Eisner, and S. Haselbeck, "Design, fabrication and testing of microlens arrays for sensors and microsystems," Pure and Applied Optics, vol. 6, no. 6, pp. 617–636, 1997

29. R. A. Stern, L. Shing, N. Waltham, H. Mapson-Menard, A. Harris, and P. Pool, "EUV and soft X-ray quantum efficiency measurements of a thinned back-illuminated CMOS active pixel sensor," IEEE Electron Device Letters, vol. 32, no. 3, pp. 354–356, 2011.

30. S. N. Vasan, P. Sharma, C. N. Ionita, et al., "Image acquisition, geometric correction and display of images from a 2x2 x-ray detector array based on electron multiplying charge coupled device (EMCCD) technology," in Medical Imaging 2013: Physics of Medical Imaging, vol. 8668 of Proceedings of SPIE, Lake Buena Vista, Fla, USA, 2013.

31. A. G. Farman and T. T. Farman, "A comparison of 18 different x-ray detectors currently used in dentistry," Oral Surgery, Oral Medicine, Oral Pathology, Oral Radiology and Endodontics, vol. 99, no. 4, pp. 485–489, 2005.

32. http://www.sensorgen.info/CanonEOS_5D_MkIII.html.

33. R. G. van Silfhout and A. S. Kachatkou, "Fibre-optic coupling to high-resolution CCD and CMOS image sensors," Nuclear Instruments and Methods in Physics Research Section A: Accelerators, Spectrometers, Detectors and Associated Equipment, vol. 597, no. 2-3, pp. 266–269, 2008.

34. H. Fan, H. L. Durko, S. K. Moore, et al., "DR with a DSLR: digital radiography with a digital single-lens reflex camera," Proceedings of the Society of Photo-Optical Instrumentation Engineers, vol. 15, p. 7622, 2010.

35. D. Uzbelger Feldman, "Real time imaging technologies, LLC, assignee. Imaging apparatus and methods," International patent application PCT/US13/49504, July 2013.

36. W. K. Chu, C. Smith, R. Bunting, P. Knoll P, R. Wobig, and R. Thacker, "Application of the CCD camera in medical imaging," in Sensors, Cameras, and Systems for Scientific/Industrial Applications, 121, vol. 3649 of Proceedings of SPIE, San Jose, Calif, USA, January 1999.

37. M. K. Shrout, J. Weaver, B. J. Potter, and C. F. Hildebolt, "Spatial resolution and angular alignment tolerance in radiometric analysis of alveolar bone change," Journal of Periodontology, vol. 67, no. 1, pp. 41–45, 1996.

38. M. K. Shrout, C. M. Russell, and B. J. Potter, "Spatial resolution in radiometric analysis of enamel loss A pilot study," Oral Surgery, Oral Medicine, Oral Pathology, Oral Radiology, and Endodontics, vol. 81, no. 2, pp. 245–250, 1996.

39. A. Wenzel, F. Haiter-Neto, and E. Gotfredsen, "Influence of spatial resolution and bit depth on detection of small caries lesions with digital receptors," Oral Surgery, Oral Medicine, Oral Pathology, Oral Radiology and Endodontology, vol. 103, no. 3, pp. 418–422, 2007.

40. A. C. Mörner-Svalling, Digital intraoral radiography: determination of technical properties and application

evaluations [Dissertation], Department of Oral Radiology and Oral and Maxillofacial Surgery, Institute of Odontology, Karolinska Institute, Stockholm, Sweden, 2002.

41. H. S. Cho, S. G. Choi, B. S. Lee, and S. Kim, "Image quality evaluation of a highly-integrated digital X-ray imaging sensor for dental intraoral-lmaging applications," Journal of the Korean Physical Society, vol. 51, no. 1, pp. 30–34, 2007.

CHAPTER 9

STUDY OF QCL LASER SOURCES FOR THE REALIZATION OF ADVANCED SENSORS

Giuseppe de Risi, Lorenzo Luigi Columbo and Massimo Brambilla

Keywords: Quantum cascade laser, optical feedback interferometry, nano displacement sensing, laser sensors

INTRODUCTION

In the last decade, quantum cascade lasers (QCLs), which represent compact, high power, highly coherent, widely tunable laser sources in the mid-infrared to terahertz range of the electromagnetic spectrum, have been extensively used in a number of sensing applications, like imaging, medical diagnosis and spectroscopy [1].

With respect to conventional bipolar semiconductor lasers, continuous wave (CW) emission in QCLs is much more stable against strong optical feedback provided by an external target in the so-called self-mixing (SM) configuration. As we recently demonstrated, this follows from the absence of relaxation oscillations due to the ultra-fast carriers recombination and to the small value of the linewidth enhancement factor (or α factor) [2,3].

Then, QCLs can be exploited to realize robust, detectorless, real-time sensors when an external target provides back-reflected radiation, which induces changes in the emitter properties (field intensity, compliance voltage at laser contacts). The modified signal hence carries information about corresponding variations of the target complex reflectivity and its displacement with respect to the laser exit facet [4]. Fields of applications range from coherent imaging [5] to motion tracking [6] and material processing [7].

The search for displacement sensors based on optical interferometry with nanometric resolution has been recently triggered by the enormous development of nanoscale technology, even in systems working at longer wavelengths (from mid-infrared to terahertz). So far, most of the systems showing these performances employ "off-line" signal post-processing to overcome the half wavelength intrinsic resolution of standard "on-line" fringe counting techniques [8,9,10]. A prototype system able to measure "on-line" nanometer-size amplitude displacements has been proposed in [11], based on a modification of the standard self-mixing scheme, namely a differential optical feedback interferometry in a two-laser configuration.

Here, we provide a proof-of-principle demonstration of a nanoscale position-sensing system based on the nonlinear dynamical response of a QCL subject to strong optical feedback. We refer in particular to the collinear double-arm configuration sketched in Figure 1 where optical feedback is provided by a slow object target (OT) with constant and unknown speed and a fast reference target (RT) with constant, controlled speed [12]. In the strong feedback regime, we show that the fast switching fringes in the SM signal typically associated with the RT motion are characterized by additional sub-features carrying information about the OT motion. As illustrated in detail in Section 2 and Section 3, this allows for a denser sampling of the OT fringes, which, in turn, leads to a displacement measure with a resolution much smaller than λ/2λ/2. With respect to our first results reported in [13], we present here a radically-improved version of the OT motion retrieval algorithm that allows us to reduce the estimated sensors resolution down to a few nanometers. We also provide an evaluation of the role of the feedback strength and discuss the role of the QCL fluctuations as a source of errors in the sensing procedure. We believe that the a super-resolved displacement sensor, like the one proposed here, may be of interest also in other applicative fields, such as QCL-based 3D imaging systems for medical and

Citation: Giuseppe de Risi, Lorenzo Luigi Columbo and Massimo Brambilla, Study of QCL Laser Sources for the Realization of Advanced Sensors, doi:10.3390/s150819140.

material processing, where the measurement of the separation between layers orthogonal to the optical axis is essential [14].

A theoretical analysis of the QCL subject to optical feedback from two independent targets is presented in Section 2. In Section 3, we describe in detail a numerical fitting procedure able to extract in real time the information about the object target displacement with a resolution of a few nanometers, which corresponds to ~λ/1000. The numerical limits in the attainable resolution and the range of application of the above algorithm are discussed in Section 3.2 and Section 3.3. In Section 3.4, we estimate the influence of the finite laser linewidth on the proposed sensor accuracy. Finally, Section 4 is devoted to the conclusions and a discussion of a possible extension of our approach.

THEORETICAL SECTION: SELF-MIXING IN QCL WITH A DOUBLE EXTERNAL CAVITY

The sensor scheme is sketched as in Figure 1, where a partially-transparent reference target moving with a known constant velocity v_r is inserted in the external cavity formed by the QCL and the object target that translates with an unknown constant velocity $v_o < v_r$.

In the single longitudinal mode, slowly varying envelope approximations and single reflection in the external cavity, the behavior of a QCL under optical feedback from two independent targets can be described by an extended version of the Lang–Kobayashi (LK) equations [4] that account for an external double cavity [15]:

$$\frac{dE(t)}{dt} = \frac{1}{2}(1+i\alpha)(N(t)-1)E(t) + \frac{k_o\tau_p}{\tau_c}E(t-\tau_o)e^{-i\omega_0\tau_o} + \frac{k_r\tau_p}{\tau_c}E(t-\tau_r)e^{-i\omega_0\tau_r} \quad (1)$$

$$\frac{dN(t)}{dt} = \gamma\left(I_p - N(t)(1+|E(t)|^2)\right) \quad (2)$$

where the adimensional field E, carriers' density N and pump intensity I_p are scaled as in [2], and the time t is expressed in units of the photon lifetime τ_p. Moreover, α is the linewidth enhancement factor; τ_c is the laser cavity round trip time; ω_0 is the solitary laser frequency (equal to the laser cavity resonance taken as the reference frequency); γ is the photon-to-carrier lifetime ratio. The feedback strength parameters k_i with i=o, r depend on the effective fraction of the field back-reflected from OT and RT that re-enters the laser cavity and the reflectivities of the external targets and of the laser exit facet. The delays τ_i change in time due to the target motions, so that $\tau_i = 2L_i/c = 2(L_{0,i} + v_i t)/c$ where $L_{0,i}$ represents the target initial position.

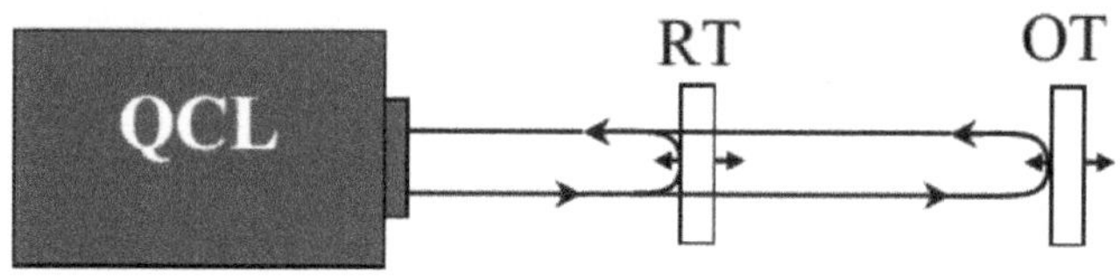

Figure 1. Scheme of the proposed sensor.

Looking for CW solutions of Equations (1) and (2), we get the following expressions for the CW QCL frequency ω_F and the associated difference ΔNΔN between the carriers' density in the presence of feedback and its value in the free running laser case [12]:

$$\omega_F = \omega_0 - \frac{k_o\tau_p}{\tau_c}\sqrt{1+\alpha^2}\sin(A_o + \omega_o t + \arctan\alpha) - \frac{k_r\tau_p}{\tau_c}\sqrt{1+\alpha^2}\sin(A_r + \omega_r t + \arctan\alpha) \quad (3)$$

$$\Delta N = -2\frac{k_o\tau_p}{\tau_c}\cos(A_o + \omega_o t) - 2\frac{k_r\tau_p}{\tau_c}\cos(A_r + \omega_r t) \quad (4)$$

where $A_i = 2L_{0,i}\omega_F/c$ and $\omega_i = 2v_i\omega_F/c$. Hence, the temporal evolution of the quantity ΔN=ΔN(t), which is proportional to the voltage offset at the QCL terminals and, thus, represents the experimentally-accessible SM signal [13], contains information about speed (and consequently, position) of both targets. From this, it follows that an explicit solution of Equations (3) and (4) would allow one, in principle, to determine the value of the OT velocity and, thus, the OT displacement. Due to the highly implicit character of the transcendent Equation (3), this is analytically impossible, and numerical methods have to be used to recover the information about the OT from the ΔN time trace.

For this purpose, we simulate the evolution of the SM function ΔN, by numerically solving Equations (3) and (4) for the values of v_r and v_o indicated by dots in the parameter space depicted in Figure 2. The other parameters used in our simulations are typical for a mid-infrared QCL and are reported in Table 1.

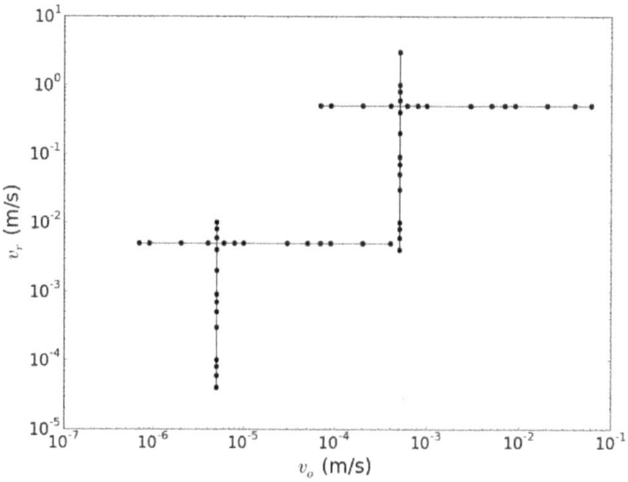

Figure 2. Two-dimensional parameter space of the simulations. Dots indicate the values of v_r and v_o used in the various simulations.

Table 1. Physical parameters used in the simulations of the Lang–Kobayashi (LK) equations.

τ_p	τ_c	α	ω_0	$L_{0,r}$	$L_{0,o}$
100 ps	35.6 ps	2	302 THz	1.5×10^{-3} m	2.5×10^{-3} m

Nonlinear Frequency Mixing

An example of the carrier density difference ΔN(t) obtained from our simulations is represented in Figure 3. Its temporal trace exhibits two distinct modulations, which show the peculiar interference fringes of the lasers operating in SM configuration: fast fringes (from Mark A to Mark B in Figure 3a) on the scale of a few 10^{-1} ms, modulated by slower fringes (see also Figure 2a in [12]). While one could expect that they correspond to the fast and slow motions of the RT and OT, respectively, the inspection of the Fourier transform of ΔN(t), shown in Figure 4, reveals the nonlinear nature of the SM signal. In fact, the first peak, at frequency $\omega_o = 4\pi v_o/\lambda_0 = 100$ Hz, is associated with the slow periodicity due to the movement of the OT; another peak at $\omega_r = 4\pi v_r/\lambda_0 = 10$ kHz, is associated with the motion of the RT, but the dominant peak in the spectrum occurs at $\omega_r-\omega_o$. This shows that the feedback fields provided by the targets cause a nonlinear response in the laser, due to the intrinsic nonlinearity of the LK Equations (3) and (4). The presence of super harmonics of the mixed frequencies are another proof of this phenomenon and will be briefly discussed in Section 3.3. In the time domain, this amounts to saying that there should exist temporal features of the time trace, associated with a fast time scale (being $2\pi/(\omega_r - \omega_o) \approx 2\pi/\omega_r$), which carry information about the slow time scale. Since the former is associated with the RT motion and the latter to the OT motion, the introduction of the RT should allow one to gather information about the OT motion by sampling the system at a fast rate. This overcomes the limit λ/2 for the appreciable displacement when only one target is considered.

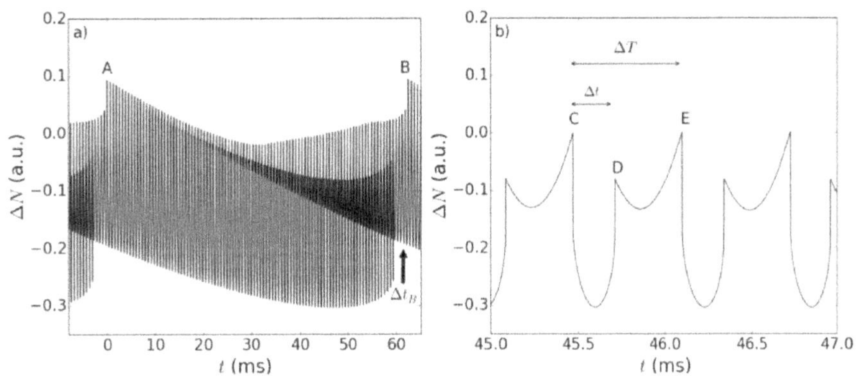

Figure 3. Typical temporal trace of the carrier density difference ΔN as a function of time. Reference and object target velocities are $v_r=5.0\times10^{-3}$ m/s and $v_o=5.0\times10^{-5}$ m/s, respectively, while feedback parameters are $k_r=0.029$ and $k_o=0.025$. (a) A complete slow fringe, delimited by Points A (at t=0ms) and B (at t=62ms), is shown, so that the faster modulation induced by the reference target is evident; in addition, the time interval of sub-feature disappearance (Δt_B), introduced in Section 3.3, is represented. (b) A close up, for a shorter time interval, of the temporal trace shown in (a), where the fast fringe can be identified (between Points C and E), with its associated time interval, denoted by ΔT. The sub-feature is delimited by Points D and E, and the time interval Δt between two consecutive sub-features is the physical quantity on which the sensing scheme is based. The other parameters are reported in Table 1.

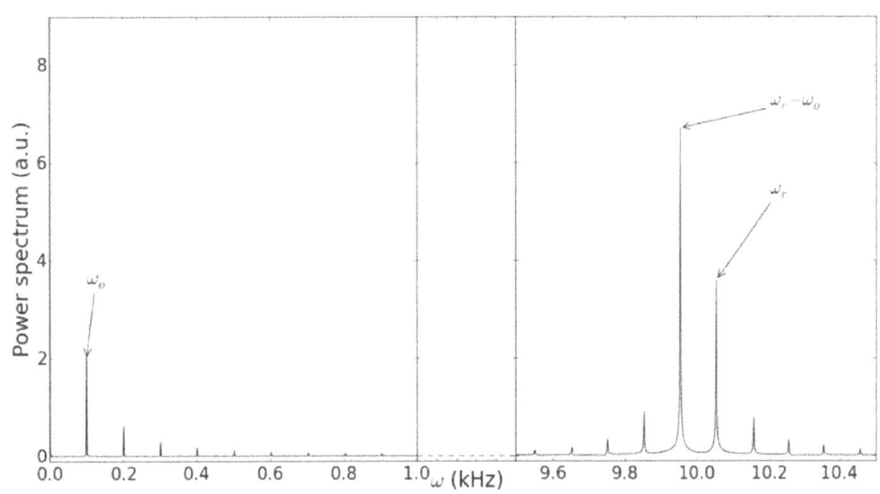

Figure 4. Fourier transform of ΔN(t) represented in Figure 3a.

More interestingly, again, if the values of the feedback parameters are high enough [13], novel sub-features can be seen in the time trace (see Figure 3); such sub-features are the fingerprint of the two-target scheme, and their temporal characteristics are linked to the frequency combination arising from the nonlinear response described above. The study of these sub-features will be the main focus of the next section, so it will be worth sketching an interpretation about their origin and evolution. For moderate and strong feedback levels, multiple solutions of Equation (3) may exist, resulting in different possible continuous wave solutions (or modes), and following the well-established literature [4], we assume that the actual lasing mode is the "maximum gain mode" (MGM), i.e., the one that minimizes ΔN [4] (simulations of the dynamical LK equations and comparison with experiments [13] show that this assumption is correct in the operating conditions that we considered). During the target evolution, it is possible that a new mode with higher gain becomes available, thus causing the lasing frequency to switch on this new mode. In the case of feedback from a single target, this leads to the typical sawtooth behavior of the ΔN temporal trace: its discontinuity represents the switching of the lasing frequency from one mode to another. In the double target setup we are considering, because of the frequency mixing described by Equation (3), the appearance of a new MGM is influenced by both target feedbacks. In particular, during the movement of the RT on a timescale comparable to the fast fringe period, there can be a time lapse (typically shorter than the fast fringe period itself) in which the new MGM exists. During this time, the laser frequency switches between the two different modes, inducing a sudden transition in the frequency and carrier density, leaving a peculiar and easily recognizable fingerprint in the time trace, which we call the "sub-feature". The duration of these sub-features, and, conversely, the time interval between two subsequent sub-features, also depends on the OT motion, as indicated by the fact that it varies from one fast fringe to another, within the periodicity of the slow fringe. This is precisely the temporal feature on the basis of which the slow motion of OT can be tracked in the fast fringes, provided that one is able to extract the dependence on the OT velocity encoded in the duration of the sub-features. Referring to Figure 3, the slow fringe is the part of the temporal trace between Points A and B in Figure 3a, while the fast fringe is the part between Points C and E in Figure 3b (its duration is indicated by ΔT), and the sub-feature is the part between Points D (the left cusp) and E (the right cusp); finally, with Δt, we denote the time interval between the edges of two consecutive sub-features, namely between Points C and D in Figure 3b.

SUPER-RESOLVED DISPLACEMENT SENSOR

In this section, we first describe the procedure for analyzing ΔN(t) and for identifying the relevant temporal marks of the sub-features described above. Then, we present the numerical scheme used to relate such data to the OT displacement.

As we have discussed in the previous section, information about the movement of RT and OT is linked together in the features of the temporal trace via Equation (4). We performed an extensive fitting procedure, to relate the position of the OT to the time interval Δt between the cusps of two subsequent sub-features. The foundations of this method have been presented in [13]. Here, we radically improve it and correlate it with an analysis of the achievable precision in relation to the calibration stage. We will make use of the increased number and accuracy of the simulations and of the broader parameter space (see Figure 2).

Numerical Approach

All of the simulations show that the sub-features emerge at the beginning of the slow fringe. Their duration $t_E - t_D = \Delta T - \Delta t$, when compared to the fast fringe duration ΔT, is very small at the beginning, and it grows as the OT moves, until, near the end of the slow fringe, it becomes of the same magnitude as ΔT. This actually results in the vanishing of the sub-features for a short period of time (indicated by Δt_B in Figure 3a) that terminates at the slow fringe jump, after which they reappear in the next slow fringe; of course, the dynamical behavior of Δt across the slow fringe is complementary). We now illustrate how it is possible to relate the change in the time lapse Δt to the position of the OT. In our simulation, the OT velocity v_o is fixed, so we can define a "theoretical" position of the OT as given by the simple formula:

$$S_{th} = v_o t \qquad (5)$$

The time trace is thus sampled to extract the values of Δt at different instants of time; at each time, we assume that the position of the OT is given by S_{th}, and we will perform a best fit analysis to find a function that relates the position S_{th} to Δt. Since, as we mentioned above, there can be a short period across the slow fringe change in which the sub-features disappear, the fitting procedure can be performed in principle only along a single slow fringe. In the next subsection, we will tackle the problem of overcoming this limit, because an actual sensor must be capable of operating on an (ideally) arbitrarily long period, while at this stage, we will limit our analysis to a single slow fringe.

We have developed an algorithm capable of identifying and classifying (potentially in real time) all of the critical points of the temporal trace labeled in Figure 3, so that we can register the time lapses between them. We set the origin of time at the first left cusp identified by the analysis algorithm, and we take the reference times t_n at each subsequent left cusp, while, as already mentioned before, we register the time lapse Δt_n that occurs between the right cusp of the (n-1)-th sub-feature and the left cusp of the n-th one. Upon varying the fit on a broad basis of polynomial and transcendent functions, we found that a quadratic dependence works surprisingly well in approximating such dependence, and the candidate test function can be cast as:

$$S_{fit}(\Delta t_n) = C_2 \Delta t_n^2 + C_1 \Delta t_n + C_0 \qquad (6)$$

The least squares method can be applied to evaluate the coefficients C_i for all of the simulations. Of course, coefficients C_i change for simulations at different v_r, because a larger v_r implies a reduced ΔT and, thus, Δt. A proper interferometric sensor cannot suffer from such "reference arm" dependence, so the next step is to make the v_r dependence explicit in Equation (6). For such a purpose, a new best fit procedure was carried out, resorting to the sets of simulations with varying v_r. The coefficients C_2 and C_1 have been found to have a quadratic and a linear dependence on v_r, respectively, while C_0 is independent of v_r. The general formula we sought, which is supposed to hold for a wide range of the (v_r, v_o) plane, could then be cast as:

$$S_{phen}(\Delta t_n) = -\gamma_2 \frac{v_r^2}{\lambda_0} \Delta t_n^2 - \gamma_1 v_r \Delta t_n + \gamma_0 \frac{\lambda_0}{2} \quad (7)$$

In this formula, we insert the solitary laser wavelength $\lambda_0 = 2\pi c/\omega_0$ to make the dimensions of the coefficients clear. To determine the parameters γ_i from the entire set of simulations, we have evaluated the parameters C_i for all of the complete slow fringes in each simulation, and we have extracted the values of the γ_i, along with their errors, by using their expressions as functions of the C_i. Each value obtained in this way has been treated as an independent detection of the "true" γ_i value, so that its best estimate is the weighted means evaluated from all of the occurrences. We found:

$$\begin{aligned} \gamma_2 &= 1.0197 \pm 0.0017 \\ \gamma_1 &= 0.6696 \pm 0.0009 \\ \gamma_0 &= 0.9375 \pm 0.0002 \end{aligned} \quad (8)$$

and these parameters are constant within the assumptions described so far.

The Sensing Procedure: Methods, Sensitivity and Limits

Having obtained the relation defined in Equation (7), it can be used to determine the displacement of the OT for any velocity pair (v_r, v_o) compatible with the limits of validity, which are going to be discussed here. First of all, the model was built on the stationary solutions of the LK equations, so we must ensure that the evolution of the target is adiabatic, i.e., the laser system has enough time to reach a stationary state as the targets move. The slowest time on which the system evolves is of the order of tens of nanoseconds [2, 3], while the evolution timescale of the carrier density difference associated with the motion of the RT is of the order of the fast fringe period, $\Delta T \simeq \lambda_0/2v_r$. Therefore, adiabaticity requires that $v_r << v_{r,max}$ with $v_{r,max} \simeq 100$ m/s. In our simulations, the maximum value for the RT velocity was thus set to 5 m/s.

The reliability of the phenomenological relation Equation (6) can be assessed by appreciation of the normalized root mean square deviation for each fitting procedure:

$$\text{RMS} = \frac{1}{\mathcal{N}-3} \sum_n \left(\frac{S_{th}(\Delta t_n) - S_{fit}(\Delta t_n)}{S_{th}(\Delta t_n)} \right)^2 \quad (9)$$

where $\mathcal{N}$ is the number of sub-features identified by the algorithm. We observed that its value was very small (of the order of a few 10^{-4}) as long as the ratio v_o/v_r was well above 10^{-4}. Near this threshold, it jumped abruptly to values of the order of 10^{-1}, thus indicating that the assumption of a quadratic dependence is no longer valid. In order to be conservative, in the evaluation of the parameters γ_i in Equation (8), we have used only simulations with $v_o/v_r \geq 0.5 \times 10^{-3}$. The smallest displacement that can be observed depends on the time between two subsequent observations of Δt, which is of the order of the fast fringe periodicity, $\Delta T \simeq \lambda_0/2v_r$, so that, in principle,

$$\Delta S \simeq \frac{v_o}{v_r} \lambda_0 \quad (10)$$

Since, as we discussed above, the sensing procedure allows $v_o/v_r \simeq 10^{-3}$, the scheme we are proposing should be capable of reaching a sensitivity of $\lambda_0/1000$, which is of order of a few nanometers. An example of the results that can be obtained with the sensing procedure described so far is given in Figure 5.

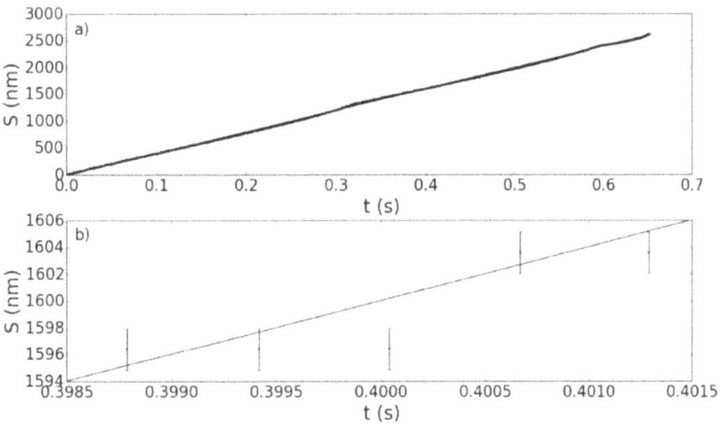

Figure 5. Plot of the position of the object target (OT) target S(t) versus time, for a simulation with v_r=5×10^{-3} m/s and v_o=4×10^{-6} m/s. The other parameters are as in Figure 3. The solid line represents the theoretical position S_{th} (see Equation (5)). The dots mark the position S_{phen} as predicted by the phenomenological Equation (7) and the associated errors σ_S as given by Equation (11). (a) The representation over the entire period of a slow fringe; (b) a close-up in which the accuracy of S_{phen} in retrieving S_{th} can be appreciated.

It is evident by inspection of the close-up, Figure 5b, that the potential measurements of the OT are well within the 10-nm deviation from the theoretically-expected values (the solid line). Of course, this result depends crucially on the accuracy in determining the coefficients γ_i of Equation (8). In fact, we can assume that the error on the determination on the OT displacement due to uncertainties in the γ_i parameters is given by applying the error propagation law to the phenomenological Equation (7):

$$\sigma_S = \sqrt{\sum_i \left(\frac{\partial S}{\partial \gamma_i}\right)^2 \sigma_{\gamma_i}^2} = \sqrt{\left(\frac{v_r^2}{\lambda_0}\Delta t_n^2\right)^2 \sigma_{\gamma_2} + (v_r \Delta t_n)^2 \sigma_{\gamma_1} + \left(\frac{\lambda_0}{2}\right)^2 \sigma_{\gamma_0}} \quad (11)$$

and this error must be smaller than the expected sensitivity given by Equation (10). Increasing the number of simulated slow fringes, on which the determination of the γ_i parameters is performed, will result in better accuracy. To check this, we have evaluated the mean error $\overline{\sigma_S}$ as a function of the number of simulations. The result is shown in Figure 6.

As we can see, the curve falls quite rapidly below 10 nm and then tends to saturate around 5 nm. This confirms that the γ_i parameters can be evaluated with sufficient precision for our sensing scheme to reach nanometric sensitivity. The values reported in Equation (8) have been obtained using a set of 100 simulated slow fringes.

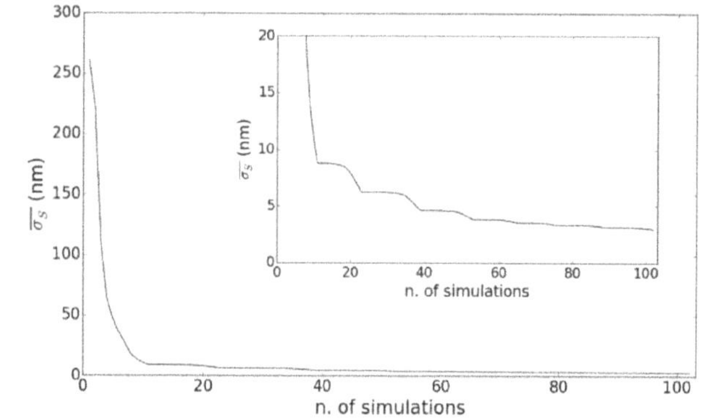

Figure 6. Plot of the mean error on the determination of the OT position $(\overline{\sigma_S})$ as a function of the number of simulations used to determine the γ_i coefficients. The inset focuses on the relevant range of values for $\overline{\sigma_S}$ and shows that it actually reaches the nanometer range.

In a realistic sensor, the determination of the γ_i parameters and their uncertainties is based on a calibration procedure where targets with known velocities are employed. The calibration should be performed for a set of pairs (v_r, v_o) conveniently distributed in the parameter space according to the guidelines discussed in the first part of this subsection. This calibration needs to be done only once, since the parameters determined in this way should remain valid for a range of at least three orders of magnitude in both target velocities. The number of calibration runs determines the accuracy of the value obtained for the γ_i parameters and, consequently, the sensitivity of the device. A curve representing the error on the determination of the position of the target as a function of the numbers of calibration runs should have the behavior reported in Figure 6, but other factors related to the experimental conditions may hinder the capability of achieving the predicted nanometric sensitivity. Just as an example, mechanical vibrations of the set-up, irregularities in the motion provided by the step motor driving the RT and/or OT and current fluctuations in the power supply of the laser are all obvious sources of errors in the measure. While such sources can be tamed in principle, we will consider the effect of a more intrinsic limitation in Section 3.4.

Another issue to be addressed is how to circumvent the detection limitation to a single slow fringe due to the disappearance of the sub-feature discussed above. Our simulations proved that the number of fast fringes not exhibiting a sub-feature occurs invariably at the end of a slow fringe, and it decreases with increasing feedback strengths k_o and k_r. In any case, in the time lapse Δt_B, our algorithm is "blind" to the target motion. A sufficiently high feedback should limit this interval to a small fraction of the slow fringe period, and during this period, the OT displacement can be extrapolated via the simple formula $\Delta S_{blind} = v_l \Delta t_B$, where v_l is the last detected OT velocity, just before sub-feature disappearance. This simple method has been experimentally tested in [13] and proven to be sufficiently accurate within experimental uncertainties. In the next subsection, we will discuss in detail this and other aspects of the sensing procedure related to the feedback level.

Finally, let us stress that, in principle, the sensitivity could be pushed further beyond, since the limiting ratio $v_o/v_r \simeq 0.5 \times 10^{-3}$ is set by the failure of the quadratic fit proposed in Equation (7). For lower values, new sets of simulations and a more refined fitting procedure might yield a reliable relation $S_{phen}(\Delta t_n)$. In any case, the necessity to detect a sufficient number of slow fringes and of densely sampling each fast fringe might increase the requirement in terms of bandwidth and the buffer memory of the electronics reading out the voltage at the QCL contacts.

The Role of the Feedback Parameters in the Sensing Scheme

As was shown in Section 2.1, the key element for the sensor proposed in this work is the high feedback level, which ensures, in the spectral domain, the nonlinear frequency coupling and, in the time domain, the appearance of the sub-features, whence our algorithm extracts the information of the OT motion with nanometric accuracy. We stress that this feature is critical of the nonlinear dynamics of the laser in providing the coupling and of the QCL in particular, since this emitter can sustain large feedback without entering chaotic regimes [2]. We will now illustrate in some detail the role of the feedback strength on the system dynamics.

We analyzed the system behavior when the feedback levels k_r and k_o are changed, while keeping the ratio k_r/k_o fixed. The values of the target velocities were set to $v_r=5\times10^{-2}$ m/s and $v_o=5\times10^{-4}$ m/s. Decreasing the feedback results in a reduction of the time fraction of the slow fringe in which the sub-features are present, as it is possible to see in Figure 7a. At even lower feedback levels, the sub-features disappear completely, as shown in Figure 7b.

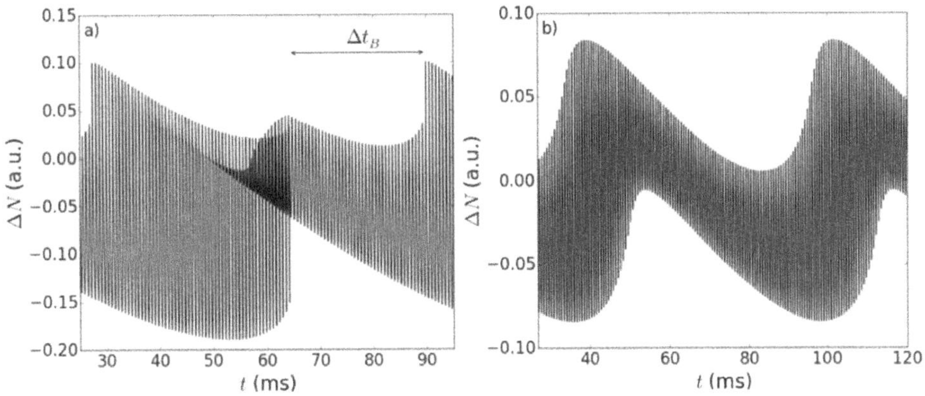

Figure 7. Plot of ΔN(t) for two different levels of feedback with fixed ratio k_r/k_o. The other parameters are as in Figure 3. (a) The feedback values are k_r=0.0180 and k_o=0.01575. Notice that the sub-features are present only in (approximatively) the second half of the slow fringe, as indicated by the time lapse t_B. The temporal trace (b), obtained using k_r=0.0080 and k_o=0.007, has no sub-features at all.

On the contrary, as the feedback parameters increase, new temporal features appear in the fast fringes, in the form of additional pairs of cusps with even shorter duration (see Figure 8a). While such "sub-sub-features" may be the subject of further investigations to enhance even further the potential sensitivity of our scheme, another interesting insight for this phenomenon can be gathered from inspecting the Fourier transform of ΔN(t) for different values of the feedback parameters.

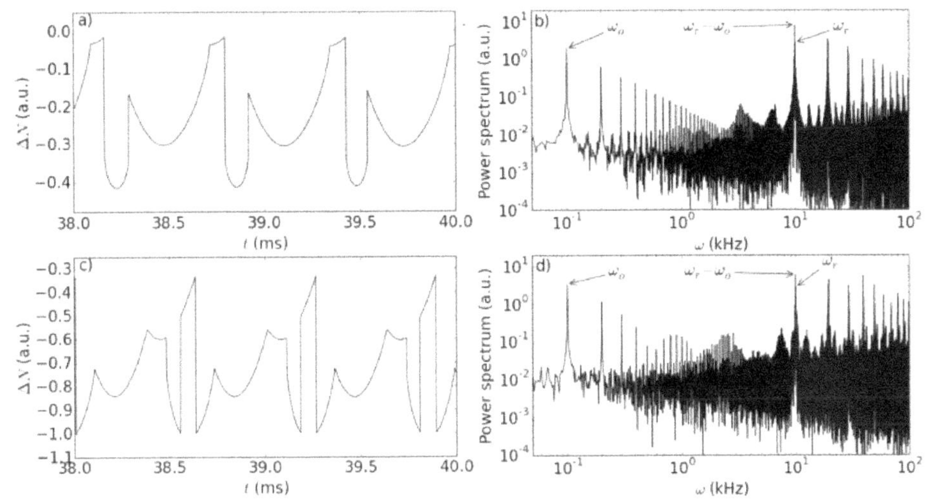

Figure 8. Plot of ΔN as a function of t (a, c) along with their corresponding Fourier transforms plotted in a log-log plane (b, d); upper part: k_r=0.0404; lower part: k_r=0.1010. The ratio k_r/k_o remains fixed. The other parameters are as in Figure 3.

Figure 8b shows the Fourier transform of the temporal trace corresponding to Figure 8a. One observes a strong peak at the lowest frequency ω_o and its higher harmonics, while at higher frequencies, as in the case of Figure 4, the main peak occurs at the frequency difference $\omega_r - \omega_o$. Remarkably, there are several higher harmonics of this dominant note, whose peaks decrease in intensity with a power law, indicating once again the strong nonlinear character of the interaction, brought about by the laser dynamics. Interestingly, the time trace is still regularly periodic in this case, although the peak pattern is complicated by the appearance of the new cusps. Upon further increase of the feedback strength, we observe (see Figure 8d) that the background increases considerably, while the peaks at the dominant frequencies become more intense and do not decrease linearly any more; accordingly, the time trace (see Figure 8c) still exhibits regular, though complex features on the short time scale, while at longer timescales, comparable with the slow fringe period, it appears irregular, and we cannot expect to recover a relation formally similar to Equation (7).

Summing up, as concerns the proposed sensing scheme, in the general dynamical scenario of the retro-injected QCL, the most suitable feedback level one should try to set as the operational point for the sensor is the one close to the threshold of the appearance of the sub-sub-features. In this condition, in fact, the time lapse Δt_B in which the sub-features disappear is the shortest possible, thus reducing the error implied by the extrapolation procedure described above.

Influence of the QCL Linewidth on the Sensitivity

As we mentioned in Section 3.2, several other sources of error may worsen the accuracy of Equation (7), as given by the errors on its coefficients in Equation (8), which are solely determined by the sample set on which the calibration is performed. We consider now the effect of the finite linewidth of the QCL emission. The latter is associated with phase variation induced by spontaneous emission, carrier-induced refractive index change and injection current fluctuations [16,17]. Moreover, it is known that the presence of optical feedback leads to linewidth broadening or narrowing depending on the external cavity phase shift [18]. A rigorous theoretical approach that describes all of these phenomena would consist of adding Langevin noise sources in the LK Equations (1) and (2), as is described in [17]. Here, in order to provide a simple estimation of the role of the QCL linewidth in limiting the proposed sensor accuracy, we suppose that the stochastic fluctuations of the free running laser frequency, denoted now as ω_{QCL}, which follows a normal distribution centered in ω0ω0 with amplitude $\Delta\omega_{QCL}$, affect the frequency ω_F of the reinjected QCL trough Equation (3) and, in turn, the values of ΔN as given by Equation (4). We can thus expect the fringe jumps and the sub-feature duration to fluctuate accordingly; this will induce additional uncertainties in the determination of the target displacement as evaluated in Equation (7). In an ergodic hypothesis, we consider the ensemble average of a large number of simulations with fixed ω_{QCL} as representative of the time average in the temporal evolution of fluctuating variables, as provided by the integration of Equations (1) and (2) with the inclusion of Langevin noise sources [17].

In particular, for the study case v_r=5×10^{-3} m/s and v_0=5×10^{-5} m/s and for values of $\Delta\omega_{QCL}$ ranging from 100 kHz to 10 MHz, which are in agreement with estimations reported in recent literature [19,20,21], a set of 50 determinations of ω_{QCL} was randomly generated and used to obtain 50 ΔN(t) traces, via Equation (3) (where, of course, ω_0 was replaced by ω_{QCL}). The immediate visual effect of the ω_{QCL} fluctuations introduced in this way is the jittering of the cusps delimiting the sub-features (see Figure 9). This amounts to saying that one can detect a set of $t_{C,i}, i = 1, \ldots 50,$ corresponding to left cusps initiating a sub-feature, and another set $t_{D,j}, j = 1, \ldots 50$, corresponding to right cusps ending a sub-feature (see Figure 3b), which define a 50×50 array of detectable sub-feature time lapses $\Delta t_{i,j} = t_{D,j} - t_{C,i}$. Indicated as Δt_{max} and Δt_{min}, the maximum and minimum values of $\Delta t_{i,j}$, respectively (see Figure 9), the dispersion $\sigma_{\Delta t} = (\Delta t_{max} - \Delta t_{min})$ is considered as the error on the determination of Δt induced by the finite linewidth.

The uncertainty on the OT position has been derived with the error propagation law:

$$\sigma_{S,lw} = \left(2\gamma_2 \frac{v_r^2}{\lambda_0^2} \Delta t_n + v_r\right) \sigma_{\Delta t_n} \tag{12}$$

In Figure 10, we represent the behavior of the mean error $\overline{\sigma_{S,lw}}$ (obtained by averaging $\sigma_{S,lw}$ the set of sub-features considered) as a function of the QCL linewidth. As we can see, the errors grow with linewidth according to a power law, and for $\Delta\omega_{QCL} \leq 5$ MHz, they are below 10 nm, thus comparable to those intrinsic to our deterministic method.

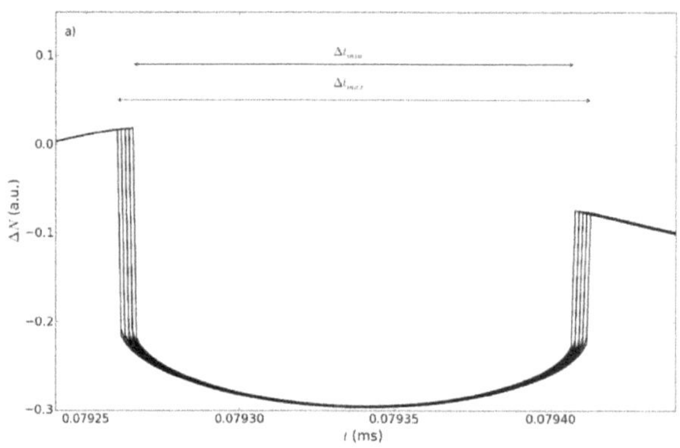

Figure 9. Broadening of the temporal trace ΔN(t) for a linewidth $\Delta\omega_{QCL}$=10 MHz. The plot focuses on the time lapse between two consequent sub-features, showing the maximum (Δt_{max}) and the minimum (Δt_{min}) values of Δt considered in the evaluation of the uncertainty $\sigma_{S,lw}$. The other parameters are as in Figure 3.

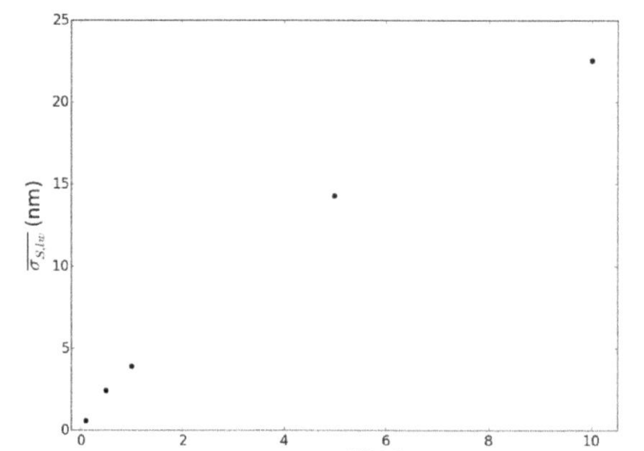

Figure 10. Plot of the uncertainty $\overline{\sigma_{S,lw}}$ against the quantum cascade laser (QCL) linewidth associated here with quantity $\Delta\omega_{QCL}$.

CONCLUSIONS AND PERSPECTIVES

In conclusion, we have studied the dynamics of a QCL with retro-injection provided by two translating collinear targets, showing that the nonlinear behavior of the emitter provides a nontrivial coupling of the two feedback fields, which allows one to extract information on the slower target on a time basis linked to the interferometric fringes associated with the fast target translation. Thus, we have proposed a scheme for a real-time, nanometric displacement sensor with a resolution on the order of $\lambda/1000$, a calibration procedure for its use in a wide range of OT velocities, and we have provided an estimation of the accuracy in dependence of several factors, among which is the intrinsic linewidth of the emitter.

While this scheme could be exported in principle to any range of wavelengths, it must be noted that high levels of feedback strength are necessary to ensure a significant nonlinear response and, thus, the occurrence of the sub-features upon whose durations we based the sensor scheme. Conventional diode lasers appear to be unsuited for such scheme, since the undamped amplification of the relaxation oscillations frequency causes the emitter to enter a chaotic regime at high feedback levels [4]. On the other hand, the QCL is quite appealing, not only for the wavelength range, at which several materials of interest are transparent, but also because of the narrow linewidth, which is beneficial for the sensitivity of the present sensor scheme. We believe that our work shows how, quite generally, the introduction of an additional interferometric element (here, the moving RT) provides a second, controlled and fast "clock tick", which allows fast sampling of an independent process (the OT translation), while the nonlinear dynamics of the laser provides the coupling between the two. It will be of course interesting to extend the present scheme to analyze arbitrary OT motions and to model more convenient RT dynamics, such as vibrations of the RT etalon, as they may be provided by a piezo controller. Other than faster RT displacements and, thus, improved basic resolution (see Equation (10)), such an extension may provide a simpler device with a reduced footprint. In this respect, entirely new knowledge must be acquired concerning the relation among RT and OT spectral features appearing in the QCL output. Work is in progress in this direction. Finally, and on a more fundamental note, it has been shown that a QCL with feedback can exhibit a multimode regime of regular oscillations, corresponding to coherent locking of modes of the external cavity provided by a reflector [3]. The inclusion of two reflectors, thus the existence of two sets of independently-tunable, external modes, might provide an "engineered" modal competition (e.g., when the two sets have free spectral range ratios in the rational or in the irrational domain) and reveal novel features in the coherent QCL dynamics.

The authors also wish to thank Francesco Paolo Mezzapesa for useful discussions.

ACKNOWLEDGMENTS

The authors acknowledge financial support from the research projects MIUR-PON 02-0576 "INNOVHEAD–Tecnologie innovative per riduzione emissioni, consumi e costi operativi di motori Heavy Duty" and "MASSIME–Sistemi di sicurezza meccatronici innovativi (cablati e wireless) per applicazioni ferroviarie, aerospaziali e robotiche", PON 01-02238 EURO6 "Elettronica di controllo, sistema d'iniezione, strategie di combustione, sensoristica e tecnologie di processo innovativi per motori diesel a basse emissioni inquinanti", COST Action BM 1205 "European Network for Skin Cancer Detection using Laser Imaging" and the Fondazione Caripuglia research project "Studio di sorgenti laser QCL per la realizzazione di sensori avanzati".

AUTHOR CONTRIBUTIONS

Lorenzo Luigi Columbo performed the numerical solution of Equations (3) and (4) for different OT and RT velocities. Giseppe De Risi developed the theoretical analysis of the resulting carriers density functions $\Delta N(t)$ and the fitting procedure leading to Equation (7). Massimo Brambilla supervised all of the theoretical and numerical work. Giuseppe De Risi, Lorenzo Luigi Columbo and Massimo Brambilla wrote the manuscript.

REFERENCES

1. Faist, J. Quantum Cascade Lasers; Oxford University Press: London, UK, 2013.
2. Mezzapesa, F.P.; Columbo, L.L.; Brambilla, M.; Dabbicco, M.; Borri, S.; Vitiello, M.S.; Beere, H.E.; Ritchie, D.A.; Scamarcio, G. Intrinsic stability of quantum cascade lasers against optical feedback. Opt. Express 2013, 21, 13748–13757.
3. Columbo, L.L.; Brambilla, M. Multimode regimes in quantum cascade lasers with optical feedback. Opt. Express 2014, 22, 10105–10118.
4. Kane, D.M.; Shore, K.A. Unlocking Dynamical Diversity-Optical Feedback Effects on Semiconductor Diode Lasers; J. Wiley and Sons: New York, NY, USA, 2005.
5. Dean, P.; Lim, Y.L.; Valavanis, A.; Kliese, R.; Nikolić, M.; Khanna, S.P.; Lachab, M.; Indjin, D.; Ikonić, Z.; Harrison, P.; et al. Terahertz imaging through self-mixing in a quantum cascade laser. Opt. Lett. 2011, 36, 2587–2589.
6. Lim, Y.L.; Dean, P.; Nikolic, M.; Kliese, R.; Khanna, S.P.; Lachab, M.; Valavanis, A.; Indjin, D.; Ikonić, Z.; Harrison, P.; et al. Demonstration of a self-mixing displacement sensor based on terahertz quantum cascade lasers. Appl. Phys. Lett. 2011, 99, 081108.
7. Mezzapesa, F.P.; Spagnolo, V.; Ancona, A.; Scamarcio, G. Detection of ultrafast laser ablation using quantum cascade laser-based sensing. Appl. Phys. Lett. 2012, 101, 171107.
8. Zabit, U.; Bony, F.; Bosch, T.; Rakić, A.D. A self-mixing displacement sensor with fringe-loss compensation for harmonic vibrations. IEEE Photon. Technol. Lett. 2010, 22, 410–412.
9. Fan, Y.; Yu, Y.; Xi, J.; Chicharo, J.F. Improving the measurement performance for a self-mixing interferometry-based displacement sensing system. Appl. Opt. 2011, 50, 5064–5072.
10. Hast, J.; Okkonen, M.; Heikkinen, H.; Krehut, L.; Myllyla, R. Nanometer-scale displacement sensing using self-mixing interferometry with a correlation-

based signal processing technique. Opto-Electron. Rev. 2013, 14, 129–134.

11. Azcona, F.J.; Atashkhooei, R.; Royo, S.; Astudillo, J.M.; Jha, A. A nanometric displacement measurement system using differential optical feedback interferometry. IEEE Photon. Technol. Lett. 2013, 25, 2074–2077.
12. Mezzapesa, F.P.; Columbo, L.L.; Dabbicco, M.; Brambilla, M.; Scamarcio, G. QCL-based nonlinear sensing of independent targets dynamics. Opt. Express 2014, 22, 5867–5874.
13. Mezzapesa, F.; Columbo, L.L.; de Risi, G.; Brambilla, M.; Dabbicco, M.; Spagnolo, V.; Scamarcio, G. Nanoscale displacement sensing based on nonlinear frequency mixing in quantum cascade lasers. 2015; arXiv:1505.00027.
14. Keeley, J.; Dean, P.; Valavanis, A.; Bertling, K.; Lim, Y.L.; Alhathlool, R.; Taimre, T.; Li, L.H.; Indjin, D.; Rakić, A.D.; et al. Three-dimensional terahertz imaging using swept-frequency feedback interferometry with a quantum cascade laser. Opt. Lett. 2015, 40, 994–997.
15. Mezzapesa, F.P.; Columbo, L.L.; Dabbicco, M.; Brambilla, M.; Ancona, A.; Sibillano, T.; de Lucia, F.; Lugará, P.M.; Scamarcio, G. Simultaneous measurement of multiple target displacements by self-mixing interferometry in a single laser diode. Opt. Express 2011, 19, 16160–16173.
16. Henry, C.H. Phase Noise in Semiconductor Lasers. J. Lightwave Technol. 1986, Lt-4, 298–311.
17. Agrawal, G.P.; Roy, R. Effect of injection-current fluctuations on the spectral linewidth of semiconductor lasers. Phys. Rev. A 1988, 37, 2495–2501.
18. Agrawal, G.P. Line Narrowing in a Single-Mode Injection Laser Due to External Optical Feedback. IEEE J. Quantum Electron. 1984, QE-20, 468–471.
19. Yamanishi, M.; Edamura, T.; Fujita, K.; Akikusa, N.; Kan, H. Theory of the Intrinsic Linewidth of Quantum-Cascade Lasers: Hidden Reason for the Narrow Linewidth and Line-Broadening by Thermal Photons. IEEE J. Quantum Electron. 2008, 44, 12–29.
20. Bartalini, S.; Borri, S.; Cancio, P.; Castrillo, A.; Galli, I.; Giusfredi, G.; Mazzotti, D.; Gianfrani, L.; de Natale, P. Observing the Intrinsic Linewidth of a Quantum-Cascade Laser: Beyond the Schawlow-Townes Limit. Phys. Rev. Lett. 2010, 104, 083904.
21. Vitiello, M.S.; Scalari, G.; Williams, B.; de Natale, P. Quantum cascade lasers: 20 years of challenges. Opt. Express 2015, 23, 5167–5182.

CHAPTER 10 QUANTUM SPECTROSCOPY OF PLASMONIC NANOSTRUCTURES

Dmitry A. Kalashnikov, Zhenying Pan, Arseniy I. Kuznetsov, and
Leonid A. Krivitsky

INTRODUCTION

New insights into understanding the behavior of materials at the nanoscale and progress in nanofabrication capabilities have triggered great interest in the study of plasmonic nanostructures. One practical application of these structures is the development of sensing technology. Plasmonic sensors benefit from their biocompatibility and high sensitivity [1–4]. With the sensitivity of the sensors approaching the atto-Molar scale, as well as sensing of photosensitive substances, it is important that the analyte is not damaged or modified by the probing light. Thus, we need to use extremely small doses of light, possibly down to the level of single photons, and highly sensitive equipment for the readout of the sensor response. For sensing with low-photon counts, the main practical issue is careful suppression of the influence of the background noise. This is challenging with conventional spectroscopy, but quantum optics can be exploited to effectively address this issue [5–7].

Interfacing quantum optics with plasmonics is an area of active research. Plasmonic nanostructures are used as efficient optical antennas for increasing the collection efficiency of light, emitted by solid-state single-photon sources [8–16]. They are also shown to preserve quantum entanglement [17–19] and squeezing [20,21], thus making it possible to use them in subwavelength quantum optical networks [22].

In this work, we demonstrate advantages of using frequency-entangled photons, generated via spontaneous parametric down conversion (SPDC), in revealing the spectroscopic response of a plasmonic refractive-index sensor under noisy conditions. The technique operates at a single-photon level and is more robust to environmental noise, compared to conventional transmission spectroscopy.

QUANTUM SPECTROSCOPY

In the SPDC, a pump photon traveling in a medium with quadratic susceptibility occasionally splits into a pair of photons, referred to as the signal and idler [23, 24]. Signal and idler photons are created almost simultaneously (within tens of fs) with their energies and momenta obeying the conservation laws:

$$\omega_p = \omega_s + \omega_i \qquad \vec{k}_p = \vec{k}_s + \vec{k}_i, \tag{1}$$

where $\omega_{p,s,i}$ and $\vec{k}_{p,s,i}$ are the energies and the wave vectors of the pump, signal, and idler photons, respectively. Signal and idler photons are detected with two single-photon avalanche photodiodes (APD). Their pairwise correlation is measured by an electronic coincidence circuit, which produces a signal (coincidence) if APD counts arrive within the fixed time window, typically of about several ns. Dedicated techniques allow one to obtain the widths of the SPDC spectra in a given spatial mode in the range of several hundreds of nanometers [25–28].

The sample under investigation is placed in the path of one of the photons (idler), and the frequency selection is performed over another photon (signal) by a monochromator [29–33] [see Fig. 1(a)]. Since the sum of energies of the down-converted photons is equal to the energy of the pump, spectral selection over the signal photon defines the frequency of the correlated idler photon. Following Ref. [29], the number of coincidences

Citation: Dmitry A. Kalashnikov, Zhenying Pan, Arseniy I. Kuznetsov, and Leonid A. Krivitsky, Quantum Spectroscopy of Plasmonic Nanostructures, DOI:http://dx.doi.org/10.1103/PhysRevX.4.011049.

between two APDs is proportional to the spectral function of the sample T(ω) at the frequency of the idler photon,

$$R(\omega_s) \sim T(\omega_p - \omega_s). \quad (2)$$

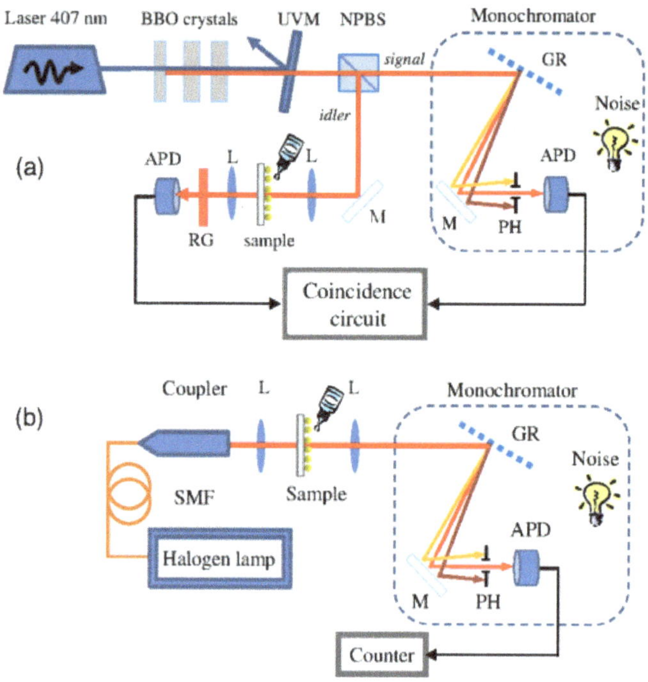

Figure. 1. (a) Quantum spectroscopy setup. Diode laser at 407 nm is used as a pump for a set of three BBO crystals cut for type-I SPDC; the UV mirror (UVM) reflects the pump beam and transmits the SPDC signal; a nonpolarizing beam splitter (NPBS) splits the SPDC signal into two arms; in one arm, there is a monochromator, which consists of a mirror (M), a diffraction grating (GR), a pinhole, and an APD. In another arm, there is a sample, which represents an array of gold nanoparticles assembled into a cell. The SPDC signal is focused onto the sample by achromatic lenses (L), filtered by a red glass filter (RG), and detected by the APD. Counts of the two APDs are sent to a coincidence circuit; the source of the noise is provided by the external APD, illuminated by the lamp (not shown), whose signal is mixed electronically with the signal of the APD in the signal channel. (b) Transmission spectroscopy setup. A halogen lamp is used as a source of the broadband light; the light after a single mode fiber (SMF) is collimated by a coupler, focused by lenses onto the sample, filtered by the monochromator, and detected by the APD.

In contrast to the conventional transmission spectroscopy, the spectral selection is performed in the channel without the structure under test. Following the analogy with imaging experiments, we refer to the method as quantum (ghost) spectroscopy.

The remarkable feature of the quantum spectroscopy is its robustness against the environmental noise, which allows one to conduct measurements at extremely low-photon fluxes. The noise typically includes background optical noise and electronic noise of the APDs. Let us denote the number of SPDC photocounts as $S_{s,i} = \eta_{s,i}P$, where P is the number of photon pairs, produced by the SPDC source, and $\eta_{s,i}$ is the quantum efficiency of the signal and idler channel, respectively. The number of coincidences is proportional to the probability of joint detection by the APDs in two channels:

$$R(\omega_s) = \eta_s\eta_i P. \quad (3)$$

Assuming that noise photocounts of the two APDs are uncorrelated, the number of noise coincidences RN is given by accidental overlap of photocounts within the time window of the coincidence circuit Δt. It also includes components due to accidental overlap of SPDC and noise photocounts:

$$R_N = N_sN_i\Delta t + S_sN_i\Delta t + N_sS_i\Delta t, \quad (4)$$

where N_s and N_i are the number of noise photocounts of APDs in signal and idler channels, respectively. The total number of coincidences is given by

$$R_{\text{Total}} = R(\omega_s) + R_N. \quad (5)$$

Even though $N_s, N_i \gg S_s, S_i$, the number of noise coincidences R_N is strongly suppressed compared to $R(\omega_s)$ because of a narrow coincidence time window Δt. Thus, the spectral response of the structure can be revealed from R_{Total} even under excessive noise situations.

COMPARISON TO THE CONVENTIONAL TRANSMISSION SPECTROSCOPY

In conventional transmission spectroscopy, the sample is placed between the light source and the monochromator; the transmission spectrum is obtained directly from APD photocounts [see Fig. 1(b)]. The signal-to-noise ratio (SNR) for the transmission spectroscopy is given by $\text{SNR}_T = \eta_s P/N_s$. For the quantum spectroscopy, the SNR is given by $\text{SNR}_Q = R(\omega_s)/R_N = \eta_i\eta_s P/N_sN_i\Delta t$, see Eqs. (3) and (4), where it is assumed that $N_sN_i \gg S_sN_i; N_sS_i$. The ratio of the two SNRs is given by

$$\text{SNR}_Q/\text{SNR}_T = \eta_i/N_i\Delta t. \quad (6)$$

The advantage in the SNR for the quantum spectroscopy is provided by the use of high-quantum efficiency, low-noise APD in the idler channel, and the narrow coincidence time window Δt.

THE PLASMONIC ARRAY SENSOR

We use an array of metal nanoparticles which has a narrow Fano-type plasmon resonance, due to diffractive coupling of localized surface plasmons [34–39]. The array acts as a metallic grating, and when the wavelength of incident light gets close to the Rayleigh cutoff wavelength, the diffracted wave propagates along the surface [40]. This effect is referred to as Wood-Rayleigh's anomaly. If the anomaly wavelength is close to the plasmon resonance of individual nanoparticles, then the collective plasmonic mode is excited. Excitation of the mode results in a narrow Fano-type resonance in the transmission spectrum of the array. The position and shape of the Fano resonance depend on the shape of the nanoparticles, distance between them, and refractive index of the surrounding medium. High sensitivity to the local change of the refractive index, as well as the high-quality (Q) factor of the nanoparticle array resonance, constitutes the basis for its sensing applications [37,41].

The nanoparticle array is fabricated using a combined method of nanosphere lithography with femtosecond laser-induced transfer, which allows production of large-scale periodic arrays of spherical nanoparticles [37]. A hexagonal array with dimensions of approximately 1×1 mm² is fabricated with a particle diameter of 130 nm and a hexagonal lattice period of 1.1 μm [see Figs. 2(a) and 2(b)]. The fabricated particles are partially embedded (around two-thirds) into a polydimethylsiloxane (PDMS) substrate, with a part (around one-third) above the substrate surface (this is accessible to the local environment). With the sample covered by a second PDMS layer, the nanoparticles are in a homogeneous surrounding and the array provides a Fano resonance centered at 806 nm with a quality factor Q≈27, where Q is defined as the ratio of the resonance wavelength to the resonance width, defined from the fit using the Fano formula (see details below).

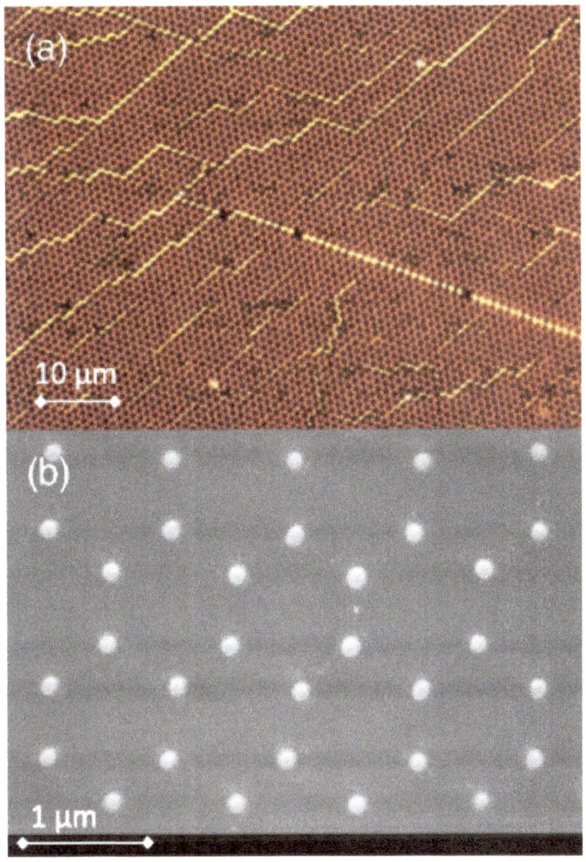

Figure. 2. Images of the plasmonic nanoparticle array on a PDMS substrate. (a) Dark-field microscope image at 10×magnification. (b) Scanning electron microscope (SEM) image.

The refractive-index sensitivity is measured by removing the cover PDMS layer and adding testing glycerin-water solutions with different concentrations on top of the array. The sensitivity is calculated as a ratio of the resonance shift in nm to the change of the refractive index of the testing solution.

EXPERIMENTAL SETUP

The quantum spectroscopy setup is shown in Fig. 1(a). Three bulk BBO crystals (Dayoptics) with thicknesses of 0.3 mm, 0.5 mm, and 0.5 mm are pumped by a continuous-wave vertically polarized diode laser at 407 nm (Omicron PhoxX-405-60). The BBOs are cut for type-I collinear SPDC. One of the BBOs is set for the frequency-degenerate regime, while two others are detuned from the degenerate regime by tilting their optical axis [26]. The resulting SPDC light source has a spectral width of around 250 nm and is shown in Fig. 3. The pump is eliminated by a UV mirror, and the SPDC is split by a nonpolarizing beam splitter (NPBS). In one arm of the NPBS, we inserted a home-built diffraction grating monochromator (Thorlabs GR-1205) with resolution 1.5 nm. The light after the monochromator is detected by an avalanche photodiode (APD, Perkin Elmer SPCM-AQRH-14FC).

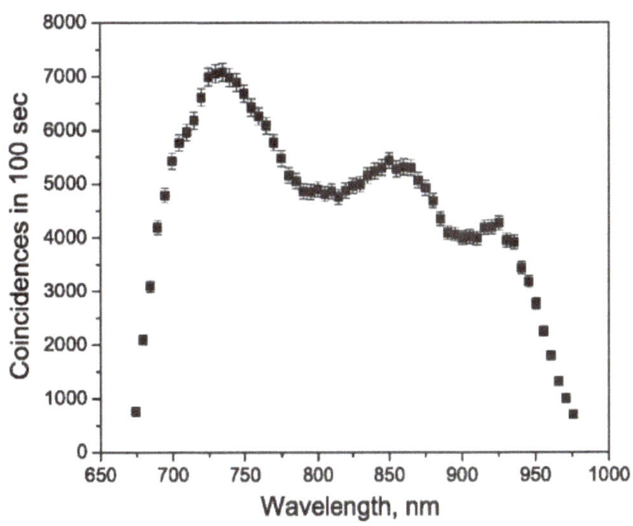

Figure. 3. Spectrum of coincidences of SPDC from three BBO crystals. The error bars are the standard deviations.

The sample is attached to a glass substrate and mounted in another arm of the NPBS in the focal plane of two confocal achromatic lenses with focus length f=60 mm. A mechanical translation stage (not shown) is used to position the sample. The focused light has a diameter of 300 μm on the sample surface. The sample area with the array is surrounded by

PDMS walls and covered by a thin glass plate forming a cell. The refractive index of the medium surrounding the array can be changed by adding different solvents inside the cell. The SPDC light passes through the broadband red glass filter (RG), and then it is coupled into the multimode fiber and detected by the APD (Perkin Elmer SPCM-AQRH-14FC). The signals from the two APDs are sent to the coincidence circuit with the time window of $\Delta t=5$ ns (Ortec 567 TAC/SCA). The number of coincidences is measured as a function of wavelength, selected by the monochromator. The step size of the wavelength selection is 3 nm.

Noisy conditions are emulated by artificially adding photocounts of an additional APD (not shown), illuminated by a lamp, to the signal of the APD in the monochromator. This approach allows one to exclude the spectral dependence of the noise. The level of noise is controlled by changing the driving current of the lamp.

Results of the quantum spectroscopy are compared with those obtained by the transmission spectroscopy [see Fig. 1(b) for the experimental setup]. The light from the lamp after the single mode fiber (SMF) is sent through the sample under investigation, using the same optical system as described above, and then coupled to the monochromator. APD photocounts are recorded versus the wavelength, selected by the monochromator.

The experimental procedure in both cases includes transmission measurements of the sample substrate and the array. The resulting transmission spectrum is calculated as the ratio of the array and the substrate spectra.

RESULTS AND DISCUSSION

To ensure that the SPDC bandwidth fully covers the range of the structure's Fano resonance, we measure the sample transmission with the quantum spectroscopy setup. The nanoparticle array is covered by a layer of PDMS to create homogeneous surroundings. The refractive index of the PDMS is about 1.4 [42]. The obtained spectrum is shown in Fig. 4. It reveals the Fano-type resonance with the distinctive asymmetrical shape. The gray solid curve in Fig. 4 shows the fitting of the experimental results with the Fano formula [38,39], with the following parameters: resonance wavelength $\lambda_R=806$ nm, resonance width $\Delta\lambda=30$ nm, and Fano (asymmetry) parameter $F=-16$. This gives the resonance quality factor $Q=\lambda_R/\Delta\lambda\approx 27$.

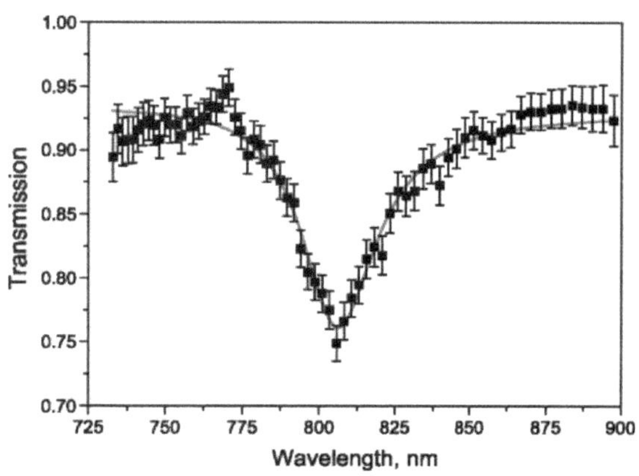

Figure. 4. Transmission spectrum of the nanoparticle array in homogeneous PDMS surroundings measured via quantum spectroscopy. The gray solid line is a fit with the Fano formula ($R^2=0.94$). The error bars are the standard deviations.

Refractive-index sensing is demonstrated by depositing a few drops (around 10 μl) of the testing solution of glycerin in water on the sample surface. In two different measurements, the glycerin concentrations are 40% and 50%, with the corresponding refractive indexes of the solutions of 1.384 and 1.398, respectively (the refractive-index values are taken from Ref. [43] for 20 °C). The refractive index of the solution of 50% glycerin in water is close to the refractive index of PDMS. Between subsequent measurements, the sample surface is washed with methanol and dried to remove the previous testing solution. Transmission spectra obtained at different glycerin concentrations using the conventional transmission spectroscopy are shown in Fig. 5(a). They demonstrate the redshift of the Fano resonance minima of about 8 nm when the glycerin concentration changes from 40% to 50%. This corresponds to the sensitivity value of 570 nm/RIU, which is a characteristic value for refractive-index sensing with plasmonic nanoparticles. Measurements performed by the quantum spectroscopy show similar results [see Fig. 5(b)].

It is important to emphasize that for the comparison of the two methods, all the measurements are performed with the same average flux of photons in the channel with the monochromator (in the vicinity of the resonance) equal to 103 photocounts/sec. The measurements are also performed for the same acquisition time: Each point of the spectrum is obtained from the average of 20 measurements, each acquired for 20 s.

Results of the measurements performed under noisy conditions for the case of the conventional transmission spectroscopy are shown in Figs. 5(a), 5(c), and 5(e). The noise levels are equal to 10^2 photocounts/s in Fig. 5(a), 2×10^4 photocounts/s in Fig. 5(c), and 7×10^4 photocounts/s in Fig. 5(e). With an increasing noise level, the resonance profile gets shallower and less pronounced. The maximum amount of noise at which it is feasible to perform sensing of the refractive-index change is given by the interplay of the spectral function of the sample, the sensitivity of the sensor, the average photon flux, the acquisition time, and the efficiency of the detection system [5]. From the experimental data in Fig. 5(e), one can see that when the noise is about 70 times larger than the signal, the resonance positions for both glycerin concentrations completely overlap, making it impossible to observe any refractive-index change.

The results obtained via the quantum spectroscopy are shown in Figs. 5(b), 5(d), and 5(f). The noise levels are equal to 10^3 photocounts/s in Fig. 5(b), 2×10^4 photocounts/s in Fig. 5(d), and 7×10^4 photocounts/s in Fig. 5(f). In contrast to the transmission spectroscopy, the shape and the depth of the resonance do not significantly change. The resonance position remains distinguishable, even when the noise is 70 times larger than the signal in the monochromator [see Fig. 5(f)]. The only effect of the noise is a slight decrease in the accuracy of the resonance measurements due to the contribution of noise coincidences. However, the difference between the resonance minima remains the same as in the case of low noise. This shows that quantum spectroscopy has high resistance to noise and can be used with extremely low-photon fluxes, which are not possible with conventional transmission spectroscopy.

This makes it very promising for sensing of photosensitive biological substances. Note that slight discrepancies in the shapes of Fano resonances in Figs. 5(a) and 5(b)are most likely caused by experimental uncertainties in setting the angle of incidence of the sensing light on the sample.

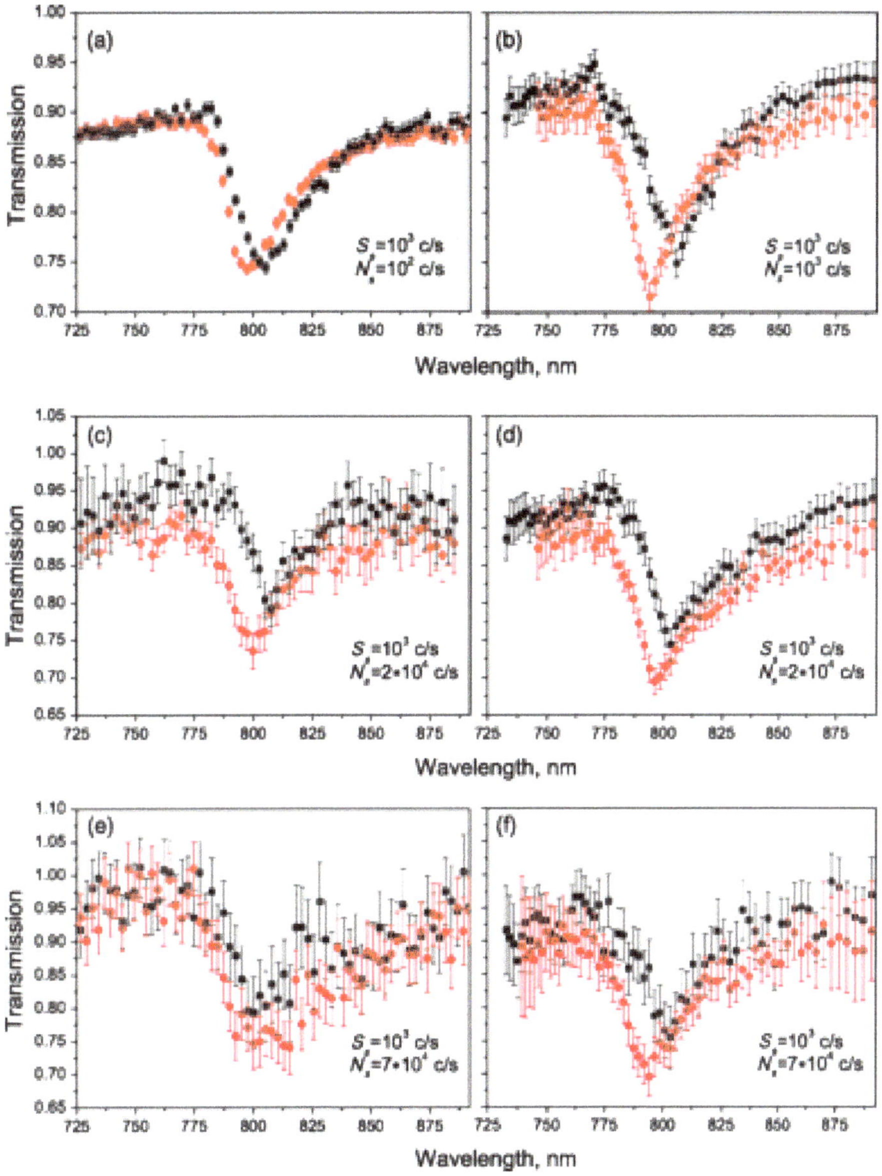

Figure. 5. Transmission spectrum of the nanoparticle array with different concentrations of glycerin-water solution on top, measured at different noise conditions. Red circles correspond to 40% glycerin concentration and black squares are assigned to 50% glycerin concentration. Results in (a), (c), and (e) are obtained by the conventional transmission spectroscopy with different noise levels of 10^2 photocounts/s, 2×10^4 photocounts/s, and 7×10^4 photocounts/s, respectively. Results in (b), (d), and (f) are obtained by the quantum spectroscopy with different noise levels: 10^3 photocounts/s, 2×10^4 photocounts/s, and 7×10^4 photocounts/s, respectively. The signal in the channel with the monochromator is the same for all the experiments in (a)-(f) and is equal to 10^3 photocounts/s. The error bars are the standard deviations.

Let us estimate the improvement in the SNR by using the quantum spectroscopy compared to the transmission spectroscopy from Eq. (6). In our experiment, the width of the coincidence time window is set to Δt=5 ns. It is limited by the jitter of the APD (around 400 ps) and the finite time resolution of the coincidence circuit (around 3 ns). The noise of the APD in the idler channel is $N_i=10^5$ photocounts/s. It is contributed by an optical noise and photocounts from the detection of broadband idler photons at wavelengths that are not correlated with the wavelength of a signal photon, selected by the monochromator. Finally, the quantum efficiency in the idler channel is approximately η_i=5%. It accounts for the quantum efficiency of the APD (η_{APD}=50%) and optical losses in the idler channel. From Eq. (6), we estimate the resulting improvement in the SNR using quantum spectroscopy $SNR_Q/SNR_T\approx100$.

For experiments with actual light-sensitive structures, it would be advantageous to place the sample just after the monochromator in the signal beam. In this case, the sample will be illuminated by a spectrally filtered light with fewer photons, compared to the present configuration. However, since the total number of detected photon pairs will remain the same, there will be no improvement in the resulting SNR.

The ratio SNR_Q/SNR_T can be significantly increased by using an additional monochromator in the idler arm. The two monochromators should select exactly conjugated frequencies of signal and idler photons, given by Eq. (1). It would be feasible to decrease N_i by at least 2 orders of magnitude. Furthermore, by using commercially available low-jitter APDs (around 50 ps) and time-to-digital converters with timing resolution on the order of around 10 ps [44], it would be possible to set the coincidence time window to about 100 ps. Given this, it is feasible to reach a ratio of SNR_Q/SNR_T as high as 5×10^5.

CONCLUSION

In conclusion, a new method for refractive-index sensing at extremely low-photon fluxes and under excessive noisy conditions is demonstrated. The method combines the use of entangled photon pairs, generated via spontaneous parametric down conversion, and plasmonic nanoparticle arrays with Fano resonance. Pairwise correlation between the photons allows spectral measurements with cancellation of uncorrelated background optical noise, and electronic noise of the detectors.

Using this method, detection of a refractive-index change of 0.014 is demonstrated with a noise level 70 times higher than the signal. The measured resonance shift of 8 nm corresponds to the relatively high refractive-index sensitivity of around 570 nm/RIU provided by the Fano resonance of the nanoparticle array. It is shown that conventional spectroscopy fails to provide any sensing results in such excessive noisy conditions.

The developed approach will contribute to further progress in the application of plasmonic sensors. In particular, it will allow nondisturbing measurements of ultrasmall concentrations of chemicals (at atto-Molar scale), and sensing of photosensitive substances, which can be affected by the probing light.

ACKNOWLEDGMENTS

We acknowledge Manuel R. Gonçalves and Othmar Marti from the Institute of Experimental Physics, Ulm University for preparation of nanoparticle arrays on glass substrates by nanosphere lithography. We acknowledge Carl Zeiss Pte Ltd (Singapore) for assistance with the SEM imaging of the sample. Fruitful discussions with Boris Luk'yanchuk and Reuben Bakker are greatly appreciated.

REFERENCES

1. J.N. Anker, W.P. Hall, O. Lyandres, N.C. Shah, J. Zhao, and R.P. Van Duyne, Biosensing with Plasmonic Nanosensors, Nat. Mater. 7, 442 (2008).
2. A.V. Kabashin, P. Evans, S. Pastkovsky, W. Hendren, G.A. Wurtz, R. Atkinson, R. Pollard, V.A. Podolskiy, and A.V. Zayats, Plasmonic Nanorod Metamaterials for Biosensing, Nat. Mater. 8, 867 (2009).
3. P.K. Jain, X. Huang, I.H. El-Sayed, and M.A. El-Sayed, Noble Metals on the Nanoscale: Optical and Photothermal Properties and Some Applications in Imaging, Sensing, Biology, and Medicine, Acc. Chem. Res. 41, 1578 (2008).
4. Y. Choi, Y. Park, T. Kang, and L.P. Lee, Selective and Sensitive Detection of Metal Ions by Plasmonic Resonance Energy Transfer-Based Nanospectroscopy, Nat. Biotechnol. 4, 742 (2009).
5. E. Jakeman and J.G. Rarity, The Use of Pair Production Processes to Reduce Quantum Noise in Transmission Measurements, Opt. Commun. 59, 219 (1986).
6. M.M. Hayat, A. Joobeur, and B.E.A. Saleh, Reduction of Quantum Noise in Transmittance Estimation Using Photon-Correlated Beams, J. Opt. Soc. Am. A 16, 348 (1999).
7. C.K. Hong, S.R. Friberg, and L. Mandel, Optical Communication Channel Based on Coincident Photon Pairs, Appl. Opt. 24, 3877 (1985).
8. D.E. Chang, A.S. Sorensen, P.R. Hemmer, and M.D. Lukin, Quantum Optics with Surface Plasmons, Phys. Rev. Lett. 97, 053002 (2006).
9. A.G. Curto, G. Volpe, T.H. Taminiau, M.P. Kreuzer, R. Quidant, and N.F. van Hulst, Unidirectional Emission of a Quantum Dot Coupled to a Nanoantenna, Science 329, 930 (2010).
10. S. Kumar, A. Huck, and U.L. Andersen, Efficient Coupling of a Single Diamond Color Center to Propagating Plasmonic Gap Modes, Nano Lett. 13, 1221 (2013).
11. R. Kolesov, B. Grotz, G. Balasubramanian, R.J. Stöhr, A.A.L. Nicolet, P.R. Hemmer, F. Jelezko, and J. Wrachtrup, Wave-Particle Duality of Single Surface Plasmon Polaritons, Nat. Phys.5, 470 (2009).
12. A. Huck, S. Kumar, A. Shakoor, and U.L. Andersen, Controlled Coupling of a Single Nitrogen-Vacancy Center to a Silver Nanowire, Phys. Rev. Lett. 106, 096801 (2011).
13. J.T. Choy, B.J.M. Hausmann, T.M. Babinec, I. Bulu, M. Khan, P. Maletinsky, A. Yacoby, and M. Lončar, Enhanced Single-Photon Emission from a Diamond-Silver Aperture, Nat. Photonics 5, 738 (2011).
14. R.M. Bakker, V.P. Drachev, Z. Liu, H.-K. Yuan, R.H. Pedersen, A. Boltasseva, J. Chen, J. Irudayaraj, A. V. Kildishev, and V.M. Shalaev, Nanoantenna Array-Induced Fluorescence Enhancement and Reduced Lifetimes, New J. Phys. 10, 125022 (2008).
15. S. Schietinger, M. Barth, T. Aichele, and O. Benson, Plasmon-Enhanced Single Photon Emission from a Nanoassembled Metal-Diamond Hybrid Structure at Room Temperature, Nano Lett. 9, 1694 (2009).
16. A.V. Akimov, A. Mukherjee, C.L. Yu, D.E. Chang, A.S. Zibrov, P.R. Hemmer, H. Park, and M.D. Lukin, Generation of Single Optical Plasmons in Metallic Nanowires Coupled to Quantum Dots,Nature (London) 450, 402 (2007).
17. E. Alterwischer, M.P. van Exter, and J.P. Woerdman, Plasmon-Assisted Transmission of Entangled Photons, Nature (London) 418, 304 (2002).
18. S. Fasel, F. Robin, E. Moreno, D. Erni, N. Gisin, and H. Zbinden, Energy-Time Entanglement Preservation in Plasmon-Assisted Light Transmission, Phys. Rev. Lett. 94, 110501 (2005).
19. G. Fujii, T. Segawa, S. Mori, N. Namekata, D. Fukuda, and S. Inoue, Preservation of Photon Indistinguishability After Transmission through Surface-Plasmon-Polariton Waveguide, Opt. Lett.37, 1535 (2012).

20. A. Huck, S. Smolka, P. Lodahl, A. Sørensen, A. Boltasseva, J. Janousek, and U. Andersen, Demonstration of Quadrature-Squeezed Surface Plasmons in a Gold Waveguide, Phys. Rev. Lett.102, 246802 (2009).
21. B. J. Lawrie, P. G. Evans, and R. C. Pooser, Extraordinary Optical Transmission of Multimode Quantum Correlations via Localized Surface Plasmons, Phys. Rev. Lett. 110, 156802 (2013).
22. R.W. Heeres, L.P. Kouwenhoven, and V. Zwiller, Quantum Interference in Plasmonic Circuits,Nat. Nanotechnol. 8, 719 (2013).
23. B.Y. Zel'dovich and D.N. Klyshko, Statistics of Field in Parametric Luminescence, Sov. Phys. JETP Lett. 9, 40 (1969).
24. D.C. Burnham and D.L. Weinberg, Observation of Simultaneity in Parametric Production of Optical Photon Pairs, Phys. Rev. Lett. 25, 84 (1970).
25. M.B. Nasr, S. Carrasco, B. Saleh, A. Sergienko, M. Teich, J. Torres, L. Torner, D. Hum, and M. Fejer, Ultrabroadband Biphotons Generated via Chirped Quasi-Phase-Matched Optical Parametric Down-Conversion, Phys. Rev. Lett. 100, 183601 (2008).
26. D.A. Kalashnikov, K.G. Katamadze, and S.P. Kulik, Controlling the Spectrum of a Two-Photon Field: Inhomogeneous Broadening Due to a Temperature Gradient, JETP Lett. 89, 224 (2009).
27. K. G. Katamadze and S. P. Kulik, Control of the Spectrum of the Biphoton Field, JETP 112, 20 (2011).
28. M. Okano, R. Okamoto, A. Tanaka, S. Subashchandran, and S. Takeuchi, Generation of Broadband Spontaneous Parametric Fluorescence Using Multiple Bulk Nonlinear Crystals, Opt. Express 20, 13977 (2012).
29. G. Scarcelli, A. Valencia, S. Gompers, and Y. Shih, Remote Spectral Measurement Using Entangled Photons, Appl. Phys. Lett. 83, 5560 (2003).
30. A. Yabushita and T. Kobayashi, Spectroscopy by Frequency-Entangled Photon Pairs, Phys. Rev. A 69, 013806 (2004).
31. A.A. Kalachev, D.A. Kalashnikov, A.A. Kalinkin, T.G. Mitrofanova, A.V. Shkalikov, and V.V. Samartsev, Biphoton Spectroscopy of YAG: Er3+ Crystal, Laser Phys. Lett. 4, 722 (2007).
32. A.A. Kalachev, D.A. Kalashnikov, A.A. Kalinkin, T.G. Mitrofanova, A.V. Shkalikov, and V.V. Samartsev, Biphoton Spectroscopy in a Strongly Nondegenerate Regime of SPDC, Laser Phys. Lett. 5, 600 (2008).
33. O. Slattery, L. Ma, P. Kuo, Y.-S. Kim, and X. Tang, Frequency Correlated Biphoton Spectroscopy Using Tunable Upconversion Detector, Laser Phys. Lett. 10, 075201 (2013).
34. V.G. Kravetz, F. Schedin, and A.N. Grigorenko, Extremely Narrow Plasmon Resonances Based on Diffraction Coupling of Localized Plasmons in Arrays of Metallic Nanoparticles, Phys. Rev. Lett. 101, 087403 (2008).
35. G. Vecchi, V. Giannini, and J.G. Rivas, Shaping the Fluorescent Emission by Lattice Resonances in Plasmonic Crystals of Nanoantennas, Phys. Rev. Lett. 102, 146807 (2009).
36. G. Vecchi, V. Giannini, and J.G. Rivas, Surface Modes in Plasmonic Crystals Induced by Diffractive Coupling of Nanoantennas, Phys. Rev. B 80, 201401(R) (2009).
37. A.I. Kuznetsov, A.B. Evlyukhin, M.R. Gonçalves, C. Reinhardt, A. Koroleva, M.L. Arnedillo, R. Kiyan, O. Marti, and B.N. Chichkov, Laser Fabrication of Large-Scale Nanoparticle Arrays for Sensing Applications, ACS Nano 5, 4843 (2011).
38. B. Luk'yanchuk, N.I. Zheludev, S.A. Maier, N.J. Halas, P. Nordlander, H. Giessen, and C.T. Chong, The Fano Resonance in Plasmonic Nanostructures and Metamaterials, Nat. Mater. 9, 707 (2010).
39. A.E. Miroshnichenko, S. Flach, and Y.S. Kivshar, Fano Resonances in Nanoscale Structures,Rev. Mod. Phys. 82, 2257 (2010).
40. Lord Reyleigh, On the Dynamical Theory of Gratings, Proc. R. Soc. A 79, 399 (1907).
41. V.G. Kravets, F. Schedin, A.V. Kabashin, and A.N. Grigorenko, Sensitivity of Collective Plasmon Modes of Gold Nanoresonators to Local Environment, Opt. Lett. 35, 956 (2010).
42. J.E. Mark, *Polymer Data Handbook* (Oxford University Press, New York, 1999).
43. L.F. Hoyt, Determination of Refractive Index of Glycerols by the Immersion Refractometer, Oil Soap 10, 43 (1933).
44. http://www.picoquant.com/products/category/photon-counting-detectors;http://www.picoquant.com/products/category/tcspc-and-time-tagging-modules.

CHAPTER 11 INFRARED PHOTODETECTORS

Chandler Downs and Thomas E. Vandervelde

Keywords: Infrared photodetectors, IRPDs, bulk detectors, QWIPs, QDIPs, DWELLs, superlattices

INTRODUCTION

Infrared photodetectors (IRPDs) are a technology with wide-ranging and rapidly expanding applications in the modern world. Ever since Friedrick William Herschel discovered the presence of infrared radiation in sunlight in the early 19th century, people have tried various means to detect and analyze this spectrum of light invisible to the naked eye [1]. The earliest practical IR detectors were developed by Macedonio Melloni in the mid-19th century [2]. These detectors were thermopiles that functioned by thermal conduction, typically by relying on the differences in thermal expansion of two dissimilar metals. In 1917, Case developed what could be considered the first modern photodetector, when in his search for materials which exhibited variable resistances depending on whether light was shined on them [3]. In this research he noted a number of materials, such as lead sulfide, exhibited responses out into the IR regime. These were the first IR detectors to operate using quantum effects rather than conductive ones, and it was this technology that fathered the field of IRPDs as we know them today [4].

Applications currently utilizing IRPDs span military (e.g., navigation, night vision, weapons detection), commercial (e.g., communications, aerospace, medical imaging), public (e.g., atmospheric sounding, pollution control, meteorology, environmental monitoring), and academic (e.g., astronomy) domains-with new uses constantly arising as the various IRPD technologies become more established [5–11]. As such, researchers have invested tremendous time and resources into developing and improving various IRPD technologies to further serve these applications. Of particular note are the advances since the new millennium. Within the past twelve years, established technologies have grown into commercial successes, nascent technologies have grown into thriving hubs of research, and new technologies have been discovered and begun to be investigated.

The world around us is a large source of infrared radiation and IRPDs can be useful in a wide array of applications utilizing it. This ubiquity is due to the fact that all objects will emit an IR spectrum based on their temperature. This emission spectrum can be approximated by wavelength λ as blackbody radiation, which can be characterized according to the blackbody's temperature T by Equation (1) [12]:

$$e_B(\lambda, T)\, d\lambda = \frac{2\pi hc^2 d\lambda}{\lambda^5[e^{hc/\lambda kT} - 1]} \quad (1)$$

This equation illustrates why IRPDs have received so much interest of late. It implies that an object at room temperature will emit IR radiation with a peak intensity of around 9.5 μm, with detectable IR emissions for wavelengths microns away from this center value. While not every portion of the IR spectrum is ordinarily present due to natural blackbody radiation (a blackbody would need to be at a temperature of over 4,000 K to have peak light emissions at the edge of the visible spectrum), these sources do emit across large portions of the IR spectrum.

One complication for IRPDs is atmospheric absorption. As IR radiation propagates through the atmosphere, some wavelengths are readily absorbed due to the molecules present in the air. Of greatest concern to IR radiation are the absorption spectra of water vapor and various hydrocarbons [13], whose absorption spectra cut large infrared windows of the

Citation: Chandler Downs and Thomas E. Vandervelde, Infrared Photodetectors, doi:10.3390/s130405054.

atmospheric transmission spectrum. These infrared windows (seen in Figure 1) do not necessary preclude the use of IRPDs at these wavelengths, but they can significantly interfere with device performance at those wavelengths [14,15].

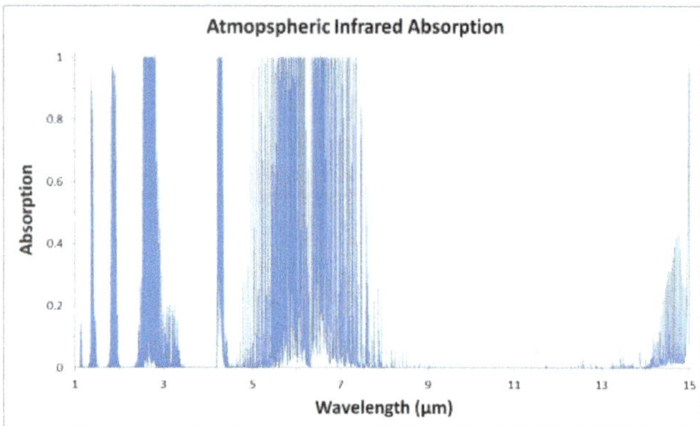

Figure 1. A plot of the atmosphere's absorption spectrum [16]. Note that light emitted in the 2.5–3.5 and 5–7 μm ranges are rapidly absorbed by the atmosphere.

The wavelengths of light associated with IRPDs are those longer than visible light (longer than 700 nm). This is a truly massive range, one no single detector technology is able to completely span. Comparisons between the performance of devices at vastly different absorption wavelengths can be difficult to make, given that performance metrics may vary by orders of magnitude across the IR spectrum. In hopes of encouraging reasonable comparisons between devices, the IR spectrum has been divided (as best as is possible) into a number of sub-regions. Where possible, these divisions have been made at convenient or recognizable points to aid in distinguishing between the regions.

Closest to the visible spectrum is the near infrared (NIR) regime, ranging from 700 nm to about 1.0 μm, corresponding to the cutoff for absorption by silicon. From the 1.0 μm cutoff to 3 μm is the short-wave infrared (SWIR) regime, with cutoff defined by one of the atmospheric windows. This is followed by the mid-wave infrared (MWIR) regime, ranging from 3 μm to 5 μm, again with the cutoff defined by the atmospheric window. The long-wave infrared (LWIR) regime ranges from 8 μm to 12 μm, while the very long-wave infrared regime (VLWIR) amasses everything beyond 12 μm [5]. There is a range of wavelengths, from 5 μm to 8 μm, which does not fall into any of these categories. This range of wavelengths corresponds to a large region of atmospheric IR absorption. This absorption does not preclude development of IRPDs at these wavelengths provided they are detecting nearby sources of radiation in that range, but it can make the transmission of those wavelengths more difficult, decreasing the effective range of the detectors [5]. In this paper, we will attempt to constrain our device comparisons to within the same wavelength regime, even within the same technology, so as to avoid any improper comparisons.

Over the years, a large number of metrics have been developed to measure the performance of IRPDs. In this review, we will focus on the most commonly reported device metrics. Peak absorption wavelength will be used to separate detectors into the different regimes listed above. For single pixel detectors, the most common metrics are dark current density, peak detectivity, peak responsivity, and operating temperature. Lower dark current densities allow lower response signals to be detected, a desired trait in all devices. Comparisons of dark currents at a given temperature within a single technology are a valid means of determining superior performance, but comparisons between IRPD technologies must be taken with a grain of salt. Different technologies (or even the same technology operating in a different wavelength regime) may produce dramatically different signal current densities operating at similar conditions. The amount of signal current generated for a given input power of IR radiation (measured for a specific wavelength) is the responsivity of the device and is defined in Equation (2):

$$R = \frac{e\eta g}{hv} \tag{2}$$

where η is the quantum efficiency (the percentage of generated carriers that are extracted from the device) of the detector and g is the photoconductive gain (the number of carriers that are generated by the device structure and applied bias for every carrier generated by an absorbed photon) [17]. As with dark current density, comparisons of responsivity are valid within specific technologies under comparable conditions, but are deceptive outside of those limitations. In general, higher responsivities will result in better device performance. For the purposes of this paper, comparisons concerning responsivity will be done according to peak responsivity values reported independent of wavelength. Comparisons of responsivity values will also be limited to devices of the same technology under similar operating conditions.

Another favored metric used to delineate IRPD performance is specific detectivity (D*). Specific detectivity incorporates aspects of both the dark current density and responsivity of a device to provide a comparison of the amount of signal current generation for a given amount of noise at a specific wavelength, defined in Equation (3):

$$D^* = \frac{R_p\sqrt{A\Delta f}}{i_n} \tag{3}$$

where R_p is the peak responsivity, A is the cross-sectional area of the IRPD, Δf is the bandwidth of the device, and i_n is the noise current [17]. High specific detectivities indicate a larger signal current generated for a given amount noise, which allows for better signal detection. Further, comparisons of specific detectivities across different technologies and operating conditions are reasonable, making this one of the most useful analytic metrics. Closely related to the three previous metrics is the maximum operating temperature of the device. The higher the temperature an IRPD is operated at, the more noise will be present. This in turn increases dark current and reduces specific detectivity. Once the size of the dark current is large enough to make the signal current indistinguishable, the device will no longer function. Photodetectors with higher maximum operating temperatures require less cooling to maintain functionality. The need for less cooling can dramatically reduce cost of operation for these devices and allow the devices to be used in environments and situations where liquid cryogens are not available.

Focal plane arrays (FPAs) necessitate a handful of additional performance metrics in addition to those used to describe single pixel detectors in order to address overall image quality. The most obvious is simply the number of pixels present in the FPA. A larger pixel count enables more precise imaging, generally a desirable feature. Similarly, pixel size is important, with smaller pixels allowing for more precise imaging. However, since most FPAs are fabricated using standard read out integrated circuit (ROIC) packaging, there is more a series of benchmarks to be met in terms of pixel size and pixel count. A more useful metric for FPAs is the noise equivalent temperature difference (NETD), which can be used to approximate the minimum resolvable temperature difference of the FPA, defined in Equation (4):

$$NETD = \frac{i_n}{R}\frac{dP}{dT_B} = \frac{i_n h\nu P}{k_B T_B^2} \tag{4}$$

where P is the incident power of light on the detector, T_B is the blackbody radiation temperature, and k_B is the Boltzmann constant [18]. A number of device parameters, from pixel size to specific detectivity, are taken into account in this parameter. A smaller NETD is preferable, indicating a capability for finer, more precise imaging. We will additionally make note of any multicolor FPAs which incorporate multiple different wavelengths of IR detection, particularly if the wavelengths of detection are significantly different. Such multispectral imaging has many advantages for a multitude of applications.

In this review, the progress made in each of the most researched IRPD technologies since 2000 will be examined in its own section. Each section will begin with a brief description of the technology in question and highest performing devices in 2000. This will be followed with a chronicle of the major trends in the technology over the ensuing twelve years, e.g., popular material choices or inclusion of specific structures in devices. A current state of the technology will also be provided for each section. If applicable, closely related technologies will be examined in a subsection.

BULK DETECTORS

As of 2000, bulk detectors were the most common IRPD technology. A wide variety of materials were utilized in various applications, but most popular were HgCdTe (MCT) and InSb. The active portions of these detectors are generally photodiodes, though there is great variety in the structure and fabrication of the specific photodiodes. There are numerous fabrication methods, each having various strengths and weakness based on the specific IRPD design (e.g., front- or back-illuminated, epitaxial or implanted layers, operation in the photoconductive or photovoltaic regime, etc.), but these are beyond the scope of this review [7,19]. Bulk detectors were uniformly the most popular detectors at the outset of the 21st century for a number of reasons, largely due to the relative maturity of the technologies. Bulk detectors, because of their relative ease of fabrication, were first developed significantly earlier than other technologies.

HgCdTe (MCT)

Since the development of MCT in 1959 and the creation of the first detector using it in 1962, MCT has been one of the most popular materials for IRPDs [20–22]. By 2000, the technology was fairly mature. Single pixel and focal plane array (FPA) detectors had been extensively studied in laboratory settings, with the research from those devices being used as the groundwork for a number of commercial ventures. As such, published research in this area was actually not as widespread as in other IRPD technologies over this timeframe. Much of the research performed was focused less on direct improvement in performance of devices, but instead on addressing specific difficulties encountered by the technology.

The scope of this research becomes much clearer when examining the strengths and weaknesses of the MCT technology as a whole. Some of its greatest strengths have been touched on already, namely the established history of the technology and the relative simplicity of fabrication compared to a number of the later technologies. There are a number of methods utilized to grow MCT devices, including metalorganic vapor phase epitaxy (MOCVD), liquid phase epitaxy (LPE), and molecular beam epitaxy (MBE), which allows for more flexibility in fabrication [19–25]. MCT detectors tend to have very high responsivities and competitive detectivities when compared with quantum structures utilized in other technologies [26]. As MCT is a ternary material, by varying the composition of the material it is possible to span all of the wavelengths between the constituent binary materials, as can be seen in Figure 2. With MCT, this is a large range, spanning all wavelengths between 0.7 μm and 25 μm and three atmospheric windows [19,23,27–29]. However, despite the wide range of wavelengths MCT spans, there is very little variation in the value of the lattice constant across that range [23]. This allows for simplification of many of the enabling techniques used in IRPDs (substrate selection, ROIC usage, etc.) to be used across all MCT compositions. MCT devices also tend to have very short integration times and very low junction capacitances compared to other bulk detectors due their low dielectric constant, both of which enhance its capability to detect fast moving objects [4,19,30].

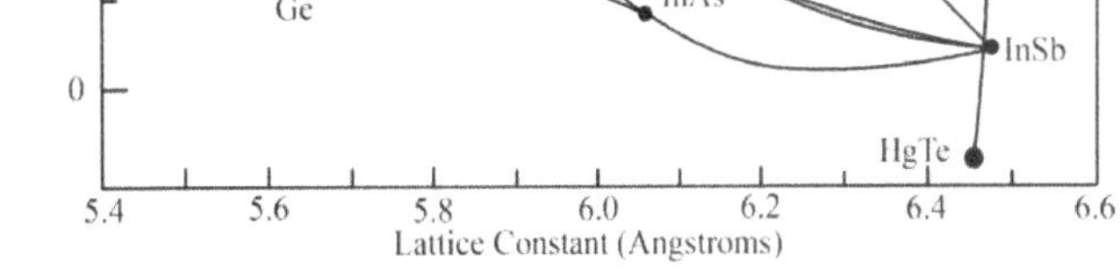

Figure 2. A lattice constant vs. band gap plot of common semiconductor materials.

With these advantages come a number of disadvantages. The most noteworthy of these is that dark currents in MCT devices tend to be extremely large, much larger than in other devices, influenced heavily by band-to-band tunneling [31]. MCT devices also have growth related issues due to differences in the densities and melting points of the constituent materials, the relative softness of the final material, and a number of crystalline properties of the material (e.g., high interdiffusion coefficients, low bond strength, and a high rate of dislocation formation) [4,15,28]. These issues can make precision growths of MCT extremely difficult, which lead to a number of other device issues. One of the major issues, particularly as the band gap of the material decreases, is finely controlling the composition of the material [28,32]. As the band gap goes down towards the LWIR and VLWIR regimes, variation of the mole composition of MCT by 0.001 can have drastic effects on the performance of a detector. That level of control is extremely difficult, regardless of the growth method. Additionally, it is hard to achieve the kind of uniformity across a wafer that is necessary for large area FPAs, which has led some to question MCTs suitability for some applications [15,26,32,33]. This problem is only exacerbated when the issue of multicolor FPAs is considered [30].

As mentioned previously, by 2000 MCT IRPDs were already an established and mature technology. Much of the research performed since then has focused on addressing some of the issues mentioned above or expanding the commercial utility of the technology, particularly with FPAs. This is partially due to the increased research performed by commercial organizations, as opposed to government or academic organizations. Companies such as Raytheon, BAE, Sofradir, and many others have published numerous papers on their progress with commercial products. In the early part of the decade, these papers generally focused on small- to mid-sized FPAs, generally operating in the NIR, SWIR, MWIR regimes. The majority of these devices were operated around liquid nitrogen temperatures (77–100 K), though there were a couple of noteworthy exceptions [19,34–37]. A few standout devices operating in the NIR and SWIR regimes, where dark currents are minimized with the corresponding larger bandgaps, were capable of functioning at elevated temperatures, up to around 250 K [24,38,39]. The fabrication method most favored for these devices was LPE, but MBE growth was growing in popularity, particularly with devices requiring large cross-sectional areas due to MBE's finer growth control. As the decade progressed, commercial interests pushed towards larger FPAs (with the largest being a 2,560 × 512 pixel FPA using a double layer heterojunction design grown by MBE by Raytheon [40]) and longer detection wavelengths, pushing detection out to as far as 13 μm [41]. Published data also showed attempts to incorporate multicolor detection, starting with dual-color detection, into the technology with reasonably successful results [40,42]. Other trends showed an increased reliance on MBE growth for what should be considered prototype devices, with the likely intention to eventually shift to faster processes after the initial structures are completed [43–45].

Research performed in various academic and governmental laboratories has had a slightly different focus than commercial research. As early as 2000, MBE was regularly used in the growth of MCT detectors. This is not wholly unexpected, as MBE does allow for better control of material quality for the small numbers of cells that are generally grown in these settings. In the early half of the decade, the bulk of published research centered mainly on two subjects: LWIR detectors and growth of MCT on silicon substrates. LWIR detection was being researched as an extension of the progress made in SWIR and MWIR previously. These devices were usually small to mid-sized FPAs (e.g., 384 × 288) operated at liquid nitrogen temperatures and focused on using MBE growth to achieve the level of composition control necessary for precise LWIR regime absorption [28,46,47]. Research focused on silicon substrates looked to address cost concerns that could arise from typical MCT substrates (most commonly CdZnTe). Growth of MCT on silicon in the early part of the decade was performed exclusively by MBE, as the lattice mismatch between various MCT alloys and the silicon substrate can be as large as 19% [25,48–51]. Research was also performed to enable improved room temperature performance of MCT detectors. While as of 2000 room temperature operation was feasible, performance was lackluster, necessitating further research [52].

Moving through the later part of the decade, research on MCT detectors by universities and government agencies had less of a unified focus. Research on deposition of MCT onto Si substrates slowed to a near halt [53]. Research continued improving the performance of high operating temperature devices, both in terms of detectivity as well as extending the wavelengths at which this could be achieved. Attempts were made to improve the detectivity of devices (e.g., by using multiple absorbing layers) [23] as well as decrease the dark current density (e.g., including Auger suppression layers) [54]. Moreover, research continued extending the operational wavelengths for MCT devices, with functional detectors pushing into the VLWIR regime up to 14 μm [7] and D* of LWIR detectors reaching up to 1.9×10^{11} cm $Hz^{1/2}/W$ [55].

Despite these local areas of focus, there was also a great range of research on individualized topics [56,57]. Further research was performed on multicolor detectors, focusing on moving into the MWIR and LWIR regimes utilizing multiple absorbing layers [58]. Amorphous MCT was investigated as a detector material, as it is easier to grow on silicon. The performance of the detectors exhibited lower detectivities (on the order of 10^8 $cmHz^{1/2}/W$) than crystalline MCT, but did exhibit the possibility of near-room temperature operation [32]. Experiments were also performed using resonant cavity enhancements for improved absorption of incident light at specific wavelengths. This design does have some special considerations, as one of the characteristics of MCT detectors is the wide absorption bands, whereas the resonant cavity enhancements tend to have narrow absorption bands. These detectors have shown improved absorption at specific absorber wavelengths, but reduced absorption at other wavelengths [59].

InSb

InSb photodetectors have been fabricated commercially since the late 1950's. InSb photodetectors, like MCT photodetectors, are bulk detectors. Most detectors utilize some form of a p-n junction (including p-i-n, p^+-p-n^+, etc.), though many incorporate a metal-insulator-semiconductor (MIS) contact structure [4,60]. In the early part of the decade, InSb was the material of choice for NIR and some MWIR applications, in part because it was an established technology and in part because the material's band gap falls squarely in that range. Further, the material is often combined with various other column IV or VI materials, most commonly arsenic. Attempts have been made to utilize thallium or bismuth to achieve a narrower band gap, but including small amounts of these materials in InSb (<5%) led to great difficulty in material fabrication and dramatically limited device detectivity and responsivity [61,62]. InSb FPAs were capable of pixel counts roughly similar to MCT FPAs at the start of the decade [4].

InSb detectors have several notable strengths. Its band gap (0.17 eV (~7.3 μm) at 300 K, 0.23 eV (~5.4 μm) at 80 K), allows it to function comfortably as a binary material in the MWIR regime. Its utility in the MWIR is expanded when arsenic is included in a ternary material. InAsSb is capable of spanning all wavelengths between InSb and InAs (band gap of 0.35 eV (~3.5 μm)). This flexibility allows InAsSb to be tuned to most wavelengths in the MIR regime. Further, because the difference in band gaps between InSb and InAs is much less than that between HgTe and CdTe, the tolerances in terms of material composition when designing specific operating wavelengths are eased [60]. This also limits the utility of these devices, as it can be very difficult for InSb or InAsSb devices to operate competitively outside the MWIR regime. InSb-based detectors also tend to have high device stability, lower fragility, and better device yield than MCT devices [60,63,64].

The use of InSb for photodetectors has also been associated with a number of fabrication difficulties. As can be seen in Figure 2, InAsSb spans a wide range of lattice constants. As such, while there are larger tolerances for material composition in terms of absorption wavelength, there are much stricter tolerances in terms of lattice matching to a substrate. At the beginning of the decade, the substrates for most of these devices were InSb, as the lattice mismatch between even InSb and GaSb was large enough to cause issues. This issue was exacerbated with more common substrates like GaAs or silicon. When attempts were made to utilize GaAs or Si substrates, the InSb/substrate interface suffered from high dislocation densities despite the use of a number of advanced techniques (e.g., inclusion of buffer layers) to ease the strain. The high dislocation densities resulted in large leakage currents, which in turn led to lower detectivities [60]. As mentioned above, while InSb and InAsSb do span a number of useful wavelengths, the precision gained in wavelength selectivity does come at a cost to the range of operating wavelengths. Additionally, the little response that InSb materials have in the LWIR regime is further reduced when operated at low temperatures, which is often necessary due to dark current considerations, due to reduced minority carrier lifetime in the n-type region of the material [4].

As InSb was a well-established technology as of 2000, numerous InSb photodetectors and FPAs had been built to this point. Most of the research performed in the decade was aimed at addressing specific concerns and problems that the technology faced. The two most common problems that this research addressed in the early part of the decade were increasing FPA performance and attempting to utilize different substrate materials in the devices. Universities typically focused on smaller FPAs with other performance improving features (e.g., an InAlSb current blocking layer in the active region of the detector) while companies, most notably Raytheon, focused on expanding pixel counts in FPAs beyond the megapixel range, out as far as 2K × 2K pixels. Whereas many single pixel devices were capable of operating at near room temperature, the FPAs were typically operated at liquid nitrogen (LN_2) temperatures or lower to achieve the levels of performance desired [65–70].

Despite the difficulties growers faced when using substrates other than InSb, research continued on InSb IRPDs fabricated on these substrates. The most popular substrates (for reasons of cost, availability, and integration with other technologies) were silicon (a = 5.431), which has a lattice mismatch with InSb of 19%, and GaAs (a = 5.653), which has a lattice mismatch of 15%. Growth on these substrates was accomplished using a variety of means, most commonly MBE growth with the incorporation of buffer layers to reduce the strain at each interface. A number of devices were fabricated on both silicon and GaAs substrates, but these devices uniformly had higher dark current densities, lower responsivities, lower detectivities, and lower maximum operating temperatures than those fabricated on InSb substrates [60,61,63,71]. Other attempts were made at utilizing GaSb substrates, which does have the benefit of being much closer to a lattice match to InSb, with a lattice constant of 6.096 Å and a mismatch of 6%. The performance of these devices were generally more favorable than those on silicon or GaAs substrates, but operating temperatures were limited to 250 K and detectivities were still reduced [62].

As other technologies rose to prominence in the middle of the decade, research on InSb, particularly at universities, slowed. The Kuze group, one of the few groups that still performed research on these devices, focused on miniaturization of InSb sensors. These devices were constructed from a number of photodiodes connected in series with the aim of weak IR detection at room temperature. The final operational area of the devices was 600 × 600 μm, with final packaging 2.2 × 2.7 × 2.7 mm [72–75]. Other research looked into incorporating nitrogen into InSb. It was found that incorporating small amounts of nitrogen (<1%) led to improved D* at wavelengths as far out as 10 μm. While its incorporation did lead to a decrease in performance at shorter wavelengths, this device had among the highest performance in the LWIR of any InSb device [76]. Research into InAsSb devices has been fairly modest to this point. Most of the research has been focused on achieving the quality of growth required for devices, typically by limiting the mole fraction of As in the devices and including buffer layers. A handful of MWIR and LWIR IRPDs have been grown, achieving D* as high as 4.03×10^{11} cm $Hz^{1/2}/W$ at 170 K at a wavelength of 3.7 μm [76–78].

QUANTUM WELL INFRARED PHOTODETECTORS

Quantum well infrared photodetectors (QWIPs) operate on a much different principle than the bulk detectors that came before it. First demonstrated in 1987 by Levine et al. [79], QWIPs rely on quantum scale physical effects, whereas bulk detectors operate on larger scale effects. QWIPs in their most basic form are a periodic repetition of layers of two materials with dissimilar band gaps. The material with the lower energy band gap is commonly referred to as the well layer, while the higher energy band gap is referred to as the barrier layer. This structure approximates a series of quantum mechanical wells, with the well layer deposited thinly enough (generally on the order of nanometers to tens of nanometers) such that discretized energy states form within the well instead of the typical band of energy states. An example quantum well structure is provided in Figure 3. The energy states within the well will be characterized by Equation (5):

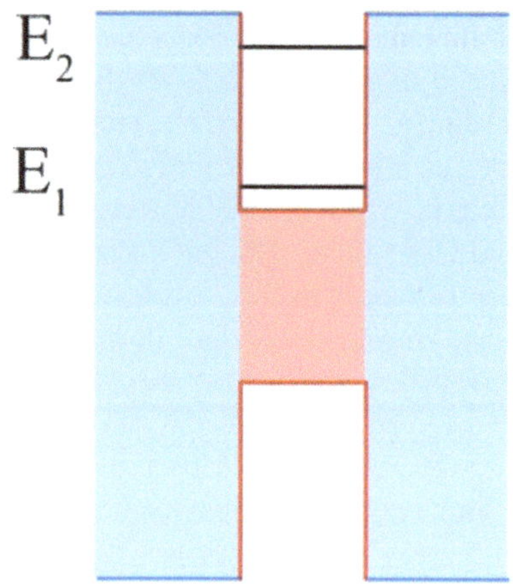

Figure 3. A band diagram of a quantum well in a QWIP. Carriers are excited by incident photons from the ground energy state of the well (E_1) to the excited energy state of the well (E_2).

$$E_n = \frac{\hbar^2 \pi^2 n^2}{2m^* L_w^2} \quad (5)$$

where m* is the effective mass of the charge carrier, L_w is the width of the well layer, and n is an integer identifying the energy level in the well [15]. The well layers are doped such that without illumination there will be carrier electrons present in the ground energy state. These carriers are then excited by incident photons to an energy state near the conduction band edge of the barrier material, where the applied voltage will sweep the carrier out of the well to the contacts. When the higher energy state is above the barrier conduction band edge (>10 meV), there is a higher escape probability from the well. This generally results in higher detectivities, and is referred to as a bound to continuum transition.

When the higher energy state is below the barrier conduction band edge (<10 meV) there is higher absorption efficiency; this is a bound to bound transition. Between these two values is a bound to quasibound transition and mixes a balance of the above traits. QWIP devices typically utilize an absorbing superlattice stack sandwiched between emitter and absorber layers, which aid in transferring carriers to and from the contacts and can be doped to assist with carrier separation. Due to the precise tolerances required for the layer thicknesses in the matrix of materials, these devices are typically grown using MBE, though other methods have been attempted with varying success [5,6,80].

QWIPs have a number of positive and negative performance characteristics that distinguish them from other IRPD technologies. In comparison to MCT devices, QWIPs are capable of lower dark currents, higher detectivities and higher NETDs. They make use of commonly used III-V material processing techniques, making fabrication of the devices easier than with MCTs. QWIPs also typically have greater radiation tolerances than narrower band gap materials like MCT and InSb. A case can be made that QWIPs have the greatest ease of material growth of multispectral imagers, as neither lattice mismatching (due to small layer thicknesses) or surface irregularities (due to a lack of nanostructures) are common hindrances for these devices [30,81–83].

QWIPs also face a number of challenges. The most distinct disadvantage arises from QWIPs' use of quantum confinement. Due to constraints defined by quantum mechanics, a QWIP can only absorb light that is incident upon it when there is quantum confinement in one of the perpendicular axes. This leads to a large problem: due to common growth techniques, most QWIPs are incapable of the absorption of normally incident light. In order to mitigate this problem, most QWIP devices use some form of a frontside scattering filter, which redirects normally incident light such that it can be absorbed [84].

It should also be noted that the maximum energy photon that a QWIP can absorb is limited by the energy difference between the conduction band edges of the materials. This means that for operation at shorter wavelengths, there must be a very large difference between the band edges. Finding materials that can satisfy this condition can be extremely difficult. For this reason, QWIPs have been used sparingly in the SWIR regime [85–87]. The scattering filters required for the absorption of normally incident light must be on a similar scale as the wavelength of the incident light. This complicates SWIR QWIPs by increasing the difficulty of fabrication as wavelength decreases. Those devices that attempt to operate at higher frequencies typically incorporate various nitride materials. There are also similar issues when the absorbed wavelength gets very long, into the VLWIR range [10,80].

QWIPs have significantly narrower absorption widths due to the discretized nature of the energy states transitions occur between. This can be advantageous if detection of a specific wavelength is required, but is a hindrance for wide band detection [81]. The narrow absorption bands also reduce photogenerated current, and hence the detectivity of QWIPs tends to be much lower than in bulk detectors. Lower detectivities in turn make QWIPs more sensitive to the magnitude of dark current density. As such, QWIPs have had significantly greater difficulty achieving high temperature operation [5,6,88,89]. Compounding the dark current issue is a short carrier lifetime in these devices, generally on the order of a handful of picoseconds. While a low carrier lifetime can have the added benefit of fast operating speeds, it also leads to increased sensitivity to dark current [90,91]. There are also a small number of issues that arise when considering QWIP FPAs. Due to the lower detectivities and responsivities of these devices, integration times tend to be longer when compared with bulk detectors (up to the order of milliseconds). This can make detection and identification of quickly moving objects more difficult [30].

Similar to bulk technologies, QWIPs had been the subject of research for some time as of 2000. Functional devices were common place and FPAs had recently begun to be developed. The vast majority of research conducted over the ensuing decade was performed by research labs at various universities. This section will focus first on single pixel QWIP designs, followed by various FPA designs, and finally with notes on a handful of technologies closely related to QWIPs, but which do not easily fit within the normal framework of QWIP devices.

One of the largest debates in the early portion of the 2000s was which material system held the most promise moving forward. Under consideration primarily were three material compositions (utilizing various mole fractions of the different ternary materials): (Al)GaAs/InGaAs, (In)GaAs/(Ga)InP, and AlGaAs/GaAs. Of primary interest during the earliest portion of the 2000s was the GaAs/InGaAs regime. It was hoped that this material system could help extend the functionality of QWIP devices out to the LWIR and VLWIR regimes ($\lambda_c > 10$ μm). Performance of these devices was generally not promising in these wavelength ranges, with required operating temperature often below LN_2 levels (77 K). Even with these lower than normal operating temperatures, device performance was still lackluster, with responsivities regularly below 0.5 A/W and detectivities on the order of 10^9 $cmHz^{1/2}/W$. Performance of these devices was marginally improved with the incorporation of aluminum into the GaAs layer, with operating temperatures in ranges of LN_2 cooling. Responsivities and detectivities were also higher after incorporation of aluminum, particularly at lower operating temperatures, peaking at around 1.96 A/W and 1.59×10^{10} $cmHz^{1/2}/W$, respectively, by Lee et al. [92]. However, when quantum dot infrared photodetectors (QDIPs) began receiving attention as a viable alternative to QWIPs, particularly at these longer wavelengths, research interest in the GaAs/InGaAs material system petered out [93–98].

The next material system under serious consideration was InGaAs/(Ga)InP. This material regime also received the most attention at the beginning of the decade, with interest fading with the introduction of QDIPs. The material system was investigated across a wider range of wavelengths than the (Al)GaAs/InGaAs system, ranging from MWIR to VLWIR. InGaAs/(Ga)InP QWIPs met with similar difficulties in the LWIR and VLWIR regimes, with detectivities on the order of 1×10^9 $cmHz^{1/2}/W$ around LN_2 temperatures. These devices did exhibit stronger performance in the MWIR regime, feature operating temperatures over 100 K, responsivities as high as 2.2 A/W, and detectivities as high as 5×10^{10} $cmHz^{1/2}$ [83,99–102]. There were also attempts made to adapt this material system for deposition on Si substrates, however performance of these devices is generally poorer than the normal performance of these devices, usually by an order of magnitude and sometimes more. This poor performance was attributed to the high density of defects and dislocations at the interface between the silicon substrate and the active QWIP stack [103,104].

By far the most commonly used material system over the decade was AlGaAs/GaAs. Whereas the other material systems often tried to extend operation out to the VLWIR regime, research with AlGaAs/GaAs typically focused on the MWIR and LWIR regimes. Dual band absorption was incorporated into a number of the structures, with performance of these devices often similar to single color devices. Performance of AlGaAs/GaAs devices outstripped that of the (Al)GaAs/InGaAs and (In)GaAs/(Ga)InP in the early portion on the decade, encouraging further research in the later portion of the decade [102,105–109]. There has been an effort to try and extend the operational wavelengths of AlGaAs/GaAs towards the VLWIR regime, but these efforts only met very limited success. Some reasonable responsivity values were achieved, but only at operating temperatures of 20 K. Performance at even LN_2temperature degrades rapidly [110–112]. Some have suggested the GaSb/AlGaSb material regime for use in the VLWIR wavelength range, but it remains sparsely investigated to this point [113].

Perhaps the most interesting results in the AlGaAs/GaAs material systems are those devices which incorporated some form of current blocking layer into the QWIP. A number of QWIP devices included such a layer, which consist of a significantly higher energy band gap intended to reduce the dark current that reaches the contacts. Such a layer can be implemented either within the active region or can be added between the complete material matrix and the other layers of the device. These devices generally exhibit improved dark current performance, but was not a topic of much interest. A number of the devices, while called QWIPs, can be seen as precursors to dots-in-a-well detectors (DWELLs), discussed later in this paper, particularly devices which incorporated the current blocking layer or layers into the material matrix. DWELLs have become increasingly interesting subjects of research in recent years, and seeing how they may have evolved from QWIPs is a fascinating occurrence [114–116].

In addition to single pixel QWIPs, a great amount of research was also performed on FPAs. Some of the research was focused on the InGaAs/InP material system. These devices, while showing some improvement over their single pixel predecessors (which can be attributed to improvement in MBE technology over the decade), still exhibited some of the same issues, namely dark current as much as an order of magnitude larger than AlGaAs/(In)GaAs QWIPs [117,118]. For this reason and the general superior performance over other material choices, most of the QWIPs FPAs researched utilized AlGaAs/(In)GaAs.

At the beginning of the decade, the most common size for these FPAs was 640 × 512 pixels. These devices were generally operated in the MWIR and LWIR regimes, and exhibited similar performance to single pixel devices. They were generally operated at LN_2 temperatures with the strongest results produced by Gunapala et al., with responsivities between 0.5–1 A/W and detectivities upwards of 2×10^{11} $cmHz^{1/2}/W$. Spectral width, where reported, was generally between 10%–15% and NETDs between 30–50 mK [91,119–123]. As the decade progressed, efforts were made to increase the amount of pixels in the FPAs and their operating temperature. FPA sizes increased to over a megapixel in size, generally 1,024 × 1,024. While accomplishing this, operating temperatures were also raised, but the amount of improvement was fairly small, with the highest performance at just 105 K. NETD across these devices also improved through the decade, reaching below 25 mK for the highest performing devices [124–128].

Quantum Cascade Detectors

One interesting technology closely related to QWIPs is the quantum cascade detector (QCD). Like a QWIP, a QCD is a matrix of two alternating material layers: a well material and a barrier material. Whereas a period of a QWIP contains only two layers, one each of the well and barrier, each period of a QCD contains many layers. The first two layers are much like a period of a QWIP, with a long barrier layer followed by a collecting well layer. Much as in a QWIP, this collection well is doped such that electrons are present in the ground state of the well without illumination and upon illumination move to an excited energy state in the well, however unlike QWIPs this excited energy state is not near the conduction band edge of the barrier material. After these two layers are an additional series of thin layers, alternating between barrier and well materials. The thicknesses of these layers are chosen so as each successive well layer has an incrementally lower energy state than the excited energy state in the well. The carrier is able to tunnel through the adjacent barrier layer into the adjacent incrementally lower energy state. This continues through the rest of the period of the QCD until the carrier finally relaxes to the ground state of the absorbing well in the next period of the detector. The requirement for a number of excitations to produce a single carrier in the conduction band has a dramatic effect on the performance of the device. The band diagram of a QCD can be seen in Figure 4. This further reduces the detectivity and responsivity of the device, but with the benefit of also dramatically reducing the dark current generated [5,129].

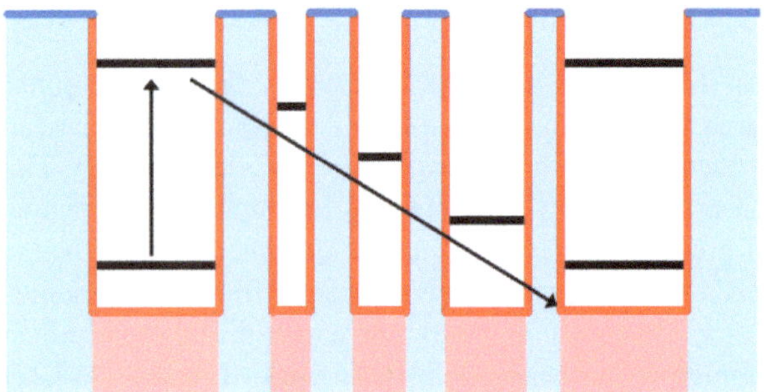

Figure 4. A band diagram of a quantum cascade detector. Incident light excites a carrier from the ground energy state of the absorbing well to an excited energy state. The carrier then tunnels through the barrier layers into the adjacent wells. This continues until the carrier reaches the ground state of the next period of the superlattice.

STRAINED-LAYER SUPERLATTICES

Since the first strained-layer superlattice (SLS) photodetector was grown from InGaAs/GaAs in 1984, superlattice (SL) and SLS structures have been of interest to the IRPD community [130]. Note: for brevity, in this publication, SL and SLS photodetectors will both be referred to as SLS, whether they are strained or not. SLS, while initially appearing to have a similar structure to QWIPs, actually operate using dramatically different physical principles. The misleading similarities result from a superlattice with extremely thin layer thicknesses (on the order of single nanometers) as an active absorbing layer. Additionally, while QWIPs generally utilize relatively thick barrier layers (on the order of tens of nanometers), in SLS the active layers have thicknesses on the same order of magnitude [131,132]. Like a QWIP, the most common structure for the superlattice is alternating layers of two different materials, but some more recent devices utilize more complex structures [30,133].

Similarities between the two technologies begin and end at the inclusion of a superlattice. QWIPs utilize a type-I band structure with explicit barrier and well materials, where the band edges of the smaller energy band gap are straddled by the band edges of the larger energy band gap. This structure allows QWIPs to utilize intraband transitions within the smaller band energy material in both the valence and conduction bands. SLS devices typically utilize a type-II band structure, where the band edges of the superlattice are staggered [134]. Due to this staggering of the band edges, each material acts as a well in one band (valence or conduction) and as a barrier for adjacent layers of the other material. An example SLS band structure can be seen in Figure 5. The thinness of the SLS layers allows for tunneling between wells in each band. Generally the barriers are thin enough that the available energy states in the wells of each band are able to effectively blur together despite being separated spatially in the different material layers of the superlattice, leading to what are commonly referred to as minibands or quasi-bands. It is the interband energy gap transitions between these minibands that determines the absorption wavelength of the device, rather than the band gap of either material (as in bulk detectors) or the spacing of energy levels within a single well (as in QWIPs and QDIPs). The devices are then biased to extract the charge carriers [133,135].

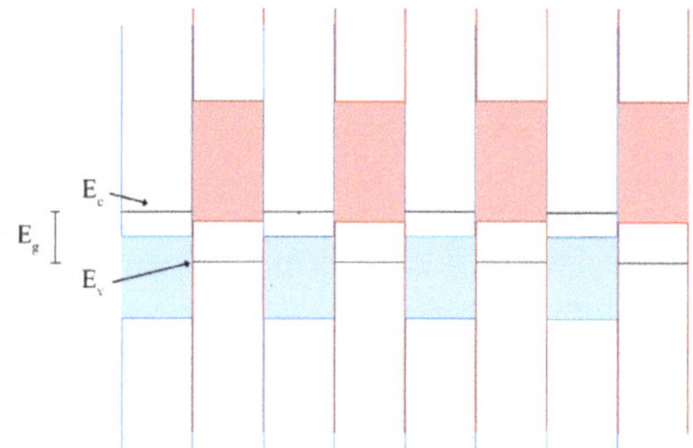

Figure 5. A band diagram of a InAs(blue)/GaSb(red) SLS. Carrier electrons are excited from the valence band of the GaSb layers to the conduction band of the InAs layers.

SLS have a number of performance strengths and weaknesses. SLS devices can be grown using many of the same techniques used for other III-V devices, but MBE growth is generally favored for its precise control of layer thicknesses and interfaces. Control of layer thickness, and particularly material interfaces, is vital for high performing devices. Poor quality interfaces can lead to increased band-to-band and defect-assisted tunneling currents. As SLS depends heavily on the spatial separation of charge for its performance, these tunneling currents are especially detrimental [136]. Materials chosen for use in SLS generally fall around the 6.1 Å lattice constant (InAs/GaSb/AlSb/etc.) with the most common superlattices comprised of InAs (electron capture) and GaSb (hole capture) layers [30,135]. Gallium-free versions also exist, using the same InAs layer for electron capture, but using an InAsSb for the hole capture. Use of InAsSb layers in place of GaSb would allow for longer wavelength absorption, but to this point research has been limited to characterization of superlattices comprised of these materials as opposed to fabrication of functioning photodetectors [137–139]. Due to the structure of SLS, dark currents tend to be fairly low for these devices. This in turn allows for higher operating temperatures compared to bulk detectors and QWIPs, particularly at longer absorption wavelengths. These detectors also allow normal incidence absorption, eschewing the need for frontside scattering filters. There have also been studies concluding higher values for the effective electron mass in these devices, which can also lead to reduced Auger recombination rates [133,140–142]. SLS devices also tend to be more tolerant to growth variances than MCT, particularly in terms of mole fraction composition [143].

Unfortunately, due to the overlap of the band structure of different materials in the devices, the effective band gap between the minibands tends to be fairly narrow. This tends to make it more difficult to use these devices for SWIR and shorter wavelength applications. It also makes SLS well suited for applications in the LWIR regime and longer wavelengths [133,144,145]. Due to high trap assisted tunneling rates in the active region of the device, there can be high amounts of non-thermal noise in these devices [146]. Additionally, the native oxides for many of the materials used in these devices are conductive. In FPAs, the sides of mesas can accumulate these oxides, which can enhance shunt current. Should adjacent pixels come into contact with each other due to oxide growth, those pixels will not perform as intended [133]. There can also be difficulty in achieving high quantum efficiency levels, as the thickness of the active regions can be small compared to other technologies, ranging from as small as hundreds of nanometers to as thick as a handful of microns [30,147].

In the early portion of the decade the majority of research in SLS was focused on InAs/GaSb devices grown on GaSb substrates. This material system had already been shown to function in previous work and various research groups sought to refine their utility. These devices were focused primarily on two miniband gaps: 150 meV (λ = 8.3 μm) and 80 meV (λ = 15.5 μm). These miniband gaps were chosen in part to fall squarely in the LWIR and VLWIR regimes, respectively, and also in part due to straightforward deposition ratios (42Å/42Å for the former, 20 ML (monolayers)/20 ML for the latter) [148,149]. Like QWIPs, some of these devices chose to incorporate current blocking layers into the superlattice. These devices showed promising performance in both regimes, capable of LN_2 temperature operation even at this early stage. At these temperatures, the LWIR SLS with the best performance achieved detectivity on the order of 1×10^{11} cmHz$^{1/2}$/W and responsivity of 1.2 A/W by Razeghi et al. Similarly, the best VLWIR device was capable of a D* and responsivity of 4.5×10^{10} and 4 A/W, respectively [148–153].

Moving forward, FPAs were created using SLS technologies. These FPAs began with small pixel counts, 256 × 256, and at this point were only utilized in the LWIR regime, but those produced were capable of similar performance metrics as the single pixel devices. As processing steps were refined, the imaging of these devices improved and an NETD of about 10 mK was achieved at LN_2 temperatures [45,154,155]. Research on single pixel devices over this time frame focused on expanding the operation wavelengths SLS in both directions. SLS operational down to 5 μm were fabricated, maintaining the strong performance of previous tests with detectivities as high as 1.5×10^{13} cmHz$^{1/2}$/W and responsivities remaining at 1 A/W by Krishna et al. [156–158]. It was also over this time frame that the first room temperature operation of an SLS IRPD, with a D* of 4.6×10^{9} cmHz$^{1/2}$/W and responsivity of 2.2 A/W at 5.25 μm [144]. Also during this time frame the mechanics of this material system was researched, revealing information on its optical properties. It was discovered by Brown, Haugan, et al. that variation of the InAs layer thicknesses has a stronger effect on the band gap than variation of the GaSb layer thicknesses, and the dependence of the amount of interfacial strain on the number of periods in the superlattice [147,159,160].

Moving towards the end of the decade, research began to focus on production of larger and higher quality FPAs. The majority of these devices had cutoff wavelengths in the LWIR regime and were operated at LN_2 temperatures. While none of the devices are capable of room temperature operation, a handful were capable of operation between 200 and 240 K. The expansion of FPA size, and perhaps the inclusion of other features such as current blocking layers into these devices, led to a small decrease in NETD, usually in the 20–30 mK range. These figures may seem low, however they are still more than sufficient for the purposes of an FPA [161–164]. By the end of the decade, a megapixel FPA had been fabricated, and its responsivities and detectivities remained similar to those found in previous FPAs. However, this FPA did suffer somewhat due to its size in the form of a higher NETD (over 50 mK) [165]. In addition to this FPA research, extension of previous work was performed, including optimization of growth conditions and further extending operational wavelengths. SLS were fabricated out to wavelengths of 4.3 μm. Other research focused on reducing the thickness of a single SLS period to 10 ML, which may enable this class of device to achieve noteworthy SWIR response, but this device resulted in lower detectivities at room temperature compared to other devices [136,145,162,164,166–168]. There were also some advancements in performance in recent years, with some single pixel devices with detectivities as high as 1.9×10^{13} cmHz$^{1/2}$/W and responsivities as high as 1.33 A/W [169].

There was also some research performed on other material systems over this timeframe. The most common alternative to the InAs/GaSb system was InAs/InGaSb. Incorporation of indium into the GaSb layer allowed for increased flexibility in defining the energy level of the valence miniband. Performance of InAs/InGaSb devices was usually similar to devices utilizing InAs/GaSb structures. However, GaSb was usually the preferred due to more convenient material growth [143,170–172].

M/W Strained-Layer Superlattices

A pair of sub-technologies of SLS spawned due to the implementation of current blocking layers. Each of these technologies incorporates a different manner of a current blocking layer. The first, and arguably more complex, manner was first proposed in 2001, but not implemented until 2006 [173,174]. This technology incorporates a current blocking layer into one of the active superlattice layers, forming an ABCB superlattice as opposed to the traditional AB superlattice. This barrier layer typically has a significantly larger energy band gap than the surrounding material, leading to the distinctive band structure that gives these devices their name. An example band diagram can be seen in Figure 6.

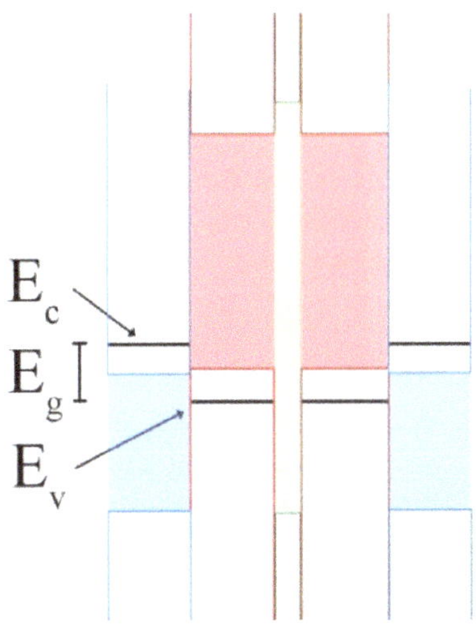

Figure 6. A band diagram of a InAs(blue)/GaSb(red)/AlSb(green) M-SLS. The inclusion of AlSb into the device gives this structure its eponymous band structure.

The current blocking barrier, typically an aluminum bearing material, can be incorporated into either the InAs or GaSb layer, giving rise to the eponymous band structure in a period of the superlattice, resembling either a W or an M. The barrier placement in W- and M-superlattices is crucial in these devices. Inclusion of the barrier in the well shifts the two highest energy levels from the original well towards each other. When the barrier is precisely in the middle of the well the energy states become nearly degenerate. This in turn increases the number of energy states available at the desired energy level [164,175].

Even by the end of the decade, W- and M- superlattices were still a nascent technology. As such, much of the research performed is in the very early stages. Most single pixel devices have only been operated at LN_2temperatures. Only at the very end of the decade was an M-superlattice operated as high as 150 K. The results from that device, led by Razeghi et al., with detectivities as high as 1.05×10^{12} cmHz$^{1/2}$/W at 150 K in the MWIR regime, do show promise for this technology. Further, a 320 × 256 pixel FPA was constructed using this design, with imaging possible up to 170 K [131,174,176–178].

Unipolar/Monovalent Barrier Strained-Layer Superlattices

The other main technology incorporating current blocking layers in SLS are unipolar or monovalent barrier devices. In these devices, a single barrier layer is incorporated in the SLS structure immediately above or below (depending on whether the barrier is in the conduction or valence band) the active superlattice stack. The barrier material is chosen such that there is a relatively large barrier in the chosen band, but as little discontinuity as possible in the other band. This barrier allows the device to conduct photocurrent and diffusion current, but block both generation-recombination current and trap-assisted tunneling current [178–182]. Because of unavoidable variabilities in the growth process, most unipolar/monovalent SLS are operated at a slightly higher bias voltage to overcome the unintentional barrier than other devices, in case there is a discontinuity at the band edge between the barrier layer and the rest of the device [131].

As with M- and W-superlattices, the first devices incorporating unipolar barriers did not arrive until the later part of the decade, 2006 [174]. With the explicit concern of limiting dark current, this class of devices quickly achieved room temperature operation. Monovalent barrier devices have been operated at a range of wavelengths from the MWIR to LWIR. One of the highest performing single pixel devices is capable of a D* of 1×10^9 for a cutoff wavelength of 4.3 μm at room temperature. Barrier layers were investigated for both the conduction band (nBn structures) and the valence band (pBp structures), but nBn (Figure 7) were far more popular in the early stages of research [146,161,183–185]. A small FPA (320 × 256 pixels) was fabricated using nBn technology, having comparable performance to single pixel devices in the MWIR regime. Responsivities of 1.5 A/W, detectivities of 6.4×10^{11}cmHz$^{1/2}$/W, and NETD of 24 mK were also achieved with nBn devices [163]. Recently, pBn have drawn interest for use in photodetectors, but very few devices have been successfully fabricated. To this point, pBn devices (Figure 8) have remained an exercise for theory and simulation, but promising results from early devices, included low dark current densities as low as 3×10^7 A/cm^2 at 150 K could lead to increased research with these devices [186–189].

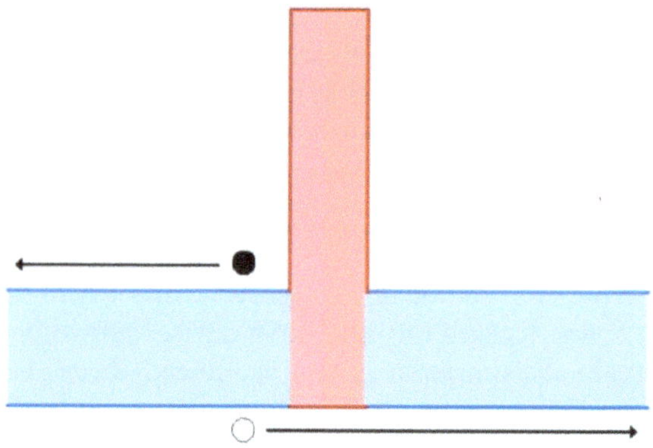

Figure 7. A band diagram of an nBn IRPD without applied bias. With the barrier in the middle, the left hand side acts as the photo-absorber. Absorbed photons create carrier pairs, which are unimpeded by the barrier. Thermally generated pairs on the right-hand side are blocked by the conduction band barrier, however.

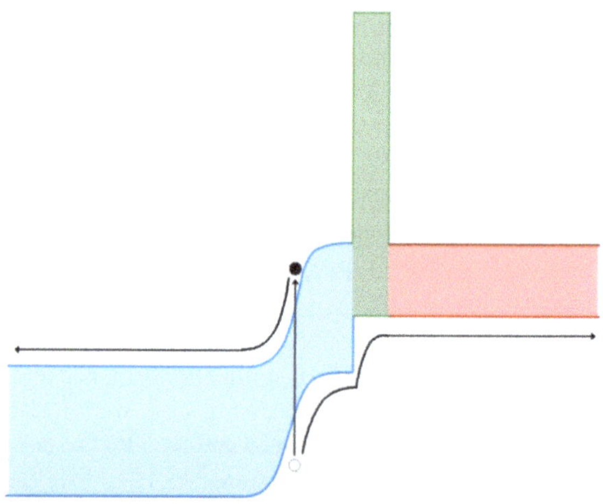

Figure 8. A band diagram of a pBn photodiode. A carrier electron is excited from the valence band of the n-type material and swept away from the barrier. The carrier hole is swept past the barrier layer, as it does not produce a potential barrier in the valence band of the p-n junction.

QUANTUM DOT INFRARED PHOTODETECTORS (QDIPS)

Quantum dots were initially posited in 1982 as a structure for use in quantum well lasers as a means to reduce or remove the temperature dependence from a laser's performance [190]. They remained a theoretical exercise for over ten years until 1993, when the first demonstrated growth of quantum dots was performed by MBE [191]. By 2000, research was well underway for quantum dot infrared photodetectors (QDIPs). In a QDIP, operation is very similar to that in a QWIP. A layer of dots, typically with diameters on the order of nanometers, is deposited on a well material, most commonly by MBE. These dots typically have narrower band gaps than the barrier material, creating a highly localized quantum well. A number of iterations are then deposited to improve the absorption of these devices [82,84,192,193–196]. Dots are typically fabricated using the Stranski-Krastanow growth mode, which requires a certain amount of lattice mismatch (between 1%–10%) between the dot material and the well material to ensure sufficient strain for islanding to occur. The thickness of the dot layer must also be precisely monitored, as excessive thickness (>2–3 ML for some systems) in the layer will change the surface morphology from quantum dots to a solid layer with large densities of dislocations [197]. Like QWIPs, QDIPs are doped such that there are electrons present in the conduction bands of the quantum dots. These electrons are then excited by incident photons to the conduction band of the well layers, where they can diffuse to the contacts of the device [9,15].

While much of the operation of QDIPs is reminiscent of that of QWIPs, there are a number of features which distinguish the former from the latter. The most notable difference is that, due to the quantum dot's confinement in all directions, QDIPs are capable of absorbing normally incident photons. The higher confinement also leads to lower dark currents, which can in turn lead to higher detectivities and operating temperatures [26,80,84,88,192,198,199]. QDIPs have longer carrier lifetimes than QWIPs, generally on the order of hundreds of picoseconds. However, the carrier lifetimes are shorter than in bulk detectors. This is believed to derive both from decreased amounts of phononic scattering in QDIPs and a phenomenon commonly referred to as phononic bottleneck. In this bottleneck, carriers fill the excited states of the dots. If these carriers are not extracted from the dots quickly, other carriers which would otherwise be excited to the higher energy level will not undergo that transition due to the lack of an available energy state [84,89,192,200]. Under ideal growth conditions, QDIPs also have higher selectivity and narrower spectral width absorption than QWIPs [8,192].

QDIPs do come with some disadvantages. While under ideal growth conditions QDIPs should have superior selectivity to QWIPs, in practice it is extremely difficult to have the kind of quantum dot size uniformity required for high levels of performance. Due to uniformity variance in the quantum dots, oftentimes QDIPs have wider spectral absorption than QWIPs [84,193]. For similar reasons, tuning QDIPs to a specific wavelength with great accuracy can be extremely difficult [15,201]. The dot layers in a QDIP are by their nature very uneven, leading to issues with strain and dislocations with an increasing number of absorption layers. As such, QDIPs often exhibit fewer absorbing layer iterations than QWIPs, leading to lower absorbances. To combat this effect, some QDIPs have strain reducing layers incorporated into their design. While this raises the complexity of the fabrication, it can lead to higher device performance. Quantum efficiencies (QEs) for QDIPs also tend to be much lower than QWIPs. This low QE is due to the relative paucity of energy states in QDIPs (a few monolayers of scattered dots in QDIPs vs. several nanometers of a solid layer in QWIPs), lower absorption of incident light, and lower capture probabilities by individual dots due to more sharply defined energy states than in QWIPs [84,193,202].

The vast majority of research related to QDIPs over the course of the decade was related to four main material regimes: InAs dots in GaAs barriers, InGaAs dots in GaAs barriers, InGaAs dots in InGaP barriers, and InAs dots in InGaAs barriers. Of these, InAs dots in GaAs barriers was the most common device in the early part of the decade. QDIPs using InAs dots in GaAs barriers typically operate in the SWIR and MWIR regimes, with a handful out as far as 7 μm. In the devices of the early part of the decade there was significant difficulty in growing numerous iterations of absorbing layers due to a buildup of strain, exacerbated by the irregular surface morphology of the dots. As such, detectors rarely exceed ten layers. As might be expected, the low number of absorbing layers led to fairly low detectivities, as high as 2.94×10^9 $cmHz^{1/2}/W$ at LN_2 temperatures by Madukar et al. [15,203–205]. It was in the early part of the decade that some of the earliest InAs/GaAs FPAs were fabricated. While the imaging of these devices was not superlative, as they were intended as proofs of concept more than final designs, they do produce recognizable images. Due to the low dark current densities present, even the initial devices were capable of operating up to 150 K in the MWIR regime [88,206]. As the decade moved forward, progress addressed a number of QDIP issues. Most notably was the number of quantum dot layers which could be deposited without dislocations. Whereas in the early part of the decade, layer numbers were limited to ten or less, by 2005 layer numbers InAs/GaAs QDIPs were in 50–70 range. Detectivities rose some with this increase, to a high of 1.2×10^{11} $cmHz^{1/2}/W$ [207]. It should be noted that even in these later devices, responsivities were still low, with a high value of 0.3 A/W [193,208,209]. Advances were also made with FPAs, with a more refined device fabricated in the middle of the decade. A 256×256 pixel FPA with similar performance metrics was fabricated with significantly improved imaging capabilities over previous devices [210]. However, following the middle part of the decade research on InAs/GaAs QDIPs slowed, replaced by research into the other material systems.

A different pair of the material systems had been under limited research over the past decade. The first of these systems was InGaAs dots on GaAs barriers. These materials offer increased design flexibility, as dot composition can be varied along with dot size. This system was researched more towards the later portion of the decade, as interest in the InAs/GaAs system waned. Operation temperatures InGaAs/GaAs QDIPs were comparable to InAs/GaAs devices. Early devices also had a number of the same issues as InAs/GaAs, including a low number of quantum dot layers and low detectivities. By the end of the decade, FPAs using InGaAs/GaAs were outperforming InAs/GaAs FPAs, capable of detectivities of 1.01×10^{11} $cmHz^{1/2}/W$ and responsivities of 2.1 A/W by Jagadish et al. [15,211–213]. The other lesser used material system involved InGaAs dots in InGaP barriers. However, these devices generally had poor performance, including low quantum efficiencies and fill factors. These low metrics have been attributed to the lower lattice constant mismatch (there is a mismatch of about 3.2% between InGaAs and InGaP, whereas InAs/GaAs devices have a mismatch of about 7%) [214–217].

The last common material system that has been investigated was InAs dots in InGaAs barriers. As with other material systems for QDIPs, research for this system did not begin getting traction until the middle portion of decade when researchers were looking to improve performance from InAs/GaAs devices. These InAs/InGaAs devices functioned in the MWIR and LWIR regimes and operated at temperatures as high as 190 K. QDIPs of this system which operated at longer wavelengths did tend to suffer in terms of D*, only reaching as high as 4.6×10^9 cmHz$^{1/2}$/W at LN_2 temperatures. However, when operated in the MWIR, these structures achieved comparable performance metrics to other material systems. Some of these devices, as well as some of the InGaAs/GaAs and InGaAs/InGaP devices, were grown by MOCVD, as opposed to MBE, with little hindrance on performance. MOCVD growth has an advantage to potentially make larger scale fabrication of QDIP devices much easier [218–221]. A 350 × 256 pixel FPA QDIP was fabricated near at the end of the decade. While operation of this device was constrained to LN_2temperatures, D* at that temperature was as high 1×10^{10} cmHz$^{1/2}$/W by Lai et al., which was among the highest for a QDIP FPA of any type [222].

A material system that may be interesting for future QDIP devices is GaAs dots on GaSb substrates. While this material system has not been used in a functioning IRPD, it has been used in other optical devices effectively, including lasers, solar cells, and LEDs, most notably by the Huffaker group [223–227]. For these devices, both Stranski-Krastanov and, more commonly due to the size of the lattice mismatch, interfacial misfit growth modes have been utilized. Most QD devices utilizing this material system have shown peak photoluminescence in the 1–2 μm range, which could make it suitable for NIR applications [228,229].

Another interesting subset of QDIPs utilized a pair of resonant tunneling layers incorporated into the periodic absorbing stack. In these structures, a quantum dot is deposited on a barrier material. This is followed by deposition of a very thin layer of a higher energy band gap material, a thin layer of the original barrier material, and a second thin layer of the higher energy band gap material. This creates two energy wells in each layer of the absorbing stack of the device, albeit at a much higher energy level than the quantum dot well. The intention is that by tuning the energy level of this second energy well, the well will transmit photocurrent excited from the initial well. However, the well won›t transmit a large fraction of the dark current. These devices, while reducing dark current and maintaining respectable detectivities, encountered great difficulty in terms of overall responsivity, not capable of achieving a responsivity of 0.1 A/W at LN_2 temperatures. This low responsivity and the additional complexity of the structure has limited the appeal of these devices [230,231].

Another interesting subset of QDIP devices is that which incorporate periodic current blocking layers. The current blocking layers (usually composed of either AlGaAs or InGaAs) were mainly incorporated at the early portion of the decade and were typically incorporated as capping layers on the quantum dots. The unintended consequence, at least in early devices, was the effect they had on the energy states in the quantum dots, particularly with InGaAs layers. These current blocking layers, if thick enough, could change the level of the energy states in the quantum dot or create entirely new energy levels in the barrier or well materials. While not having the explicit aim or the ultimate polish of later devices, it is easy to see these devices as precursor to quantum dot-in-a-well photodetectors (DWELLs). These early DWELL devices were plagued by many of the same issues as resonant tunneling barrier QDIPs, including low responsivities and low counts of quantum dot layers. However, they were able also to achieve detectivities up to 1.1×10^{12} cmHz$^{1/2}$/W. Ultimately, a 320 × 256 pixel FPA was fabricated using these techniques, which was capable of imaging at 120 K, albeit with an NETD of 344 mK [213,220,232–235].

QUANTUM DOTS-IN-A-WELL PHOTODETECTORS (DWELL-IPS)

Utilization of structures called DWELLs in IRPDs began in 2002, though the structure was previously used in other applications, such as semiconductor lasers [33]. While 2002 is the earliest direct reference to a DWELL structure, other structures did utilize various capping layers in QDIPs. These structures unintentionally made the introduction of DWELLs to IRPDs harder to pin down. Over the next couple years there was a gradual delineation between the DWELL structures and QDIPs. Once that delineation had been made, research on devices specifically labeled DWELLs picked up.

The physical structure of a DWELL is much akin to a combination of a QWIP and a QDIP. A layer of the absorbing stack begins with deposition of a wide band gap material, the barrier material. This deposition is followed by growth of a narrower band gap material, the well material, much as in a QWIP. Growth is then followed with deposition of a layer of quantum dots with a band gap that is narrower still, known as the dot material. At this point, some devices will add a capping layer, sometimes composed of the well material [236]. Because these devices typically require precise tolerances, MBE deposition is typically the method of growth chosen for DWELLs. The absorbing layers will then be repeated a number of times to improve the absorption of incident light [8].

The structure of the DWELL is designed to address some of the weaknesses of both QWIPs and QDIPs. One of the most notable flaws of the QWIP is the inability to absorb normally incident light. The inclusion of a quantum dot layer allows DWELLs to utilize light at normal incidence. QDIPs can have difficulty in precisely controlling the energy levels in the quantum dots, as those energy levels depend heavily on dot size. The inclusion of a well layer in addition to the quantum dot layer adds significantly greater control over the energy level spacing. The well and dot layers, being adjacent, form what amounts to a combined energy well. For DWELLs which utilize two energy states in the well, the ground state of the well occurs in the dot, with the absorbing layer doped to the point such that electrons populate the lowest energy state in the dot at rest (Figure 9). The next energy level is designed to be located in the quantum well layer. The location of this energy level can be adjusted by varying the thickness of the well layer. This can allow for easier wavelength selectivity, as that selectivity can be determined by varying layer thicknesses instead of quantum dot diameters. The amount of flexibility in the design, in terms of dot size, material choices, well thickness, etc., has also led to alternative methods of multispectral imaging. While multispectral imaging is available in a number of different technologies, it is generally implemented by stacking detectors for two different wavelengths monolithically. In DWELLs, there is enough flexibility such that two or three energy transitions can exist within the well. Each transition then corresponds to a different absorption wavelength [8,33,237].

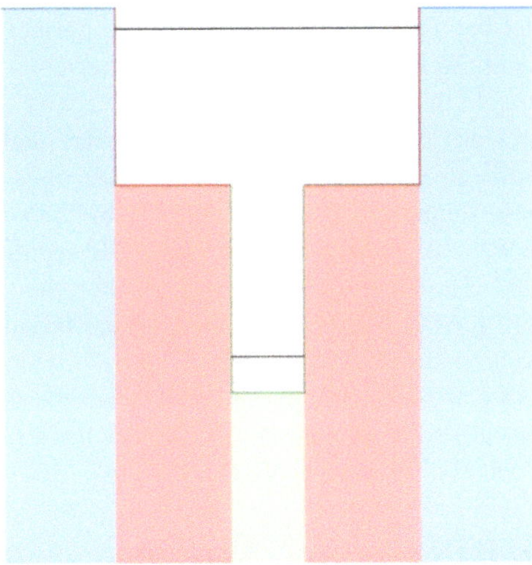

Figure 9. A band diagram of a DWELL structure, with a barrier layer (blue), well layer (red), and dot layer (green).

While DWELLs do address a number of the weaknesses present in QWIPs and QDIPs, the technology still has some of the faults present in other devices. One of the major difficulties DWELLs have in common with QDIPs is a low number of absorbing layers which can be grown on a single device. As with QDIPs, use of the quantum dots can lead to strain build-up. As more layers are deposited, the strain accrues, and if too many layers are deposited defects can form. Accumulations of defects limit the amount of absorbing layers that can be deposited in a DWELL, which can in turn lead to low absorbances and quantum efficiencies. Much like in QDIPs, this buildup of strain can be mitigated with the inclusion of strain reducing layers at the expense of device complexity [238].

Many of the earliest devices which utilized DWELLs existed either as proofs of concept or confirmation of theoretical predictions. Characteristics, such as the effect of variation of quantum dot size on absorption wavelength (smaller dots absorbed shorter wavelengths), the effect of variation of well layer thickness on absorption wavelength (thicker layers led to longer absorbed wavelengths), and the effect of variation of well layer mole composition (narrower band gaps led longer absorption wavelengths), were all investigated. Other growth characteristics were also characterized in this time frame. The effect of overall deposition speed on device performance was examined, with slower growth speeds associated with higher photoluminescent response. The effect of growth speed proved to be most important with the quantum dot layer, as the lower growth speed allows for greater uniformity of quantum dot size, which in turn leads to better performance at the desired absorption wavelength. Similarly, higher quantum dot densities were also associated with stronger response, but were much easier to achieve with slower deposition speeds [8,82,239,240].

Most of the early DWELL devices had a similar structure. Operating temperatures were typically around that of LN_2, though some early designs did achieve operation near 100 K. These devices utilized a structure of GaAs barriers, some composition of InGaAs for the wells, and InAs quantum dots. The barriers were typically tens of nanometers thick, had well layers under ten nanometers, and contained quantum dots 1–3 ML thick. The variations in layer thicknesses were enough to dramatically alter devices absorption characteristics. Early DWELL-IPs spanned a wide range of absorption wavelengths, from the MWIR to VLWIR. A number of early devices also exhibited multispectral absorption from multiple available energy transitions within individual wells [33,201,241]. By the middle portion of the decade, a small 320 × 256 pixel FPA was produced, though for this early device responsivity and quantum efficiency was very low [8]. Dark current levels in these early devices were generally lower than those in QWIPs, but still higher than those in QDIPs. Responsivities varied significantly depending on the absorption wavelength, but responsivities in the MWIR regime as high as 3.58 A/W were achieved by Krishna et al. At longer wavelengths, this value decreased dramatically, with some of the VLWIR regime devices having responsivities as low as 0.01–0.02 A/W by Krishna et al. [33].

Moving through the decade, research continued on DWELLs at universities nationwide. Operating wavelengths were extended out further into the VLWIR regime. Often, long wavelength detection occurred as the secondary or tertiary absorption wavelength in the DWELL. As long as the devices were operated at low temperatures (sometimes as low as 5 K) detection of VLWIR radiation was still possible [242,243]. A number of groups continued characterizing the effect of various growth conditions and deposition characteristics on DWELL performance. Variation in the doping density for DWELLs was investigated by Attaluri, Krishna, et al, with higher doping densities leading to both larger spectral responses and large dark current densities. It was also theorized that doping DWELLs to the point that there was an average of two electrons per quantum dot (opposed to the lower numbers normally used) could lead to a decrease in photocurrent due to overfilling energy states in the well [89]. Perera, Krishna et al. investigated the effect of annealing on DWELL performance. Annealing was found to increase the responsivity of the device, but also increased the dark current and shifted the absorption peak by up to 3.5 µm in wavelength. The change in performance is attributed to the effect of annealing on the quantum dots. It is believed that annealing increased material diffusion between the dot and the capping layer and, therefore, reduces the confinement of the quantum dots in devices. Despite these advances, the operating temperatures of DWELL devices remained fairly low, rarely operating significantly above LN_2 temperatures and even then barely above them [244–246]. Double well DWELLs, where one period of the DWELL absorbing incorporates a barrier layer, two well layers, and a dot layer, were also developed over this time period, led by the Krishna group. In these structures, the dot layer is generally InAs, the first well is a thin layer of InGaAs to aid in the formation of the quantum dots, the second well layer is GaAs, and the barrier layer is AlGaAs. Less strain develops in a DWELL using this structure, allowing for more iterations of the absorbing layer to be deposited in a single device [247–249]. Using this structure, devices with up to 80 periods have been fabricated. These devices operated in both the MWIR and LWIR regimes and performed similarly to normal DWELL devices with a handful of minor distinctions. These devices tended to have marginally lower responsivities (around half that of single well DWELLs), while have improved NETDs in FPAs, reaching as low as 105 mK [250]. These devices also were capable of D* as high as 3×10^{10} $cmHz^{1/2}/W$ at LN_2 temperatures by Krishna et al. [251–258].

Other growth methods besides MBE were attempted for DWELLs, with the intention of easing fabrication of the devices. These alternative methods still required fine control of deposition speeds, to accommodate quantum dot growth, and the ability to deposit a variety of materials without breaking vacuum. The first alternative method was MOCVD. Working DWELL devices were fabricated using MOCVD, though it should be noted that the devices produced had notably lower responsivities and detectivities at LN_2 temperatures. These devices performed in the range of 0.05–15 A/W and 1–6 × 10^9 $cmHz^{1/2}/W$, respectively. Additionally, a small change to the typical DWELL structure was made in these devices. Instead of using InGaAs for the well material and GaAs for the barrier material, these devices used GaAs for the well material and AlGaAs for the barrier material. The effect in terms of the band diagram of the DWELL structure is effectively the same, with only slightly different differences in the band edge locations. The use of these two materials does have one particular advantage. Compositions of AlGaAs are closely lattice-matched to GaAs, whereas InGaAs (particularly compositions with high indium contents) may not necessarily have the same trait. This trait is of particular significance when changing to a lower quality growth method, such as from MBE to MOCVD, as the larger mismatch can lead to high densities of defects [259,260].

A number of DWELL designs utilized similar resonant structures that have been mentioned previously for other IRPD technologies. These designs implemented a resonant tunneling barrier, typically located in the well layer of the DWELL. These layers are typically a pair of wide band gap layers of AlGaAs with a layer of InGaAs in between. Notably, the AlGaAs layer has a wider band gap than the barrier of the normal DWELL structure and the InGaAs layer has a narrower band gap than the band gap of the well material. As in the previous examples, these layers are included to attempt to reduce dark current and improve detectivity. The DWELL structures utilizing the resonant tunneling structures did exhibit strong D*, as high as 2.9 × 10^{10} $cmHz^{1/2}/W$ in the LWIR regime at LN_2 temperatures. However, they also suffered in terms of device responsivity [261–263]. There was additional progress made in the development of DWELL FPAs during this time frame. FPAs as large as 640 × 512 pixels were fabricated for MWIR applications at LN_2 temperatures. This device has a D* as high as 1 × 10^{10} $cmHz^{1/2}/W$ and an NETD of 40 mK [264].

DISCUSSION

Tables 1, 2, 3, 4, 5 and 6 list the highest performing devices of different IRPD technologies across different wavelength regimes. In these Tables, a number of the most common IRPD performance metrics are examined. Results which stray too far from these standard conditions and metrics, or utilize wholly different metrics for characterization, are not included in these Tables. Detectivity (D*), both at LN_2 and at the highest temperature with successful results are reported. Similar results are reported for responsivity (R). The highest temperature at which a device has been successfully operated is also reported. The lowest dark current density (J_{dark}) at LN_2 temperatures is included in the table. Finally, to compare FPA performance, a comparison of noise equivalent temperature difference (NETD) is also included.

Table 1. Highest reported performance metrics for HgCdTe IRPDs across different wavelength regimes.

Metric	SWIR	MWIR	LWIR	VLWIR
$D^*_{LN_2}\ [cmHz^{1/2}/W]$	1 × 10^{14} [265]	2 × 10^{13} [265]	2 × 10^{13} [265]	X
$D^*_{T_{high}}[cmHz^{1/2}/W](T_{high}\ [K])$	1 × 10^{12} (300) [265]	7 × 10^{10} (210) [25]	1 × 10^{7} (300) [52]	X
$R_{LN_2}\ [(A/W)]$	X	X	1 [266]	X
$R_{T_{high}}(T_{high})[A/W]$	1 [24] (300)	X	X	X
$T_{max}\ [K]$	300	210 [25]	300	100 [35]
$J_{dark}\ [A/cm^2]$	X	X	2.5 × 10^{-4} [36]	2.7 × 10^{-7} [41]
$NETD\ [mK]$	20 [24]	10 [267]	9 [55]	22 [41]

Table 2. Highest reported performance metrics for InSb IRPDs across different wavelength regimes.

Metric	SWIR	MWIR	LWIR	VLWIR
$D^*_{LN_2}\ [cmHz^{1/2}/W]$	X	1 × 10^{10} [63]	7.6 × 10^{8} [64]	X
$D^*_{T_{high}}[cmHz^{1/2}/W](T_{high}\ [K])$	X	2.8 × 10^{8} (300) [73]	X	X
$R_{LN_2}\ [(A/W)]$	X	1.4 [60]	X	X
$R_{T_{high}}(T_{high})[A/W]$	X	X	X	X
$T_{max}\ [K]$	X	300 [73]	300	X
$J_{dark}\ [A/cm^2]$	X	X	X	X
$NETD\ [mK]$	X	2.2 [72]	X	X

Table 3. Highest reported performance metrics for QWIPs across different wavelength regimes.

Metric	SWIR	MWIR	LWIR	VLWIR
$D^*_{LN_2}\ [cmHz^{1/2}/W]$	X	2 × 10^{11} [116]	2 × 10^{11} [91]	5 × 10^{9} [121]
$D^*_{T_{high}}[cmHz^{1/2}/W]$	X	1 × 10^{11} (210) [107]	3.2 × 10^{10} (110) [99]	5 × 10^{9} (70) [121]
$R_{LN_2}\ [(A/W)]$	X	1 [268]	1.4 [118]	0.38 [119]
$R_{T_{high}}(T_{high})[A/W]$	X	0.7 (290) [107]	0.3 (120) [92]	0.25 (70) [121]
$T_{max}\ [K]$	X	290 [107]	120	70 [121]
$J_{dark}\ [A/cm^2]$	X	1 × 10^{-7} [116]	1 × 10^{-9} [99]	1.7 × 10^{-3} [109]
$NETD\ [mK]$	X	14 [123]	19 [118]	48 [122]

Table 4. Highest reported performance metrics for QDIPs across different wavelength regimes.

Metric	*SWIR*	*MWIR*	*LWIR*	*VLWIR*
$D^*_{LN_2}$ $[cmHz^{1/2}/W]$	1×10^9 [203]	1×10^{12} [235]	$1x10^{11}$ [213]	X
$D^*_{T_{high}}[cmHz^{1/2}/W]$	1×10^9 (77) [203]	2.4×10^8 (250) [269]	1.2×10^{11} (77) [207]	2×10^8 (40) [15]
R_{LN_2} $[(A/W)]$	1 [203]	2.5 [270]	5.3 [232]	X
$R_{T_{high}}(T_{high})[A/W]$	1 (77) [203]	0.012 (300) [270]	5.3 (77)[232]	4 (40) [15]
T_{max} $[K]$	77 [203]	300 [270]	77	40 [15]
J_{dark} $[A/cm^2]$	X	1×10^{-13} [193]	3×10^{-8} [232]	X
$NETD$ $[mK]$	X	87 [214]	X	X

Table 5. Highest reported performance metrics for DWELL IRPDs across different wavelength regimes.

Metric	*SWIR*	*MWIR*	*LWIR*	*VLWIR*
$D^*_{LN_2}$ $[cmHz^{1/2}/W]$	X	1×10^{10} [271]	3×10^{10} [256]	X
$D^*_{T_{high}}[cmHz^{1/2}/W]$	X	1×10^8 (250) [272]	1×10^8 (200) [256]	X
R_{LN_2} $[(A/W)]$	X	0.12 [259]	3.58 [33]	X
$R_{T_{high}}(T_{high})[A/W]$	X	0.12 (77) [259]	0.2 (110) [244]	0.025 (4.6) [8]
T_{max} $[K]$	X	77	110 [244]	4.6 [8]
J_{dark} $[A/cm^2]$	X	1×10^{-5} [271]	5×10^5 [261]	X
$NETD$ $[mK]$	X	X	40 [264]	X

Table 6. Highest reported performance metrics for SLS IRPDs across different wavelength regimes.

Metric	*SWIR*	*MWIR*	*LWIR*	*VLWIR*
$D^*_{LN_2}$ $[cmHz^{1/2}/W]$	X	1.9×10^{13} [169]	1×10^{12} [155]	4.5×10^{10} [150]
$D^*_{T_{high}}[cmHz^{1/2}/W](T_{high}$ $[K])$	1×10^8 (300) [166]	4.6×10^9 (300) [144]	X	4.5×10^{10} (80) [150]
R_{LN_2} $[(A/W)]$	X	1.33 [169]	3.2 [162]	4 [150]
$R_{T_{high}}(T_{high})[A/W]$	X	2.2 (300) [144]	X	4(80) [150]
T_{max} $[K]$	300 [166]	300	180 [162]	80 [150]
J_{dark} $[A/cm^2]$	X	1×10^{-7} [163]	1×10^{-3} [171]	3.3×10^{-3} [184]
$NETD$ $[mK]$	X	10 [178]	270 [162]	X

CONCLUSIONS AND FUTURE PROSPECTS

The progress that the field of infrared photodetection has made in recent history is tremendous. What started as primarily academic research has transformed into a budding industry, including a number of new technologies in the pipeline. IRPDs will see increasing use in a number of fields in the present and near future, ranging from military target detection to environmental sensing. With the number of applications for these technologies rapidly increasing, there is a need to determine which technology is best suited to each application.

The number of different IRPD technologies has led to segmentation of which is currently the most viable depending on the application. Bulk detectors have achieved the most commercial penetration due to ease of large scale fabrication, long track record of growth, flexible absorption wavelengths (which can span from SWIR to LWIR), and high detectivity at cryogenic temperatures, but there has been difficulty in raising the operating temperatures of these devices while maintaining performance. QWIPs and QDIPs have also drawn significant attention due to the former's capability to perform in the VLWIR regime and the latter's high temperature performance, currently the best among current devices. DWELL-IPs and SLS devices have shown noteworthy promise in terms of performance, range of absorption range, and operating temperature. It may well be that these devices could soon overtake QWIPs and QDIPs in terms of performance.

Moving forward, progress in the field of IRPDs may come from dramatically different sources than research to date. In recent years, there has been a shift in the source of funding for some IRPD technologies. The increase in research conducted by private businesses in these technologies has led to a change in project selection from many funding agencies, viewing research in these fields increasingly as industrial development as opposed to academic research. That academic research that is funded has a greater focus on producing front-end devices, which may eventually lead to commercial ventures, as opposed to investigating novel applications or phenomenon in the established technology. This has primarily occurred with bulk detectors so far, primarily with MCT and InSb, but as other technologies become more prevalent this trend will likely spread across the field.

ACKNOWLEDGMENTS

We would like to thank Tufts University for the John Mellon Fellowship and The Alexander Von Humboldt Foundation of Germany for the Humboldt Research Fellowship for Experienced Researchers. Their support enabled his work.

REFERENCES

1. Herschel, W. Experiments on the refrangibility of the invisible rays of the sun. Phil. Trans. Royal Soc. Lond. 1800,90, 284–292.
2. Barr, E.S. The infrared pioneers—II. Macedonio melloni. Infrared Phys. 1962, 2, 67–73.
3. Case, T.W. Notes on the change of resistance of certain substances in light. Phys. Rev. 1917, 9, 305–310.

4. Rogalski, A. Infrared detectors: Status and trends. Prog. Quantum Electron. 2003, 27, 59–210.
5. Buffaz, A.; Carras, M.; Doyennette, L.; Nedelcu, A.; Bois, P.; Berger, V. State of the art of quantum cascade photodetectors. Proc. SPIE 2010.
6. Barve, A.; Krishna, S. Advances in Infrared Photodetectors; Elsevier: San Diego, CA, USA, 2011.
7. Xiaohui, X.; Hua, H.; Guangyin, Q.; Qingjun, L.; Xiaoning, H. Study of the characteristics of VLWIR HgCdTe photovoltaic detectors in variable-area diode test structures. Proc. SPIE 2011.
8. Krishna, S. Quantum dots-in-a-well infrared photodetectors. J. Phys. D 2005, 38, 2142–2150.
9. Stiff-Roberts, A.D. Quantum-dot infrared photodetectors: A review. J. Nanophon. 2011.
10. Jdidi, A.; Sfina, N.; Abdi-Ben Nassrallah, S.; Saïd, M.; Lazzari, J.L. A multi-color quantum well photodetector for mid- and long-wavelength infrared detection. Semicond. Sci. Technol. 2011.
11. Bundas, J.; Patnaude, K.; Dennis, R.; Burrows, D.; Cook, R.; Reisinger, A.; Sundaram, M.; Benson, R.; Woolaway, J.; Schlesselmann, J.; et al. Two-color quantum well infrared photodetector focal plane arrays. Proc. SPIE 2006.
12. Chubb, D.L. Fundamentals of Thermophotovoltaic Energy Conversion; Elsevier: Amsterdam, The Netherlands, 2007; p. 515.
13. Daniel, J.S.; Solomon, S.; Kjaergaard, H.G.; Schofield, D.P. Atmospheric water vapor complexes and the continuum. Geophys. Res. Lett. 2004, 31, 1–4.
14. Richards, A.A. Spectral selection in infrared imaging. Proc. SPIE 2004, 5405, 288–293.
15. Towe, E.; Pan, D. Semiconductor quantum-dot nanostructures: Their application in a new class of infrared photodetectors. IEEE J. Sel. Top. Quantum Electron. 2000, 6, 408–421.
16. Mikhailenko, S.N.; Babikov, Y.L.; Golovko, V.F. Information-calculating system spectroscopy of atmospheric gases. Atmos. Ocean. Opt. 2005, 18, 685–695.
17. Liu, H.; Zhang, F.; Zhang, J.; He, G. Performance analysis of quantum dots infrared photodetector. Proc. SPIE2011.
18. Schneider, H.; Walther, M.; Schonbein, C.; Rehm, R.; Fleissner, J.; Pletschen, W.; Braunsteina, J.; Koidl, P.; Weimann, G.; Ziegler, J.; Cabanski, W. QWIP FPAs for high-performance thermal imaging. Physica E 2000, 7, 101–107.
19. Reine, M.B. HgCdTe photodiodes for IR detection: A review. Proc. SPIE 2001, 4288, 266–277.
20. Kohn, A. N.; Schlickman, J. J. 1-2 micron (Hg, Cd)Te photodetectors. Electron Devices, IEEE Transactions on 1969,16, 885–890.
21. Lawson, W.D.; Nielsen, S.; Putley, E.H.; Young, A.S. Preparation and properties of HgTe and mixed crystals of HgTe-CdTe. J. Phys. Chem. Solid. 1959, 9, 325–329.
22. Kruse, P.W.; Blue, M.D.; Garfunkel, J.H.; Saur, W.D. Long wavelength photoeffects in mercury selenide, mercury telluride and mercury telluride-cadmium telluride. Infrared Phys. 1962, 2, 53–60.
23. Piotrowski, J.; Orman, Z.; Nowak, Z.; Pawluczyk, J.; Pietrzak, J.; Piotrowski, A.; Szabra, D. Uncooled long wave infrared photodetectors with optimized spectral response at selected spectral ranges. Proc. SPIE 2005, 5783, 616–624.
24. Chorier, P.; Tribolet, P. High performance HgCdTe SWIR detectors development at Sofradir. Proc. SPIE 2001,4369, 698–712.
25. Vasilyev, V. V; Ovsyuk, V.N.; Sidorov, Y.G. IR photodetectors based on MBE grown MCT layers. Proc. SPIE2003, 5065, 39–46.
26. Krishna, S. Quantum dots-in-a-well infrared photodetectors. Infrared Phys. Technol. 2005, 47, 153–163.
27. Pal, E.; Towe, D. Uniformity assessment of key characteristics of quantum-dot infrared detectors: A prerequisite for focal plane arrays. J. Appl. Phys. 2006.
28. Stafeev, V. Mercury cadmium telluride-main semiconductor material of modern IR photoelectronics. Proc. SPIE2000, 4340, 240–243.
29. Shi, Y.-L.; Lei, S.-Q.; Yang, T.-F.; Zhuang, J.-S.; Cai, Y. Short wavelength Hg1-xCdxTe infrared detectors prepared by loophole technology. Proc. SPIE 2005, 5964, 596401–596409.
30. Rogalski, A. New material systems for third generation infrared detectors. Proc. SPIE 2009.
31. Razeghi, M.; Hoffman, D.; Nguyen, B.-M.; Delaunay, P.-Y.; Huang, E.K.-W.; Tidrow, M.Z.; Nathan, V. Recent advances in LWIR Type-II InAs/GaSb superlattice photodetectors and focal plane arrays at the center for quantum devices. Proc. IEEE 2009, 97, 1056–1066.
32. Wang, J.; Chen, X.; Hu, W.; Wang, L.; Lu, W.; Xu, F.; Zhao, J.; Shi, Y.; Ji, R. Amorphous HgCdTe infrared photoconductive detector with high detectivity above 200 K. Appl. Phys. Lett.
33. Raghavan, S.; Rotella, P.; Stintz, A.; Fuchs, B.; Krishna, S.; Morath, C.; Cardimona, D.A.; Kennerly, S.W. High-responsivity, normal-incidence long-wave infrared (lambda~7.2 um) InAs/In(0.15)Ga(0.85)As dots-in-a-well detector. Appl. Phys. Lett. 2002, 81, 3.
34. Cabanski, W.A.; Breiter, R.; Koch, R.; Mauk, K.; Rode, W.; Ziegler, J.; Eberhardt, K.; Oelmaier, R.; Schneider, H.; Walther, M. High-resolution focal plane array IR detection modules and digital signal processing technologies at AIM. Proc. SPIE 2000, 4028, 113–121.
35. D'Souza, A.I.; Dawson, L.C.; Staller, C.O.; Reekstin, J.P.; Wijewarnasuriya, P.S.; DeWames, R.E.; McLevige, W.V; Arias, J.M.; Edwall, D.E.; Hildebrandt, G. VLWIR Hg(1-x)Cd(x)Te photovoltaic detectors performance. Proc. SPIE 2000, 4028, 343–352.
36. Wijewarnasuriya, P.S.; Zandian, M.; Phillips, J.; Edwall, D.; DeWames, R.E.; Hildebrandt, G.; Bajaj, J.; Arias, J.M.; D'Souza, A.I.; Moore, F. Advances in large area

Hg(1-x)Cd(x)Te photovoltaic detectors for remote sensing applications. Proc. SPIE 2001, 4454, 10–18.

37. Grudzien, M.; Piotrowski, J. Infrared detectors: View from a high-tech company. Proc. SPIE 2003, 5124, 15–21.

38. Vydyanath, H.R.; Tobin, S.P.; Norton, P.W.; Odette, P.; Nathan, V. HgCdTe photodiodes with state-of-the-art performance in the near infrared spectral region. Proc. SPIE 2001, 4288, 316–334.

39. D'Souza, A.I.; Stapelbroek, M.G.; Robinson, J. HgCdTe and silicon detectors and FPAs for remote sensing applications. Proc. SPIE 2004, 5234, 262–271.

40. Smith, E.P.G.; Bornfreund, R.E.; Kasai, I.; Pham, L.T.; Patten, E.A.; Peterson, J.M.; Roth, J.A.; Nosho, B.Z.; De Lyon, T.J.; Jensen, J.E.; et al. Status of two-color and large format HgCdTe FPA technology at Raytheon vision systems. Proc. SPIE 2006, 6127, 61271.

41. Haiml, M.; Bauer, A.; Bitterlich, H.; Bruder, M.; Hofmann, K.; Lutz, H.; Mai, M.; Nothaft, H.-P.; Ruhlich, I.; Wendler, J.; et al. AIM results for space-qualified HgCdTe photovoltaic detectors from 0.9-um to 13-um spectral range. Proc. SPIE 2006, 6361, 636111–636119.

42. Vuillermet, M.; Pistone, F.; Reibel, Y. Latest developments in MCT for next generation of infrared staring arrays. Proc. SPIE 2009.

43. Crawford, S.; Skivington, T.; Craig, R.; Haining, A.; Costard, E.; Belhaire, E.; Bois, P. Beyond 3rd generation MCT: SXGA QWIP. Proc. SPIE 2005, 5783, 777–788.

44. Rochette, L.; Smithson, T.; Gurgenian, R.; St-Pierre, G. Long wave focal plane array detector: Development performance assessment of an 8×8 sterling cooled photovoltaic MCT detector module. Proc. SPIE 2007, 6661, 666107.

45. Cabanski, W.A.; Eberhardt, K.; Rode, W.; Wendler, J.C.; Ziegler, J.; Fleissner, J.; Fuchs, F.; Rehm, R.H.; Schmitz, J.; Schneider, H.; et al. Third-generation focal plane array IR detection modules and applications. Proc. SPIE2004, 5406, 184–192.

46. Bovina, L.A.; Boltar, K.O.; Bourlakov, I.D.; Golovin, S. V; Ivanov, V.I.; Saginov, L.D.; Stafeev, V.I.; Akimov, V.M.; Iakovleva, N.I.; Sidorov, Y.G. 384 × 288 HgCdTe staring focal plane array. Proc. SPIE 2000, 4340, 23–26.

47. Boltar, K.O.; Bourlakov, I.D.; Golovin, S.V; Lognikov, V.E.; Mansvetov, N.G. LWIR heterodyne HgCdTe detectors. Proc. SPIE 2000, 4340, 81–84.

48. Varavin, V.S.; Gutakovsky, A.K.; Dvoretsky, S.A.; Kartashev, V.A.; Latyshev, A.V; Mikhailov, N.N.; Pridachin, D.N.; Remestnik, V.G.; Rukhlitsky, S.V; Sabinina, I.V; et al. The state of the art and prospects of CdxHg1-xTe molecular beam epitaxy. Proc. SPIE 2003, 5126, 398–406.

49. Lee, T.-S.; Zhao, J.; Chang, Y.; Ashokan, R.; Sivananthan, S.; Boieriu, P.; Chen, Y.; Brill, G.; Wijewarnasuriya, P.S.; Dhar, N.K. MWIR detectors on HgCdTe grown by MBE on 3-in. diameter silicon substrates. Proc. SPIE2004, 5564, 113–122.

50. Bajaj, J. State-of-the-art HgCdTe infrared devices. Proc. SPIE 2000, 3948, 42–54.

51. Wijewarnasuriya, P.S.; Brill, G.; Chen, Y.P.; Dhar, N.K.; Velicu, S. LWIR MBE HgCdTe photovoltaic detectors grown on Si composite substrates. Proc. SPIE 2004, 5406, 323–331.

52. Piotrowski, J.; Rogalski, A. Uncooled long-wavelength infrared photon detectors. Proc. SPIE 2004, 5359, 10–22.

53. Yakushev, M.V; Brunev, D.V; Varavin, V.S.; Vasilyev, V.V; Dvoretskii, S.A.; Marchishin, I.V; Predein, A.V; Sabinina, I.V; Sidorov, Y.G.; Sorochkin, A.V. HgCdTe heterostructures on Si (310) substrates for midinfrared focal plane arrays. Semiconductors 2011, 45, 385–391.

54. Wijewarnasuriya, P.S. Nonequilibrium operation of long wavelength HgCdTe photo detectors for higher operating temperatures. Proc. SPIE 2010.

55. Sizov, F.F.; Vasiliev, V.V; Gumenjuk-Sichevska, J.V; Kamenev, Y.E.; Dvoretsky, S.A.; Mikhailov, N.N. MCT detectors: From IR to sub-mm and mm wavelength bands. Proc. SPIE 2008.

56. Aqariden, F.; Elsworth, J.; Zhao, J.; Grein, C.H.; Sivananthan, S. MBE HgCdTe for HDVIP Devices: Horizontal integration in the US HgCdTe FPA industry. J. Electron. Mater. 2012, 41, 2700–2706.

57. Wenisch, J.; Eich, D.; Lutz, H.; Schallenberg, T.; Wollrab, R.; Ziegler, J. MBE growth of MCT on GaAs substrates at AIM. J. Electron. Mater. 2012, 41, 2828–2832.

58. Wei, Q.; Zhou, W.; Wang, W.; Fu, X.; Yu, M.; Chen, L.; Wu, Y.; He, L. Characterization of multilayered HgCdTe for MW/LW two-color application. Proc. SPIE 2007, 6835, 68351.

59. Wehner, J.G.A.; Musca, C.A.; Sewell, R.H.; Dell, J.M.; Faraone, L. Mercury cadmium telluride resonant-cavity-enhanced photoconductive infrared detectors. Appl. Phys. Lett. 2005, 87, 211101–211104.

60. Besikci, C. III-V infrared detectors on Si substrates. Proc. SPIE 2000, 3948, 31–39.

61. Kimukin, I.; Biyikli, N.; Kartaloglu, T.; Aytur, O.; Ozbay, E. High-speed InSb photodetectors on GaAs for mid-IR applications. IEEE J. Sel. Top. Quantum Electron. 2004, 10, 766–770.

62. Joullie, A.F.; Christol, P.; Rodriguez, J.B.; Ait-Kaci, H.; Chevrier, F.; Nieto, J.; De Anda, F. A survey of new laser and detector structures for 3–5 um mid-infrared spectral range. Proc. SPIE 2004, 5582, 211–221.

63. Ozer, S.; Besikci, C. Assessment of InSb photodetectors on Si substrates. J. Phys. D 2003, 36, 559–563.

64. Razeghi, M. Novel Sb-based alloys for uncooled infrared photodetector applications. Proc. SPIE 2001, 4288, 304–315.

65. Ashley, T.; Baker, I.M.; Burke, T.M.; Dutton, D.T.; Haigh, J.A.; Hipwood, L.G.; Jefferies, R.; Johnson, A.D.; Knowles, P.; Little, J.C. InSb focal plane array (FPAs) grown by molecular beam epitaxy (MBE). Proc. SPIE 2000,4028, 398–403.

66. Akimov, V.M.; Chishko, V.F.; Dirochka, A.I.; Kasatkin, I.L.; Klimanov, Y.A.; Kravtchenko, N.V; Lopukhin, A.A.; Pasekov, V.F. The 128 × 128 and 288 × 384 InSb array photodetective assemblies for 3–5-um spectral range. Proc. SPIE 2003, 5126, 91–97.

67. Hoffman, A.W.; Love, P.J.; Ando, K.J.; Rosbeck, J.P.; Lum, N.A.; Holcombe, R.; Durkee, J.M.; McMurtry, C.W. Development of 2k × 2k FPA InSb modules for the NGST mission. Proc. SPIE 2003, 4850, 836–846.

68. Shkedy, L.; Markovitz, T.; Calahorra, Z.; Hirsh, I.; Shtrichman, I. Megapixel digital InSb detector for midwave infrared imaging. Optical Engineering 2011, 50, 61008.

69. Hoffman, A.W.; Corrales, E.; Love, P.J.; Rosbeck, J.; Merrill, M.; Fowler, A.; McMurtry, C. 2K × 2K InSb for Astronomy. SPIE USE 2004, 5499, 59–67.

70. Nesher, O.; Pivnik, I.; Ilan, E.; Calalhorra, Z.; Koifman, A.; Vaserman, I.; Oiknine Schlesinger, J.; Gazit, R.; Hirsh, I. High resolution 1,280 × 1,024, 15 um pitch compact InSb IR detector with on-chip ADC. Proc. SPIE2009.

71. Kimukin, I.; Biyikli, N.; Ozbay, E. InSb high-speed photodetectors grown on GaAs substrate. J. Appl. Phys. 2003,94, 5414–5416.

72. Camargo, E.G.; Ueno, K.; Morishita, T.; Sato, M.; Endo, H.; Kurihara, M.; Ishibashi, K.; Kuze, N. High-sensitivity temperature measurement with miniaturized InSb mid-IR sensor. IEEE Sens. J. 2007, 7, 1335–1339.

73. Kuze, N.; Camargo, E.G.; Ueno, K.; Morishita, T.; Sato, M.; Kurihara, M.; Endo, H.; Ishibashi, K. High performance miniaturized InSb photovoltaic infrared sensors operating at room temperature. J. Cryst. Growth2007, 301-302, 997–1000.

74. Ueno, K.; Gomes Camargo, E.; Morishita, T.; Sato, M.; Yokoyama, A.; Endo, H.; Yanagita, Y.; Goto, H.; Kuze, N. Miniaturized InSb mid-IR photovoltaic sensor for room temperature operation. Physica Status Solidi C 2008, 5, 2819–2821.

75. Camargo, E.G.; Ueno, K.; Kawakami, Y.; Moriyasu, Y.; Nagase, K.; Satou, M.; Endo, H.; Ishibashi, K.; Kuze, N. Miniaturized InSb photovoltaic infrared sensor operating at room temperature. Opt. Eng. 2008, 47, 14401–14402.

76. Lim, K.P.; Pham, H.T.; Yoon, S.F.; Tan, K.H.; Ngo, C.Y. InSb1-xNx/InSb/GaAs alloys by thermal annealing for midinfrared photodetection. Appl. Phys. Lett. 2010.

77. Lackner, D.; Pitts, O.J.; Martine, M.; Cherng, Y.T.; Mooney, P.M.; Thewalt, M.L.W.; Plis, E.; Watkins, S.P. InAsSb and InPSb Materials for Mid Infrared Photodetectors. Proceedings of the International Conference on Indium Phosphide and Related Materials, Kagawa, Japan, 31 May–4 June 2010; pp. 1–4.

78. Shafir, I.; Katz, M.; Sher, A.; Raizman, A.; Zussman, A.; Nathan, M. Suppression of leakage currents in InAsSb MWIR photodiodes by chemical treatment and illumination. Semicond. Sci. Technol. 2010.

79. Levine, B.F.; Choi, K.K.; Bethea, C.G.; Walker, J.; Malik, R. New 10 um infrared detector using intersubband absorption in resonant tunneling GaAlAs superlattices. Appl. Phys. Lett. 1987, 50, 1092–1094.

80. Ryzhii, V.; Ryzhii, M.; Khmyrova, I.; Suris, R.; Mitin, V.; Shur, M. Quantum well and quantum dot infrared photodetectors: Physics of operation and modeling. Proc. SPIE 2003.

81. Berurier, V.G.A.; Huet, O.; Manissadjian, A.; Facoetti, H.; Marcadet, X.; Carras, M.; Trinité, V.; Nedelcu, A. Quantum well infrared photodetectors: Present and future. Opt. Eng. 2011.

82. Sakoglu, U.; Tyo, J.S.; Hayat, M.M.; Raghavan, S.; Krishna, S. Spectrally adaptive infrared photodetectors with bias-tunable quantum dots. J. Opt. Soc. Amer. B 2004, 21, 7–17.

83. Razeghi, M.; Erdtmann, M.; Jelen, C.; Guastavinos, F.; Brown, G.J.; Park, Y.S. Development of quantum well infrared photodetectors at the center for quantum devices. Infrared Phys. Technol. 2001, 42, 135–148.

84. El-Tokhy, M.; Mahmoud, I.; Konber, H. Comparative study between different quantum infrared photodetectors. Opt. Quantum Electron. 2009, 41, 933–956.

85. Hofstetter, D.; Schad, S.-S.; Wu, H.; Schaff, W.J.; Eastman, L.F. GaN/AlN-based quantum-well infrared photodetector for 1.55 m. Appl. Phys. Lett. 2003, 83, 572–574.

86. Guillot, F.; Monroy, E.; Leconte, S.; Bellet-Amalric, E.; Baumann, E.; Giorgetta, F.; Hofstetter, D.; Nevou, L.; Tchernycheva, M.; Doyennette, L.; et al. MBE growth of nitride-based photovoltaic intersubband detectors. Superlattice. Microstruct. 2006, 40, 418–425.

87. Sherliker, B.; Halsall, M.; Kasalynas, I.; Seliuta, D.; Valusis, G.; Vengris, M.; Barkauskas, M.; Sirutkaitis, V.; Harrison, P.; Jovanovic, V.D.; et al. Room temperature operation of AlGaN/GaN quantum well infrared photodetectors at a 3–4 um wavelength range. Semicond. Sci. Technol. 2007, 22, 1240–1244.

88. Stiff-Roberts, K.S.; Bhattacharya, P.; Kennerly, S. Low-bias, high-temperature performance of a normal incidence InAs/GaAs vertical quantum dot infrared photodetector with a current blocking barrier. J. Vac. Sci. Technol. B 2002, 20, 3.

89. Attaluri, R. S.; Annamalai, S.; Posani, K. T.; Stintz, A.; Krishna, S. Influence of Si doping on the performance of quantum dots-in-well photodetectors. J. Vac. Sci. Technol. B 2006, 24, 3.

90. Tidrow, M.Z. Device physics and state-of-the-art of quantum well infrared photodetectors and arrays. Mater. Sci. Eng. B 2000, B74, 45–51.

91. Gunapala, S.D.; Bandara, S.V; Liu, J.K.; Luong, E.M.; Stetson, N.; Shott, C.A.; Bock, J.J.; Rafol, S.B.; Mumolo, J.M.; McKelvey, M.J. Long-wavelength 256 × 256 GaAs/AlGaAs quantum well infrared photodetector (QWIP) palm-size camera. IEEE Trans. Electron. Devices 2000, 47, 326–332.

92. Li, S.S.; Kim, S.-H.; Moon, J.-H.; Lee, J.H. A two-stack, multi-color quantum well infrared photodetector for mid- and long-wavelength infrared detection. Infrared Phys. Technol. 2003, 44, 235–241.

93. Perera, A.G.U.; Matsik, S.G.; Liu, H.C.; Gao, M.; Buchanan, M.; Schaff, W.J.; Yeo, W. GaAs/InGaAs quantum well infrared photodetector with a cutoff wavelength at 35 m. Appl. Phys. Lett. 2000, 77, 741–743.

94. Perera, A.G.U.; Matsik, S.G.; Liu, H.C.; Gao, M.; Buchanan, M.; Schaff, W.J.; Yeo, W. 35 um cutoff

bound-to-quasibound and bound-to-continuum InGaAs QWIPs. Infrared Phys. Technol. 2001, 42, 157–162.

95. Lee, J.-H.; Li, S.S.; Tidrow, M.Z.; Liu, W.K. Investigation of multi-color, broadband quantum well infrared photodetectors with digital graded superlattice barrier and linear-graded barrier for long wavelength infrared applications. Infrared Phys. Technol. 2001, 42, 123–134.

96. Alves, F.D.P.; Santos, R.A.T.; Nohra, L.F.M.; Magalhaes, L.B.; Karunasiri, G. Tunable NIR Quantum Well Infrared Photodetector Using Interband Transitions. Proceedings of the 2009 SBMO/IEEE MTT-S International Microwave and Optoelectronics Conference (IMOC), Belem, Brazil, 3–6 November 2009; pp. 303–306.

97. Rehm, R.; Schneider, H.; Schwarz, K.; Walther, M.; Koidl, P.; Weimann, G. Responsivity and gain of InGaAs/GaAs-QWIPs and GaAs/AlGaAs-QWIPS: A comparative study. Proc. SPIE 2001, 4288, 379–385.

98. Budkin, I.V.; Bulaev, P.V.; Vacilevskay, L.M.; Zalevsky, I.D.; Kuznetsov, U.A.; Kulikov, V.B.; Marmalyuk, A.A.; Nikitin, D.B.; Padalitsa, A.A.; Petrovsky, A.V. 3–5 um InGaAs/AlGaAs QWIP heterostructure growth by metal organic chemical vapor deposition (IR photodetector applications). Proc. SPIE 2007, 5126, 178–180.

99. Bahir, G.; Gusakov, Y.; Maimon, S.; Cohen, G.; Ritter, D.; Finkman, E. InP based QWIPs for long- and mid-wavelength band detection. Proc. SPIE 2003, 4820, 663–670.

100. Arora, B.M.; Majumdar, A.; Shah, A.P.; Gokhale, M.R.; Ghosh, S.; Bhattacharya, A.; Sengupta, D. Characteristics of high responsivity 8.5 um InGaAs/InP QWIPs grown by metalorganic vapour phase epitaxy. Infrared Phys. Technol. 2007, 50, 206–210.

101. Zhang, D.H.; Sun, L.; Yuan, K.H.; Yoon, S.F.; Radhakrishnan, K. InGaAsP/InP long wavelength quantum well infrared photodetectors. Thin Solid Films 2007, 515, 4450–4453.

102. Liu, H.C.; Song, C.Y.; Shen, A.; Gao, M.; Dupont, E.; Poole, P.J.; Wasilewski, Z.R.; Buchanan, M.; Wilson, P.H.; Robinson, B.J. Dual-band photodetectors based on interband and intersubband transitions. Infrared Phys. Technol. 2001, 42, 163–170.

103. Razeghi, M.; Jiang, J.; Jelen, C.; Brown, G.J. GaInAs/InP quantum-well infrared photodetectors grown on Si substrates. Proc. SPIE 2001, 4454, 78–84.

104. Jiang, J.; Jelen, C.; Razeghi, M.; Brown, G.J. High detectivity GaInAs-InP quantum-well infrared photodetectors grown on Si substrates. IEEE Photon. Technol. Lett. 2002, 14, 372–374.

105. Fu, L.; Tan, H.H.; Jagadish, C.; Li, N.; Li, N.; Liu, X.; Lu, W.; Shen, S.C. Tuning of detection wavelength of quantum-well infrared photodetectors by quantum-well intermixing. Infrared Phys. Technol. 2001, 42, 171–175.

106. Hsu, M.C.; Kuan, C.H.; Wang, S.Y. Multicolor infrared detection using two stacks of superlattice structures in a back-to-back configuration. Appl. Phys. Lett. 2000, 77, 3.

107. Oogarah, T.; Liu, H.C.; Dupont, E.; Wasilewski, Z.R.; Byloos, M.; Buchanan, M. High absorption GaAs/AlGaAs quantum well infrared photodetectors. Semicond. Sci. Technol. 2002, 17, 41–43.

108. Sim, S.K.H.; Liu, H.C.; Shen, A.; Gao, M.; Lee, K.F.; Buchanan, M.; Ohno, Y.; Ohno, H.; Li, E.H. Effect of barrier width on the performance of quantum well infrared photodetector. Infrared Phys. Technol. 2001, 42, 115–121.

109. Lee, J.-H.; Chang, C.-Y.; Li, C.-H.; Lin, S.-Y.; Lee, S.-C. Performance improvement of AlGaAs/GaAs QWIP by NH3 plasma treatment. IEEE J. Quantum Electron. 2012, 48, 922–926.

110. Liu, H.C.; Song, C.Y.; Spring Thorpe, A.J.; Cao, J.C. Terahertz quantum-well photodetector. Appl. Phys. Lett.2004, 84, 4068–4070.

111. Mei, T.; Li, H.; Karunasiri, G.; Fan, W.J.; Zhang, D.H.; Yoon, S.F.; Yuan, K.H. Normal incidence silicon doped p-type GaAs/AlGaAs quantum-well infrared photodetector on (111) A substrate. Infrared Phys. Technol. 2007, 50, 119–123.

112. Graf, M.; Dupont, E.; Luo, H.; Haffouz, S.; Wasilewski, Z.R.; Thorpe, A.J.S.; Ban, D.; Liu, H.C. Terahertz quantum well infrared detectors. Infrared Phys. Technol. 2009, 52, 289–293.

113. Ferrini, R.; Guizzetti, G.; Patrini, M.; Franchi, S. Optical study of intersubband transitions in GaSb/AlGaSb systems for QWIPs. Opt. Mater. 2001, 17, 351–354.

114. Chen, C.C. Performance and application of a superlattice infrared photodetector with a blocking barrier. J. Appl. Phys. 2002, 91, 6.

115. Costard, E.; Nedelcu, A.; Marcadet, X.; Belhaire, E.; Bois, P. QWIP development status at thales research and technology. Proc. SPIE 2006, 6206, 62060–62061.

116. Gueriaux, V.; Nedelcu, A.; Carras, M.; Huet, O.; Marcadet, X.; Bois, P. Mid-wave QWIPs for the (3–4.2 m) atmospheric window. Infrared Phys. Technol. 2009, 52, 235–240.

117. Cellek, O.O.; Ozer, S.; Besikci, C. High responsivity InP-InGaAs quantum-well infrared photodetectors: Characteristics and focal plane array performance. IEEE J. Quantum Electron. 2005, 41, 980–985.

118. Eker, S.U.; Arslan, Y.; Onuk, A.E.; Besikci, C. High conversion efficiency InP/InGaAs strained quantum well infrared photodetector focal plane array with 9.7 um cut-off for high-speed thermal imaging. IEEE J. Quantum Electron. 2010, 46, 164–168.

119. Gunapala, S.D.; Bandara, S.V; Singh, A.; Liu, J.K.; Rafol, B.; Luong, E.M.; Mumolo, J.M.; Tran, N.Q.; Ting, D.Z.-Y.; Vincent, J.D. 640 × 486 long-wavelength two-color GaAs/AlGaAs quantum well infrared photodetector (QWIP) focal plane array camera. IEEE Trans. Electron. Devices 2000, 47, 963–971.

120. Gunapala, S.D.; Bandara, S.V; Singh, A.; Liu, J.K.; Luong, E.M.; Mumolo, J.M.; LeVan, P.D. Recent developments and applications of quantum well infrared photodetector focal plane arrays. Physica E 2000, 7, 108–111.

121. Bandara, S.V; Gunapala, S.D.; Liu, J.K.; Rafol, S.B.; Ting, D.Z.; Mumolo, J.M.; Reininger, F.M.; Fastenau,

J.M.; Liu, A.K. Large format, broadband and multicolor GaAs/AlGaAs quantum well infrared photodetector (QWIP) focal plane arrays. Proc. SPIE 2001, 4454, 30–39.

122. Rafol, S.B.; Gunapala, S.D.; Bandara, S.V; Liu, J.K.; Ting, D.Z.; Mumolo, J.M.; Reininger, F.M. Characterization of QWIP (10–16 um) broadband FPA. Proc. SPIE 2003, 4820, 273–281.

123. Schneider, H.; Fleissner, J.; Rehm, R.; Walther, M.; Pletschen, W.; Koidl, P.; Weimann, G.; Ziegler, J.; Breiter, R.; Cabanski, W. High-resolution QWIP FPAs for the 8–12 um and 3–5 um regimes. Proc. SPIE 2003, 4820, 297–305.

124. Jhabvala, M.; Choi, K.; Goldberg, A.C.; La, A.T.; Gunapala, S.D. Development of a 1K × 1K GaAs QWIP far IR imaging array. Proc. SPIE 2004, 5167, 175–185.

125. Ozer, S.; Tumkaya, U.; Besikci, C. Large format AlInAs-InGaAs quantum-well infrared photodetector focal plane array for midwavelength infrared thermal imaging. IEEE Photon. Technol. Lett. 2007, 19, 1371–1373.

126. Jhabvala, M.; Choi, K.K.; Monroy, C.; La, A.; Adams, J.; Devitt, J.; Forrai, D.; Endres, D. Development of a 1 megapixel long IR QWIP focal plane array. Proc. SPIE 2007, 6542, 65420–65421.

127. Gunapala, S.D.; Bandara, S.V; Liu, J.K.; Mumolo, J.M.; Hill, C.J.; Ting, D.Z.; Kurth, E.; Woolaway, J.; LeVan, P.D.; Tidrow, M.Z. Development of megapixel dual-band QWIP focal plane array. Proc. SPIE 2008, 6940, 69401–69402.

128. Arslan, Y.; Çolakoglu, T.; Torunoglu, G.; Aktas, O.; Besikci, C. Enhanced performance QWIP FPAs. Infrared Phys. Technol. 2012. in press.

129. Hinds, S.; Buchanan, M.; Dudek, R.; Haffouz, S.; Laframboise, S.; Wasilewski, Z.; Liu, H.C. Near-room-temperature mid-infrared quantum well photodetector. Adv. Mater. 2011, 23, 5536–5539.

130. Bulman, G.E.; Myers, D.R.; Wiczer, J.J.; Dawson, L.R.; Biefeld, R.M.; Zipperian, T.E. Ion implanted lateral In(.2)Ga(.8)As/GaAs strained-layer superlattice photodetector. IEEE Electron. Devices Meet. 1984, 30, 719–722.

131. Nguyen, B.-M.; Chen, G.; Hoang, A.M.; Abdollahi Pour, S.; Bogdanov, S.; Razeghi, M. Effect of contact doping in superlattice-based minority carrier unipolar detectors. Appl. Phys. Lett. 2011, 99, 033501.

132. Rehm, R.; Masur, M.; Schmitz, J.; Daumer, V.; Niemasz, J.; Vandervelde, T.; DeMeo, D.; Luppold, W.; Wauro, M.; Wörl, A.; et al. InAs/GaSb superlattice infrared detectors. Infrared Phys. Technol. 2013. in press.

133. Brown, G.J. Type-II InAs/GaInSb superlattices for infrared detection: An overview. Proc. SPIE 2005, 5783, 65–77.

134. Haugan, H.J.; Szmulowicz, F.; Brown, G.J.; Ullrich, B.; Munshi, S.R.; Wickett, J.C.; Stokes, D.W. Short-period InAs/GaSb superlattices for mid-infrared photodetectors. Physica status solidi 2007, 4, 1702–1706.

135. Nathan, V.; Anselm, A.; Lin, C.H.; Johnson, J. InAs/InGaSb type-2 strained layer superlattice IR detectors. Proc. SPIE 2002, 4650, 185–198.

136. Khoshakhlagh, A.; Plis, E.; Myers, S.; Sharma, Y.D.; Dawson, L.R.; Krishna, S. Optimization of InAs/GaSb type-II superlattice interfaces for long-wave (~8 mm) infrared detection. J. Cryst. Growth 2009, 311, 1901–1904.

137. Lackner, D.; Pitts, O.J.; Najmi, S.; Sandhu, P.; Kavanagh, K.L.; Yang, A.; Steger, M.; Thewalt, M.L.W.; Wang, Y.; McComb, D.W.; et al. Growth of InAsSb/InAs MQWs on GaSb for mid-IR photodetector applications. J. Cryst. Growth 2009, 311, 3563–3567.

138. Lackner, D.; Pitts, O.J.; Steger, M.; Yang, A.; Thewalt, M.L.W.; Watkins, S.P. Strain balanced InAs/InAsSb superlattice structures with optical emission to 10 m. Appl. Phys. Lett. 2009, 95, 081906.

139. Lackner, D.; Steger, M.; Thewalt, M.L.W.; Pitts, O.J.; Cherng, Y.T.; Watkins, S.P.; Plis, E.; Krishna, S. InAs/InAsSb strain balanced superlattices for optical detectors: Material properties and energy band simulations. J. Appl. Phys. 2012, 111, 034507.

140. Rodriguez, J.B.; Plis, E.; Bishop, G.; Sharma, Y.D.; Kim, H.; Dawson, L.R.; Krishna, S. NBn structure based on InAs/GaSb type-II strained layer superlattices. Appl. Phys. Lett. 2007.

141. Lu, J.-H.; Wu, K.-J.; Hsieh, K.-J.; Kuan, C.-H.; Feng, J.-Y.; Lay, T.-S.; Yang, C.-W.; Tu, S.-L. A superlattice infrared photodetector integrated with multiple quantum wells to improve the performance. IEEE J. Quantum Electron 2007, 43, 72–77.

142. Vurgaftman, I.; Canedy, C.L.; Jackson, E.M.; Nolde, J.A.; Affouda, C.A.; Aifer, E.H.; Meyer, J.R.; Hood, A.; Evans, A.J.; Tennant, W.T. Analysis and performance of type-II superlattice infrared photodetectors. Opt. Eng.2011, 50, 061007.

143. Brown, G.J.; Szmulowicz, F. InAs/InGaSb superlattices for very long wavelength infrared detection. Proc. SPIE2001, 4288, 200–208.

144. Plis, E.; Lee, S.J.; Zhu, Z.; Amtout, A.; Krishna, S. InAs/GaSb superlattice detectors operating at room temperature. IEEE J. Sel. Top. Quantum Electron. 2006, 12, 1269–1274.

145. Zhang, Y.; Ma, W.; Cao, Y.; Huang, J.; Wei, Y.; Cui, K.; Shao, J. Long wavelength infrared InAs/GaSb superlattice photodetectors with InSb-Like and mixed interfaces. IEEE J. Quantum Electron. 2011, 47, 1475–1479.

146. Plis, E.; Rodriguez, J.B.; Balakrishnan, G.; Sharma, Y.D.; Kim, H.S.; Rotter, T.; Krishna, S. Mid-infrared InAs/GaSb strained layer superlattice detectors with NBn design grown on a GaAs substrate. Semicond. Sci. Technol. 2010.

147. Haugan, H.J.; Brown, G.J.; Grazulis, L.; Mahalingam, K.; Tomich, D.H. Optimization of InAs/GaSb type-II superlattices for high performance of photodetectors. Physica E 2004, 20, 527–530.

148. Mohseni, H.; Tahraoui, A.; Wojkowski, J.S.; Razeghi, M.; Mitchel, W.; Saxler, A. Growth and characterization

of very long wavelength type-II infrared detectors. Proc. SPIE 2000, 3948, 145–152.

149. Mohseni, H.; Wojkowski, J.S.; Tahraoui, A.; Razeghi, M.; Brown, G.; Mitchel, W. Growth and characterization of type-II nonequilibrium photovoltaic detectors for long wavelength infrared range. Proc. SPIE 2000, 3948, 153–160.

150. Razeghi, M.; Wei, Y.; Gin, A.; Brown, G.J.; Johnstone, D. Type II InAs/GaSb superlattices and detectors with lambda c >18 um. Proc. SPIE 2002, 4650, 111–116.

151. Wei, Y.; Gin, A.; Razeghi, M.; Brown, G.J. Type II InAs/GaSb superlattice photovoltaic detectors with cutoff wavelength approaching 32 um. Appl. Phys. Lett. 2002, 81, 3675–3677.

152. Wong, A.F.; Nelson, M.J.; Plis, E.A.; Skrutskie, M.F.; Yao, L.; Vandervelde, T.; Krishna, S.; Kim, H.; Khoshakhlagh, A.; Myers, S.A. Characterization of multicolor type-II InAs/GaSb strained-layer superlattice photodetectors for use in astronomical observation. Proc. SPIE 2010.

153. Zhang, Y.; Ma, W.; Wei, Y.; Cao, Y.; Huang, J.; Cui, K.; Guo, X. Narrow-band long-/very-long wavelength two-color type-II InAs/GaSb superlattice photodetector by changing the bias polarity. Appl. Phys. Lett. 2012.

154. Razeghi, M.; Wei, Y.; Bae, J.; Gin, A.; Hood, A.; Jiang, J.; Nah, J. Type II InAs/GaSb superlattices for high-performance photodiodes and FPAs. Proc. SPIE 2003, 5246, 501–511.

155. Wei, Y.; Hood, A.; Gin, A.; Yazdanpanah, V.; Razeghi, M.; Tidrow, M. High performance LWIR Type II InAs/GaSb superlattice photodetectors and infrared focal plane arrays. Proc. SPIE 2005, 5732, 309–315.

156. Rodriguez, J.B.; Christol, P.; Chevrier, F.; Joullié, A. Optical characterization of symmetric InAs/GaSb superlattices for detection in the 3–5 μm spectral region. Physica E 2005, 28, 128–133.

157. Wei, Y.; Hood, A.; Yau, H.; Gin, A.; Razeghi, M.; Tidrow, M.Z.; Nathan, V. Uncooled operation of type-II InAs/GaSb superlattice photodiodes in the midwavelength infrared range. Appl. Phys. Lett. 2005, 86, 1–3.

158. Zhu, Z.M.; Bhattacharya, P.; Plis, E.; Su, X.H.; Krishna, S. Low dark current InAs/GaSb type-II superlattice infrared photodetectors with resonant tunnelling filters. J. Phys. D. 2006, 39, 4997–5001.

159. Haugan, H.J.; Brown, G.J.; Szmulowicz, F.; Grazulis, L.; Mitchel, W.C.; Elhamri, S.; Mitchell, W.D. InAs/GaSb type-II superlattices for high performance mid-infrared detectors. J. Cryst. Growth 2005, 278, 198–202.

160. Haugan, H.J.; Brown, G.J.; Mahalingam, K.; Elhamri, S.; Mitchel, W.C.; Grazulis, L.; Shank, J.M.; Houston, S. Growth of short-period InAs/GaSb superlattices for mid-infrared photodetectors. Proc. SPIE 2006.

161. Rodriguez, J.B.; Plis, E.; Lee, S.J.; Kim, H.; Bishop, G.; Sharma, Y.D.; Dawson, L.R.; Krishna, S.; Jones, C. Type-II InAs/GaSb strain layer superlattice detectors for higher operating temperatures. Proc. SPIE 2007, 6542, 654201–654208.

162. Delaunay, P.-Y.; Nguyen, B.M.; Hoffman, D.; Razeghi, M. 320 × 256 infrared focal plane array based on type II InAs/GaSb superlattice with a 12 um cutoff wavelength. Proc. SPIE 2007, 6542, 654204.

163. Plis, E.; Kim, H.S.; Rodriguez, J.B.; Bishop, G.D.; Sharma, Y.D.; Khoshakhlagh, A.; Dawson, L.R.; Bundas, J.; Cook, R.; Burrows, D. NBn based infrared detectors using type-II InAs/(In,Ga)Sb superlattices. Proc. SPIE 2008,6940, 69400–69401.

164. Huang, E.K.-W.; Delaunay, P.-Y.; Nguyen, B.-M.; Pour, S.A.; Razeghi, M. Photovoltaic MWIR type-II superlattice focal plane array on GaAs substrate. IEEE J. Quantum Electron. 2010, 46, 1704–1708.

165. Ting, D.Z.; Soibel, A.; Nguyen, J.; Hoglund, L.; Khoshakhlagh, A.; Rafol, S.B.; Keo, S.A.; Liao, A.; Mumolo, J.M.; Liu, J.K. Type II superlattice barrier infrared detector. Proc. SPIE 2011.

166. Guo, J.; Peng, Z.; Sun, W.; Xu, Y.; Zhou, Z.; Niu, Z. InAs/GaSb superlattices for photodetection in short wavelength infrared range. Infrared Phys. Technol. 2009, 52, 124–126.

167. Chen, J.; Xu, Q.; Zhou, Y.; Jing, J.; Lin, C.; He, L. Growth and fabrication of InAs/GaSb type II superlattice mid-wavelength infrared photodetectors. Nanoscale Res. Lett. 2011, 6, 1–14.

168. Jianliang, H.; Ma, W.; Wei, Y.; Zhang, Y.; Cui, K.; Cao, Y.; Guo, X.; Shao, J. How to use type II InAs/GaSb superlattice structure to reach detection wavelength of 2–3 μm. IEEE J. Quantum Electron. 2012, 48, 1322–1326.

169. Salihoglu, O.; Muti, A.; Kutluer, K.; Tansel, T.; Turan, R.; Kocabas, C.; Aydinli, A. Atomic layer deposited Al2O3 passivation of type II InAs/GaSb superlattice photodetectors. J. Appl. Phys. 2012.

170. Brown, G.J.; Szmulowicz, F.; Linville, R.; Saxler, A.; Mahalingham, K.; Lin, C.-H.; Kuo, C.H.; Hwang, W.Y. Type-II superlattice photodetector on a compliant GaAs substrate. IEEE Photon. Technol. Lett. 2000, 12, 684–686.

171. Anselm, K.A.; Ren, H.; Vilela, M.; Zheng, J.; Lin, C.H.; Nathan, V.; Brown, G. Long-wavelength infrared InAs/InGaSb type-II superlattice photovoltaic detectors. Proc. SPIE 2001, 4288, 183–190.

172. Patrashin, M.; Hosako, I.; Akahane, K. Type-II InAs/GaInSb superlattices for terahertz range photodetectors. Proc. SPIE 2011.

173. Fuchs, F.; Burkle, L.; Hamid, R.; Herres, N.; Pletschen, W.; Sah, R.E.; Kiefer, R.; Schmitz, J. Optoelectronic properties of photodiodes for the mid- and far-infrared based on the InAs/GaSb/AlSb materials family. Proc. SPIE 2001, 4288, 171–182.

174. Aifer, E.H.; Tischler, J.G.; Warner, J.H.; Vurgaftman, I.; Bewley, W.W.; Meyer, J.R.; Kim, J.C.; Whitman, L.J.; Canedy, C.L.; Jackson, E.M. W-structured type-II superlattice long-wave infrared photodiodes with high quantum efficiency. Appl. Phys. Lett. 2006.

175. Razeghi, M.; Nguyen, B.-M. Band gap tunability of Type II Antimonide-based superlattices. Phys. Procedia 2010,3, 1207–1212.

176. Nguyen, B.-M.; Razeghi, M.; Nathan, V.; Brown, G.J. Type-II "M" structure photodiodes: An alternative material design for mid-wave to long wavelength infrared regimes. Proc. SPIE 2007, 6479, 64790S.
177. Nguyen, B.-M.; Hoffman, D.; Delaunay, P.-Y.; Razeghi, M. Dark current suppression in type II InAsGaSb superlattice long wavelength infrared photodiodes with M-structure barrier. Appl. Phys. Lett. 2007.
178. Razeghi, M.; Pour, S.A.; Huang, E.K.-W.; Chen, G.; Haddadi, A.; Nguyen, B.-M.B.-M.; Delaunay, P.-Y.; Abdollahi Pour, S. High-operating temperature MWIR photon detectors based on Type II InAs/GaSb superlattice. Proc. SPIE 2011, 8012, 80122Q.
179. Savich, G.R.; Pedrazzani, J.R.; Maimon, S.; Wicks, G.W. Use of epitaxial unipolar barriers to block surface leakage currents in photodetectors. Physica Status Solidi C. 2010, 7, 2540–2543.
180. Savich, G.R.; Pedrazzani, J.R.; Sidor, D.E.; Maimon, S.; Wicks, G.W. Dark current filtering in unipolar barrier infrared detectors. Appl. Phys. Lett. 2011, 99, 121112.
181. DeMeo, D.F.; Vandervelde, T.E. Simulations of Gallium Antimonide (GaSb) p-B-n Thermophotovoltaic Cells. MRS Proc. 2011.
182. Savich, G.R.; Pedrazzani, J.R.; Sidor, D.E.; Wicks, G.W. Benefits and limitations of unipolar barriers in infrared photodetectors. Infrared Phys. Technol. 2013. in press.
183. Khoshakhlagh, A.; Rodriguez, J.B.; Plis, E.; Bishop, G.D.; Sharma, Y.D.; Kim, H.S.; Dawson, L.R.; Krishna, S. Bias dependent dual band response from InAsGa (In) Sb type II strain layer superlattice detectors. Appl. Phys. Lett. 2007, 91, 263504.
184. Nguyen, B.-M.; Bogdanov, S.; Pour, S.A.; Razeghi, M. Minority electron unipolar photodetectors based on type II InAs/GaSb/AlSb superlattices for very long wavelength infrared detection. Appl. Phys. Lett. 2009, 95, 183502.
185. Khoshakhlagh, A.; Myers, S.; Kim, H.; Plis, E.; Gautam, N.; Lee, S.J.; Noh, S.K.; Dawson, L.R.; Krishna, S. Long-wave InAs/GaSb superlattice detectors based on nBn and pin designs. IEEE J. Quantum Electron. 2010, 46, 959–964.
186. Wicks, G.W.; Savich, G.R.; Pedrazzani, J.R.; Maimon, S. Infrared detector epitaxial designs for suppression of surface leakage current. Proc. SPIE 2010, 7608, 760822.
187. Klipstein, P.; Klin, O.; Grossman, S.; Snapi, N.; Lukomsky, I.; Yassen, M.; Aronov, D.; Berkowitz, E.; Glozman, A.; Magen, O. High operating temperature XBn-InAsSb bariode detectors. Proc. SPIE 2012.
188. Klipstein, P. "XBn" barrier photodetectors for high sensitivity and high operating temperature infrared sensors. Proc. SPIE 2008.
189. Kim, H.S.; Cellek, O.O.; Lin, Z.-Y.; He, Z.-Y.; Zhao, X.-H.; Liu, S.; Li, H.; Zhang, Y.-H. Long-wave infrared nBn photodetectors based on InAs/InAsSb type-II superlattices. Appl. Phys. Lett. 2012, 101, 161114.
190. Arakawa, Y.; Sakaki, H. Multidimensional quantum well laser and temperature dependence of its threshold current. Appl. Phys. Lett. 1982, 40, 939–941.
191. Leonard, D.; Krishnamurthy, M.; Reaves, C.M.; Denbaars, S.P.; Petroff, P.M. Direct formation of quantum-sized dots from uniform coherent islands of InGaAs on GaAs surfaces. Appl. Phys. Lett. 1993, 63, 3203–3205.
192. Kim, S.M. Review on recent development of quantum dots: From optoelectronic devices to novel biosensing applications. Proc. SPIE 2003, 4999, 1117.
193. Asano, T.; Chong, H.; Yi, Z.; Mingguo, L.; Campbell, J.C.; Madhukar, A. Design consideration and demonstration of resonant-cavity-enhanced quantum dot infrared photodetectors in mid-infrared wavelength regime. IEEE J. Quantum Electron. 2010, 46, 1484–1491.
194. Kubis, A.J.; Vandervelde, T.E.; Bean, J.C.; Dunn, D.N.; Hull, R. Analysis of the three-dimensional ordering of epitaxial Ge quantum dots using focused ion beam tomography. Appl. Phys. Lett. 2006, 88, 263101–263103.
195. Vandervelde, T.E.; Sun, K.; Merz, J.L.; Kubis, A.; Hull, R.; Pernell, T.L.; Bean, J.C. The effect of two-temperature capping on germanium/silicon quantum dots and analysis of superlattices so composed. J. Appl. Phys. 2006, 99, 124301.
196. Vines, P.; Tan, C.H.; David, J.P.R.; Attaluri, R.S.; Vandervelde, T.E.; Krishna, S. Multiple stack quantum dot infrared photodetectors. Proc. SPIE 2008.
197. Ting, D.Z.-Y.; Bandara, S.V; Gunapala, S.D.; Mumolo, J.M.; Keo, S.A.; Hill, C.J.; Liu, J.K.; Blazejewski, E.R.; Rafol, S.B.; Chang, Y.-C. Submonolayer quantum dot infrared photodetector. Appl. Phys. Lett. 2009, 94, 111107.
198. Shao, J.; Vandervelde, T.E.; Barve, A.; Jang, W.; Stintz, A.; Krishna, S. Barrier engineered superlattice and quantum dot detectors for HOT operation. Proc. SPIE 2011.
199. Vines, P.; Tan, C.H.; David, J.P.R.; Attaluri, R.S.; Vandervelde, T.E.; Krishna, S.; Jang, W.-Y.; Hayat, M.M. Quantum dot infrared photodetectors with highly tunable spectral response for an algorithm-based spectrometer. Proc. SPIE 2010.
200. Benisty, H.; Sotomayor-Torres, C.M.; Weisbuch, C. Intrinsic mechanism for the poor luminescence properties of quantum-box systems. Phys. Rev. B. 1991, 44, 10945–10948.
201. Krishna, S.; Raghavan, S.; Von Winckel, G.; Rotella, P.; Stintz, A.; Morath, C.P.; Le, D.; Kennerly, S.W. Two color InAs/InGaAs dots-in-a-well detector with background-limited performance at 91 K. Appl. Phys. Lett. 2003, 82, 2574–2576.
202. Ryzhii, V.; Khmyrova, I.; Mitin, V.; Stroscio, M.; Willander, M. On the detectivity of quantum-dot infrared photodetectors. Appl. Phys. Lett. 2001.
203. Chen, Z.H.; Baklenov, O.; Kim, E.T.; Mukhametzhanov, I.; Tie, J.; Madhukar, A.; Ye, Z.; Campbell, J.C. InAs/AlxGa1-xAs quantum dot infrared photodetectors with undoped active region. Infrared Phys. Technol.2001, 42, 479–484.
204. Lo, M.C.; Wang, S.Y.; Ling, H.S.; Lee, C.P. Vertically coupled quantum-dot infrared photodetectors. IEEE Photon. Technol. Lett. 2010, 22, 796–798.

205. Lin, W.-H.; Lin, S.-Y. Broadband InGaAs-capped InAs/GaAs quantum-dot infrared photodetector with Bi-modal dot height distributions. J. Appl. Phys. 2012, 112, 034508.

206. Stiff-Roberts, A.D.; Chakrabarti, S.; Pradhan, S.; Kochman, B.; Bhattacharya, P. Raster-scan imaging with normal-incidence, midinfrared InAs/GaAs quantum dot infrared photodetectors. Appl. Phys. Lett. 2002, 80, 3265–3267.

207. Sengupta, S.; Kim, J.O.; Barve, A.V.; Adhikary, S.; Sharma, Y.D.; Gautam, N.; Lee, S.J.; Noh, S.K.; Chakrabarti, S.; Krishna, S. Sub-monolayer quantum dots in confinement enhanced dots-in-a-well heterostructure. Appl. Phys. Lett. 2012, 100, 191111.

208. Chakrabarti, S.; Stiff-Roberts, A.D.; Bhattacharya, P.; Gunapala, S.; Bandara, S.; Rafol, S.B.; Kennerly, S.W. High-temperature operation of InAs-GaAs quantum-dot infrared photodetectors with large responsivity and detectivity. IEEE Photon. Technol. Lett. 2004, 16, 1361–1363.

209. Aivaliotis, P.; Wilson, L.R.; Zibik, E.A.; Cockburn, J.W.; Steer, M.J.; Liu, H.Y. Enhancing the dot density in quantum dot infrared photodetectors via the incorporation of antimony. Appl. Phys. Lett. 2007.

210. Tang, S.-F.; Chiang, C.-D.; Weng, P.-K.; Gau, Y.-T.; Luo, J.-J.; Yang, S.-T.; Shih, C.-C.; Lin, S.-Y.; Lee, S.-C. High-temperature operation normal incident 256 × 256 InAs-GaAs quantum-dot infrared photodetector focal plane array. IEEE Photon. Technol. Lett. 2006, 18, 986–988.

211. Drozdowicz-Tomsia, K.; Goldys, E.M.; Fu, L.; Jagadish, C. Doping effect on dark currents in In(0.5)Ga(0.5)As/GaAs quantum dot infrared photodetectors grown by metal-organic chemical vapor deposition. Appl. Phys. Lett. 2006, 89, 1023–1025.

212. Fu, L.; Li, Q.; Kuffner, P.; Jolley, G.; Gareso, P.; Tan, H.H.; Jagadish, C. Two-color InGaAs/GaAs quantum dot infrared photodetectors by selective area interdiffusion. Appl. Phys. Lett. 2008, 93, 13501–13504.

213. Chakrabarti, S.; Adhikary, S.; Halder, N.; Aytac, Y.; Perera, A.G.U. High-performance, long-wave (~10.2 um) InGaAs/GaAs quantum dot infrared photodetector with quaternary In0.21Al0.21Ga0.58As capping. Appl. Phys. Lett. 2011, 99, 181102.

214. Jiang, J.; Tsao, S.; O'Sullivan, T.; Zhang, W.; Lim, H.; Sills, T.; Mi, K.; Razeghi, M.; Brown, G.J.; Tidrow, M.Z. High detectivity InGaAs/InGaP quantum-dot infrared photodetectors grown by low pressure metalorganic chemical vapor deposition. Appl. Phys. Lett. 2004, 84, 2166–2168.

215. Jiang, J.; Mi, K.; Tsao, S.; Zhang, W.; Lim, H.; O'Sullivan, T.; Sills, T.; Razeghi, M.; Brown, G.J.; Tidrow, M.Z. Demonstration of a 256 × 256 middle-wavelength infrared focal plane array based on InGaAs/InGaP quantum dot infrared photodetectors. Appl. Phys. Lett. 2004, 84, 2232–2234.

216. Zhang, W.; Lim, H.-C.; Taguchi, M.; Tsao, S.; Szafraniec, J.; Movaghar, B.; Razeghi, M.; Tidrow, M. High performance InAs quantum dot infrared photodetectors (QDIP) on InP by MOCVD. Proc. SPIE 2005, 5732, 326–333.

217. Chakrabarti, S.; Adhikary, S.; Halder, N.; Aytac, Y.; Perera, A.G.U. Demonstration of high responsivity(~2.16 A/W) and detectivity(~10∧11 Jones) in the long wavelength (~10.2 um) from InGaAs/GaAs quantum dot infrared photodetector with quaternary InAlGaAs capping. Proc. SPIE 2012, 8353, 835338.

218. Ouattara, L.; Mikkelsen, A.; Lundgren, E.; Hoglund, L.; Asplund, C.; Andersson, J.Y. A cross-sectional scanning tunneling microscopy study of a quantum dot infrared photodetector. J. Appl. Phys. 2005.

219. Lim, H.-C.; Tsao, S.; Zhang, W.; Razeghi, M. Self-assembled semiconductor quantum dot infrared photodetector operating at room temperature and focal plane array. Proc. SPIE 2007, 6542, 65420–65421.

220. Tsao, S.; Lim, H.; Zhang, W.; Razeghi, M. High-performance InAs quantum-dot infrared photodetectors grown on InP substrate operating at room temperature and high operating temperature focal plane array. Proc. SPIE2007, 6593, 65931.

221. Lu, X.; Vaillancourt, J.; Meisner, M.J. Temperature-dependent photoresponsivity and high-temperature (190 K) operation of a quantum dot infrared photodetector. Appl. Phys. Lett. 2007.

222. Chen, T.-C.; Lai, J.-H. Investigation for optoelectronic characteristics and imaging performance of InAs quantum dot covered with In(0.1)Ga(0.9)As/GaAs multilayer based focal plane array. Opt. Eng. 2011.

223. Huffaker, D.L.; Mehta, M.; Balakrishnan, G.; Huang, S.; Khoshakhlagh, A.; Patel, P.; Kutty, M.N.; Dawson, L.R. GaSb QW-Based "Buffer-Free" Vertical LED Monolithically Embedded within a GaAs Cavity Using Interfacial Misfit Arrays. Proceedings of the 19th Annual Meeting of the IEEE Lasers and Electro-Optics Society, Montreal, QC, Canada, 29 October–2 November 2006.

224. Jallipalli, A.; Kutty, M.N.; Balakrishnan, G.; Tatebayashi, J.; Nuntawong, N.; Huang, S.H.; Dawson, L.R.; Huffaker, D.L.; Mi, Z.; Bhattacharya, P. 1.54 m GaSb/AlGaSb multi-quantum-well monolithic laser at 77 K grown on miscut Si substrate using interfacial misfit arrays. Electron. Lett. 2007, 43, 1198–1199.

225. Laghumavarapu, R.B.; Moscho, A.; Khoshakhlagh, A.; El-Emawy, M.; Lester, L.F.; Huffaker, D.L. GaSbGaAs type II quantum dot solar cells for enhanced infrared spectral response. Appl. Phys. Lett. 2007.

226. Tatebayashi, J.; Jallipalli, A.; Kutty, M.N.; Huang, S.H.; Rotter, T.J.; Balakrishnan, G.; Dawson, L.R.; Huffaker, D.L. Device characteristics of GaInSb/AlGaSb quantum well lasers monolithically grown on GaAs substrates by using an interfacial misfit array. J. Electron. Mater. 2008, 37, 1758–1763.

227. Yarborough, J.M.; Lai, Y.-Y.; Kaneda, Y.; Hader, J.; Moloney, J.V; Rotter, T.J.; Balakrishnan, G.; Hains, C.; Huffaker, D.; Koch, S.W.; Bedford, R. Record pulsed power demonstration of a 2 um GaSb-based optically pumped semiconductor laser grown lattice-mismatched on an AlAs/GaAs Bragg mirror and substrate. Appl. Phys. Lett. 2009, 95, 081112.

228. Balakrishnan, G.; Tatebayashi, J.; Khoshakhlagh, A.; Huang, S.H.; Jallipalli, A.; Dawson, L.R.; Huffaker, D.L. III/V ratio based selectivity between strained Stranski-Krastanov and strain-free GaSb quantum dots on GaAs. Appl. Phys. Lett. 2006.

229. Tatebayashi, J.; Liang, B.; Bussian, D.A.; Htoon, H.; Huang, S.; Balakrishnan, G.; Klimov, V.; Dawson, L.R.; Huffaker, D.L. Formation and optical characteristics of type-II strain-relieved GaSb/GaAs quantum dots by using an interfacial misfit growth mode. IEEE Trans. Nanotechnol. 2009, 8, 269–274.

230. Su, X.; Chakrabarti, S.; Bhattacharya, P.; Ariyawansa, G.; Perera, A.G.U. A resonant tunneling quantum-dot infrared photodetector. IEEE J. Quantum Electron. 2005, 41, 974–979.

231. Bhattacharya, P.; Su, X.H.; Chakrabarti, S.; Ariyawansa, G.; Perera, A.G.U. Characteristics of a tunneling quantum-dot infrared photodetector operating at room temperature. Appl. Phys. Lett. 2005, 86, 1–3.

232. Kim, J.-W.; Oh, J.-E.; Hong, S.-C.; Park, C.-H.; Yoo, T.-K. Room temperature far infrared (8~10 m) photodetectors using self-assembled InAs quantum dots with high detectivity. IEEE Electron. Device Lett. 2000,21, 329–331.

233. Wang, S.Y. Low dark current quantum-dot infrared photodetectors with an AlGaAs current blocking layer. Appl. Phys. Lett. 2001, 78, 1023–1025.

234. Kim, E.-T.; Chen, Z.; Ho, M.; Madhukar, A. Tailoring mid- and long-wavelength dual response of InAs quantum-dot infrared photodetectors using In(x)Ga(1-x;As capping layers. J. Vac. Sci. Technol. B 2002, 20, 1188–1191.

235. Szafraniec, J.; Tsao, S.; Zhang, W.; Lim, H.; Taguchi, M.; Quivy, A.A.; Movaghar, B.; Razeghi, M. High-detectivity quantum-dot infrared photodetectors grown by metalorganic chemical-vapor deposition. Appl. Phys. Lett. 2006.

236. Vandervelde, T.E.; Lenz, M.C.; Varley, E.; Barve, A.; Shao, J.; Shenoi, R.V; Ramirez, D.A.; Jan, W.; Sharma, Y.D.; Krishna, S. Quantum dots-in-a-well focal plane arrays. IEEE J. Sel. Top. Quantum Electron 2008, 14, 1150–1161.

237. Vandervelde, T.E.; Krishna, S. Progress and prospects for quantum dots in a well infrared photodetectors. J. Nanosci. Nanotechnol. 2010, 10, 1450–1460.

238. Ling, H.S.; Wang, S.Y.; Lee, C.P.; Lo, M.C. High quantum efficiency dots-in-a-well quantum dot infrared photodetectors with AlGaAs confinement enhancing layer. Appl. Phys. Lett. 2008, 92.

239. Liu, H.Y.; Alia, E. Optimizing the growth of 1.3 um InAs/InGaAs dots-in-a-well structure. J. Appl. Phys. 2003,93, 2931–2936.

240. Barve, A.V; Sengupta, S.; Kim, J.O.; Montoya, J.; Klein, B.; Shirazi, M.A.; Zamiri, M.; Sharma, Y.D.; Adhikary, S.; Godoy, S.E. Barrier selection rules for quantum dots-in-a-well infrared photodetector. IEEE J. Quantum Electron. 2012, 48, 1243–1251.

241. Rotella, P.; Raghavan, S.; Stintz, A.; Fuchs, B.; Krishna, S.; Morath, C.; Le, D.; Kennerly, S.W. Normal incidence InAs/InGaAs dots-in-well detectors with current blocking AlGaAs layer. J. Cryst. Growth 2003, 251, 787–793.

242. Hoglund, L.; Asplund, C.; Wang, Q.; Almqvist, S.; Malm, H.; Petrini, E.; Andersson, J.Y.; Holtz, P.O.; Petersson, H. Origin of photocurrent in lateral quantum dots-in-a-well infrared photodetectors. Appl. Phys. Lett. 2006.

243. Ariyawansa, G.; Perera, A.G.U.; Raghavan, G.S.; von Winckel, G.; Stintz, A.; Krishna, S. Effect of well width on three-color quantum dots-in-a-well infrared detectors. IEEE Photon. Technol. Lett. 2005, 17, 1064–1066.

244. Aivaliotis, P.; Zibik, E.A.; Wilson, L.R.; Cockburn, J.W.; Hopkinson, M.; Airey, R.J. Tuning the photoresponse of quantum dot infrared photodetectors across the 8–12 um atmospheric window via rapid thermal annealing.Appl. Phys. Lett. 2007.

245. Shao, J.; Vandervelde, T.E.; Jang, W.-Y.; Stintz, A.; Krishna, S. High operating temperature quantum-dot infrared photodetector using advanced capping techniques. IEEE Trans. Nanotechnol. 2011, 10, 1010–1014.

246. Shao, J.; Vandervelde, T.E.; Jang, W.-Y.; Stintz, A.; Krishna, S. High Operating Temperature InAs Quantum Dot Infrared Photodetector Via Selective Capping Techniques. Proceedings of the 8th IEEE Conference on Nanotechnology, Arlington, TX, USA, 18– 21 August 2008; pp. 112–115.

247. Shenoi, R. V; Attaluri, R.S.; Siroya, A.; Shao, J.; Sharma, Y.D.; Stintz, A.; Vandervelde, T.E.; Krishna, S. Low-strain InAsInGaAsGaAs quantum dots-in-a-well infrared photodetector. J. Vacu. Sci. Technol. B 2008, 26, 1136–1139.

248. Vines, P.; Tan, C.H.; David, J.P.R.; Attaluri, R.S.; Vandervelde, T.E.; Krishna, S. Noise, gain, and responsivity in low-strain quantum dot infrared photodetectors with up to 80 dot-in-a-well periods. IEEE J. Quantum Electron.2011, 47, 607–613.

249. Shenoi, R.V; Hou, J.; Sharma, Y.; Shao, J.; Vandervelde, T.E.; Krishna, S. Low strain quantum dots in a double well infrared detector. Proc. SPIE 2008, 7082, 708207.

250. Andrews, J.; Restaino, S.R.; Vandervelde, T.E.; Brown, J.S.; Sharma, Y.D.; Lee, S.L.; Teare, S.W.; Reisinger, A.; Sundaram, M.; Krishna, S. Comparison of long-wave infrared quantum-dots-in-a-well and quantum-well focal plane arrays. IEEE Trans. Electron. Devices 2009, 56, 512–516.

251. Jang, W.-Y.; Hayat, M.M.; Tyo, J.S.; Attaluri, R.S.; Vandervelde, T.E.; Sharma, Y.D.; Shenoi, R.; Stintz, A.; Cantwell, E.R.; Bender, S.C. Demonstration of bias-controlled algorithmic tuning of quantum dots in a well (DWELL) MidIR detectors. IEEE J. Quantum Electron 2009, 45, 674–683.

252. Shenoi, R.V; Rosenberg, J.; Vandervelde, T.E.; Painter, O.J.; Krishna, S. Multispectral quantum dots-in-a-well infrared detectors using plasmon assisted cavities. IEEE J. Quantum Electron. 2010, 46, 1051–1057.

253. Shao, J.; Vandervelde, T.E.; Barve, A.; Jang, W.-Y.; Stintz, A.; Krishna, S. Enhanced normal incidence

photocurrent in quantum dot infrared photodetectors. J. Vacu. Sci. Technol. B 2011, 29, 3–123.

254. Vines, P.; Tan, C.H.; David, J.P.R.; Attaluri, R.S.; Vandervelde, T.E.; Krishna, S.; Jang, W.; Hayat, M.M. Versatile spectral imaging with an algorithm-based spectrometer using highly tuneable quantum dot infrared photodetectors. IEEE J. Quantum Electron. 2011, 47, 190–197.

255. Barve, A.V; Sharma, Y.D.; Montoya, J.; Shao, J.; Vandervelde, T.; Rotter, T.; Krishna, S. Engineering the barrier of quantum dots-in-a-well (DWELL) infrared photodetectors for high temperature operation. Int. J. High Speed Electron. Syst. 2011, 20, 549–555.

256. Andrews, J.R.; Restaino, S.R.; Teare, S.W.; Sharma, Y.D.; Jang, W. Comparison of quantum dots-in-a-double-well and quantum dots-in-a-well focal plane arrays in the long-wave infrared. IEEE Trans. Electron. Devices2011, 58, 2022–2027.

257. Vandervelde, T.E.; Lenz, M.C.; Varley, E.; Barve, A.; Shao, J.; Shenoi, R.; Ramirez, D.A.; Jang, W.; Sharma, Y.D.; Krishna, S. Multicolor quantum dots-in-a-well focal plane arrays. Proc. SPIE 2008, 6940, 694003.

258. Krishna, S. Barrier engineering in quantum dots in a well detector. Proc. SPIE 2012.

259. Jolley, G.; Fu, L.; Tan, H.H.; Jagadish, C. Effects of well thickness on the spectral properties of In(0.5)Ga(0.5)As/GaAs/Al(0.2)Ga(0.8)As quantum dots-in-a-well infrared photodetectors. Appl. Phys. Lett. 2008.

260. Jolley, G.; Fu, L.; Tan, H.H.; Jagadish, C. Influence of quantum well and barrier composition on the spectral behavior of InGaAs quantum dots-in-a-well infrared photodetectors. Appl. Phys. Lett. 2007.

261. Barve, A.V; Shah, S.Y.; Shao, J.; Vandervelde, T.E.; Shenoi, R.V; Jang, W.-Y.; Krishna, S. Reduction in dark current using resonant tunneling barriers in quantum dots-in-a-well long wavelength infrared photodetector. Appl. Phys. Lett. 2008, 93, 131115.

262. Barve, A.; Shao, J.; Sharma, Y.D.; Vandervelde, T.E.; Sankalp, K.; Lee, S.J.; Noh, S.K.; Krishna, S. Resonant tunneling barriers in quantum dots-in-a-well infrared photodetectors. IEEE J. Quantum Electron. 2010, 46, 1105–1114.

263. Rosenberg, J.; Shenoi, R.V; Vandervelde, T.E.; Krishna, S.; Painter, O. A multispectral and polarization-selective surface-plasmon resonant midinfrared detector. Appl. Phys. Lett. 2009, 95, 161101.

264. Gunapala, S.D.; Bandara, S.V; Hill, C.J.; Ting, D.Z.; Liu, J.K.; Rafol, S.B.; Blazejewski, E.R.; Mumolo, J.M.; Keo, S.A.; Krishna, S. 640 × 512 pixels long-wavelength infrared (LWIR) quantum-dot infrared photodetector (QDIP) imaging focal plane array. IEEE J. Quantum Electron. 2007, 43, 230–237.

265. Kozlowski, L.J. HgCdTe focal plane arrays for high performance infrared cameras. Proc. SPIE 1997, 3179, 200–211.

266. Mollard, L.; Destefanis, G.; Bourgeois, G.; Ferron, A.; Baier, N.; Gravrand, O.; Barnes, J.P.; Papon, A.; Milesi, F.; Kerlain, A.; Rubaldo, L. Status of p-on-n arsenic-implanted HgCdTe technologies. J. Electron. Mater. 2011, 40, 1830–1839.

267. Kozlowski, L.J.; Kosonocky, W.F. Chapter 23—Infrared Detector Arrays. In Handbook of Optics; McGraw-Hill: New York USA, 1995; p. 1664.

268. Costard, E.; Truffer, J.P.; Huet, O.; Dua, L.; Nedelcu, A.; Robo, J.A.; Marcadet, X.; De l'Isle, N.B.; Bois, P. QWIP from 4 um up to 18 um. Proc. SPIE 2006, 6361, 636111–636117.

269. Tang, S.F.; Lin, S.Y.; Lee, S.C. Near-room-temperature operation of an InAs/GaAs quantum-dot infrared photodetector. Appl. Phys. Lett. 2001, 78, 2428–2430.

270. Chakrabarti, S.; Stiff-Roberts, A.D.; Su, X.H.; Bhattacharya, P.; Ariyawansa, G.; Perera, A.G.U. High-performance mid-infrared quantum dot infrared photodetectors. J. Phys. D 2005, 38, 2135–2141.

271. Krishna, S.; Raghavan, S.; von Winckel, G.; Stintz, A.; Ariyawansa, G.; Matsik, S.G.; Perera, A.G.U. Three-color ($\lambda p1 \sim 3.8$ μm, $\lambda p2 \sim 8.5$ μm, and $\lambda p3 \sim 23.2$ μm) InAs/InGaAs quantum-dots-in-a-well detector. Appl. Phys. Lett. 2003.

272. Shao, J.; Vandervelde, T.E.; Jang, W.; Stintz, A.; Krishna, S. Improving the operating temperature of quantum dots-in-a-well detectors. Proc. SPIE 2010.

CHAPTER 12

QUANTUM DOT INFRARED PHOTODETECTORS: PHOTORESPONSE ENHANCEMENT DUE TO POTENTIAL BARRIERS

Vladimir Mitin, Andrei Antipov, Andrei Sergeev, Nizami Vagidov, David Eason, Gottfried Strasser

Keywords: Quantum dots, Infrared detectors, Photoresponse, Doping, Potential barriers, Capture processes

INTRODUCTION

Many optoelectronic devices are based on the phenomenon of photoconductivity in which a material becomes more electrically conductive due to photocarriers created by electromagnetic radiation. Photocarriers contribute to the electric current until they are trapped by impurities and/or defects. Long photocarrier lifetime would substantially improve the operation of optoelectronic devices, such as IR and THz detectors and solar cells. New nanostructured materials that provide long photocarrier lifetime at room temperatures would significantly increase the commercial market for infrared and terahertz technologies.

Initial hopes related to QD nanostructures were associated with the "phonon bottleneck" concept, which assumes that the phonon-assisted bound-to-bound transitions in QDs are prohibited, unless the energy between two discrete levels matches the phonon energy [1]. According to this concept, the intrinsic electron relaxation in quasi-1D nano-objects, such as QDs, was anticipated to be significantly slower than in 2D and 3D structures. However, the phonon bottleneck model completely ignores interaction between electrons and corresponding modification of electron states. It is not surprising that the experimentally measured phonon-mediated electron relaxation turned out to be much faster than it is expected in the phonon bottleneck concept [2]. Recent investigations [3] unambiguously demonstrated that the actual intradot kinetics is completely opposite to what can be expected for weakly interacting electrons and phonons. In reality, strong coupling between electrons and longitudinal optical (LO) phonons leads to the formation of the polaron states, which decay due to the interaction of LO phonons with acoustical phonons. Such kinetics results in strong energy and temperature dependences of the electron relaxation. For example, for 14 meV transition, the relaxation time reduced from 1.5 ns at 10 K to 560 ps at 30 K, and further to 260 ps at 50 K. At room temperatures, the polaron decay time is observed in the range of 2–30 ps, depending on the electron energy [3]. Thus, after numerous experiments with various QD structures, no true phonon bottleneck has been found [3, 4, 5].

Another possibility to suppress photoelectron relaxation and to increase a photoelectron lifetime is related with the interdot kinetics. In theoretical works [6, 7, 8, 9], we proposed to suppress the capture processes by means of potential barriers in specially engineered QD structures. Potential barriers are always created, when electrons populating the dots are taken from the specific areas located relatively far from the dots. Changing the position of dopants and doping level, one can manage the potential barriers around dots and control the photoelectron capture processes [8, 9].

In this work we present direct experimental demonstration of strong effects of potential barriers on photoresponse of QD structures. The paper is organized in the following way. In the next section we describe the fabrication of the QD structures with different positions of dopants and various doping levels. After that we present results of measurements of the photoresponse and noise characteristics of these structures. Then, using the nextnano^{3}software we calculate the dot population and potential barriers around dots as a function of dopant positions and doping levels. Finally, we discuss the photoresponse enhancement in terms of potential barriers around QDs.

Citation: Vladimir Mitin, Andrei Antipov, Andrei Sergeev, Nizami Vagidov, David Eason, Gottfried Strasser, Quantum Dot Infrared Photodetectors: Photoresponse Enhancement Due to Potential Barriers, DOI: 10.1007/s11671-010-9767-y.

DEVICE FABRICATION

Several standard quantum dot infrared photodetector (QDIP) structures have been grown by the Riber Compact 21 MBE (molecular beam epitaxy) in AlGaAs matrix materials. For a given amount of In, adding aluminum to the nucleation layer and the surrounding matrix decreases the surface mobility of In and, thus, increases the dot density and decreases the dot size, when compared to InAs QDs grown onto GaAs surfaces and embedded in the binary compound GaAs only. In addition to the dot size and density distribution, an increase in the aluminum concentration also increases the conduction band offset. This shifts the center energy (decreases the wavelength) of the detector from the THz to the MID-IR region. In addition, the dark current is further reduced by increasing the aluminum concentration of the matrix material.

Growth temperatures on the substrate surface are monitored by infrared pyrometer. The surface temperature affects the density, quality, and size of the quantum dots that are formed and has to be constant, reproducible, and adjustable to control the QD formation. Typical growth temperatures are 500 ± 10°C as read by the pyrometer.

The doping for the active layers of the QDIPs can be done in several manners (see Table 1). The QDIP structures with two types of doping have been grown: intra-QD layer doping and inter-QD layer doping. Samples with intra-QD layer doping B44 and B52 have been grown with Si doping in the InAs dots (Figure 1a). The equivalent doping sheet concentration is 2.7 and 5.4 × 10^{11} cm^{-2} for B44 and B52, respectively. Samples with inter-QD layer doping B45, B53, and B54 have been grown with the Si doping directly in the middle of each AlGaAs barrier layer (Figure 1b). The thickness of doped layer is 6.4 nm. The doping sheet concentration is 2.7, 5.4, and 8.1 × 10^{11} cm^{-2} for B45, B53, and B54, respectively. Also, samples with inter-QD layer doping, B46, have been grown with the modulation doping. Each Si-doped 6.4-nm thick layer moves down 2.144 nm in AlGaAs barrier layer. Each repetition ends with 2.5, 4.644, 6.788, 8.932, 11.076, 13.22, 15.364, 17.508, 19.652, and 21.8 nm below AlGaAs barrier layer. The doping sheet concentration is 2.7 × $10^{11}cm^{-2}$.

Table 1 Devices

Device	Dopant position	Dopant concentration ($\times 10^{11}$ cm^{-2})	Number of electrons per QD	Barrier height (meV)
B44	QD layer	2.7	2.4	25
B45	Middle of AlGaAs layers	2.7	2.8	70
B46	Modulation doping	2.7	2.8	60
B52	QD layer	5.4	4.7	79
B53	Middle of AlGaAs layers	5.4	6.1	130
B54	Middle of AlGaAs layers	8.1	9	200

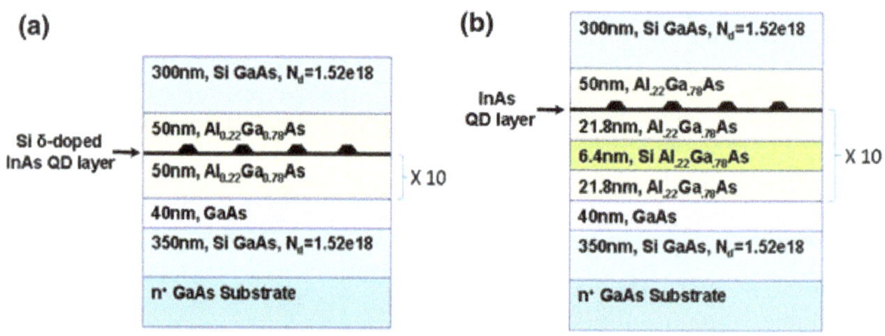

Figure 1 QDIP structures with *n* -type intra-QD layer doping (a) and inter-QD layer doping (b).

InAs dots grown on GaAs and AlGaAs surfaces form in approximately 2.2 monolayers of InAs growth. During the normal growth of layers, the substrate is rotated at 30 RPMs to insure uniform thickness of layers.

Typically 10 layers of quantum dots are grown in the active layer of the structure. Also, a layer of InAs dots is grown on the final top surface. All the QDIP structures are grown on *n*–type doped GaAs epi-ready substrates. Room-temperature images of surface quantum dots taken at ambient conditions by AFM measurements have been used to calibrate and control the quantum dot size and density. A typical AFM result for InAs quantum dots grown on a GaAs surface is shown in Figure 2: the substrate rotation is stopped during the growth of the QDs to get a density distribution over the 2 in. (or 3 in.) wafer. This gives one side of the wafer, closest to the indium source, a higher density of dots and the other side of the wafer, away from the indium source, a lower density of dots.

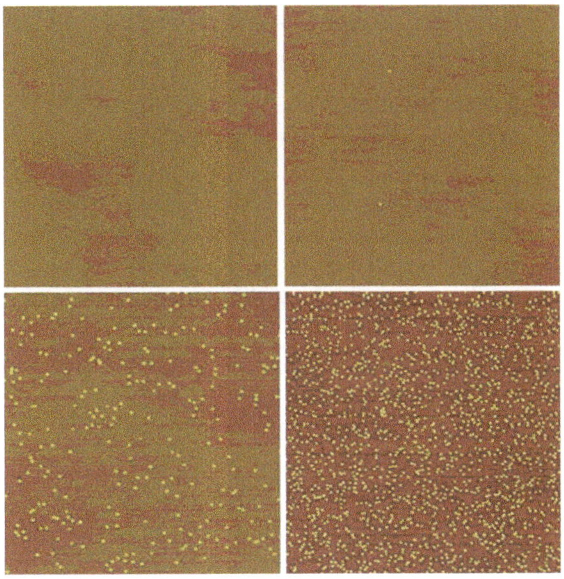

Figure 2 AFM images of a typical surface quantum dot sample; different positions on a 2 in. wafer; the deposited In increases from *left* to *right* and from *top* to *bottom* going from just a wetting layer (*left upper image*) to a concentration of 4 × 10 [10] dots/cm [2] in the *right lower image*.

The shown images are taken at different positions on a 3-inch wafer with respect to the indium effusion cell. The image size is 3 × 3 μm per image; the distance between images is half an inch, moving closer to the indium cell by going from left to right and from top to the bottom. Further adjustments to growth rates and times allow the properties of the quantum dots to be adjusted. The size of the dots and the matrix material used to embed the QDs determine the wavelength at which the QDIPs operate.

The vertical QDIP structure was processed by standard optical lithography, etching, and metallization techniques. To investigate the electron transport in QDIP structures, square 100 × 100 μm^2 mesas with alloyed Ni/Ge/Au/Ni/Au high-quality ohmic contacts were formed. Top and bottom contacts were deposited on the highly Si-doped GaAs layers and followed by rapid thermal annealing at 430°C for 40 s. Positive bias polarity corresponds to a positive voltage, which is applied to the top contacts.

PHOTORESPONSE AND DARK CURRENT

For low-temperature optical measurements, each sample was mounted inside a helium continuous-flow cryostat. The current–voltage characteristics were recorded with Keithley 2602 Multimeter. Our measurements have been done with 10^{-7} A/cm^2 accuracy. The spectral response of our QDIP structures was measured using a Bruker Optics Vertex 70 Fourier transform infrared (FTIR) spectrometer and a low noise current mode 7265 DSP Lock-in Amplifier.

Normal incidence photoresponse spectra of the B44, B45, and B46 structures at $T = 80$ K and bias voltage -5 V are presented in Figure 3. We observed the maximum response at ~329 meV. Full width at half maximum is ~40 meV. Also, there is an additional small spike that can be clearly observed at the high-energy tail of the spectra. The energy corresponding to the spike is ~380 meV. The valley on the low-energy tail of the spectra at 300 meV corresponds to absorption of IR light by CO_2 gas in the air. As seen from Figure 3, the photoresponse of B45 is several times larger then that for samples B44 and B46.

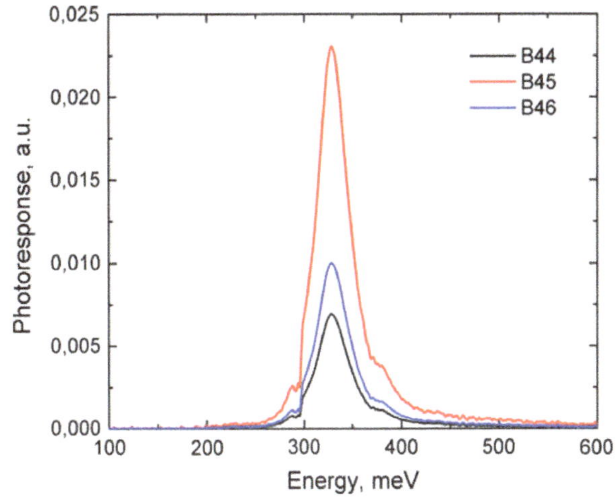

Figure 3 Normal incidence spectral photoresponse of samples B44–B46 at $T = 80$ K and bias voltage -5 V.

Photoresponse spectra of the samples B52, B53, and B54 are shown in Figure 4. Spectra of all these samples have the same position of a maximum at ~335 meV at negative bias voltages 3.2 V for B52, 2 V for B53, and 0.8 V for B54, correspondingly. The photoresponse of B54 is several times bigger than that of B52. Each spectrum exhibits also local maximums on the low-energy tail at ~300 meV and at the high-energy tail at ~380 meV. Full width at half maximum is ~45 meV. Let us note that at low voltages the photoresponse increases exponentially as well as the dark current density (see next paragraph). At high voltages, photoresponse sharply decreases. The same shapes and positions of maximum in the spectra provide strong evidence that QDs in all our structures are nearly identical.

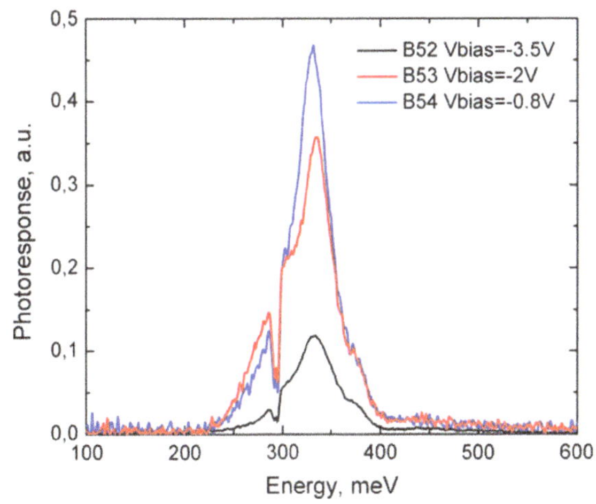

Figure 4 Normal incidence maximum spectral photoresponse of samples B52–B54 at $T = 80$ K.

Dark current densities for samples B52, B53, and B54 are presented in Figure 5. The dark current density in the sample B54 is higher by two orders of magnitude than that in the sample B53 and by four orders of magnitude higher than that in B52. Let us note that the current–voltage characteristic for the sample B52 is more symmetrical than that for samples B53 and B54. The asymmetric characteristic of this sample is due to 40 nm GaAs undoped layer at the substrate side.

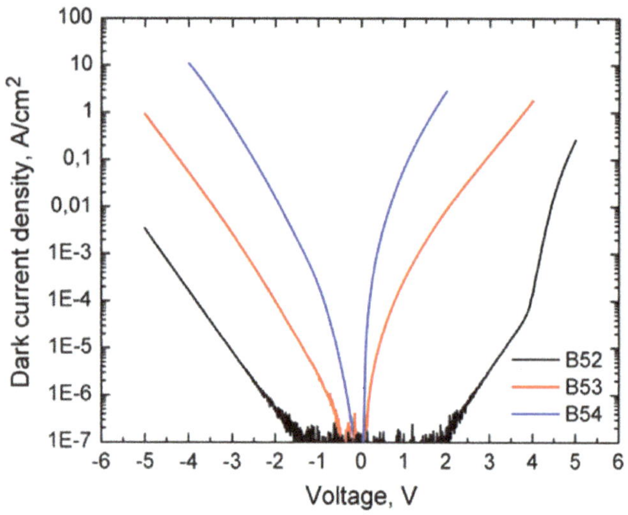

Figure 5 Dark current density of samples B52–B54 at $T = 80$ K.

We also found that the dark current density for sample B46 is higher by two orders of magnitude than that for samples B44 and B45. The sample B46 with the modulation doping has the highest dark current density and the lowest gain. The sample B44 with the silicon Delta doping just before InAs QD layer and the sample B45 with Si doping directly in the middle of each AlGaAs barrier layer show almost the same values of the dark currents.

We investigated the photocurrent in our samples as a function of optical pumping. In these measurements, we used a red light-emitting diode (LED) as a source of the optical radiation. The corresponding energy of photons, ~2 eV, is higher than QDIP's intersubband transitions and InAs energy gap. The energy of IR photoexcitations was tuned to the resonance absorption, i.e. to 380 meV in accordance with the data in Figure 3. The optical power of LED was calibrated by using a silicon power sensor.

Photocurrent densities for sample B45 at different power of LED's radiation are shown in Figure 6. A photocurrent response increases linearly with a background power and finally saturates at high power. The optical pumping allowed us to increase the photoresponse by three orders of magnitude. This effect is mainly due to an increase in the electron population in the quantum dot energy levels.

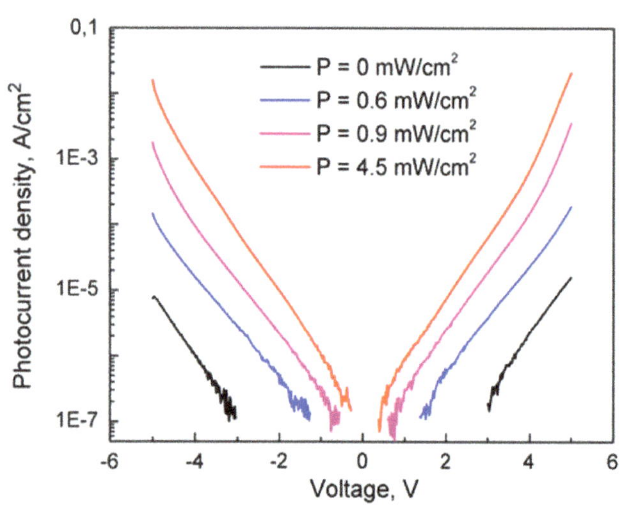

Figure 6 Photocurrent density of the sample B45 at $T = 80$ K as a function of optical pumping from LED.

MODELING: POTENTIAL BARRIERS IN QD STRUCTURES

The analysis of optical and electrical properties of the grown samples was done using nextnano[3] software [10]. This versatile software allows for simulation of multilayer structures combined of different materials with realistic geometries in three dimensions. The simulation tool solves self-consistently Schrödinger, Poisson, and current equations. The conduction and valence bands are defined within single-band or multiband k.p envelope. In this modeling, we used well-established material parameters for the simulated structures: the effective mass of electron, m^*, in γ-valley of GaAs is 0.067 and in $Al_{0.22}Ga_{0.78}As$, m^*, is 0.085. Effects of strain [11, 12] were included in simulations.

Using this software, the three-dimensional bandstructures of the grown samples were obtained. Analyzing two-dimensional slices of conduction band profiles, the heights of potential barriers that divide neighboring QDs were defined.

The height of potential barriers for electrons located in interdot area is defined as the difference between the maximum of the conduction band in QDs and minimum in the depletion regions that occur as a result of interdot doping, as it is shown in Figure 7.

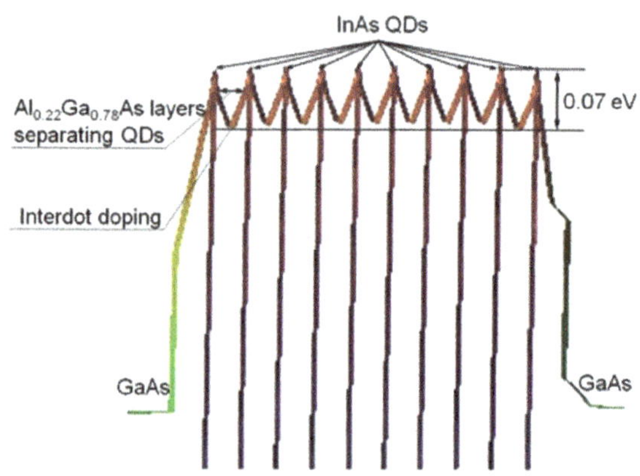

Figure 7 Two-dimensional slice of the calculated using nextnano[3] software conduction band structure of B45 sample.

The calculated heights of these potential barriers at $T = 80$ K in B44, B45, B52, B53, and B54 samples are shown in Figure 8. Point B45 in Figure 8 corresponds to the potential barrier height of 0.07 eV shown in Figure 7.

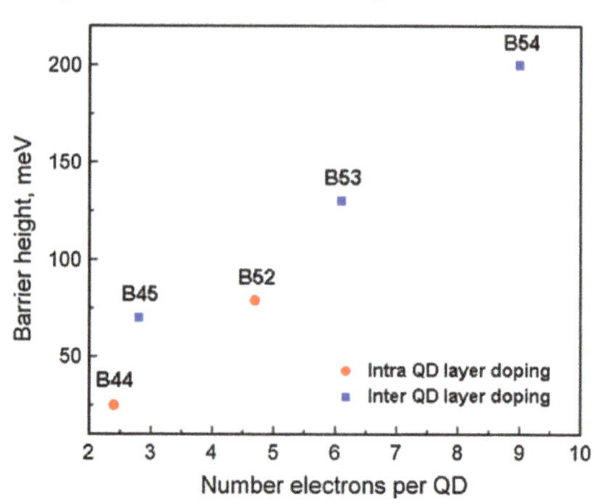

Figure 8 Barrier height versus number electrons per QD of samples B44, B45, B52, B53, and B54 at $T = 80$ K.

CONCLUSIONS

In this work we investigated the effects of the potential barriers around QDs on the photoresponse of QDIPs. We found that in accordance with our theoretical conclusions [6, 7, 8, 9], the potential barriers substantially suppress photoelectron capture and enhance the photoresponse.

In Figure 9 we summarize our main results. As seen, when the height of the potential barriers in the selected devices changes from 28 to 130 meV, the photoresponse increases exponentially as a function of the barrier height [6, 7, 8, 9]. At high doping level (the sample B54), the effect is saturated. We observed ~30 times enhancement of the photoresponse due to potential barriers around dots. Our results show that QDIPs can substantially outperform quantum-well photodetectors due to the manageable photocarrier kinetics, which can be controlled by potential barriers.

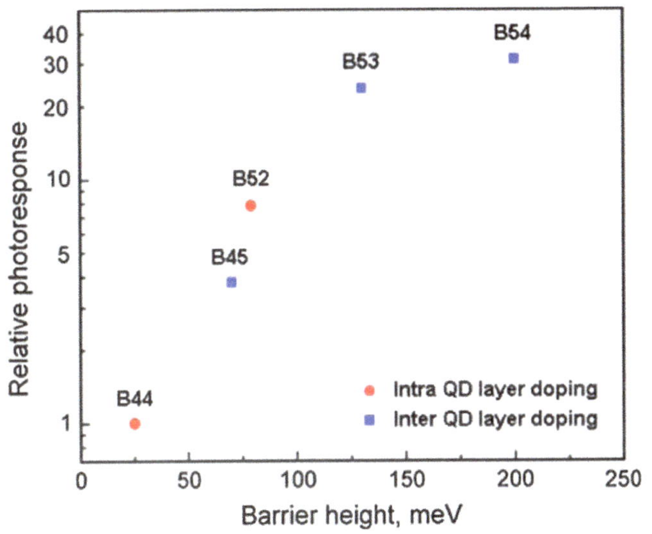

Figure 9 Normal incidence relative spectral photoresponse versus barrier height of samples B44, B45, B52, B53, and B54 at $T = 80$ K.

DECLARATIONS

Acknowledgements

This work was supported by AFOSR, the research of Antipov was also supported by NSF under Grant No DMR 0907126.

REFERENCES

1. Bockelmann U, Bastard G: Phys Rev B 1990, 42:8947.
2. Toda Y, Moriwaki O, Nishioka M, Arakawa Y: Phys Rev Lett 1999, 82:4114.
3. Zibik EA, Grange T, Carpenter BA, et al: Nat Mater 2009, 8:403.
4. Razeghi M, Lim H, Tsao S, Szafraniec J, Zhang W, Mi K, Movaghar B:
5. Nanotechnology 2005, 16:219.
6. Lim H, Movaghar B, Tsao S, et al: Phys Rev B 2006, 74:205321.
7. Mitin VV, Pipa VI, Sergeev AV, Dutta M, Stroscio M: Infrared Phys Technol
8. 2001, 42:467.
9. Sergeev A, Mitin V, Stroscio M: Physica B 2002, 316–317:369.
10. Chien LH, Sergeev A, Vagidov N, Mitin V: Selected Topics in Electronics and Systems. World Scientific, New Jersey; 2009:50:255.
11. Sergeev A, Chien L-H, Vagidov N, Mitin V: Future Trends in Microelectronics: From Nanophotonics to Sensors to Energy. Edited by: Luryi S, Xu J, Zaslavsky A. Wiley, Hoboken, NJ; 2010.
12. [http://www.nextnano.de/nextnano3].
13. Li SS, Xia JB, Yuan ZL, Xu ZY: Phys Rev B 1996, 54:11575.
14. Li SS, Xia JB: Phys Rev B 1997, 55:15434

CHAPTER 13 PHOTO-DETECTORS INTEGRATED WITH RESONANT TUNNELING DIODES

Bruno Romeira, Luis M. Pessoa, Henrique M. Salgado, Charles N. Ironside and José M. L. Figueiredo

Keywords: Double barrier quantum well structures, clock recovery, microwave-photonics, optical injection locking, optoelectronic circuits, oscillators, photo-detectors, radio-over-fiber, resonant tunneling diodes

INTRODUCTION

Semoconductor photo-detectors are devices that can convert optical signals into electronic signals, and they are extensively used in optical communication systems [1]. Ideally a photo-detector should detect all the incident light, respond to the fastest changes in the incoming signal and not introduce additional noise. In practice, photo-detectors have limited bandwidth with finite response times, they have associated noise sources and the probability of detecting an individual photon of light is less than one. Furthermore, photo-detectors have tradeoffs particularly between speed (bandwidth) and quantum efficiency, whereby designs are aimed to increase their bandwidth-efficiency product [1].

Several types of detectors can be used in fiber optic systems for communications, sensing and measurement systems. Some of the most common types include p-i-n photodiodes, metal semiconductor metal photodiodes and avalanche photodiodes [1]. Alternatives to traditional photo-detectors include optical control of microwave devices for generation of high-speed low-phase noise oscillations. Injection locking of optically controlled microwave oscillators allows synchronizing the frequency and phase of free-running oscillators to optical modulated signals and can find extensive applications, ranging from wireless communications [2] to high-speed optical communications such as radio-over-fiber networks [3,4], from bio-imaging for identifying cancers [5] to test instruments. Direct optical control of microwave devices has been investigated in Gunn oscillators [6], tuning and power modulation of trapped plasma avalanche and transit time (TRAPATT) oscillators [7], and tuning of impact avalanche and transit time (IMPATT) oscillators [8]. However, these devices are limited in bandwidth and not amenable to integrated circuit realization and, because of nature of the processes involved, they require inconveniently high bias voltages, for example the avalanche effect is noisy and can introduce additional timing jitter and phase noise resulting in considerable inferior stability.

Currently, many researchers are looking for solutions to cleanly transfer the stability of high quality (Q) optical signals to microwave electronic signals. The proposed configurations employ either direct or external modulation of laser diodes ([9], and references therein), or heterojunction bipolar transistors (HBTs) [10], among others [11]. However, most of these schemes require several electronic and optoelectronic components such as phase-locked local oscillators, amplifiers, photo-detectors, light emitters, and/or amplifiers or modulators, which make them rather complex and costly for applications where compact solutions for the transmission and distribution of radio-frequency (RF) signals are needed [11].

Recently, optical injection locking of double barrier quantum well (DBQW) resonant tunneling diode (RTD) photo-detector devices to intensity modulated optical signals [12,13], and self-injection locking of a low cost, simple and compact RTD optoelectronic oscillator (OEO) configuration [14,15] have been successfully demonstrated that could substantially increase the practicality of optically controlled microwave oscillators for a wider range of microwave and photonics applications. In this article, DBQW-RTD based optical waveguide photo-detectors are described for the implementation of novel optically injection locked microwave oscillators that are controlled using signals at telecommunications wavelengths for applications in clock

Citation: Bruno Romeira, Luis M. Pessoa, Henrique M. Salgado, Charles N. Ironside and José M. L. Figueiredo, Photo-Detectors Integrated with Resonant Tunneling Diodes, doi:10.3390/s130709464.

recovery and fiber optic based communication systems. The photodetection results show the noise of the microwave injection locked RTD signal matches that of the input optical analogue and digital signals resulting in low timing jitter of the electrically detected carrier signal. Optically controlled RTD-based photo-detectors can have interesting applications in generation of microwave stable low-phase noise signals for clock recovery systems, low-power miniature atomic clocks as time references [16], and synchronization in communication networks [17]. Because of the fast response of RTDs [18], this technology has the potential to deliver Gb/s wireless transmission [19] which matches the data capability of current fiber optics, and considerably enhances radio-over-fiber communications [20].

DBQW RTD PHOTO-DETECTORS

Double barrier quantum well RTD based photo-detectors are interesting alternatives for optical detection when compared with traditional photo-detectors due to their photo detection mechanism and built in electrical gain arising from RTD's current-voltage (I-V) characteristic showing a region of negative differential resistance (NDR). With optimum epi-layer structure design that takes advantage of a photo detection mechanism based on the scenario that in biased RTDs the photo-excited electron-hole pairs become locally separated, which in turn modulates the internal electric field, RTDs photo-detectors can have sensitivities of up to 10^3 A/W [21]. By taking advantage of these effects, RTD structures can be used to implement light-by-light switching and optically switched RTD photo-detectors. Early demonstrations of such optically switched devices have shown potential for high-speed transmission communications at Gb/s rates with very low electrical power dissipation [22]. Demonstration of single photon detectors were also achieved combining DBQW-RTDs and quantum dot structures [23].

Resonant tunneling diode based photo-detectors (RTD-PDs) consist of nano-electronic structures that use vertical stacking of epitaxial layers of semiconductor alloys with the active regions consisting of 10-nm thick InGaAs/AlAs DBQW semiconductor alloys, surrounded by InGaAlAs photoconductive layers that act as like a Fabry-Pérot interferometer for the electron wave functions. The Fabry-Pérot effect gives rise to a highly nonlinear and N-shaped current versus voltage characteristic which provides the occurrence of NDR. DBQW-RTD structures for photo-detection emerged in the last decades as devices with high responsivities and gain-bandwidth-efficiency products, as well as low switching energies [23] because of their intrinsic gain and low voltage operation characteristics arising from the NDR effect. Essentially, they work as follows: a light pulse incident upon a resonant tunneling structure biased close to the peak voltage produces photo-charges that reduce the series resistance, leading to a shift of the peak and valley voltages which can induce RTD switching from the peak to the valley of I-V NDR and give rise to changes in the current flow [24].

Device Description

An example of a surface-illuminated DBQW-RTD photo-detector consisting of InGaAs/AlAs semiconductor compounds is shown in Figure 1a. As exemplified in the conduction band diagram, since the applied external voltage drops almost entirely on the thick undoped InGaAs absorption layer, the photo-generated electrons and holes in it are separated by the electric field, the holes are accumulated at the barriers of the resonant structure whereas the electrons are collected by the external circuit. As a result, the applied electric field in the undoped layer is partially screened and that within the resonant structure is enhanced. Consequently, the NDR of the resonant tunneling device shifts to lower voltages, as exemplified in Figure 1b. If the accumulated holes have a long recombination time the sensitivity of the device is high but its frequency response is low whereas for fast hole recombination times the sensitivity is lower but high-speed can be achieved.

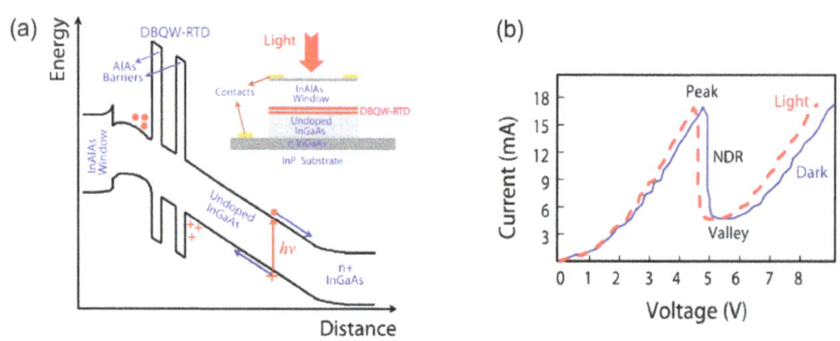

Figure 1. (a) Schematic of the conduction band diagram of a surface illuminated InGaAs/AlAs DBQW-RTD. Inset is shown the schematic cross-section of the surface RTD-PD [22]. (b) Typical dark and illuminated I-V characteristics.

Side-illuminated waveguide photo-detectors were proposed as a means to ease the trade-off between the speed and responsivity in order to achieve a large bandwidth-efficiency product [25,26]. This type of detector is generally edge-illuminated with light guided parallel to the plane of the intrinsic layer and where the light is absorbed. The photo-generated carriers and the light travel in perpendicular directions. In comparison with the surface transverse photodiode, the waveguide longitudinal photodiode responsivity is mainly determined by the waveguide length, and not by absorption layer thickness, therefore the waveguide detector can have a larger bandwidth-efficiency performance than the surface photo-detector.

The optical waveguiding structure is similar to that of the double heterostructure lasers. Figure 2shows the RTD-PD optical waveguide schematic, the wafer structure and the energy band diagram of the waveguide detector used in this work. The RTD component was a device fabricated from pre-existing material, the layer structure of which is given in [27]. The InGaAlAs RTD structure was grown by molecular bema epitaxy in a Varian Gen II system on a n$^+$ InP substrate. It essentially consisted of two 2-nm-thick AlAs barriers separated by a 6-nm-wide InGaAs quantum well, embedded in a 1-μm thick InGaAlAs optical waveguide core, see epi-layer structure in the inset of Figure 2a. Besides the epi-layer structure shown in the inset of Figure 2a, a δ-doped InGaAs cap layer was provided for formation of Au-Ge-Ni ohmic contacts. Ridges were fabricated by wet-etching after which ohmic contacts were deposited on top of the ridges. A SiO_2 layer was then deposited, and access contact windows were etched on the ridge electrodes making contact to high frequency bonding pads [coplanar waveguide (CPW) transmission line].

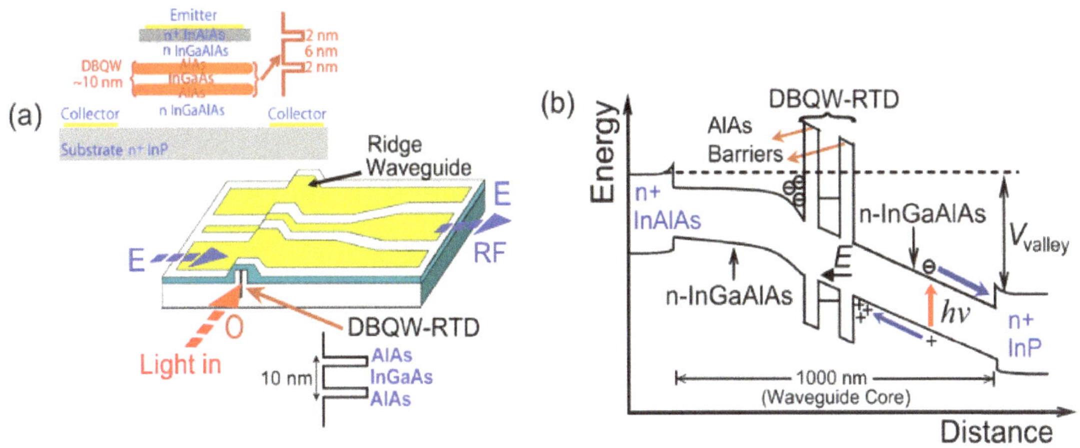

Figure 2. (a) Schematic diagram of RTD-PD device. The ridge waveguide dimensions were ~3 μm wide and 1 μm thick. Inset (on top) is shown the schematic diagram (not scaled) of the RTD epi-layer structure, (b) RTD-PD unipolar In Al As-In$_{0.53}$Ga$_{0.42}$Al$_{0.05}$ As-InP band energy diagram.

The design described here was originally used to obtain a DBQW-RTD operating as an electro-absorption modulator (EAM), but the same epitaxial structure can be used for novel photo-detector devices and applications as discussed in this article. Light is edge-coupled to the waveguide and the responsivity is mainly limited by the coupling efficiency and the scattering loss along the waveguide. The coupling can be improved through the use of lensed fibers, antireflection coating at waveguide facet, or laterally tapered waveguide sections.

Detection Mechanism in DBQW-RTD Waveguide Photo-Detectors

Inour device, the RTD operates as a waveguide photo-detector by taking advantage of DBQW-RTD photoconductive layers and optical waveguide structure. An RTD-PD optical waveguide works as a waveguide photo-detector for light with energy close or above the waveguide core bandgap energy. The ridge waveguide structure, Figure 2a, offers an advantage over traditional RTD-PDs by providing light confinement along the DBQW plane. The RTD-PD presents the typical RTD nonlinear I-V characteristic with an NDR region which leads to electrical gain. The photo-detection mechanism works as follows. For the low bias condition, electron-hole pairs are created in the depleted part of the waveguide core (that is, the 500 nm undoped In$_{0.53}$Ga$_{0.42}$Al$_{0.05}$ As layer) adjacent to the double barriers, see energy band diagram of Figure 2b. However, due to the fact that the electric field is very weak under low bias, the generated electron-hole pairs tend to recombine and neutralize after the creations. As the bias is increased, the hole accumulation density at the hetero-interface increases, resulting in an enhancement of the hole tunneling current. However, for bias voltages below the peak voltage, the current flow is still low and therefore a portion of the photogenerated carriers recombine rather than tunnel through the DBQW structure. When the bias is further increased to the valley voltage, Figure 2b, the effective conductivity of the double barriers drops as the resonant level in the quantum well becomes lower than the Fermi level of the emitter region. Consequently, the band bending on the double barriers increase, resulting in a substantially increase of the hole tunneling current. Once the hole current becomes large enough, the recombination in the photogeneration region decreases to zero, and the photogenerated carriers are collected by the electrodes.

As described previously, if the photon energy is larger than the waveguide core band-gap energy (here 0.815 eV) the RTD-PD operates as a photoconductive detector. However, if the injected photon energy is slightly lower (e.g., 0.8 eV) the Franz-Keldysh effect [28], which is associated with interband photon-assisted tunneling, plays also a role in absorption mechanism, i.e., the absorption of a photon is considerably enhanced due the presence of the electric field via the Franz-Keldysh effect.

DC Photo-Detection Characteristics

Photo-detection characterization of RTD-PDs consisted in coupling light into the waveguide cleave ends (endfire coupling) using a lensed single mode optical fiber (spot diameter of 5 ± 0.5 μm and coated on the tip). The lensed fibers and the RTD-PD devices were mounted on microblock 3-axis translation stages, in an endfire arrangement where both input optical fibers and samples could be translated and rotated independently. The light source employed for optical detection characterization was a Photonetics Tunics laser diode tunable in the wavelength region 1,457–1,599 nm, with up to 10 mW optical output power, that could be used for high frequency modulation with a bandwidth up to 1 GHz. RTD-PD devices mounted on the brass fixture were characterized directly on-chip by probing the RTD CPW contacts using a ground-signal-ground CPW probe with a pitch size of 125 μm.

Due to the polarization dependence of the RTD waveguide detectors, laser light was coupled to the waveguide using a polarization controller and the position of the input fiber was optimized to obtain the highest photocurrent from the device (and hence a high response to light modulated signals), ensuring a maximized overlap between the optical mode from the input fiber and the DBQW region. In Figure 3a is shown the I-V characteristics for both dark and illuminated conditions showing the photocurrent gradually turned on as the bias was increased to the valley voltage. Consequently, the NDR of the RTD-PD shifts to lower voltages, although the effect of the current increase shown in Figure 3a is not very pronounced because of the low optical power received by the RTD due to the small coupling efficiency as detailed next in this sub-section. Also shown is the external responsivity (the photocurrent generated divided by the optical power in fiber) as a function of the bias voltage. Note that a bias voltage of only 0.5 V was required to efficiently collect the photogenerated carriers. The results show clearly a higher responsivity close in or in the valley region which confirms the detection mechanism analyzed previously. Figure 3b presents the photogenerated valley current plotted versus the optical power in fiber. The straight line in the figure is the linear regression obtained from the data. The photogenerated valley current increases approximately linearly with the illumination power level (dI/dP_i) with a slope of 0.29 mA/mW, and up to 10 mW in-fiber optical power level. No saturation of the photogenerated valley current was observed, indicating that photoexcited RTD waveguide detectors can operate at moderate incident power light conditions.

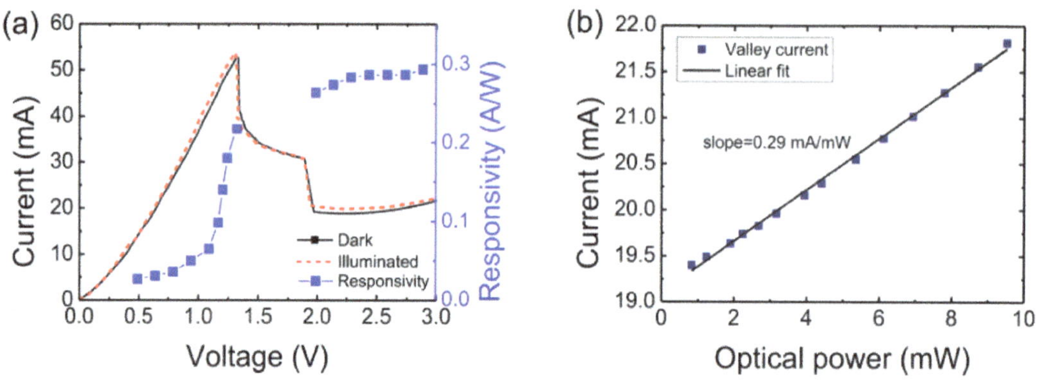

Figure 3. (a) Dark RTD-PD I-V characteristic and I-V curve with illumination at 1,550 nm for input optical power in fiber around 5 mW. The estimated responsivity as a function of the bias voltage is also plotted. (b) Photogenerated valley current as a function of illumination intensity and the linear regression.

The quantum efficiency in photo-detectors is defined as the fraction of photons creating electron-hole pairs:

$$\eta_{ph} = \frac{I_{ph}}{q\phi} \quad (1)$$

where I_{ph} is the photocurrent, and $\varphi = \lambda P(\lambda)/hc$ is the photon flux, with λ the operation wavelength, and h and c being the Planck constant and the speed of light in the vacuum, respectively. Therefore, the RTD photo-generated current I_{ph} in response to an incident optical signal P(λ) can be expressed by:

$$I_{ph} = \eta_{ph} \frac{q\lambda}{hc} P(\lambda) \quad (2)$$

If the surface reflections and the finite dimensions of the RTD waveguide photo-detector are considered, then the quantum efficiency, η_{ph} is given by light coupling factor κ multiplied by the absorbance [28]:

$$\eta_{ph} = \kappa(1 - R_{ref})\left(1 - e^{-\alpha(\lambda,V)\gamma_{ph}\Lambda}\right) \quad (3)$$

with R_{ref} is the waveguide facet reflectivity (sidewall imperfections and refractive index nonuniformity contribute to this kind of losses), α(λ,V) is the waveguide core absorption coefficient at wavelength λ and bias V, γ_{ph} is the overlap integral of the electric field and the optical field, and Λ is the waveguide photo-detector absorbing active length. This expression does not take into account with propagation losses.

In Table 1 are presented the typical physical parameters of the RTD photo-detectors. For the waveguide structure described in Figure 2a, we estimate γ_{ph} ~0.25 and κ ~0.35 at 1.55 μm, and an absorption coefficient α(λ,V), in the valley region, around 400 cm^{-1}, estimated from absorption spectra characterization reported in [11], which for devices with absorbing active length $A \geq 150$ μm yields a quantum efficiency η_{ph} ~0.2. Finally, the external responsivity is expressed by:

Table 1. Description of the typical physical parameters of the waveguide photo-detector.

Symbol	Quantity	Typical order of magnitude
λ	Operation wavelength	1.55 μm
κ	Light coupling factor	0.35
R_{ref}	Waveguide facet reflectivity	0.3
$\alpha(\lambda,V)$	Waveguide core absorption coefficient (at 1.55 μm and biased in the valley)	400 cm^{-1}
γ_{ph}	Overlap integral of the electric and optical fields	0.25
Λ	Waveguide contact length	150 μm

$$R_{\lambda} = \frac{\lambda}{1.24}\eta_{ph} \quad (4)$$

that gives a responsivity of 0.25 AAV in the valley region, which compares well with the typical values measured between 0.25 AAV and 0.3 AAV for RTD-PD devices biased in the valley region (see measured responsivity in the valley region shown in Figure 3a). According to the calculations, most of the light is not coupled in the waveguide (25% coupling efficiency which gives a coupling loss of 6 dB), which seems to be the major factor limiting the responsivity.

Although the responsivity of 0.25 AAV at these wavelengths (~1,550 nm) is low and below available commercial devices, as shown in the next section, the values reported here are sufficient to achieve an efficient optical control of the RTD microwave signals when the RTD is operating in the oscillating regime. Moreover, considering the estimated losses, we expect RTD-PD devices can provide PDs with responsivities up to 1 A/W which will further improve the efficiency of the optical control. The losses can be reduced by increasing the overlap integral of the input beam profile with the waveguide cross-section, reducing facet reflections, and design an optical waveguide with the RTD-PD active region closer to the input facet. In addition, calculations show that there would be no significant increase in the responsivity even with a much higher overlap, indicating that ~150 μm is long enough to absorb almost all light. In fact, calculations suggest there is not a substantial increase in responsivity for waveguides longer than ~100 μm. Therefore, the waveguide structure will occupy much less area, and the RTD could operate at higher frequencies. Currently, we are also investigating RTD-based surface illuminated photo-detectors which can provide advantages in terms of overlap integral of the input beam profile and coupling efficiency.

REGIMES OF OPERATION OF DBQW RTD PHOTO-DETECTORS

The RTD-PD can be operated in either non-oscillating or oscillating regimes depending on the bias voltage quiescent point. The oscillating regime of operation takes place when the RTD-PD is biased in the NDR region giving rise to electrical gain and microwave self-sustained oscillations which boosts RTD sensitivity to incident modulated light signals. In this section, we demonstrate efficient detection of GHz modulated optical carriers and optical control of RTD high-frequency oscillator when RTD-PD devices are operated in the non-oscillating and oscillating regimes.

Non-Oscillating Regime

The measurement system used for the characterization of the RTD-PD optical modulation response in the non-oscilating regime consisted of continuous wave light from a tunable single mode laser modulated by an external 10 GHz electro-optic modulator driven by an RF signal generator. A single mode optical fiber was used to guide the modulated light and to illuminate the RTD-PD waveguide by means of a lensed fiber. The RTD-PDs, the electrical GSG probe (with high frequency response up to 40 GHz), and the optical fiber were mounted on the positioners as described previously in the DC characterization setup. The variable DC bias was applied via a wide bandwith bias-T (45 MHz–26.5 GHz) connected to the high-frequency probe SMA cable. The RTD-PD electrical response to optical modulated injected signals was measured using a high-bandwidth spectrum analyzer and oscilloscope.

Figure 4a presents the RF injection locking capture level as function of the DC bias using light (λ= 1,550 nm) modulated by a sinusoidal RF reference signal at 1.2 GHz, with optical power levels in fiber of 1 mW and 6 mW.

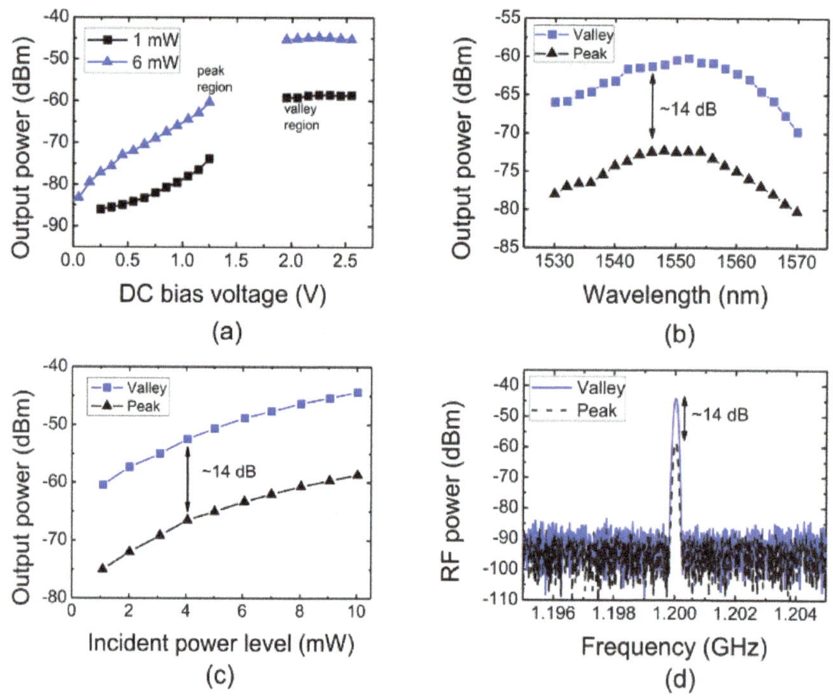

Figure 4. (a) RF injection locking capture level as function of the DC bias using light at $\lambda = 1{,}550$ nm. (b) Output power as function of wavelength with DC bias voltage as parameter for 1 mW optical power in fiber. (c) Output power as function of optical power level at $\lambda = 1{,}550$ nm. (d) Power spectra of photo-detected signal using a 10 mW optical signal X = 1,550 nm). In figures (b), (c), and (d), the RTD-PD was DC biased in the peak$V_p = 1.2$ V and valley $V_v = 2.0$ V regions. In all figures, light was modulated by a RF reference source signal at 1.2 GHz.

The figure shows the dependence of 1.2 GHz photodetection signal on bias conditions. The responsitivity increases with the transition from RTD-PD peak to valley voltage, V_p and V_v, respectively, with an improvement up to 14 dB. As discussed previously, the higher response in this region occurs because the holes photo-generated in the depletion region accumulate around the collector barrier (see Figure 2b), causing a local enhancement field which leads to higher carrier velocity and hence high current. Figure 4b presents the photo-detected RF power as function of the light wavelength with the DC bias voltage as parameter. Again, the increased responsivity under higher voltage bias has also been observed at the wavelength of 1,530-1,570 nm (C-band).

Figure 4c presents the measured RTD-PD RF power output when biased in the peak and valley regions as a function of incident optical power, and Figure 4d shows the RF power spectra of the photo-detected optical modulated signals. Up to 14 dB gain is obtained in the transition from the peak to the valley region. The results demonstrate the RTD-PD works as a high-speed optical-to-electrical converter, taking advantage of the optical waveguide design. In what follows, the optical injection locking operation employing RTD-PDs is investigated when the devices are DC biased in the NDR region, i.e., when they generate self-sustained oscillations at frequencies close to the detected microwave signals.

Oscillating Regime and Optical Control of Microwave Signals

Optical injection locking of an electrical oscillator can synchronize the frequency and phase of a free-running oscillator to the modulated optical signal. This, is very attractive in applications such as local oscillators embedded in radio base station cells and in optical transmission systems, e.g., optoelectronic clock recovery circuits, because it can lead to much simpler systems and is more suitable for higher bit-rates than a fully electrical circuit. There are two types of optically injection locked oscillators (OILOs): the direct and indirect type. A direct OILO, whose active oscillator device itself is directly illuminated for synchronization is preferred because is much simpler and is more suitable for monolithic integration than an indirect one which needs an external photo-detector. In this section, we investigate the injection locking phenomena of an RTD-PD circuit showing self-sustained oscillations up to 1.41 GHz. The influence of optical injection locking on the quality of the microwave carriers is evaluated and we achieved phase noise quality of the signals that are of interest for many microwave photonics applications, as discussed in detail in Section 4.

The RTD-PD based oscillator circuit utilized in the optical injection locking experiments was realized using RTD components attached directly onto the surface of a microwave substrate containing a 50 Ω copper microstrip transmission line laminated onto a non-conductive substrate. The transmission lines were realized in microstrip hybrid technology using Rogers substrate with a dielectric constant of 3.48 and a thickness of 0.762 mm. At the end terminals of the 50 Ω microstrip line were high frequency SMA connectors which acted as the RF output port. Gold bond wires were employed to connect the RTD-PD emitter and collector to the 50 Ω transmission line and to the ground, respectively. The RTD-PD devices were attached to the substrate with a high conductivity adhesive. The orientation of the RTD›s optical waveguide was taken into account in the layout and in the chip bonding to PCB to ensure good alignment to an optical fiber. In the oscillator circuit, a 10 Ω thin-film resistor, Figure 5, was utilized as part of the bias circuit of the oscillator to suppress the bias circuit spurious oscillations for the RF signals generated by the RTD [14]. The operation frequency of the RTD-PD is essentially determined by the RTD-PD›s parallel capacitance and the series equivalent inductance of the wire bonding connections (the wire length was around 4 mm, giving an equivalent inductance around 3.6 nH), which for the circuit reported here gives an operation frequency up to 1.4 GHz. High-frequency characterization of these devices was previously investigated [29] providing detailed analysis of RTD impedance versus frequency dependence for various bias voltage values. For a 45 MHz to 18 GHz frequency range the samples with negative resistance R ~ −15Ω. (estimated from the NDR region) showed RTD equivalent capacitance of RTD around 1.5 pF. For these circuit parameters, one can estimate the typical maximum frequency of operation of RTD devices, f_{max} ~14.15 GHz. Therefore, the RTD-PD hybrid circuit oscillators employed in our experiments could operate up to 10 GHz reducing circuit parasitics.

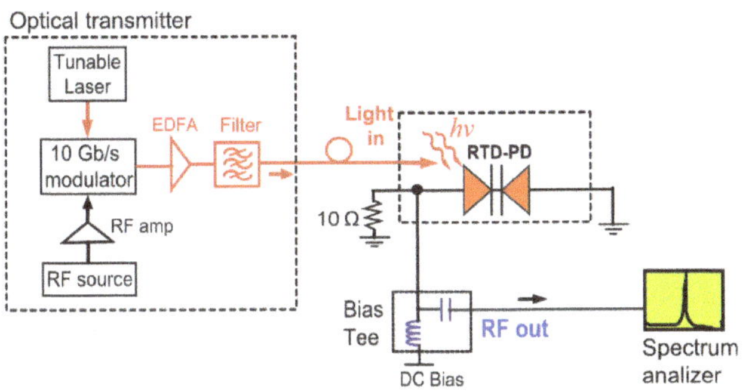

Figure 5. RTD-PD oscillator circuit schematic, and setup of the optical injection locking experiment.

The circuit diagram and experimental setup are presented schematically in Figure 5 showing the optical transmitter and measurement setups. Optical modulated signals at communications wavelengths were generated by a 10 Gb/s Mach-Zehnder modulator that encode the light (typically λ = 1,550 nm) from a tunable single mode laser. The output from the Mach-Zehnder modulator was amplified by an erbium-doped fiber amplifier (EDFA), and then filtered to eliminate spontaneous emission noise from the EDFA. Then, optical modulated signals illuminated the RTD-PD oscillator waveguide input facet through single mode lensed fiber. The free running oscillation frequency of the RTD-PD oscillator was controlled by adjusting the bias voltage provided by a power supply and through an external bias tee. Optically injection locked signals were measured in the electrical output of the circuit using a high frequency oscilloscope and spectrum analyzer.

In the following, we study the injection locking phenomena of an RTD-PD circuit showing an oscillation ranging from 1.04 GHz to 1.41 GHz as the DC bias is swept across the NDC region with voltage range ~0.4 V. The coupling of optical modulated signals to the RTD-PD waveguide locks the self-oscillations to the optically injected RF sub-carrier for in-fiber optical powers above a given threshold that depends on the waveguide characteristics, on the frequency and the RF power of the modulated light, and on the RF power of the unperturbed self-oscillations. To investigate the influence of optical injection locking on the quality of the RF carriers, signals were observed at the output RF port of the RTD-PD oscillator. The oscillator was considered injection-locked when its output spectrum matched the optical modulation reference source frequency with stable oscillation and minimum noise.

The optical injection locking investigation included measurements of a RTD-PD circuit self-oscillating at around 1.405 GHz. The circuit showed a substantial improved performance in terms of injection locking and spectral purity. The results are summarized in Figure 6. Figure 6a presents the typical broad spectrum of free-running oscillation (light off) around 1.405 GHz. When the light is on, with an average optical power level in fiber of 1 mW and employing a modulation extinction ratio close to 10 dB, the free-running signal is pulled and the spectrum collapses showing a substantial phase noise reduction. The inset of Figure 6a (top) presents the recorded spectrum surveillance confirming the spectral long-term stability of the optical injection locking. Injection locking for optical power levels in fiber as low as 0.2 mW was observed. The locking range was 23.8 MHz for 4.9 mW optical power in fiber. This corresponds to an exceptional wide value with a fractional locking bandwidth of 1.7%. The observed locking range corresponds to the largest value reported in the literature for RTD-based devices at communication wavelengths ([12] and references therein).

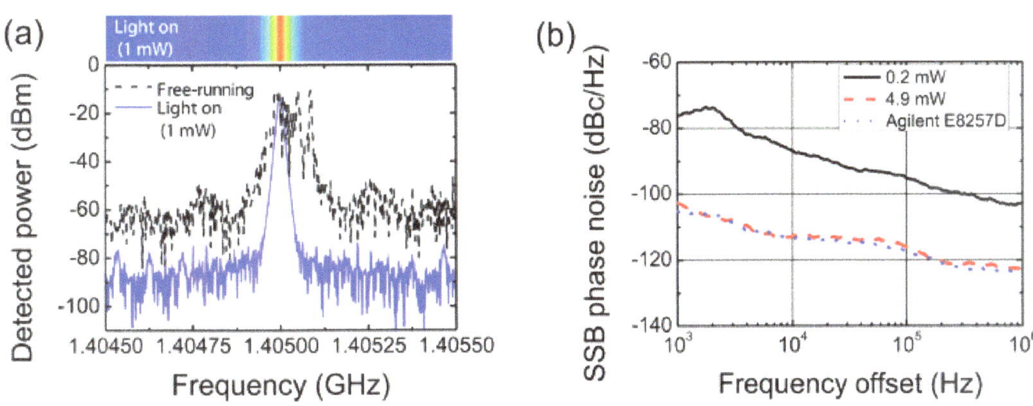

Figure 6. (a) Power spectra of the free-running and locked signals. (b) Single side band phase noise measurements.

Since injection-locking leads to a significant phase-noise and jitter reduction the respective oscillator performance has been further investigated. Figure 6b presents the phase noise measurements showing phase noise of −112.4 dBc/Hz at 10 kHz offset of the carrier frequency, a very close match to the RF source (Agilent E8257D) signal for the moderate optical power levels employed (below 5 mW), which results in a noise suppression of more than 40 dB when compared with the free-running oscillator without illumination. For the infiber optical power locking threshold the phase noise was −87.1 dBc/Hz at 10 kHz offset of the carrier frequency.

The results, although obtained using devices with small responsivities when compared with commercial available photo-detectors, show the RTD-PD can operate as a stable optically controlled microwave photonics oscillator to obtain a highly efficient O/E conversion. The RTD-PD phase noise characteristics should be sufficient for many current microwave photonic applications that use phase modulated schemes, e.g., phase shift keying (PSK) data transmission (as discussed in the next section), as long as the locking conditions are achieved and maintained.

APPLICATIONS IN CLOCK RECOVERY AND FIBER OPTIC SUPPORTED RADIO FREQUENCY COMMUNICATIONS

In this section, we demonstrate applications of injection locked RTD-PD oscillator devices in microwave photonics systems. The ability to lock to reference RF sources by optical injection locking techniques allows remote synchronization providing a large range of functions including generation, amplification and distribution of RF carriers, clock recovery, carrier recovery, modulation, and frequency synthesis. Optically controlled RTD-based photo-detectors can have interesting applications in the generation of microwave stable low-phase noise signals ranging from wireless communications [17] to high-speed optical communications such as radio-over-fiber networks [20], from clock recovery systems to low-power miniature atomic clocks as time references [16]. In what follows two main applications of RTD-PD oscillators in high-speed optical communication transmission systems are discussed: clock-recovery of Gb/s return-to-zero signals and transmission of modulated signals in wireless/photonics links.

Clock Recovery

The ability to recover the clock signal from incoming random data is a fundamental operation in digital communications. Clock recovery consists in deriving a clock signal from the received signal which is synchronized in both frequency and phase, avoiding its transmission along with the data, and providing a timing basis to sample the received signal. In this context, the ability to perform clock recovery using optical techniques is an important issue in the development of circuits for novel high speed optical communication systems.

RTD-PD oscillators are candidates for that purpose since they combine the functions of microwave oscillation and photo-detection, while offering intrinsic gain and high responsivities, being suitable for injection locking with incoming data in both optical and electrical domains. In what follows we present results on a clock recovery experiment using optical injection (O/E conversion). Clock recovery via optical injection is important because it enables the clock of a high-speed data signal in a fiber-optic system to be directly recovered without first converting the data to electrical pulses. In this experiment the free running RTD-PD oscillator is tuned to oscillate at a frequency sufficiently close to the incoming signal. The locking range is typically on the order of tens of MHz, and depends on the power of the injected clock frequency. The incoming optical signal is injected into the oscillator and the recovered signal is accessed at the electrical output port.

Figure 7 shows the experimental schematic for timing extraction and bit-error-rate (BER) measurements. The performance of the recovered clock was assessed from timing jitter, phase noise and locking bandwidth. Additionally, BER measurements were carried out using the recovered clock as a time base for sampling of the received data.

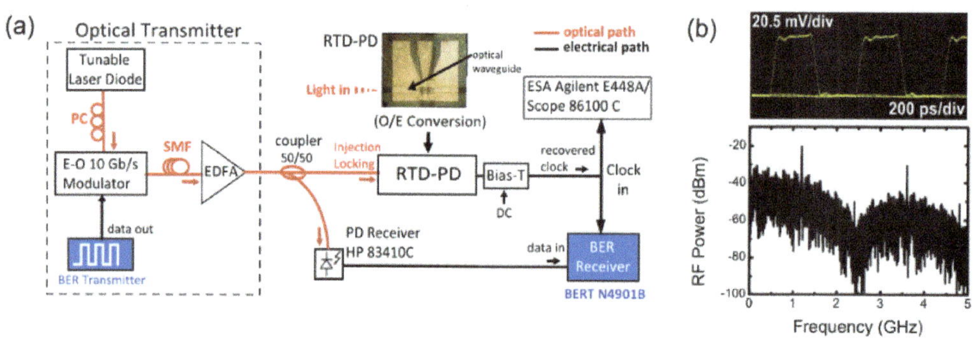

Figure 7. (a) Experimental schematic for timing extraction and BER measurements via optical injection locking. (b) RZ data stream input of 1.24 Gb/s with 100 mV amplitude ($V_{p\text{-}p}$) and 50% duty cycle using a PRBS of $2^7 - 1$; (top): eye diagram of RZ data; (bottom): RZ data signal measured in the frequency domain.

Clock recovery and error free timing extraction results are demonstrated using return-to-zero (RZ) format data signals that optically injection locks the free-running RTD-PD oscillator. A 1.24 Gb/s RZ signal generated from a bit-error-rate test (BERT) transmitter, whose spectrum together with the respective eye diagram are shown in Figure 7b, was used to modulate a tunable laser diode optical carrier at 1.550 nm using a 10 Gb/s modulator that then fed an EDFA which output was subsequently split in a 50/50 coupler. One of the coupler outputs went to a commercial photo-receiver whose output was connected to the error analyzer. The other part of the optical signal was coupled to the RTD-PD optical input, Figure 7a.

Figure 8a shows the frequency spectrum of the data after clock recovery via optical injection of a 3 mW in fiber optical power signal, showing a large frequency component at the clock frequency. Figure 8b presents the power spectra of the free-running oscillation slightly above 1.24 GHz and the injection locked output at 1.24 GHz. The output spectrum was clearly pulled to 1.24 GHz in the presence of the optical injected signal.

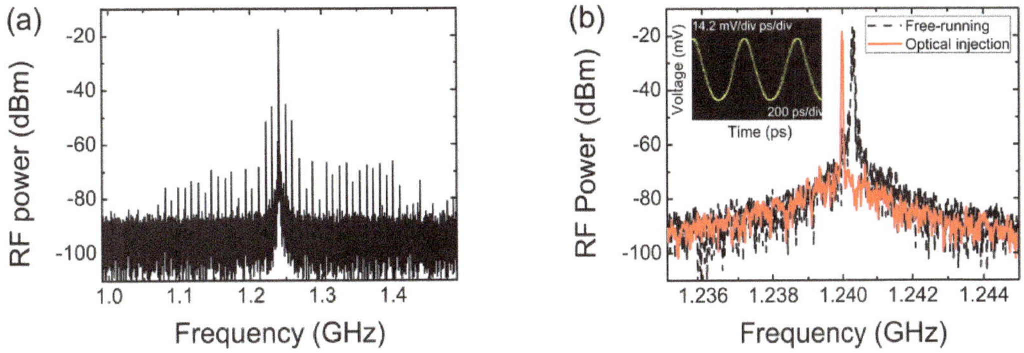

Figure 8. (a) The 1.24 GHz clock recovery in the frequency domain from an optically injected RZ data signal using an in fiber optical power level of ~3 mW. (b) Power spectra of the free-running signal and the 1.24 GHz recovered clock (span of 10 MHz). The inset shows the recovered clock in the time domain.

The inset of Figure 8b presents also the injection locked RTD-PD output signal in time domain. A phase noise reduction of around 24 dB at 10 kHz offset from the carrier frequency was achieved for an optical injection power level in fiber of around 6 mW, Figure 8a. The minimum optical power in the fiber to obtain locking was around 1.1 mW with a locking range of 0.3 MHz. The locking range and the phase noise reduction were strongly dependent on the power of the optically injected signal. The phase noise was considerably reduced as shown in Figure 9a. The RMS jitter measured by the oscilloscope was 2.16 ps, which also matches well with that from phase noise measurement. The integrated timing jitter over a 90-kHz bandwidth was below 2 ps.

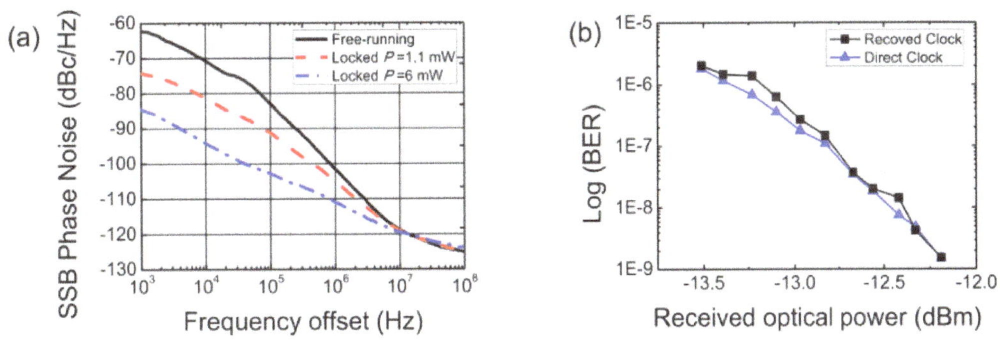

Figure 9. (a) SSB phase noise of free-running and recovered clock signals. (b) BER measurements of recovered and direct clocks.

A comprehensive way to characterize the performance of a recovered clock signal is to measure the BER. First, the quality of data detection was investigated for a direct clock situation where the clock signal from the pattern generator was used to trigger the BERT equipment. Next, without illumination, because of the drifting of the RTD oscillation frequency it was not possible to provide a clock signal to trigger the BERT instrument. Then, after illumination the optical injection locking mechanism provided a stable oscillation reference source and clock recovery was performed where the optically injection locked RTD-PD electrical output was used to trigger the BER equipment. Figure 9b shows the BER at 1.24 Gb/s as a function of the received power for both direct and recovered clocks using a data pattern length of $2^7 - 1$. The two BER curves are similar, confirming the low jitter of the extracted clock signal. It is believed that the mismatch in some of the BER points results from environmental instabilities (vibration, temperature fluctuations) which can be reduced significantly with further optimizations in the measurement setup.

In summary, clock recovery using RTD-PD oscillators was demonstrated by taking advantage of the optical injection locking capabilities of the devices. One of the attractive properties include the power of the recovered signal is independent of the input power of the signal to be recovered. Other attractive properties include frequency tunability and relatively wide tracking range.

Fiber Optic Supported Radio Frequency Communications

The demand in the coverage and speed of optical communications networks, together with rapid growth in wireless networks, has motivated an increasing attention into hybrid wireless-photonics communication systems. The integration and transmission of RF signals over fiber, RoF systems, are expected to play an important role in broadband wireless communication systems. Taking advantage of optical fiber low loss and wide bandwidth transmission characteristics, it is possible to distribute broadband data and/or high frequency signals to many wireless subscribers. Among several approaches for realizing fiber optic supported radio frequency communication systems, many efforts has been dedicated to providing low cost and simple radio access points bringing the fixed network bandwidths closer to the mobile users [3,4].

Digital standards based on phase modulation, such as Gaussian minimum shift keying (GMSK), quadrature phase shift keying (QPSK), and quadrature amplitude modulation (QAM) are the basis for communications through the latest generation of mobile telephones and wireless local area networks (LAN). In this subsection, results on the transmission of digital modulated signals are demonstrated using a wireless/photonics arrangement that take advantage of the optical injection locking characteristics of RTD-PD oscillator circuits, and the error vector magnitude (EVM) performance of the RTD-PD for different phase and frequency modulation formats are evaluated, namely phase shift keying (PSK), Gaussian minimum shift keying (GMSK) and frequency shift keying (FSK). The investigation presented here allows us to assess the implementation of RTD-PD-based oscillator transceivers as simple and compact low phase-noise oscillators for optical/electrical signal processing units in fiber optic supported radio frequency communications and cellular network base stations. Although in this work we operate at 1-2 GHz frequencies partly because the signals are easier to analyze and the data rates easier to measure with the equipment available in our laboratories, the principles we present could be replicated at higher frequencies [18].

The setup of an optical-RoF downlink is shown schematically in Figure 10. The vector signal generator (VSG) signal was used to modulate an optical carrier generated by a tunable laser diode at 1,550 nm using a 10 Gb/s Mach-Zehnder modulator. The modulated optical signal was subsequently amplified by an EDFA, then fed to a variable optical attenuator, and then split in a 90/10 coupler. One of the coupler outputs went to an optical multimeter used for controlling the power level. The other part of the optical signal was coupled to the RTD-PD ridge waveguide, which locks to the carrier modulated on the incoming optical signal and converts it to the electrical domain, with responsivity around 0.25 AAV. Finally, a power supply provides the necessary DC biasing and a spectrum analyzer down converts the received signal to be demodulated by the Agilent 89600 VSA software.

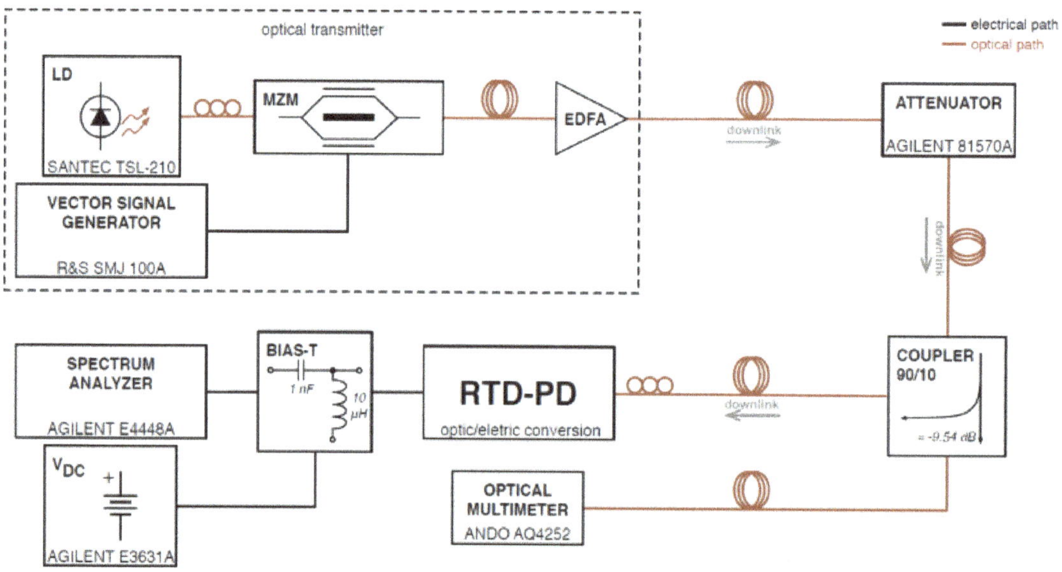

Figure 10. Description of the downlink experimental setup.

The locking range was measured as a function of the injected optical carrier power. It was found to be in average {2, 4.2, 6.5, 9} MHz, for injected optical powers of {−0.5, 3.5, 6.5, 9.5} dBm, respectively. The power of the electrical carrier on the VSG was kept at 17 dBm, and we verified that the locking range also depends significantly on this parameter. The results show that the higher the power of the modulating electrical carrier or the injected optical carrier the larger the locking range. The measured electrical output power of the RTD-PD was −7.9 dBm. The measured performance for different modulation formats is shown in Figure 11, where the EVM was obtained as a function of the power of the electrical carrier modulation, using a data-rate of 2 Msymbols/s, for a fixed injected optical power of 9.5 dBm. The PSK modulation formats provide EVM results no better than 14%, while FSK based formats achieve 0.55% and GMSK attains 1% in the best case, which is again one order of magnitude better. It can be observed that the downlink performance is generally affected by the limitations associated to the optical injection process itself, since the coupling method is still very incipient, it could be optimized in a future prototype, for example by including facets with anti-reflection coatings, or tapering the ridge end region. The poor efficiency of the optical coupling leads us to the necessity of amplifying the optical signal, which introduced noise and consequently degraded the EVM results. Furthermore, there is observable performance degradation for electrical modulated powers above 10 dBm, which results from the nonlinearity in the Mach-Zehnder modulator.

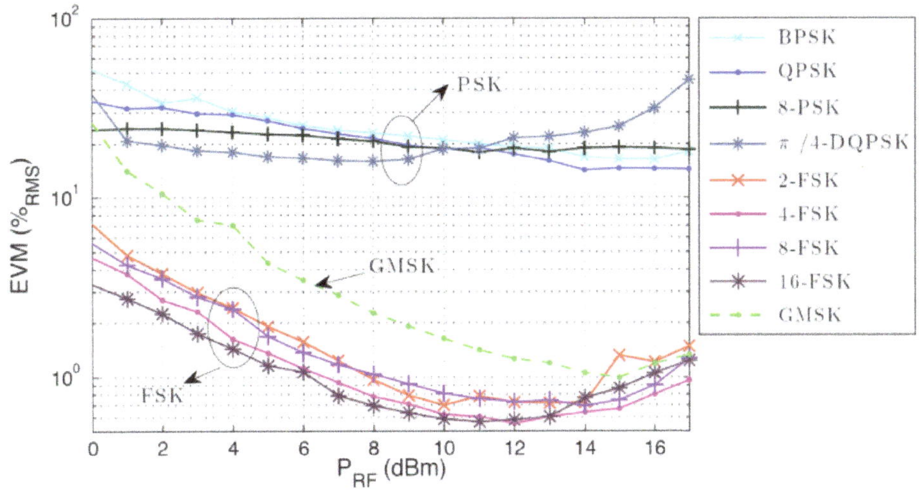

Figure 11. EVM performance comparison for PSK, FSK and GMSK modulated formats for the downlink.

As a summary, we have determined the error vector magnitude performance of different phase and frequency modulation using RTD-PD based oscillator circuits. We conclude that frequency modulated formats achieve better performance, although the Gaussian Minimum Shift Keying format also appears to be a practical alternative. Furthermore, future activities may address the performance of the device under realistic simultaneous downlink traffic, as well as the optimization of the optical injection efficiency.

CONCLUSIONS

We have presented double barrier quantum well resonant tunneling diode based optical waveguide photo-detectors for the implementation of novel optical injection locked microwave oscillators. These novel devices operate at optical communication wavelengths. We have demonstrated efficient detection of GHz modulated optical carriers and optical control of RTD high-frequency oscillator when RTD-PD devices are operated in the non-oscillating and oscillating regimes. When operating in the oscillating regimes clock recovery from return-to-zero format data by using injection locking of an RTD-PD free-running oscillator circuit was successfully demonstrated. Finally, transmission of digital modulated signals was demonstrated using a wireless/photonics arrangement that takes advantage of the optical injection locking characteristics of RTD-PD oscillator circuits. The error vector magnitude performance of the RTD-PD for different phase and frequency modulation formats was evaluated.

RTD-PDs provide simple and compact circuit configurations and can find extensive applications in high-speed optical communications such as synchronization in communication networks and radio-over-fiber links. RTDs are very high-speed devices and we anticipate that the systems and devices we have presented here could be implemented at much higher frequencies.

ACKNOWLEDGMENTS

This work was supported by FCT under the project WOWi (PTDC/EEA-TEL/100755/2008) and in the framework of programme POCTI/FEDER with grant REEQ/1272/EEI/2005. B. Romeira thanks FCT Portugal for a Postdoctoral Fellowship (Grant No. SFRH/BPD/84466/2012). The authors acknowledge the fruitful discussions with Anthony E. Kelly, and Horacio Cantú of the School of Engineering, University of Glasgow, United Kingdom, on RTD photo-detectors and optoelectronic oscillators.

REFERENCES

1. Sackinger, E. Broadband Circuits for Optical Fiber Communication; John Wiley & Sons, Inc.: Hoboken, NJ, USA, 2005; pp. 25–40.
2. Granastein, V. L. Physical Principles of Wireless Communications; CRC Press: Boca Raton, FL, USA, 2012; pp. 171–201.
3. Sauer, M.; Kobyakov, A.; George, J. Radio over fiber for picocellular network architectures. J. Lightware Technol. 2007, 25, 3301–3320.
4. Gomes, N.J.; Morant, M.; Alphones, A.; Cabon, B.; Mitchell, J.E.; Lethien, C.; Csornyei, M.; Stohr, A.; Iezekiel, S. Radio-over-fiber transport for the support of wireless broadband services. J. Opt. Netw. 2009, 8, 156–178.
5. Klemm, M.; Leendertz, J.A.; Gibbons, D.; Craddock, I.J.; Preece, A.; Benjamin, R. Microwave radar-based breast cancer detection: Imaging in inhomogeneous breast phantoms. IEEE Ant. Wirel. Propag. Lett. 2009, 8, 1349–1352.
6. Guetin, P. Interaction between a light beam and a Gunn oscillator near the fundamental edge of GaAs. J. Appl. Phys. 1969, 40, 4114–4122.
7. Kiehl, R.A. Behavior and dynamics of optically controlled TRAPATT oscillators. IEEE Trans. Electron Devices 1978, 25, 703–710.
8. Yen, H.W.; Barnoski, M.K.; Hunsperger, R.G.; Melville, R.T. Switching of GaAs IMPATT diode oscillator by optical illumination. Appl. Phys. Lett. 1977, 31, 120–122.
9. Lee, K.H.; Kim, J.Y.; Choi, W.Y.; Kamitsuna, H.; Ida, M.; Kurishima, K. Low-cost optoelectronic self-injection-locked oscillators. IEEE Photonic Technol. Lett. 2008, 20, 1151–1153.

10. Shumakher, E.; Magrisso, T.; Kraus, S.; Elias, D.C.; Gavrilov, A.; Cohen, S.; Eisenstein, G.; Ritter, D. An InP HBT-based oscillator monolithically integrated with a photodiode. J. Lightwave Technol. 2008, 26, 2679–2683.
11. Seeds, A.J.; Williams, K.J. Microwave photonics. J. Lightware Technol. 2006, 24, 4628–4641.
12. Ramond, T.; Hollberg, L.; Juodawlkis, P.W.; Calawa, S.D. Low noise optical injection locking of a resonant tunneling diode to a stable optical frequency comb. Appl. Phys. Lett. 2007, 90, 171124.
13. Romeira, B.; Figueiredo, J.M.L.; Ironside, C.N.; Kelly, A.E.; Slight, T.J. Optical control of a resonant tunneling diode microwave-photonic oscillator. IEEE Photonic Technol. Lett. 2010, 22, 1610–1612.
14. Romeira, B.; Seunarine, K.; Ironside, C.N.; Kelly, A.E.; Figueiredo, J.M.L. A self-synchronized optoelectronic oscillator based on an RTD photo-detector and a laser diode. IEEE Photonic Technol. Lett. 2011, 23, 1148–1150.
15. Romeira, B.; Javaloyes, J.; Figueiredo, J.M.L.; Ironside, C.N.; Cantú, H.I.; Kelly, A.E. Delayed feedback dynamics of liénard-type resonant tunneling-photo-detector optoelectronic oscillators. IEEE J. Quantum Electron. 2013, 49, 31–42.
16. Knappe, S.; Shah, V.; Schwindt, P.D.D.; Hollberg, L.; Kitching, J.; Liew, L.-A.; Moreland, J. Microfabricated atomic clock. Appl. Phys. Lett. 2004, 85, 1460.
17. Romeira, B.; Figueiredo, J.M.L.; Slight, T.J.; Wang, L.; Wasige, E.; Ironside, C.N.; Kelly, A.E.; Green, R. Nonlinear dynamics of resonant tunneling optoelectronic circuits for wireless/optical Interfaces. IEEE J. Quantum Electron. 2009, 45, 1436–1445.
18. Wang, J.; Wang, L.; Li, C.; Romeira, B.; Wasige, E. 28 GHz MMIC resonant tunneling diode oscillator of around 1 mW output power. Electron. Lett. 2013, 49, 816–819.
19. Ishigaki, K.; Shiraishi, M.; Suzuki, S.; Asada, M.; Nishiyama, N.; Arai, S. Direct intensity modulation and wireless data transmission characteristics of terahertz-oscillating resonant tunnelling diodes. Electron. Lett. 2012, 48, 482–483.
20. Cantú, H.I.; Romeira, B.; Kelly, A.E.; Ironside, C.N.; Figueiredo, J.M.L. Resonant tunneling diode optoelectronic circuits applications in radio-over-fiber networks. IEEE Trans. Microware Theory 2012, 60, 2903–1912.
21. Hartmann, F.; Langer, F.; Bisping, D.; Musterer, A.; Hofling, S.; Kamp, M.; Forchel, A.; Worschech, L. GaAs/AlGaAs resonant tunneling diodes with a GaInNAs absorption layer for telecommunication light sensing. Appl. Phys. Lett. 2012, 100, 172113.
22. Moise, T.S.; Kao, Y.-C.; Goldsmith, C.L.; Schow, C.L.; Campbell, J.C. High speed resonant-tunneling photodetectors with low-switching energy. IEEE Photonic Technol. Lett. 1997, 9, 803–805.
23. Li, H.W.; Kardynal, B.E.; Ellis, D.J.P.; Shields, A.J.; Farrer, I.; Ritchie, D.A. Quantum dot resonant tunneling diode single photon detector with aluminum oxide aperture defined tunneling area. Appl. Phys. Lett. 2008, 93, 153503.
24. Moise, T.S.; Kao, Y.-C.; Garrett, L.D.; Campbell, J.C. Optically switched resonant tunneling diodes. Appl. Phys. Lett. 1995, 66, 1104–1106.
25. Chaisakul, P.; Marris-Morini, D.; Isella, G.; Chrastina, D.; Rouifed, M.-S.; Le Roux, X.; Edmond, S.; Cassan, E.; Coudevylle, J.-R.; Vivien, L. 10-Gb/s Ge/SiGe multiple quantum-well waveguide photodetector. IEEE Photonic Technol. Lett. 2011, 23, 1430–1432.
26. Fidaner, O.; Okyay, A.K.; Roth, J.E.; Schaevitz, R.K.; Yu-Hsuan, Kuo; Saraswat, K.C.; Harris, James S; Miller, D.A.B. Ge–SiGe quantum-well waveguide photodetectors on silicon for the near-infrared. IEEE Photonic Technol. Lett. 2007, 19, 1631–1633.
27. Figueiredo, J.M.L.; Ironside, C.N.; Stanley, C.R. Electric field switching in a resonant tunneling diode electroabsorption modulator. IEEE J. Quantum Electron. 2001, 37, 1547–1552.
28. Chuang, S. L. Physics of Optoelectronic Devices; Wiley: NY, USA, 1995.
29. Alkeev, N.V.; Lyubchenko, V.E.; Ironside, C.N.; Figueiredo, J.M.L.; Stanley, C.R. Super high-frequency characteristics of optical modulators on the basis of InGaAlAs resonance-tunnel heterostructures. J. Commun. Technol. Electron. 2000, 45, 911–914.

CHAPTER 14

TUNNEL JUNCTION WITH PERPENDICULAR MAGNETIC ANISOTROPY

Mengxing Wang, Yue Zhang, Xiaoxuan Zhao and Weisheng Zhao

Keywords: Electronic properties and materials, Magnetic properties and materials, Metals and alloys

INTRODUCTION

Following Moore's law, the development of semiconductor industry, especially the complementary metal oxide semiconductor (CMOS) technology, has deeply impacted the world since 1970s. However, as the advanced node scaling goes below 45 nm, static power consumption increases faster than the dynamic power consumption due to intrinsic current leakage, which challenges the progress of very large scale integrated circuits (VLSI). Additionally, the long distance data traffic between memory and logic chip causes large dynamic power consumption and interconnection delay [1]. In 2015, Gordon Moore claimed that his law would reach saturation during the next decade. Novel approaches based on spintronics has opened another avenue to continuous scaling, among which magnetic tunnel junction (MTJ) has been recommended as a potential replacement since its first proposal and demonstration [1,2,3].

MTJ is basically configured with ultra-thin multilayer, *i.e.*, the free and reference layer comprising ferromagnetic metal with a sandwiched metal oxide film as tunnel barrier. Driven by external magnetic field or charge current, the magnetization direction of free layer can switch between parallel and antiparallel states (P and AP) relative to that of reference layer, thus the corresponding low or high resistance states (R_L and R_H) can be used for data storage. The margin of two resistance states is defined as tunnel magnetoresistance ratio (TMR ratio):

$$TMR\ ratio = \frac{R_H - R_L}{R_L} \qquad (1)$$

Such characteristics allow the proposal and commercialization of magnetic random access memory (MRAM) by integrating the device onto the silicon based CMOS circuit, which demonstrates non-volatility, fast access speed, and infinite endurance. However, MTJ with amorphous Al_2O_3 tunnel barrier suffers from the modest TMR ratio below 70% [4], thus the reliability of reading operation cannot be assured. In 2004, high TMR ratio of 180% was observed by Yuasa *et al.* [5,6] from MTJ using single crystal MgO tunnel barrier as particular symmetry filter. Since then, MgO has become the mainstream material for tunnel barriers.

Moreover, the property of MTJ also relies on its magnetic anisotropy, specifically when current induced magnetization direction switching, *i.e.*, the spin transfer torque (STT) switching, is taken into consideration. Tremendous work has been undertaken towards MTJ with in-plane magnetic anisotropy referring to shape anisotropy. Although having achieved giant TMR ratios up to 604% [7], in-plane magnetic anisotropy based MTJ still faces the obstacles from thermal stability and power consumption. Thermal stability is determined by energy barrier E between P and AP magnetization alignments:

$$E = \frac{M_s H_k V}{2} \qquad (2)$$

where M_s is the saturation magnetization, H_k the anisotropy field, and V the volume of free layer. It is reasonable that scaled node weakens the shape anisotropy from the ellipse nano-pillar cross section, which is also difficult to fabricate when shrinking down to 22 nm. On the other hand, the out-of-plane precession during spin transfer torque (STT) switching contributes to the

Citation: Mengxing Wang, Yue Zhang, Xiaoxuan Zhao and Weisheng Zhao, Tunnel Junction with Perpendicular Magnetic Anisotropy, doi:10.3390/mi6081023

power consumption. These disadvantages can be eliminated by introducing perpendicular magnetic anisotropy (PMA), thus MTJ with PMA is promising to enable high-density low-power VLSI. Consequently, the key point of research turns to how to achieve strong PMA for practical application [8,9].

In the rest of this review, we firstly introduce the fundamentals of PMA and STT in the memory point of view, followed with the relevant description of fabrication and characterization methods. Considering the progress of PMA, especially interfacial PMA, we discuss the crucial challenges and potential solutions with respect to material, structure, and processing, as well as other attractive research and applications. Finally, we give a brief conclusion and prospective.

PERPENDICULAR MAGNETIC ANISOTROPY PRINCIPLES

PMA in periodic Co/Pd multilayer with Co layer thinner than 0.8 nm was observed by Carcia *et al.* [10] in 1985. They attributed its origin to the anisotropy at Co/Pd interface and the strain in Co thin film. For the multilayer based on transition and noble metal such as Co/Pd (Pt), PMA, saturation magnetization, anisotropy field, and other parameters can be easily modulated by adjusting the number of bilayer periods, as well as the thickness of each layer [10,11,12]. However, the face-centered cubic fcc Co/Pd (Pt) (111) multilayers possess low spin polarization compared to CoFeB or CoFe, and cannot favor the formation of MgO (001) tunnel barriers, thereby impeding the improvement of TMR ratio [13,14]; Moreover, the free layer thickness should be reduced for lower threshold current to manipulate magnetization direction reversal. Thus, Yakushiji *et al.* [15,16] developed the ultra-thin Co/Pd (Pt) multilayers configured with 0.14–0.2 nm films, which present not only large PMA, but also a relatively high TMR ratio of 62% and an ultra-low resistance area product (RA) with adherent CoFeB film. For the MTJ consists of CoFe/Pd multilayer, MgO tunnel barrier, and CoFeB insertion between, TMR ratio has been promoted to 78% after being annealed at 200 °C [17]. Even though, this kind of MTJ still suffers from complicated structure, weak thermal stability, and large damping constant α.

To realize general application, especially the low-power spin transfer torque magnetic random access memory (STT-MRAM), numerous materials and structures have been studied, among which amorphous rare earth transition metal alloys (e.g., TbCoFe) attract interest due to their large PMA and remnant squareness. However, the employment of TbCoFe degrades the thermal endurance of MTJ, while the TMR ratio cannot be increased by post-deposition annealing because no MgO crystallization happens [18].

The milestone of PMA is the CoFeB/MgO structure based MTJ proposed by Ikeda *et al.* [19] in 2010. This work reported excellent TMR ratio of 120%, high thermal stability, and low switching current of 49 μA at 40 nm diameter, hereafter the research focus has converted to the CoFeB/MgO interface. In detail, the easy axis turns from in-plane to out-of-plane direction when the thickness of CoFeB is reduced below 1.5 nm and exhibits a 3.4 kOe anisotropy field with 1.3 nm CoFeB film. With respect to the crystal structure, the amorphous CoFeB crystalized into body-centered cubic bcc (001) texture via MgO (001) template during post-deposition annealing. Therefore, in addition to the thickness variation of MgO tunnel barrier, lattice mismatch at CoFeB/MgO interface happens during the sputtering and annealing process, resulting in failure and reliability challenges of STT-MRAM [20]. Actually, first-principles calculation has predicted the existence of interfacial PMA in Fe/MgO bilayer prior to the experimental discovery [21]. Afterwards, Yang *et al.* [22] extended the study to Co/MgO interface. The origin of interfacial PMA is attributed to the overlap between Fe-3d and O-2p orbitals at the interface, as well as the spin-orbit coupling (SOC) induced Fe d orbitals hybridizations in out-of-plane direction.

SPIN TRANSFER TORQUE: FUNDAMENTAL AND MEMORY OPERATION

STT effect was theoretically predicted by Slonczewski *et al.* [23,24] in 1996 and experimentally observed in MTJ with CoFe/Al_2O_3/CoFe structure in 2004. A charge current, instead of the external magnetic field for classic MRAM, is used to switch the free layer magnetization direction. The primary mechanism of STT switching can be illustrated with the typical one transistor one MTJ STT-MRAM cell shown in Figure 1a, where the write operation is performed by changing the polarity of charge current passing through the MTJ selected by bit and word line. The writing current can be spin polarized by the reference layer, transfer its spin angular momentum to free layer magnetic moment, and eventually reorient it by induced torque.

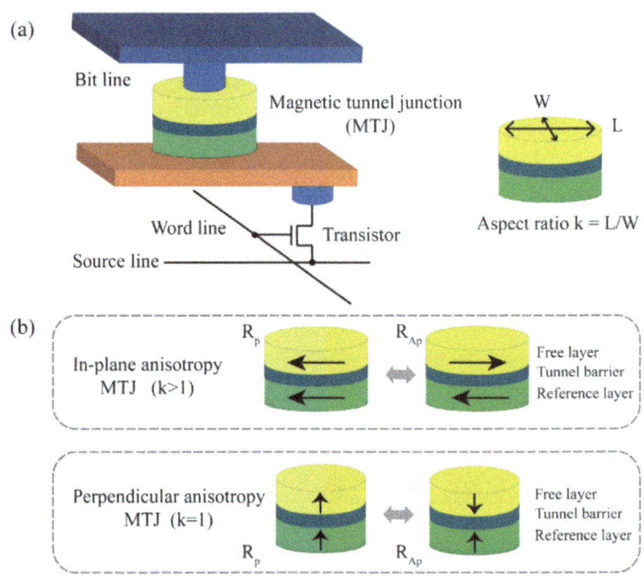

Figure 1. (a) One transistor one magnetic tunnel junction (MTJ) cell using mechanism of spin transfer switching, where the MTJ is selected by word line and transistor, and operated by bit line. (b) Comparison between in-plane and perpendicular anisotropy based MTJ nano-pillar [25].

After the significant enhancement of TMR ratio realized by replacing Al_2O_3 with MgO tunnel barrier, the first STT switching of perpendicular magnetic moment was clearly observed in MTJ with TbCoFe/CoFeB/MgO/CoFeB/TbCoFe structure fabricated at room temperature [26]. Although low threshold current density of 4.7×10^6 A/cm^2 has been achieved in comparison with in-plane magnetic anisotropy based MTJ, the TMR ratio is only 15% for the reason mentioned before. The theoretical expression of threshold current is:

$$I_{c0} = \alpha \frac{\gamma e}{\mu_B g} M_s H_k V \tag{3}$$

where the notation α is the damping constant, γ the gyromagnetic constant, e is the charge of electron, μ_B is the Bohr Magnetron, and g is a parameter determined by the spin polarization and the angle between free and reference layer magnetization directions. Another parameter should be concerned with is thermal stability factor, which is defined as:

$$\Delta = \frac{E}{k_B T} = \frac{K_{eff} V}{k_B T} \tag{4}$$

where E denotes energy barrier as Equation (2), K_{eff} the effective anisotropy energy constant defined by $K_{eff} = E/V$, k_B the Boltzmann constant, and T the absolute temperature. It is evident that both I_{c0} and Δ are in direct proportion to device dimension, which means K_{eff} should be elevated with technology node shrinking to ensure 10 year data retention against thermal fluctuation. For instance, the design of gigabit memory with 1X nm PMA MTJ requires K_{eff} larger than 10^7 erg/cm^{3}for Δ more than 40. Furthermore, the compromise between low power consumption and long retention time can be reached through α reduction, thus the ratio of Δ/I_{c0} considered without the influence introduced by Δ variation among devices.

Compared with in-plane magnetic anisotropy based MTJ, whose energy barrier E originates from the asymmetric ellipse shape, PMA based MTJ depends on the interfacial atom arrangement, thus circular device which is more manufacturable can be adopted for smaller feature size (Figure 1b). On the other hand, I_{c0} of MTJ with PMA is lower for the cancel of demagnetization field. In view of this, PMA based MTJ has been recommended as the most potential candidate to realize high-density memory. Additionally, the write and read speed of PMA based MTJ should be minimized to further enable the replacement of static random access memory (SRAM) and embedded dynamic RAM (eDRAM) [9]. Although switching speed of 1 ns using writing current as eight times as I_{c0} has been reported by Worledge *et al.* [27] in 2011, the fastest data obtained from in-plane magnetic anisotropy based MTJ is 190 ps [28]. It is worth noting that the switching speed is determined by the amplitude of writing current rather than the orientation of magnetic moment, because the switching mode transits from being processional to thermal fluctuation assisted by STT along with extended pulse duration, further indicating the importance to pursue lower I_{c0}.

FABRICATIONS AND MEASUREMENTS OF STT-MRAM DEVICE

The integration of STT-MRAM is an "above IC" process, thus the adjustment of MTJ device process to standard CMOS technology is absolutely required to realize potential applications (Figure 2a). Generally speaking, the fabrication mainly comprises substrate preparation, stack deposition, annealing, etching, dielectric encapsulation, and wiring; each procedure should be carefully examined with relevant equipment and instruments. Figure 2b depicts the typical flow of art.

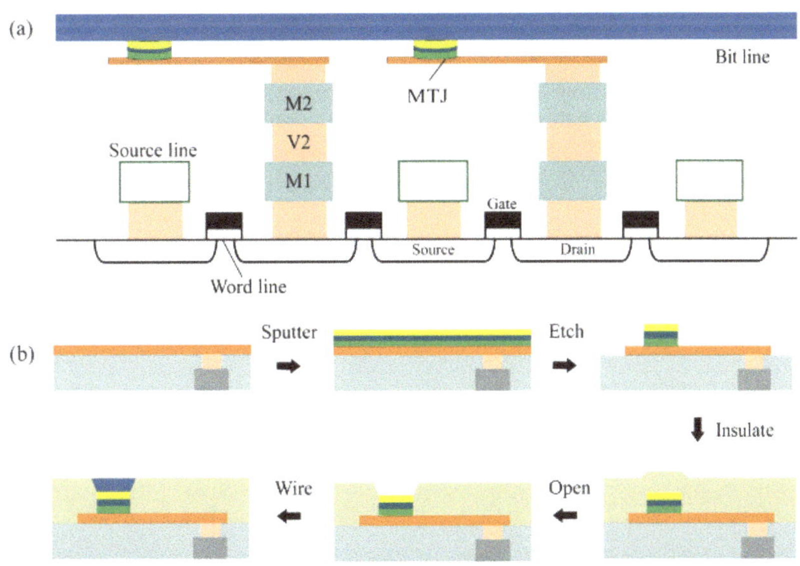

Figure 2. (a) Topology of spin transfer torque magnetic random access memory (STT-MRAM) with two cells atop the silicon based complementary metal oxide semiconductor (CMOS) front-end circuit. M1/M2 and V1 denote different level of metal layers and via, separately. (b) Typical flow of MTJ device fabrication corresponding to (a).

Magnetic Tunnel Junction Device Process Technology

Substrate Preparation

To produce STT-MRAM chip, the transistor-based circuit can be prepared in front end of line (FEOL) process by standard COMS technology. Then, an insulation layer (e.g., SiN) is deposited on the substrate, where metal contacts are formed for the interconnection between CMOS circuit and MTJ stack. Surface roughness is a critical parameter requiring optimization to avoid Neel coupling, hence the procedure usually ends up with polish by chemical mechanical planarization (CMP) or other methods. Moreover, film defects, oxidation, and CMP residual should be minimized against open/short circuit or other negative situations. For the experimental fabrication, thermally oxidized Si wafer is preferred as substrate for the favorable properties and prices. Pre-cleaning is necessary to ensure device quality and measurement accuracy, e.g., ultrasound bathed in acetone for one hour, followed by nitrogen drying and optical microscope check.

Stack Deposition

Following the preparation of smooth substrate, ultra-thin films of MTJ stack can be grown with advanced deposition tools, among which the typical ones includes sputtering or molecular beam epitaxy (MBE) systems. The base pressure in the ultra-high vacuum chamber should be lower than 10^{-8} Torr to obtain optimum interfacial PMA. Deposition rate should be controlled under 0.1 nm/s to accurately regulate the thickness of ultra-thin film, as well as the uniformity and oriented texture. Firstly, at least one nonmagnetic metal buffer layer is deposited to benefit surface flatness, film adhesion, and crystalline growth (e.g., Ta/Ru/Ta). Then, the essential part of MTJ stack (e.g., CoFeB/MgO/CoFeB) can be grown; in the case of in-plane magnetic anisotropy based MTJ, pinned layer of antiferromagnetic metal (e.g., PtMn) should be deposited adjacent to reference layer to fix the magnetization direction. Finally, at least one nonmagnetic capping layer (e.g., Ta/Ru/Ta/Ru) should be coated for protection, as well as for interfacial PMA induction in certain configuration.

How to grow perfect MgO (001) tunnel barrier is the most notable challenge of this procedure. Giant TMR ratio has been achieved using MBE [5]. With sputtering, this film is usually deposited from MgO target by radio-frequency magnetron gun under Ar atmosphere. Pre-sputtering Ta getter has been suggested as one of the effective method to remove H_2O and O_2 residual gas before MgO deposition, which improves the stoichiometry and interstitial defects in tunnel barriers. Another method is to deposit Mg metal first and then strictly oxidize it into MgO by either natural or plasma process [29]. High quality of MgO tunnel barriers enables the achievement of giant TMR ratio and acceptable RA. However, the thickness variation and lattice mismatch issues noted before still need to be solved for high reliability.

In view of throughput and quality, sputtering should be the most suitable tool to realize mass manufacture and general application, thus Canon ANELVA (Kawasaki, Japan), SINGULUS TECHNOLOGIES (Kahl am Main, Germany), and other tool vendors, have developed advanced systems capable of processing 300 mm diameter wafers with multiple sources in one chamber. Take the tool of IBM for illustration, this system is an integration of sputtering, oxidation, and pre-cleaning chambers, where wafers can be transferred with a central robot without breaking the vacuum. Nevertheless, there is still space for promotion.

Post-Deposition Annealing

After MTJ stack deposition, the sample is subsequently annealed at ~300 °C in vacuum for certain time, e.g., 1 h. Effect of this heat treatment can be illuminated from structural and natural points of view. As has been mentioned, with (001) texture MgO as template, the amorphous CoFeB can crystalized into bcc (001) texture at high temperature, which is critical to generate high TMR ratio, as well as interfacial PMA in CoFeB/MgO structure based MTJ. Moreover, the annealing temperature dependent tendency of magnetization M and K_{eff} can be observed in Figure 3 [30], where the initial increase can be attributed to the B absorption by Ta, and the later decrease arises from the formation of additional magnetic dead layer. Referring to in-plane magnetic anisotropy based MTJ, this procedure should be performed in an external magnetic field around 10 kOe to fix the magnetization direction reference layer through the exchange bias with pinned layer. However, the standard CMOS fabrication temperature is 400 °C, hence the integration of MTJ device demands robust compatibility with elevated temperature. This is a key challenge for the commercial development of MRAM or STT-MRAM.

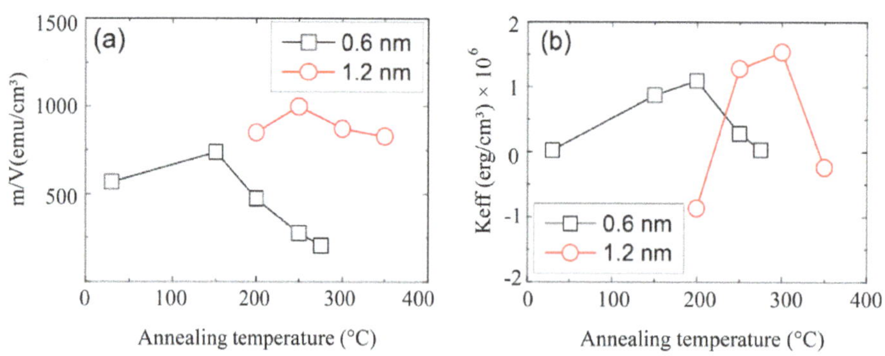

Figure 3. Annealing temperature dependence of (a) magnetic moment per unit volume (m/V); (b) K_{eff} for films Ta 1/CoFeB*t*/MgO 2/Ta 1 (units in nm): *t* is 0.6 nm for the black squares and 1.2 nm for the red circle. Reproduced with permission from Sinha *et al.* [30], Journal of Applied Physics; published by AIP Publishing, 2015.

Device Etching

After magnetic measurement certification, the blanket MTJ stack is patterned into a nanoscale device with lithography and etching process. Ultra-violet lithography (UVL) can define photomask of micron order, while the delicate MTJ shape needs to use electron beam lithography (EBL) for ultra-high resolution. Available etchers have evolved from ion beam etching (IBE) to diverse options, including reactive ion etching (RIE), inductively coupled plasma (ICP), and other systems, from which more than one technique are often chosen and mixed.

First, the bottom electrode is patterned with hard mask to bridge the metal contact below and MTJ on the top. Typical hard mask materials include SiO_2, Ta, and TiN, and the pattern is transferred from photoresist photomask by ICP/RIE. The difficulty of patterning is significantly reduced due to the circular shape of PMA based MTJ device.

Second, hard mask for MTJ device etching is defined by EBL and ICP/RIE following a similar route, with which unnecessary material is subsequently removed by IBE or other methods for nano-pillar profile. The nanoscale etching usually terminates over MgO tunnel barrier or penetrates it, leaving certain conductive films as the bottom electrode. This can be accurately monitored by secondary ion mass spectroscopy (SIMS) end point detector mounted in the process chamber. During the process, shorts across MgO tunnel barrier, which is a limit for yield, may be caused by non-volatile residuals redeposited on sidewall.

Here, it is meaningful to highlight the art of IBE/RIE/ICP. IBE has been a general apparatus for MTJ device fabrication in both laboratory and industry (Figure 4a). This technique is fully based on physical mechanism, *i.e.*, removing surface atoms with the force of incident ion beam. IBE can be universally used for most materials with little chemical damage to magnetic property, whereas the device obtained suffers redeposition and shadow effect. Wafer tilt and rotation can improve the device profile, but limits its application in high-density array patterning [31]. Besides, enlarging the size of ion source to deal with 300 mm wafer while maintaining beam uniformity and directionality remains a challenge [32,33]. In contrast, RIE/ICP are more suited for mass production. Introduced by RIE, chemical reaction provides the possibility of material dependent etching selectivity, which quite contributes to endpoint detection as well as morphology control. However, Cl based RIE requires processing temperature over 350 °C to enable product volatilization, which detrimentally impacts the magnetic property [34,35]. Other plasmas, e.g., NH_3/CO [36] and Methanol [37], also more or less present pros and cons. ICP presented in Figure 4b is another type of RIE. High-density plasma can be generated from electromagnetic induction, thus bringing enhanced anisotropy.

Third, compared with in-plane magnetic anisotropy based MTJ, the negative influence from etching on PMA based MTJ is more sensitive because smaller size is considered. Practically, MTJ device property degradation (e.g., coercivity and TMR ratio decrease) induced by either method is the major issue to be overcome. Kinoshita *et al.* [38] investigated this damage and proposed post-etching treatment using reductive He/H_2 plasma, specifically for the CoFeB/MgO structure based MTJ with interfacial PMA. In this study, Methanol gas, which exhibits high etching selectivity against Ta hard mask, was used to conduct ICP processing. After He/H_2 plasma treatment at 180 °C for 150 s, TMR ratio almost restored to the IBE level, and this trend can only be observed with pillar below 97 nm in size. Therefore, post-etching recovery is a necessary procedure for the scalability of PMA based MTJ.

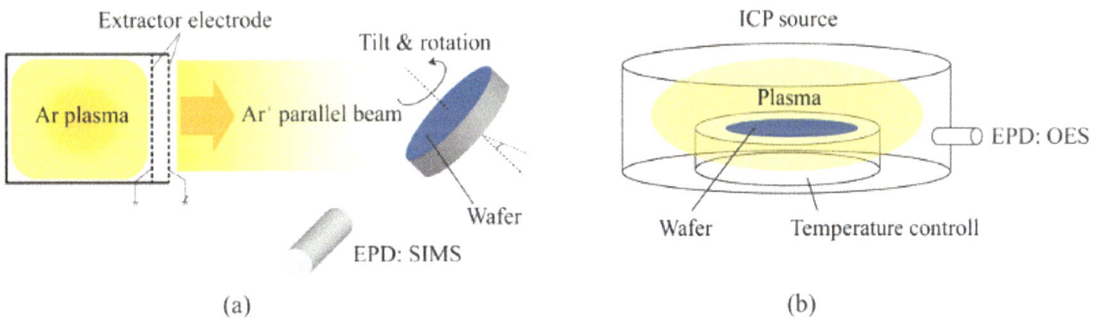

Figure 4. Schematic illustration of (a) ion beam etching (IBE) with endpoint detector (EPD), *i.e.*, secondary ion mass spectroscopy (SIMS), wafer tilt and rotation; (b) inductively coupled plasma (ICP) using an optical emission spectrometer (OES).

Dielectric Encapsulation

During the procedure, an insulation layer (e.g., SiO_2) is deposited over the bottom electrode and MTJ nano-pillar by plasma enhanced chemical vapor deposition (PECVD) or other methods for electric protection. This procedure is preferred to be performed *in-situ*, because $Mg(OH)_2$ forms instantly with MgO tunnel barrier in presence of water, leading to device property deterioration. Besides, standard temperature of PECVD for CMOS fabrication is 400 °C, where magnetic property of MTJ devices suffer. Accordingly, magnetic and dielectric material engineering should reach a compromise focusing on processing temperature. Without regard to output, a low-temperature technique for encapsulation is atom layer deposition (ALD), which demonstrates well controlled precision and isotropy.

Wiring

The final procedure is to create a connection between the MTJ device and peripheral circuit. Chemical mechanical polishing (CMP) can be undertaken at first to smooth the insulating layer surface, where contact is defined with photomask and opened until the underlying capping layer is exposed. Then, nonmagnetic metal (e.g., copper) can be deposited upon, followed with another CMP to remove the undesired wiring metal. This is the so called damascene process.

Property and Morphology Characterization

Magnetic Measuring Method

The vibrating-sample magnetometer (VSM) based on Faraday's law is the most general metrology equipment to characterize blanket MTJ ultra-thin films at room temperature. For MTJ sample with PMA, magnetic moment *versus* external out-of-plane and in-plane magnetic field (m-H) curves reveal the existence of perpendicular easy axis. For the sample with interfacial PMA, saturation magnetization M_S and magnetic dead layer thickness t_d can be given by linear fitting on m and CoFeB thickness t in the equation below:

$$m/A = M(t - t_d) \quad (5)$$

where A denotes the area of MTJ nano-pillar. Furthermore, K_{eff} is determined by bulk anisotropy constant K_b, demagnetization field $2\pi M_S^2$, and interfacial anisotropy constant K_i, as the equation shown:

$$K_{eff} = K_b - 2\pi M_S^2 + K_i/t_{eff} \quad (6)$$

here, t_{eff} is the effective CoFeB thickness determined with $t_{eff} = t - t_d$. Fitting results from Equation (6) have indicated the existence of PMA at CoFeB/MgO interface while that K_b is negligible.

Similar measurement techniques include superconducting quantum interference device (SQUID), magneto-optic Kerr effect (MOKE) magnetometer and other instruments. Each method performs differently on resolution, accuracy, sensitivity and other specifications, thus compromise should be reached according to the particular situation. Besides, TMR ratio can be quantified with current-in-plane tunneling (CIPT) method.

The film structure can be investigated by transmission electron microscope (TEM) with specially ion milled sample. As the cross-section image shown in Figure 5, free and synthetic antiferromagnetic (SAF) reference layers separated by ultra-thin 1.09 nm MgO tunnel barrier can be recognized with high resolution. During the nano-pillar etching procedure, scanning electron microscope (SEM) is an effective tool to examine device profile. In addition to critical dimension, how to achieve nearly-vertical sidewall angle (e.g., 80°) is another issue worth being concerned.

After the profile of MTJ nano-pillar is confirmed, the measurement can turn to its electrical performance, *i.e.*, the STT behavior. Four-probe method is often used at room temperature for current induced magnetization direction switching, thus resistance curve can be evaluated as a function of applied current (R-I) in absence of external magnetic field. A series of intrinsic parameters can be characterized, such as threshold current, TMR ratio, and breakdown voltage. The yield detractor is usually the fort MTJ device due to etching redeposition mentioned before. Additionally, the quality of circuit integration should be assessed by testing certain amount of MTJ device for the process variation.

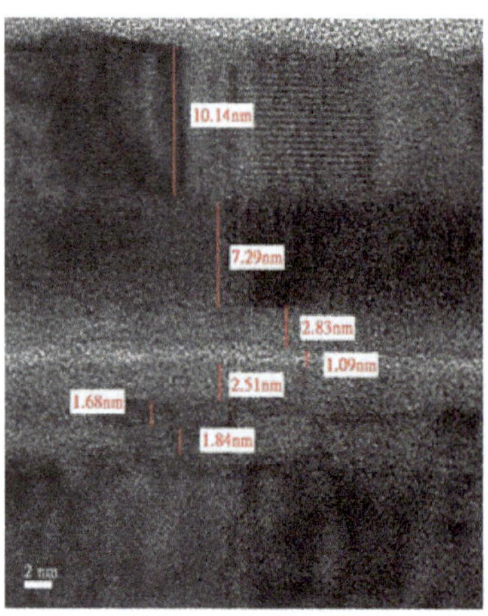

Figure 5. Cross-section image of MTJ stack, where free and synthetic antiferromagnetic (SAF) reference layers separated by ultra-thin 1.09 nm MgO tunnel barrier can be recognized respectively with high resolution.

CHALLENGES AND ENHANCEMENTS TOWARDS INTERFACIAL PMA

Material Engineering

CoFeB Behavior and Composition

The distribution of B in annealed MTJ multilayer is an interesting problem to clarify. Standing-wave (SW) hard *x*-ray photoemission spectroscopy (HXPS) analysis conducted by Greer *et al.* [39] using $Co_{20}Fe_{60}B_{20}$ showed that 19.5% of B diffuses into MgO for Mg–B–O and 23.5% forms a thin TaB-like film at Ta/CoFeB interface, and certain theory related the achievement of high TMR ration with Mg–B–O. Meanwhile, other investigations controversially claimed that B migrates into Ta film if MgO barrier without O vacancy is fabricated [40]. On the basis of these published results, Mukherjee *et al.* [41] studied series of Ta/CoFeB/MgO samples and concluded that only by covering the top surface with Ta film can avoid the presence of B oxide at CoFeB/MgO interface.

On the other hand, it is not difficult to understand that the element ratio of CoFeB also has an effect on interfacial PMA. Ikeda *et al.* [42] published the experimental results towards PMA based MTJ consisting of $(Co_{0.25}Fe_{0.75})_{100-x}B_x$ (x = 0, 15, 20, and 25), which uncovered that less B content can benefit interfacial anisotropy energy. After subtracting demagnetization energy, the K_{eff} reaches a maximum of 1.9 × 10^6 erg/cm^3 when $x = 20$ ($Co_{20}Fe_{60}B_{20}$); besides, high TMR ratio of 136% in MTJ with $x = 25$. Since then, using $Co_{20}Fe_{60}B_{20}$ has become a general method in interfacial PMA study.

Capping or Buffer Layer Material

Apart from CoFeB/MgO interface, the buffer or capping layer has been recently verified the significance to induce interfacial PMA, both by experiment and first-principles calculation [43,44,45,46,47,48,49,50]. PMA based MTJ with Ta/CoFe/MgO structure is classic, whereas the annealing temperature must be strictly controlled, for the Ta, Co, and Fe diffusion at Ta/CoFeB interface is intensified when temperature arises above 300 °C, and the diffusion induced magnetic dead layer degrades PMA dramatically. It is obvious that the tradeoff between B getter and the diffusion of other atoms should be balanced. Moreover, 300 °C is still incompatible with standard CMOS fabrication temperature, 400 °C, hence thermal endurance of PMA based MTJ has impeded its industrialization [9].

In order to solve the problems, Hf, Mo, Nb, W and other novel materials have been proposed as buffer or capping layer since 2012. For example, An *et al.* [49] investigated the properties of W/CoFe/MgO structure that withstands 450 °C. Limitation of CoFeB thickness was released from 1.3 to 1.7 nm at 300 °C, while the K_{eff} of 5 × 10^6 erg/cm^3 is extremely high in comparison with that of Ta. Considering the ability to enhance interfacial PMA, we rank the materials as: Ru < Ta ≈ Nb < W < Mo < Hf [45,46,47,48]. In addition to B getter, other characteristics of the buffer or capping layers are speculated to explain the optimization: first, the amorphous statement of Ta film sputtered on SiO_2 substrate lead to easy Ta diffusion during annealing, while the crystallized Mo and W films maintain stable; second, the surface of amorphous Hf buffer layer is atomically smooth, which benefits the following deposition; third, top stack (MgO/CoFeB /capping layer) and bottom stack (buffer layer/CoFeB/MgO) show various PMA due to different levels of magnetic dead layer formation or interface oxidation.

Miura *et al.* [44] analyzed the effect of capping or buffer layer in X/Fe/MgO (X is nonmagnetic metals, e.g., Hf, Ti, Ta, Nb, V, and Pd) structures by first-principles calculation, which revealed that X with fully occupied d states tend to execute PMA at X/Fe interface; conversely, X with partially occupied d states present in-plane easy axis. It is worth noting that Hf without fully occupied d states is an exception, because it provides unoccupied majority-spin d_z^2 states for the spin-orbit coupling with occupied minority-spin Fe d_{yz} states, the calculated result is consistent with experimental data published by Liu *et al.* [46], which promoted interfacial PMA by 35%. However, the negative magnetocrystalline anisotropy values of certain X/Fe interfaces, e.g., Ta/Fe, are opposite actual situations.

Another essential parameter involved is the thickness of capping or buffer layer, specifically the "window effect", which means that PMA may disappear due to either too thick or too thin nonmagnetic film. Cheng *et al.* [43] measured the magnetic properties in MgO/CoFe/Ta structure and concluded that the window margin is 1.1–1.7 nm in top stack. As for W/CoFe/MgO structure, the effect occurs in bottom stack rather than top stack, showing a maximum anisotropy field of 5.9 kOe with 5.6 nm W film [49]. The mechanism of window effect is still under argument. Most research indicated that increasing capping or buffer layer thickness aggravates magnetic dead layer formation. Besides, atomic force microscopy (AFM) images shows the appearance of pin hole at 7.4 nm W thickness, along with a rising surface roughness. Moreover, island growth at initial stage of sputtering may interpret the lower bound of window effect. It is important to note that our first-principles calculation agrees with this tendency, whereas the factors mentioned above are not involved.

Structure Optimization: Double CoFeB/MgO Interfaces

Since PMA mainly originates from CoFeB/MgO interface, the idea of improving PMA using multiple interfaces is reasonable. With carefully tuned MgO oxidation, the MgO/CoFeB/MgO layer fabricated by TDK-Headway Technologies Inc. (Milpitas, CA, USA) in 2012 demonstrated low RA of 6.6 Ω·μm², which is similar to that of Ta film capped structure, while retaining a TMR ratio of 102%. This is the first time that low switching current (45% decrease) and high interfacial PMA have been achieved simultaneously with double oxidation interfaces [50].

The same year, Sato *et al.* [51] investigated MTJ with MgO/CoFeB/Ta/CoFeB/MgO free layer containing double CoFeB/MgO interfaces, which exhibited high thermal stability and comparable threshold currents. The thickness of Ta spacer layer was designed as 0.4 nm for the top and bottom CoFeB layers to form large enough exchange coupling, enabling the simultaneous STT switching of the two layers. In Figure 6a, the combination between double CoFeB/MgO interfaces and ultra-thin Co/Pt multilayer based synthetic ferrimagnetic (SyF) reference layer, which has demonstrated suppressed exchange bias and high thermal stability, has further improved the properties of PMA based MTJ [52]. Thus, the junction diameter can be scaled down to 1X nm, with Δ maintaining almost constant value 85 when the diameter is larger than 30 nm, below which it starts to decline [53]. As shown in Figure 6b, threshold current decreases monotonously with scaled device area. Blanket film based calculations attribute the tendencies to the nucleation and single domain type magnetization reversals conversion, as well as the decreasing effective damping constant.

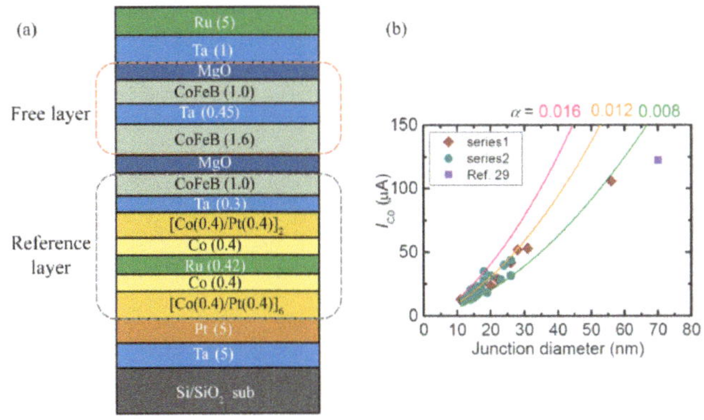

Figure 6. (a) Perpendicular magnetic anisotropy (PMA) MTJ with double CoFeB/MgO interfaces and ultra-thin Co/Pt multilayer based synthetic ferrimagnetic (SyF) reference layer, which has demonstrated suppressed exchange bias and high thermal stability. [53] (b) Average threshold current with respect to junction diameter down to 1X nm (Reproduced with permission from Ikeda *et al.* [53], International Electron Devices Meeting (IEDM); published by IEEE, 2014).

RECENT PROGRESS AND FUTURE APPLICATIONS

Spin Orbit Coupling Induced Magnetization Reversal

STT-MRAM has been through a rapid development since its first proposal and demonstration. As the MTJ scales, thinner MgO tunnel barriers are required for lower RA to remain comparable with the transistor, which contradicts the improvement of TMR ratio in the material point of view. Besides, the aging and even breakdown of ultra-thin MgO tunnel barriers caused by writing current passing through the MTJ stack leads to retention and reliability issues of conventional STT-MRAM [54].

To address these concerns, spin orbit torque (SOT), involving spin orbit coupling (SOC) such as spin Hall effect (SHE) or interface effects, has been investigated to realize magnetization rotation and novel devices recently. Figure 7 shows the three-terminal device designed from SOT (SOT-MRAM), where the in-plane writing current is injected through the nonmagnetic film (e.g., Pt, β-Ta, and β-W) on the bottom, generating a transverse spin current to induce quick magnetization reversal of adjacent ferromagnetic layer; the MTJ nano-pillar on the top is used for reading operation referring to its magnetoresistance [55,56,57]. The write path is decoupled from the read path without spin current flowing past the MgO tunnel barrier, which means the properties can be independently tuned. Since the threshold current density to perform SOT based magnetization reversal is nearly constant, competitive power consumption can be expected by reducing the cross-section area of nonmagnetic film, or increasing the material dependent spin Hall angle. The expression of spin Hall angle is $\theta_{SH} = J_s/J_e$, which characterizes the transmission efficiency from charge current density (J_e) to spin current density (J_S). Following the study of transition metals in high resistivity phase, large spin Hall angle of 0.3 has been found in β-W [57].

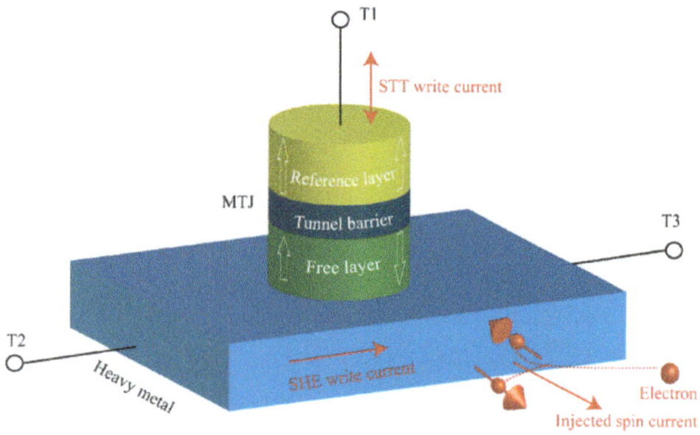

Figure 7. Mechanism of the spin Hall effect (SHE) assisted STT switching in PMA MTJ based three-terminal device. SHE write current is injected from terminal T2 and T3 into β-W strap or other heavy metal on the bottom, while the STT write current from terminal T1 penetrates the nano-pillar. Magnetization direction of the free layer can be reversed in absence of external magnetic field.

Naturally, the research upon SOT has been concentrated on its application towards MTJ devices with interfacial PMA for its advantages as previously noted. However, in this geometry, an external in-plane magnetic field must be added to assist SOT induced magnetization reversal, which brings negative influence on device fabrication and thermal stability [58]. In 2014, Yu *et al.*[59] replaced the external magnetic field with lateral structural inversion asymmetry by creating a gradient of TaO_x tunnel barrier thickness and oxidation. However, the fabrication of TaO_x wedge is not practical in application. Another possible method to cancel additional magnetic field is combining SOT (SHE) and STT, which has been theoretically validated by Brink *et al.* [60] and Wang *et al.* [61] using Landau-Lifshitz-Gilbert (LLG) equation. It has been demonstrated that a SHE writing current with 0.5 ns pulse duration allows fast switching speed (1 ns) using a relatively low STT writing current. Additional progress can be made if the interfacial PMA of MTJ with CoFeB/MgO structure is reinforced simultaneously. Similar to the discussion before, an ultra-thin spacer layer made of materials that generate strong interfacial PMA as capping or buffer layers is inserted between the nonmagnetic and magnetic films, e.g., the W/Hf/CoFeB/MgO multilayer, allowing the adjustment of both this parameter and spin Hall angle [62].

Perpendicular Magnetic Anisotropy for Magnetic Field Sensor

Magnetoresistance sensor based on MTJ with in-plane magnetic anisotropy has been applied in a wide range of areas, e.g., biosensor and magnetic field sensor. For magnetic field sensor, the magnetization direction of sensing layer, *i.e.*, free layer, should align orthogonally to that of the reference layer for large sensitivity and linear magnetoresistance response with minimum hysteresis. Thus, two successive annealing procedures at different temperatures under crossed external magnetic fields are required for fabrication, and the magnetization direction of the sensing layer can be rotated during the second procedure for its lower blocking temperature [63,64]. The complicated configuration and processing make it difficult to integrate such magnetic field sensors on Si chip. This problem is aggravated in the case of perpendicular magnetic field sensor.

Thanks to the introduction of PMA MTJ, linear response can be achieved with the coupling between out-of-plane and in-plane magnetization directions of sensing and reference layer [65]. As shown in Figure 8a, when a perpendicular magnetic field is applied, the magnetization direction of the free layer is pulled from out-of-plane to parallel or antiparallel to that of reference layer, along with the linear change of magnetoresistance. The perpendicular easy axis of sensing layer is generated from CoFe layer pinned with CoFe/Pt multilayer, while its formation is completely natural without additional thermal processing. Another kind of perpendicular magnetic field sensor based on CoFeB/MgO interface has been proposed by Lee *et al.* [66] in 2015. Because of the opposite magnetization direction situation as illustrated in Figure 8b, perpendicular magnetic field can be characterized with linear magnetoresistance response without rotating the position of sensor. It should be mentioned that its sensitivity is held to 0.3%/Oe within a dynamic range of ±25 Oe, which needs to be further improved in comparison with in-plane magnetic anisotropy-based magnetoresistance sensor (2.6%/Oe). All of the parameters, including reference layer coercivity, linear region, and TMR ratio, are independent with various device shapes or areas, thus the performance of sensor can remain designable, stable, and uniform. Additionally, the reference layer coercivity can be manipulated by adjusting the thickness of buffer layer.

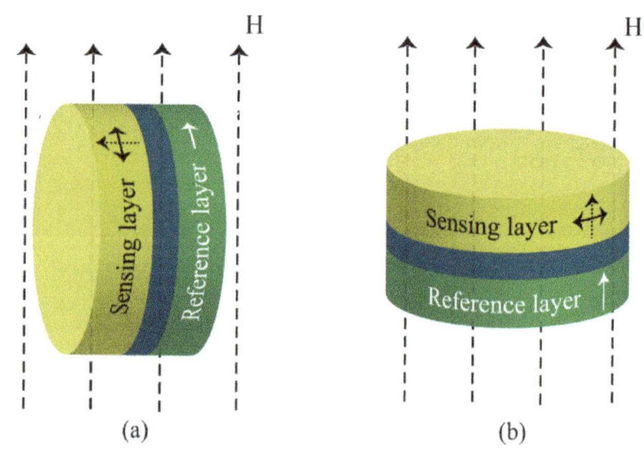

Figure 8. Perpendicular magnetic field sensor with (a) out-of-plane and in-plane magnetization direction for sensing and reference layer; (b) the opposite situation. The solid arrow indicate the easy axis, while it orientates to the dash arrow direction in the presence of external magnetic field H.

Using PMA is a very convenient method to detect perpendicular magnetic field, whereas the robust PMA is still required to achieve high sensitivity and detectivity. With shortcomings gradually being overcome, PMA MTJ based magnetoresistance sensor provides a potential candidate in magnetic field sensor industry, particularly the versatile sensors for omnidirectional detection and mobile devices.

Perpendicular Magnetic Anisotropy for Logic Circuits

The increasing static power consumption in CMOS technology beyond 45 nm, as well as the interconnection delay caused by long distance data traffic, also limit the development of transistor based logic circuits. These issues have given rise to logic circuits using the MTJ device. Further promotion is expected when concerning PMA based MTJs.

Logic-in-memory architecture built with hybrid MTJ/CMOS logic circuits, where memory devices are merged into a logic circuit, has drawn considerable attention recently [67,68]. This structure is configured with a write circuit for memory programming, a volatile logic block using MOS transistors, a sense amplifier (S.A) for logic result evaluation, and a non-volatile memory block (e.g., MRAM) for instant data storage. Because of the tricky circuit layout, interconnection distance is shortened for both high access speed and low power consumption. One typical example is the magnetic flip-flop (MFF) shown in Figure 9a [69,70]. Bidirectional current is generated by two nMOS (MN2-3) and two pMOS (MP4-5) transistors when two of them are active, thus the two complementary PMA based MTJ devices perform STT switching and store the result. MN0-1, MP0-3, and MN4 transistors constitute a precharge sense amplifier as the master register in coordination with the slave resister for output. Simulation undertaken with PMA MTJ compact model has proven the feasibility of this MFF. Logic-in-memory architecture demonstrates instant on/off capacity, zero standby power consumption, and immunity to power failure, whereas it still suffers from certain drawbacks: in comparison with MOS logic circuit, the large switching latency of MTJ device limits its computing frequency to the order of GHz; S.A mismatch and intrinsic stochastic switching due to thermal fluctuation cause the of sense reliability degradation. Therefore, progress can be reasonably expected form the further study towards based MTJ.

Actually, the MTJ in previous logic circuits is only used in memory rather than logic block for data storage, thus the extra lateral circuits offset the advantages of MTJs, e.g., scalability. A more intrinsic way to build logic-in-memory circuit is proposed as stateful logic: the MTJ serves as both logic gate and memory element [71]. For instance, the three-input logic device shown in Figure 9b can perform fundamental AND, OR, NAND, NOR and Majority functions with a single MTJ, where the current directions and resistance states denote "1" ("0") for inputs and outputs respectively [72]. Direct information communication between MTJs is realized by domain wall motion through nano-magnetic channel, enabling more complicated logic function with arithmetic logic unit (ALU).

More aggressively, when spin current is completely applied for input, logic function, data storage, information communication, and output, repeated spin-to-charge conversion can be eliminated for circuit simplification and energy reduction. This is the concept of all spin logic gates: charge current is spin polarized by magnetization direction of input magnet, then the spin current is routed through the channel towards output magnet, and determines its final statement by STT switching [73]. The channel material can comprise metal (e.g., Cu) or semiconductor, while graphene channel is preferred for its long spin diffusion length over 100 μm [74]. Taking advantage of both PMA based MTJ and graphene channel, Su *et al.* [75] proposed a novel all spin logic device, with which the AND (OR) gate is configured as Figure 10 shown. In this gate, magnetization direction of each input MTJ A, B, and C is controlled with bias voltages, while the sum of modulated spin currents decides whether the magnetic moment of output MTJ can be reversed. Finally, the resistance state is read with another bias voltage. Although simulation has confirmed, the attractive prospect of all spin logic by demonstrating its concatenability, nonlinearity, feedback elimination, gain, and a complete set of Boolean operations, much work is still required to bring this concept into reality.

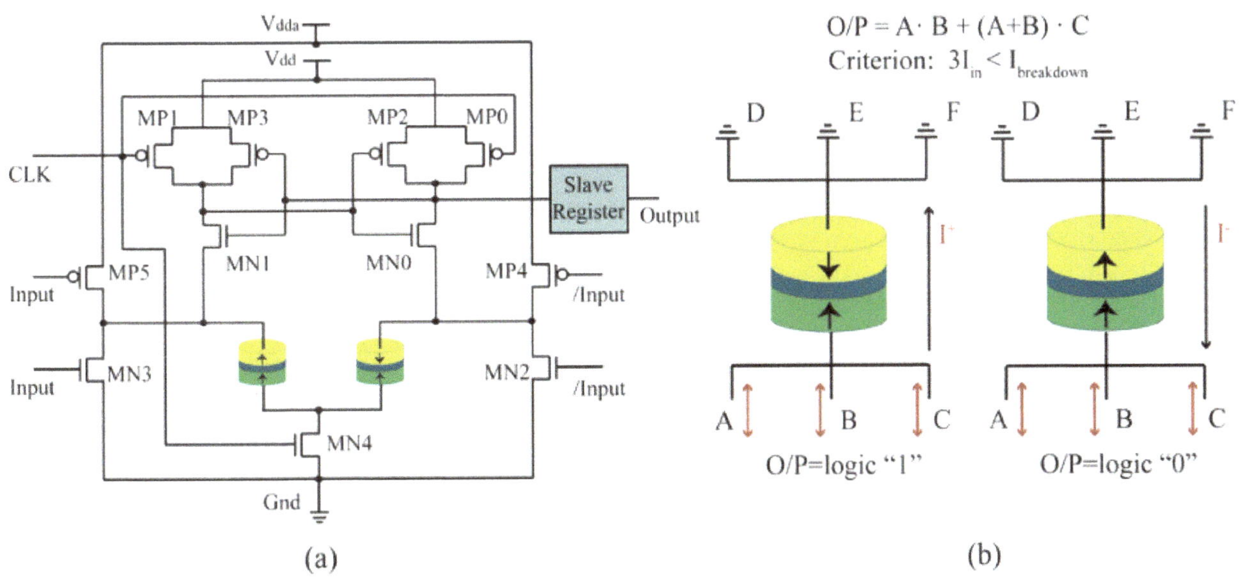

Figure 9. (a) Magnetic flip-flop (MFF) built with PMA MTJ devices and MOS transistors for logic-in-memory architecture. (b) The three-input logic device to realize stateful logic. When top electrode D, E, and F are grounded, the device executes AND/OR functions under the control of input C. Total current from input A, B, and C should be less than the breakdown current.

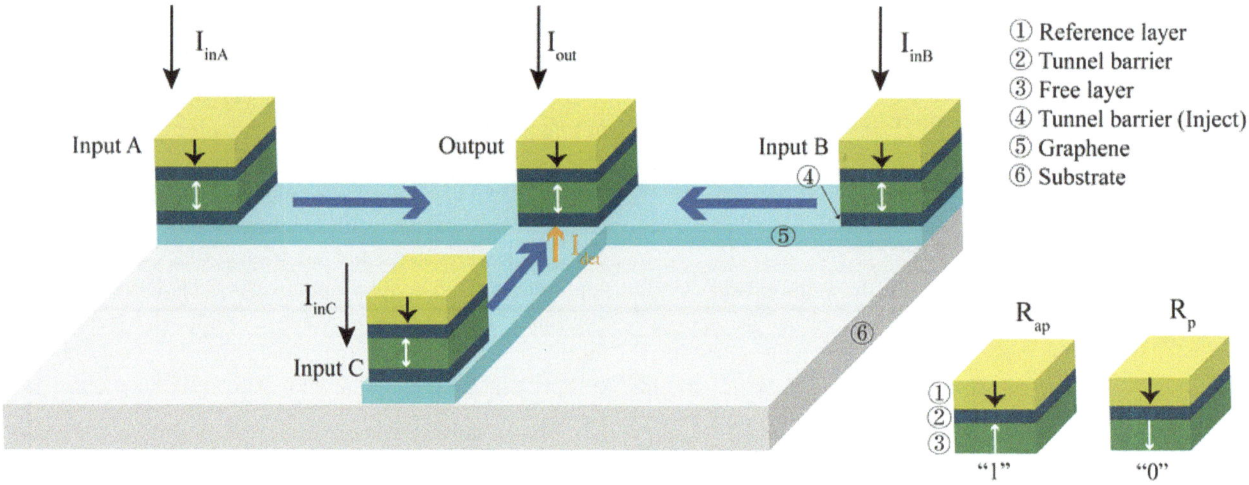

Figure 10. All spin logic gate AND/OR built with PMA-based MTJs. Data is communicated through bottom graphene channels in the form of spin current injected from input MTJ A, B, and C, of which the sum induces the STT switching in the output MTJ.

CONCLUSION AND PERSPECTIVE

This review states advantages, drawbacks and corresponding solutions with respect to PMA-based MTJs. It can be confirmed that the PMA MTJ is indeed a promising candidate to realize low-power computing and other meaningful applications. Meanwhile, great efforts have been made in the last few years for the achievement of robust interfacial PMA in CoFeB/MgO system, including the research upon material and fabrication. Considering that the progress of a PMA MTJ is still in its young stage, we can positively expect improvements in terms of thermal stability, power consumption, switching speed, and thermal endurance. In addition, other potential mechanisms (e.g., SOT) relevant to PMA MTJ and their applications are worth further developing.

ACKNOWLEDGMENTS

The authors gratefully acknowledge the International Collaboration Project 2015DFE12880 from the Ministry of Science and Technology in China, and National Natural Science Foundation of China (Grant No. 61471015) for their financial support of this work.

AUTHOR CONTRIBUTIONS

Mengxing Wang wrote the main part of manuscript. Yue Zhang performed the simulations with PMA MTJ compact model. Xiaoxuan Zhao wrote the part of magnetoresistance sensor and prepared the figures. Weisheng Zhao supervised the work and edited the manuscript.

REFERENCES

1. International Technology Roadmap for Semiconductor (ITRS), 2010 Update. Available online:http://www.itrs.net (accessed on 27 March 2014).
2. Chappert, C.; Fert, A.; Van Dau, F.N. The emergence of spin electronics in data storage. *Nat. Mater.* 2007, *6*, 813–823.
3. Nikonov, D.E.; Young, I. Overview of beyond-CMOS devices and a uniform methodology for their benchmarking. *Proc. IEEE* 2013, *101*, 2498–2533.
4. Wang, D.; Nordman, C.; Daughton, J.M.; Qian, Z.; Fink, J. 70% TMR at room temperature for SDT sandwich junctions with CoFeB as free and reference layers. *IEEE Trans. Magn.* 2004, *40*, 2269–2271.
5. Yuasa, S.; Nagahama, T.; Fukushima, A.; Suzuki, Y.; Ando, K. Giant room-temperature magnetoresistance in single-crystal Fe/MgO/Fe magnetic tunnel junctions. *Nat. Mater.* 2004,*3*, 868–871.
6. Parkin, S.S.P.; Kaiser, C.; Panchula, A.; Rice, P.M.; Hughes, B.; Samant, M.; Yang, S.H. Giant tunnelling magnetoresistance at room temperature with MgO (100) tunnel barriers. *Nat. Mater.* 2004, *3*, 862–867.
7. Ikeda, S.; Hayakawa, J.; Ashizawa, Y.; Lee, Y.M.; Miura, K.; Hasegawa, H.; Tsunoda, M.; Matsukura, F.; Ohno, H. Tunnel magnetoresistance of 604% at 300 K by suppression of Ta diffusion in CoFeB/MgO/CoFeB pseudo-spin-valves annealed at high temperature. *Appl. Phys. Lett.* 2008, *93*, 082508.
8. Thomas, L.; Jan, G.; Zhu, J.; Liu, H.; Lee, Y.J.; Le, S.; Tong, R.Y.; Pi, K.; Wang, Y.J.; Shen, D.; *et al.* Perpendicular spin transfer torque magnetic random access memories with high spin torque efficiency and thermal stability for embedded applications. *J. Appl. Phys.* 2014, *115*, 172615.
9. Kent, A.D.; Worledge, D.C. A new spin on magnetic memories. *Nat. Nanotech.* 2015, *10*, 187–191.
10. Carcia, P.F.; Meinhaldt, A.D.; Suna, A. Perpendicular magnetic anisotropy in Pd/Co thin film layered structures. *Appl. Phys. Lett.* 1985, *47*, 178–180.
11. Carcia, P.F. Perpendicular magnetic anisotropy in Pd/Co and Pt/Co thin-film layered structures. *J. Appl. Phys.* 1988, *63*, 5066–5073.
12. Hashimoto, S.; Ochiai, Y.; Aso, K. Perpendicular magnetic anisotropy and magnetostriction of sputtered Co/Pd and Co/Pt multilayered films. *J. Appl. Phys.* 1989, *66*, 4909–4916.

13. Park, J.H.; Park, C.; Jeong, T.; Moneck, M.T.; Nufer, N.T.; Zhu, J.G. Co/Pt multilayer based magnetic tunnel junctions using perpendicular magnetic anisotropy. *J. Appl. Phys.* 2008, *103*, 07A917.

14. Lim, D.; Kim, K.; Kim, S.; Jeung, W.Y.; Lee, S.R. Study on exchange-biased perpendicular magnetic tunnel junction. *IEEE Trans. Magn.* 2009, *45*, 2407–2409.

15. Yakushiji, K.; Saruya, T.; Kubota, H.; Fukushima, A.; Nagahama, T.; Yuasa, S.; Ando, K. Ultrathin Co/Pt and Co/Pd superlattice films for MgO-based perpendicular magnetic tunnel junctions. *Appl. Phys. Lett.* 2010, *97*, 232508.

16. Ishikawa, S.; Sato, H.; Yamanouchi, M.; Ikeda, S.; Fukami, S.; Matsukura, F.; Ohno, H. Magnetic properties of MgO-[Co/Pt] multilayers with a CoFeB insertion layer. *J. Appl. Phys.*2013, *113*, 17C721.

17. Mizunuma, K.; Ikeda, S.; Park, J.H.; Yamamoto, H.; Gan, H.; Miura, K.; Hasegawa, H.; Hayakawa, J.; Matsukura, F.; Ohno, H. MgO barrier-perpendicular magnetic tunnel junctions with CoFe/Pd multilayers and ferromagnetic insertion layers. *Appl. Phys. Lett.* 2009,*95*, 232516.

18. Lee, C.M.; Ye, L.X.; Hsieh, T.H.; Huang, C.Y.; Wu, T.H. Magnetic properties of TbFeCo-based perpendicular magnetic tunnel junctions. *J. Appl. Phys.* 2010, *107*, 09C712.

19. Ikeda, S.; Miura, K.; Yamamoto, H.; Mizunuma, K.; Gan, H.D.; Endo, M.; Kanai, S.; Hayakawa, J.; Matsukura, F.; Ohno, H. A perpendicular-anisotropy CoFeB–MgO magnetic tunnel junction. *Nat. Mater.* 2010, *9*, 721–724.

20. Zhao, W.S.; Zhang, Y.; Devolder, T.; Klein, J.O.; Ravelosona, D.; Chappert, C.; Mazoyer, P. Failure and reliability analysis of STT-MRAM. *Microelectron. Reliab.* 2012, *52*, 1848–1852.

21. Shimabukuro, R.; Nakamura, K.; Akiyama, T.; Ito, T. Electric field effects on magnetocrystalline anisotropy in ferromagnetic Fe monolayers. *Physica E Low-Dimens. Syst. Nanostruct.* 2010, *42*, 1014–1017.

22. Yang, H.X.; Chshiev, M.; Dieny, B.; Lee, J.H.; Manchon, A.; Shin, K.H. First-principles investigation of the very large perpendicular magnetic anisotropy at Fe|MgO and Co|MgO interfaces. *Phys. Rev. B* 2011, *84*, 054401.

23. Slonczewski, J.C. Current-driven excitation of magnetic multilayers. *J. Magn. Magn. Mater.*1996, *159*, L1–L7.

24. Huai, Y.; Albert, F.; Nguyen, P.; Pakala, M.; Valet, T. Observation of spin-transfer switching in deep submicron-sized and low-resistance magnetic tunnel junctions. *Appl. Phys. Lett.* 2004,*84*, 3118–3120.

25. Ikeda, S.; Sato, H.; Yamanouchi, M.; Gan, H.; Miura, K.; Mizunuma, K.; Kana, N.; Fukami, S.; Matsukura, F.; Kasai, N.; Ohno, H. Recent progress of perpendicular anisotropy magnetic tunnel junctions for nonvolatile VLSI. *Spin* 2012, *2*, 1240003.

26. Nakayama, M.; Kai, T.; Shimomura, N.; Amano, M.; Kitagawa, E.; Nagase, T.; Yoshikawa, M.; Kishi, T.; Ikegawa, S.; Yoda, H. Spin transfer switching in TbCoFe/CoFeB/MgO/CoFeB/TbCoFe magnetic tunnel junctions with perpendicular magnetic anisotropy. *J. Appl. Phys.* 2008, *103*, 07A710.

27. Worledge, D.C.; Hu, G.; Abraham, D.W.; Sun, J.Z.; Trouilloud, P.L.; Nowak, J.; Brown, S.; Gaidis, M.C.; O'Sullivan, E.J.; Robertazzi, R.P. Spin torque switching of perpendicular Ta/CoFeB/MgO-based magnetic tunnel junctions. *Appl. Phys. Lett.* 2011, *98*, 022501.

28. Zhao, H.; Glass, B.; Amiri, P.K.; Lyle, A.; Zhang, Y.; Chen, Y.J.; Rowlands, G.; Upadhyaya, P.; Zeng, Z.; Katine, J.A.; *et al.* Sub-200 ps spin transfer torque switching in in-plane magnetic tunnel junctions with interface perpendicular anisotropy. *J. Phys. D Appl. Phys.* 2012, *45*, 025001.

29. Choi, Y.S.; Nagamine, Y.; Tsunekawa, K.; Maehara, H.; Djayaprawira, D.D.; Yuasa, S.; Ando, K. Effect of Ta getter on the quality of MgO tunnel barrier in the polycrystalline CoFeB/MgO/CoFeB magnetic tunnel junction. *Appl. Phys. Lett.* 2007, *90*, 012505.

30. Sinha, J.; Gruber, M.; Kodzuka, M.; Ohkubo, T.; Mitani, S.; Hono, K.; Hayashi, M. Influence of boron diffusion on the perpendicular magnetic anisotropy in Ta|CoFeB|MgO ultrathin films. *J. Appl. Phys.* 2015, *117*, 043913.

31. Gokan, H.; Esho, S. Pattern fabrication by oblique incidence ion-beam etching. *J. Vac. Sci. Technol.* 1981, *18*, 23–27.

32. Kinoshita, K.; Utsumi, H.; Suemitsu, K.; Hada, H.; Sugibayashi, T. Etching magnetic tunnel junction with metal etchers. *Jpn. J. Appl. Phys.* 2010, *49*, 08JB02.

33. Kinoshita, K.; Yamamoto, T.; Honjo, H.; Kasai, N.; Ikeda, S.; Ohno, H. Damage recovery by reductive chemistry after methanol-based plasma etch to fabricate magnetic tunnel junctions. *Jpn. J. Appl. Phys.* 2012, *51*, 08HA01.

34. Vasile, M.J.; Mogab, C.J. Chemically assisted sputter-etching of permalloy using CO or Cl_2. *J. Vac. Sci. Technol. A* 1986, *4*, 1841–1849.

35. Kinoshita, K.; Yamada, K.; Matsutera, H. Reactive ion etching of Fe-Si-Al alloy for thin film head. *IEEE Trans. Magn.* 1991, *27*, 4888–4890.

36. Nakatani, I. Ultramicro fabrications on Fe-Ni alloys using electron-beam writing and reactive-ion etching. *IEEE Trans. Magn.* 1996, *32*, 4448–4451.

37. Osada, T.; Doi, M.; Sakamoto, K.; Maehara, H.; Kodaira, Y. Dry etching of magnetic thin film stacks using CO/NH_3 and CH_3OH gases for tunneling magneto-resistance devices. In Proceedings of 4th International Symposium on Dry Process, Tokyo, Japan, 30 November–1 December 2004; pp. 127–131.

38. Kinoshita, K.; Honjo, H.; Fukami, S.; Sato, H.; Mizunuma, K.; Tokutome, K.; Murahata, M.; Ikeda, S.; Miura, S.; Kasai, N.; Ohno, H. Process-induced damage and its recovery for a CoFeB–MgO magnetic tunnel junction with perpendicular magnetic easy axis. *Jpn. J. Appl. Phys.* 2014, *53*, 103001.

39. Greer, A.A.; Gray, A.X.; Kanai, S.; Kaiser, A.M.; Ueda, S.; Yamashita, Y.; Bordel, C.; Palsson, G.; Maejima, N.; Yang, S.H.; *et al.* Observation of boron diffusion in an annealed Ta/CoFeB/MgO magnetic tunnel junction with

standing-wave hard x-ray photoemission.*Appl. Phys. Lett.* 2012, *101*, 202402.

40. Kurt, H.; Rode, K.; Oguz, K.; Boese, M.; Faulkner, C.C.; Coey, J.M.D. Boron diffusion in magnetic tunnel junctions with MgO (001) barriers and CoFeB electrodes. *Appl. Phys. Lett.*2010, *96*, 262501.

41. Mukherjee, S.; Knut, R.; Mohseni, S.M.; Nguyen, T.A.; Chung, S.; Le, Q.T.; Akerman, J.; Persson, J.; Sahoo, A.; Hazarika, A.; *et al.* Role of boron diffusion in CoFeB/MgO magnetic tunnel junctions. *Phys. Rev. B* 2015, *91*, 085311.

42. Ikeda, S.; Koizumi, R.; Sato, H.; Yamanouchi, M.; Miura, K.; Mizunuma, K.; Gan, H.; Matsukura, F.; Ohno, H. Boron composition dependence of magnetic anisotropy and tunnel magnetoresistance in MgO/CoFe (B) based stack structures. *IEEE Trans. Magn.* 2012, *48*, 3829–3832.

43. Cheng, C.W.; Feng, W.; Chern, G.; Lee, C.M.; Wu, T.H. Effect of cap layer thickness on the perpendicular magnetic anisotropy in top MgO/CoFeB/Ta structures. *J. Appl. Phys.* 2011, *110*, 033916.

44. Miura, Y.; Tsujikawa, M.; Shirai, M. A first-principles study on magnetocrystalline anisotropy at interfaces of Fe with non-magnetic metals. *J. Appl. Phys.* 2013, *113*, 233908.

45. Liu, T.; Cai, J.W.; Sun, L. Large enhanced perpendicular magnetic anisotropy in CoFeB/MgO system with the typical Ta buffer replaced by an Hf layer. *Aip Adv.* 2012, *2*, 032151.

46. Lee, D.S.; Chang, H.T.; Cheng, C.W.; Chern, G. Perpendicular magnetic anisotropy in MgO/CoFeB/Nb and a comparison of the cap layer effect. *IEEE Trans. Magn.* 2014, *50*, 1–4.

47. Liu, T.; Zhang, Y.; Cai, J.W.; Pan, H.Y. Thermally robust Mo/CoFeB/MgO trilayers with strong perpendicular magnetic anisotropy. *Sci. Rep.* 2014, *4*.

48. An, G.G.; Lee, J.B.; Yang, S.M.; Kim, J.H.; Chung, W.S.; Hong, J.P. Highly stable perpendicular magnetic anisotropies of CoFeB/MgO frames employing W buffer and capping layers. *Acta Mater.* 2015, *87*, 259–265.

49. Wang, M.; Peng, S.; Zhang, Y.; Zhang, Y.; Zhang, Y.; Zhang, Q.; Ravelosona, D.; Zhao, W. A Multi-Level Cell for STT-MRAM Realized by Capping Layer Adjustment. In Proceedings of the IEEE Magnetics Conference, Beijing, China, 11–15 May 2015.

50. Jan, G.; Wang, Y.J.; Moriyama, T.; Lee, Y.J.; Lin, M.; Zhong, T.; Tong, R.Y.; Torng, T.; Wang, P.K. High spin torque efficiency of magnetic tunnel junctions with MgO/CoFeB/MgO free layer. *Appl. Phys. Express* 2012, *5*, 093008.

51. Sato, H.; Yamanouchi, M.; Ikeda, S.; Fukami, S.; Matsukura, F.; Ohno, H. Perpendicular-anisotropy CoFeB-MgO magnetic tunnel junctions with a MgO/CoFeB/Ta/CoFeB/MgO recording structure. *Appl. Phys. Lett.* 2012, *101*, 022414.

52. Sato, H.; Yamanouchi, M.; Ikeda, S.; Fukami, S.; Matsukura, F.; Ohno, H. MgO/CoFeB/Ta/CoFeB/MgO recording structure in magnetic tunnel junctions with perpendicular easy axis. *IEEE Trans. Magn.* 2013, *49*, 4437–4440.

53. Ikeda, S.; Sato, H.; Honjo, H.; Enobio, E.C.I.; Ishikawa, S.; Yamanouchi, M.; Fukami, S.; Kanai, S.; Matsukura, F.; Endoh, T.; Ohno, H. Perpendicular-anisotropy CoFeB-MgO based magnetic tunnel junctions scaling down to 1X nm. In Proceedings of the IEEE International Electron Devices Meeting, San Francisco, CA, USA, 15–17 December 2014; pp. 3321–3324.

54. Amara-Dababi, S.; Sousa, R.C.; Chshiev, M.; Béa, H.; Alvarez-Hérault, J.; Lombard, L.; Prejbeanu, I.L.; Mackay, K.; Dieny, B. Charge trapping-detrapping mechanism of barrier breakdown in MgO magnetic tunnel junctions. *Appl. Phys. Lett.* 2011, *99*, 083501.

55. Liu, L.; Moriyama, T.; Ralph, D.C.; Buhrman, R.A. Spin-torque ferromagnetic resonance induced by the spin Hall effect. *Phys. Rev. Lett.* 2011, *106*, 036601.

56. Liu, L.; Pai, C.F.; Li, Y.; Tseng, H.W.; Ralph, D.C.; Buhrman, R.A. Spin-torque switching with the giant spin Hall effect of tantalum. *Science* 2012, *336*, 555–558.

57. Pai, C.F.; Liu, L.; Li, Y.; Tseng, H.W.; Ralph, D.C.; Buhrman, R.A. Spin transfer torque devices utilizing the giant spin Hall effect of tungsten. *Appl. Phys. Lett.* 2012, *101*, 122404.

58. Cubukcu, M.; Boulle, O.; Drouard, M.; Garello, K.; Avci, C.O.; Miron, I.M.; Langer, J.; Ocker, B.; Gambardella, P.; Gaudin, G. Spin-orbit torque magnetization switching of a three-terminal perpendicular magnetic tunnel junction. *Appl. Phys. Lett.* 2014, *104*, 042406.

59. Yu, G.; Upadhyaya, P.; Fan, Y.; Alzate, J.G.; Jiang, W.; Wong, K.L.; Takei, S.; Bender, S.A.; Chang, L.T.; Jiang, Y.; Lang, M.; Tang, J.; Wang, Y.; Tserkovnyak, Y.; Amiri, P.K.; Wang, K.L. Switching of perpendicular magnetization by spin-orbit torques in the absence of external magnetic fields. *Nat. Nanotech.* 2014, *9*, 548–554.

60. Van den Brink, A.; Cosemans, S.; Cornelissen, S.; Manfrini, M.; Vaysset, A.; Van Roy, W.; Min, T.; Swagten, H.J.M.; Koopmans, B. Spin-Hall-assisted magnetic random access memory.*Appl. Phys. Lett.* 2014, *104*, 012403.

61. Wang, Z.; Zhao, W.; Deng, E.; Klein, J.O.; Chappert, C. Perpendicular-anisotropy magnetic tunnel junction switched by spin-Hall-assisted spin-transfer torque. *J. Phys. D Appl. Phys.*2015, *48*, 065001.

62. Pai, C.F.; Nguyen, M.H.; Belvin, C.; Vilela-Leão, L.H.; Ralph, D.C.; Buhrman, R.A. Enhancement of perpendicular magnetic anisotropy and transmission of spin-Hall-effect-induced spin currents by a Hf spacer layer in W/Hf/CoFeB/MgO layer structures. *Appl. Phys. Lett.* 2014, *104*, 082407.

63. Negulescu, B.; Lacour, D.; Montaigne, F.; Gerken, A.; Paul, J.; Spetter, V.; Marienb, J.; Duret, C.; Hehn, M. Wide range and tunable linear magnetic tunnel junction sensor using two exchange pinned electrodes. *Appl. Phys. Lett.* 2009, *95*, 112502.

64. Chen, J.Y.; Feng, J.F.; Coey, J.M.D. Tunable linear magnetoresistance in MgO magnetic tunnel junction

sensors using two pinned CoFeB electrodes. *Appl. Phys. Lett.* 2012, *100*, 142407.

65. Ding, Y.; Judy, J.H.; Wang, J.P. Magneto-resistive sensors with perpendicular magnetic anisotropy. *IEEE Trans. Magn.* 2005, *41*, 707–712.

66. Lee, Y.C.; Chao, C.T.; Li, L.C.; Suen, Y.W.; Horng, L.; Wu, T.H.; Chang, C.R.; Wu, J.C. Magnetic tunnel junction based out-of-plane field sensor with perpendicular magnetic anisotropy in reference layer. *J. Appl. Phys.* 2015, *117*, 17A320.

67. Matsunaga, S.; Hayakawa, J.; Ikeda, S.; Miura, K.; Hasegawa, H.; Endoh, T.; Ohno, H.; Hanyu, T. Fabrication of a nonvolatile full adder based on logic-in-memory architecture using magnetic tunnel junctions. *Appl. Phys. Express* 2008, *1*, 091301.

68. Zhao, W.; Torres, L.; Cargnini, L.V.; Brum, R.M.; Zhang, Y.; Guillemenet, Y.; Guillemenet, Y.; Sassatelli, G.; Lakys, Y.; Klein, J.O.; Etiemble, D.; Ravelosona, D.; Chappert, C. High performance SoC design using magnetic logic and memory. In Proceedings of the IEEE International Conference on Very Large Scale Integration, Hong Kong, China, 3–5 October 2011; pp. 10–33.

69. Zhao, W.; Belhaire, E.; Chappert, C. Spin-MTJ based non-volatile flip-flop. In Nanotechnology, Proceedings of the IEEE conference, Hong Kong, China, 2–5 August 2007; pp. 399–402.

70. Zhang, Y.; Zhao, W.; Lakys, Y.; Klein, J.O.; Kim, J.V.; Ravelosona, D.; Chappert, C. Compact modeling of perpendicular-anisotropy CoFeB/MgO magnetic tunnel junctions. *IEEE Trans. Electron Devices* 2012, *59*, 819–826.

71. Mahmoudi, H.; Sverdlov, V.; Selberherr, S. MTJ-based implication logic gates and circuit architecture for large-scale spintronic stateful logic systems. In Proceedings of the European Solid-State Device Research Conference, Brodeaux, France, 17–21 September 2012; pp. 254–257.

72. Yao, X.; Harms, J.; Lyle, A.; Ebrahimi, F.; Zhang, Y.; Wang, J.P. Magnetic tunnel junction-based spintronic logic units operated by spin transfer torque. *IEEE Trans. Nanotechnol.* 2012,*11*, 120–126.

73. Behin-Aein, B.; Datta, D.; Salahuddin, S.; Datta, S. Proposal for an all-spin logic device with built-in memory. *Nat. Nanotech.* 2010, *5*, 266–270.

74. Dlubak, B.; Martin, M.B.; Deranlot, C.; Servet, B.; Xavier, S.; Mattana, R.; Sprinkle, M.; Berger, C.; De Heer, W.A.; Petroff, F.; *et al.* Highly efficient spin transport in epitaxial graphene on SiC. *Nat. Phys.* 2012, *8*, 557–561.

75. Su, L.; Zhao, W.; Zhang, Y.; Querlioz, D.; Zhang, Y.; Klein, J.O.; Dollfus, P.; Bournel, A. Proposal for a graphene-based all-spin logic gate. *Appl. Phys. Lett.* 2015, *106*, 072407.

CHAPTER 15

INTEGRATION OF SPINTRONIC INTERFACE FOR NANOMAGNETIC ARRAYS

Andrew Lyle, Jonathan Harms, Todd Klein, August Lentsch, Angeline Klemm, Daniel Martens and Jian-Ping Wang

INTRODUCTION

An energy efficient logic device for the post-CMOS era has been the goal of a variety of research fields. The limits of scaling, which we expect to reach by the year 2025,[1] demand that future advances in computational power will not be realized from ever-shrinking device sizes, but rather by innovative designs and new physics, including molecular or DNA based computation,[2] organics,[3] magnonics,[4,5] or spintronics.[6] One of the key metrics that will determine the success or failure of these architectures is energy consumption. In this respect, magnetic quantum cellular automata (MQCA)[7–10] is promising based on theoretical prediction that nanomagnets can be switched using orders-of-magnitude less energy[11,12] than even the most efficient semiconductors. The switching energy of dipole coupled nanomagnet arrays for use in MQCA logic can in fact be tuned to meet the needs of the designer.

As a consequence of tunable magnetic response, the same dipole coupled nanomagnet arrays may find use as magnetic field sensors. Magnetic sensors are already utilized in applications ranging from hard disk drives to medical devices. A key feature for these sensors is the ability to tune the magnetic response and working field range to optimize the sensitivity and allow use in a range of environments and conditions.[13,14] In current technology, this is commonly done by biasing the magnetic sensor with a permanent magnet.[15–17] The permanent magnet unfortunately creates additional processing for integration with the device and increases device size. These magnetic sensors are commonly 10s or 100s of microns in size,[15–17] but advances in nanofabrication have allowed nano-scale devices to become more commonly used and applied.[18,19] These nano-scale devices are not only smaller but also incorporate new physics and concepts to improve performance and add functionality. For nano-scale sensing devices to reach their full potential, a reduction or removal of permanent magnets must be addressed. Self-biased schemes using nanomagnet arrays to tune the magnetic response of individual nanomagnets may reduce device size, add functionality, and expand the working environment to where permanent magnets cannot be used. For example, a promising attribute of the self-biased nanomagnet arrays is the feasibility of multi-axis detection with a sub-micron footprint device.

Both technologies (MQCA and field sensing) depend on reliable coupling across an array of nanomagnets. Unfortunately, the lack of a reliable electronic interface has thus far impeded real nanomagnet arrays from reaching their potential.[20] Several approaches such as spintronics, extra-ordinary Hall-effect, and optical reading have been proposed but few have been demonstrated due to difficult nanofabrication and system integration. Microscopy techniques[7,8,21] have been employed to demonstrate the requisite dipole coupling, but these methods are too time consuming to gather enough statistical data for performance and reliability measurements. Furthermore, these techniques are not compatible with integrated circuitry based on electrical input output (I/O). The extraordinary Hall-effect has been shown as a potential means of electrical I/O,[22,23] but there are several key limitations. First, the effect has only been demonstrated on submicron features down to 800 nm. Also, current in-plane (CIP) configuration leads to crowded contacts near adjacent devices and imposes design limitations. Finally, the method of programming is not easily confined to individual nanomagnets. In contrast, spintronics[24,25] interfaces well with

Citation: Andrew Lyle, Jonathan Harms, Todd Klein, August Lentsch, Angeline Klemm, Daniel Martens and Jian-Ping Wang, Integration of spintronic interface for nanomagnetic arrays, doi: 10.1063/1.3672177.

traditional I/O systems and provides not only an attractive read scheme but also a means to directly program individual nanomagnets. This has been demonstrated in devices such as magnetic random access memory (MRAM),[19] nonvolatile programmable computing,[6] and spin transfer torque (STT) diodes.[26] The use of integrated spintronics as illustrated in Figure 1(a) allows dipole coupled nanomagnet arrays to be quantifiably characterized. This is a critical achievement for both MQCA and field sensing technologies.

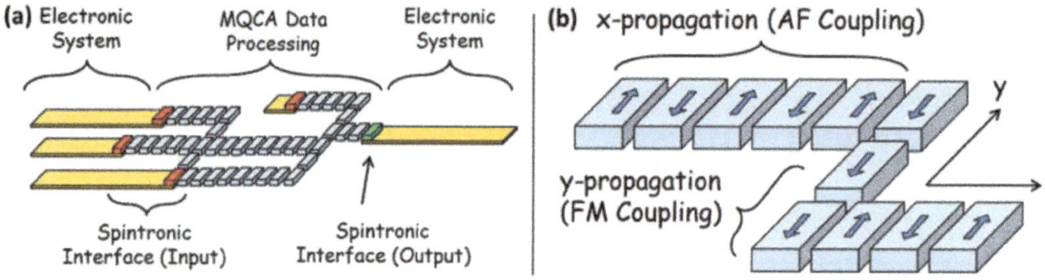

Figure. 1. Illustrations (a) showing a hybrid system which integrates the low power operation of MQCA with the high speed interface of spintronics and (b) the preferred coupling configuration for MQCA nanomagnets which is used for data propagation and computation.

SPINTRONIC MQCA

Spintronic based magnetoresistive (MR) devices store and process information based on the relative orientation of two thin magnetic layers separated by a nonmagnetic layer.[19] The nonmagnetic material is either an insulating or conductive layer for magnetic tunnel junction (MTJ) or giant magneto resistance (GMR), respectively. There are two sides to spintronics. One side is the MR effects, through which electronic signals are related with magnetization configurations of spintronic devices.[19] When the two magnetic layers are parallel (P) electrons easily flow through the device which results in a low resistance. When they are anti-parallel (AP) the electrons cannot easily pass which results in a high resistance. The other side is the manipulation and excitation of nanomagnets by STT through the transfer of angular momentum from electrons to the magnetic layer.[27,28] The manipulation of the nanomagnet using STT allows the magnetization direction to be written by only passing a current of sufficient magnitude through the MR device. We have demonstrated the potential for spintronic based nanomagnet arrays using magneto resistance (MR) read of individual 125 nm x 250 nm pillars.[29]

MQCA relies on the magnetic dipole coupling of nanomagnets sized around 100 nm or less and spaced 0 to 50 nm apart. The nanomagnets will either couple ferromagnetically (FM) or anti-ferromagnetically (AF) with neighboring elements depending on the arrangement as shown in Figure 1(b). Logic architectures use coupling competition between neighboring nanomagnets to process and transfer information without the movement of charge, which results in lower power operation. The competition between neighboring nanomagnets is controlled using a clocking magnetic field which allows the coupling to determine the state of an element based on the orientation of input pillars.

In this work, three key demonstrations are realized for the first time based on the integration of MQCA and spintronics. The first is the electrical programming of an individual nanomagnet in a MQCA device using STT. The second is the evaluation of clocking field used for logic operation and STT programming of inputs. Finally, the reliability of the coupling between nanomagnetic elements, which is utilized for data propagation and logic computation, is measured. The sensitivity and ability to gather statistical data based on these demonstrations was shown by probing the effect of the fixed magnetic layer and varying pillar size, spacing between elements, number of neighbors, and nanomagnets orientation.

EXPERIMENTAL DETAILS

The MQCA logic device relies on the dipole coupling between closely spaced nanomagnets to perform logic operations. All nanomagnets have a multilayered structure as follows: Si substrate / SiO_2 (100 nm) / bottom electrode / PtMn (20 nm) / $Co_{70}Fe_{30}$ (2.4 nm) / Ru (0.85 nm) / $Co_{40}Fe_{40}B_{20}$ (2.4 nm) / MgO (0.9 nm) / $Co_{60}Fe_{20}B_{20}$ (1.8 nm) / top electrode. The nanomagnets were patterned with a Vistec electron beam (e-beam) lithography system at 100 kV with a bi-layer of polydimethylglutarimide (PMGI) and polymethyl methacrylate (PMMA) resist followed by e-beam evaporation of 180 nm of Ti and liftoff. Ion milling was then used to mill the pillars down to the bottom electrode. Next, silicon dioxide deposition using plasma enhanced chemical vapor deposition followed by a planarization process left only the top of the MQCA elements exposed. Top contacts with a 50 nm width and 80 nm pitch at the narrowest point were defined by e-beam lithography. The narrow spacing between the top electrodes and precise alignment allowed adjacent pillars to be contacted individually. Photolithography, e-beam evaporation, and liftoff were then used to form large contact pads connecting to the e-beam leads. See supplementary material for a more detailed explanation of the fabrication process.[30]

Figure 2(a) shows the direction of the external fields and currents for our measurement setup. The positive current is defined as the electron flow from the fixed layer into the free layer, which is from the bottom to top electrode. A magnetic field in the y direction, H_y, was applied along the long axis of the nanomagnets for field switching, biasing, and setting of states. Positive H_y is defined as the direction opposite to the magnetization in the fixed layer. The clocking field applied in the x direction, H_x, is defined along the one dimensional array. For example, the clocking field for AF coupling with MTJs aligned along the long axis is applied along the minor axis of the pillars while for FM coupling the field is applied along the long axis, both of which are along the one dimensional array.

This arrangement for the clocking field is required and utilized for logic operation since both MTJ configurations are required. Figure 2(b) shows a simplified timing diagram for STT programming for MQCA operation to show the relationship between the clocking field pulse and current pulse. Figure s2 shows a full version of the timing diagram.[30]

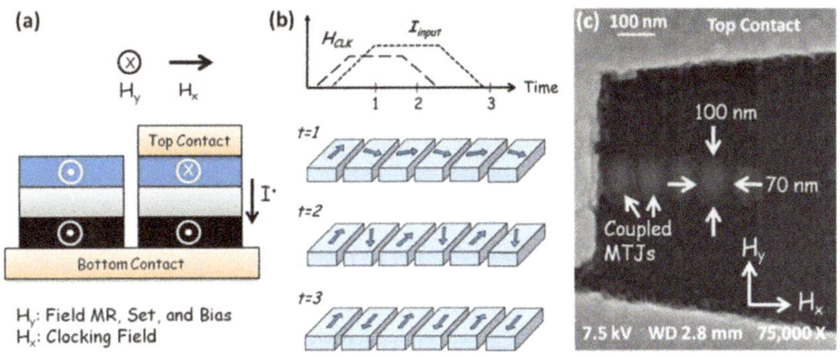

Figure. 2. (a) Cross section schematic of measurement setup and contact of pillars. (b) A timing diagram and illustrations (only free layer shown) at different time steps for STT programming for MQCA operation. (c) SEM image of a fabricated one dimensional array with 70 nm x 100 nm with 10 nm spacing between MTJ nanomagnets.

RESULTS AND DISCUSSION

Pillar sizes of 50 nm x 80 nm and 70 nm x 100 nm were fabricated with 0 to 30 nm spacing between elements. The pillars share a common bottom electrode; however, individual nanomagnets can be contacted using top electrodes defined by e-beam lithography. Figure 2(c) shows a scanning electron microscope (SEM) image of the fabricated MTJ nanomagnet one dimensional arrays for 70 nm x 100 nm pillars. Individual and adjacent nanomagnets were contacted for programming and reliability studies. The number of nanomagnets in the one dimensional array was varied from 2 to 20 nanomagnets.

One dimensional arrays with the MTJ nanomagnet arranged for AF or FM coupling were fabricated since both are required for logic application. A simplified system with only two pillars in the one dimensional arrays was fabricated to allow fundamental understanding of the results. A 20 pillar system was also fabricated where the center MTJ was probed to investigate the effect of the number of neighbors and to correlate the results to a system which is similar to an actual logic design.

Figure 3(a) shows an SEM image of a one dimensional array and top contacts setup for programming demonstrations. The array has 70 x 100 nm nanomagnets and 10 nm spacing between the 20 elements. Individual pillars are contacted to allow programming of those elements. Figure 3(b) shows the traditional field programming with the field applied along H_y, the long axis of the nanomagnets. Two clear states of high and low resistance are observed which correspond to states 1 and 0, respectively. The MR loop is well centered and only a 1.5 Oe offset is observed which is induced by the fixed layer and adjacent neighbors. For all subsequent measurements the offset field is cancelled out with an external field.

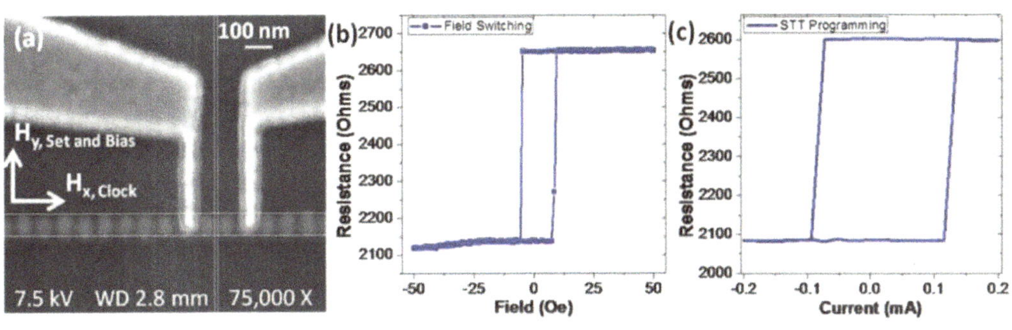

Figure. 3. (a) SEM Image showing contact of individual 70 nm x 100 nm MTJ nanomagnets with 10 nm spacing in one dimensional array of 20 elements for STT programming (b) MR from a nanomagnet for field switching with field applied along H_y (long axis of MTJ) and (c) STT programming of nanomagnets with no clocking field applied. Nanomagnets initially set P or AP with respect to fixed layer then current pulse applied. This is repeated until a switch is observed.

Magnetic field programming of nanomagnets severely limits the areal density and scalability because stray fields can write neighboring bits. A spintronic interface that utilizes STT switching overcomes this challenge, allowing for reliable control of individual input nanomagnets. Figure3(c) shows the STT switching of the same pillar which was field programmed in Figure 3(b). Clear P to AP and AP to P switching is observed. The current density, J_c, to switch from AP to P and P to AP was 1.66 and 2.42 MA/cm^2, respectively. For the measurement, the MTJ is initially set AP or P then a current pulse is applied to attempt to switch the element. This process is repeated while increasing the current pulse amplitude until switching is observed.

Figure 3(c) is the programming case where no clocking field is applied and the input's initial state and the direction of neighboring elements is set to the same configuration before the current pulse. This allows the behavior of the system to be understood for the most straight forward programming scenario which is analogous to STT programming in MRAM. The neighboring elements prefer a coupling configuration with the input nanomagnets which means that they will either help or fight the STT programming based on the initial directions. Since the nanomagnets are reset to the same direction with respect to one another and AF coupling is preferred for this arrangement, the neighboring elements are always helping to switch the element for both P to AP and AP to P switching directions. In actual operation the neighboring elements would be fighting or helping the STT switch depending on the initial coupling between the elements and a clocking field would be applied to allow adjacent nanomagnets to follow the programmed bit.

To demonstrate an actual logic write, STT and clocking field pulses were applied and a phase diagram is formed as shown in Figure 4. The black and white regions are the programmed regions. The tan region is where the state of the input nanomagnet was not properly set. The same 70 nm x 100 nm MTJ nanomagnet with 10 nm spacing and 20 pillars in the one dimensional array from Figure 3 is shown here. The nanomagnets are still reset to an initial state of P or AP with respect to the fixed layer. However, now a clocking field is applied which lowers the energy barrier for magnetic reversal and allows neighboring elements to follow the programmed bit. Then a current pulse is applied through the clocking field pulse which allows STT to set the programmed bit state and the other bits to follow. See the supplementary section for a timing diagram of the STT-H_{clock} measurements.[30]

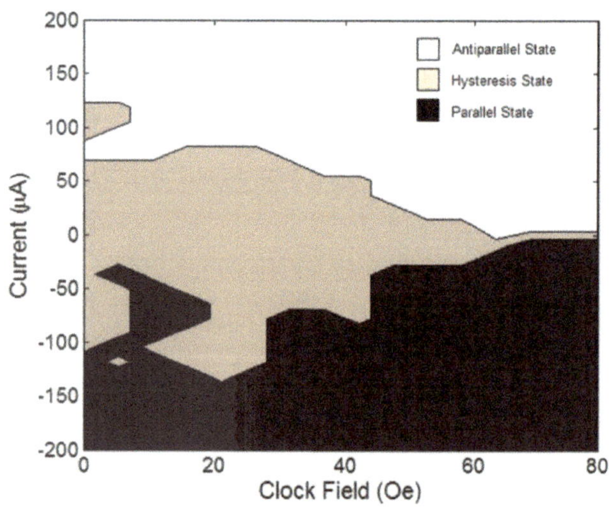

Figure. 4. The current required to properly program the input nanomagnets based on the clocking field applied for MQCA operation. The black and white regions are the properly programmed region while the tan region was not properly set. The clocking field lowers the energy barrier for magnetic reversal and allows neighboring elements to follow programmed nanomagnet. [70 nm x 100 nm elements with 10 nm spacing in a 20 pillar one dimensional array. Clocking field is applied H_x (along the one dimensional array)].

Compared to STT programming in Figure 3, if a large enough clocking field is applied the magnetic coupling between elements will reset the nanomagnets for the preferred coupling. We demonstrate this relationship of coupling to clocking field strength later in the manuscript. When the nanomagnets couple, the neighboring elements fight the STT programming. However, Figure 4 shows that at these higher fields, for this particular device 60-80 Oe, STT is dominant and the proper state is set regardless of the neighboring elements configuration.

For an optimized spacing, size and architecture, particular clocking field amplitude would be chosen which is discussed later in the manuscript. Based on this clocking field the STT current amplitude is chosen where reliable programming is achieved regardless of the initial state and coupling configuration. For example, if a 60 Oe clocking field is used for logic operation then current amplitude of 100 uA and -100 μA would reliably set the AP and P states, respectively. The observation of undesired switching at low current and clocking field amplitudes demonstrate a threshold for proper operation. The statistical nature of this phase measurement allows a means to optimize the programming of individual nanomagnets in the logic device.

Proper programming of individual nanomagnets is crucial for logic computation. Another key is the reliable coupling between neighboring elements since the interaction is used to propagate data and compute logic functions. To characterize this two MTJ nanomagnets are contacted with a single top electrode as shown in Figure 5(a) for a one dimensional array with 50 nm x 80 nm pillars and 10 nm spacing between the elements. The number of elements in the array was varied to probe the effect of neighboring elements. For simplicity a two pillar array is demonstrated first. Figure 5(b) shows the field switching loop with the field applied along H_y, the long axis of the two nanomagnets. Three states are observed corresponding to both P MTJ nanomagnets, one AP and one P, and both AP with respect to the fixed layer direction. The coercivity of the two nanomagnets is similar since they have similar aspect ratios; however the loops are shifted due to the coupling.

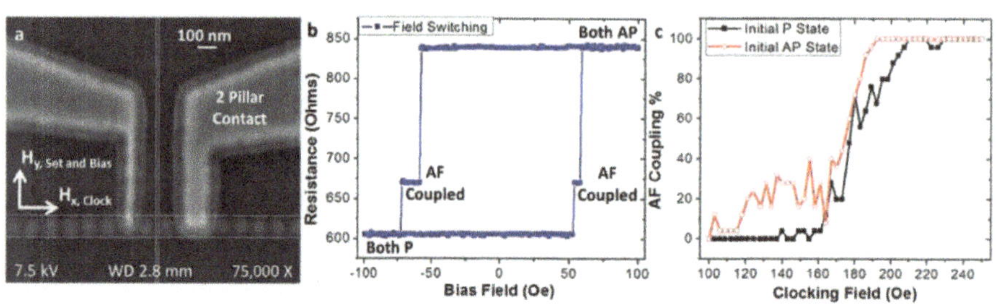

Figure. 5. (a) SEM image showing two-pillar contact used for reliability measurements on a one dimensional array with 50 nm x 80 nanomagnets and 10 nm spacing. (b) MR H_y switching using the two-pillar contact which shows three states corresponding to both nanomagnets P,

one P and one AP, and both AP with respect to the fixed layer direction. (c) Determining the required clocking field to reliably couple contacted nanomagnets based on 25 measurements at each clocking field strength. The nanomagnets are initially set P or AP with respect to the fixed layer then the clocking field is applied. Fixed layer does effect clocking field required to properly couple 100% however is not a detrimental to operation [Clocking field is applied H_x (along the one dimensional array)].

To demonstarte the proper coupling of neighboring elements, this measurement initially sets the pillars to be either P or AP with respect to the fixed layer, applies the H_x clocking field, and then checks the state of the pillars. This process is repeated while increasing the amplitude of the clocking field which tilts the magnetization direction towards the minor axis and lowers the energy barrier between the P and AP state. Once the energy barrier is low enough for the coupling to overcome it, the MTJs AF couple and set the intermediate state. Figure 5(c) shows the coupling reliability for the two-pillar array. The probability is based off of 25 measurements per clocking field amplitude. This measurement technique allows a threshold for the required clocking field amplitude to be determined and evaluation of MQCA logic devices based on statistical data.

Figure 5(c) shows the coupling of the pillars is affected by the fixed layer. Setting the pillars initially P with respect to the fixed layer requires 190 Oe to be applied to achieve 100% coupling versus over 200 Oe for the initial AP state. The fixed layer is a synthetic antiferromagnetic (SAF)[19] structure which would not apply any dipole field to the fixed layer when fully optimized. However, even when the SAF is not perfectly balanced the difference in required clocking field amplitudes is relatively small and not detrimental to the logic operation.

The effect of the nanomagnet size and spacing on the clocking field required for proper coupling is shown in Figure 6. The 50 nm x 80 nm MTJ nanomagnets with 10 nm spacing only require 80 Oe to properly couple since the dipole field between elements is proportional to V/r3. The clocking field required to get the 70 nm x 100 nm nanomagnets to properly couple for the same 10 nm spacing is 140 Oe and 200 Oe for 20 nm spacing. For the larger nanomagnets and spacing the dipole field between elements is weaker. As a result, a larger clocking field is needed to tilt the nanomagnets magnetic direction making it easier for them to couple. The reduction of dipole field accounts for a 75-150% difference in coupling field strength. Another factor that should be considered is a difference in coercivity of the nanomagnets due to the size and slight difference in aspect ratio. For the devices used, the coercivity for the 50 x 80 nm and 70 x 100 nm with 10 nm spacing is around 80 and 40 Oe, respectively. Even when the larger nanomagnets have a lower coercivity the clocking field to properly couple is still substantially larger. Therefore, the difference in dipole field strength is the clear parameter which must be considered for MQCA devices.

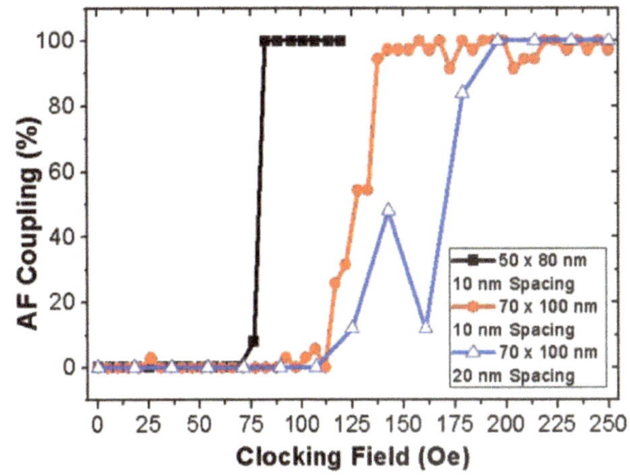

Figure. 6. Effect of pillar size and spacing on coupling reliability for 50 nm x 80 nm nanomagnets with 10 nm spacing and 70 nm x 100 nm nanomagnets with 10 and 20 nm spacing in one dimensional arrays of 20 pillars [Clocking field is applied H_x (along the one dimensional array)].

The effect of the number of neighbors on the clocking field is shown in Figure 7. Both systems have 50 nm x 80 nm nanomagnets with 10 nm spacing between elements. However, the array with 20 nanomagnets properly couples at 80 Oe compared to 190 Oe for the 2 pillar case. This can be easily understood since the addition of neighbors which prefer to be AF coupled will help to overcome the energy barrier between the P and AP state. However, only the nearest neighbors will strongly affect the MTJ nanomagnets since the dipole field falls off with distance cubed. This effect must be clearly understood since various layouts of nanomagnets are required for logic operations. The clocking field will be limited by the worst case coupling scenario of the system.

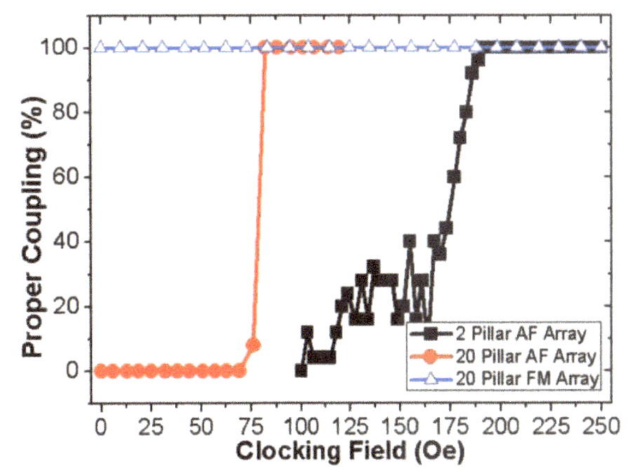

Figure. 7. Coupling behavior based on number of pillars and coupling configuration for 50 nm x 80 nm MTJ nanomagnets with 10 nm spacing with 2 and 20 pillars. [Clocking field is applied H_x].

Up to this point all data has been measured from one dimensional arrays with the MTJ arranged side by side to induce AF coupling. However, for an actual logic design the nanomagnets will also be arranged end to end where FM coupling, adjacent nanomagnets align P to one another, is preferred. Figure 7 shows the coupling reliability for a FM arrangement of 50 nm x 80 nm MTJ nanomagnets in a one dimensional array with 20 elements and 10 nm spacing. Regardless of the initial state and coupling field strength the nanomagnets properly couple 100% of the time. This result is expected since the individual MTJ nanomagnetic anisotropy and the coupling prefer the magnetization direction along the long axis of the element, so no competition is present.

TUNING THE MAGNETIC RESPONSE

Another important aspect of designing nanomagnetic arrays for logic is that the magnetic response may be different for various elements. The nanomagnets which are in transmission lines or perform the logic do not need to have a high thermal stability. So, the spacing between elements or the size can be optimized to make it easier for them to couple, compute, and propagate data. By balancing the shape anisotropy and dipole fields, the net anisotropy field could be tuned such that under the finite slew rate of the clock field, data propagation always goes from input to output. However, the input and output MTJ should be thermally stable since they need to preserve the logic data in order for the logic to be nonvolatile. Therefore, the ability to easily tune the magnetic response is an advantageous aspect of nanomagnet arrays.

The magnetic response of individual nanomagnets in this context is determined by the competition between the magnetic anisotropy and the dipole interactions with neighboring elements. Nanomagnets can have a preferred magnetization orientation along an "easy" axis, based on the anisotropy which may come from crystal structure, geometry, atomic pair ordering, and stress. For nanomagnets with an easy axis direction in the film plane, shape anisotropy is a strong contributor. The energy difference between long and short axis magnetization orientations grows with increased aspect ratio of the device. This will induce a preferred direction along the longest axis due to an energy minimum. When nanomagnets are brought within close proximity, they interact magnetostatically and modify the overall energy. This magnetostatic interaction may be approximated as a dipole field proportional to element volume and the inverse cube of the center-to-center distance ($1/r^3$). For sub 100 nm elements the dipole field will mainly be observed for nanomagnet spacing from 0 to 50 nm. When the nanomagnets are arranged near one another the dipole field may compete with the shape anisotropy of the individual nanomagnets. This allows the magnetic response to be tuned and the individual nanomagnets to be biased by neighboring elements. Nanomagnets of size 50 x 80 nm, 70 x 100 nm, and 100 x 100 nm were fabricated. The spacing between nanomagnets was varied from 0 to 20 nm.

Figure 8 shows field switching plots for different nanomagnet spacing and size in the one dimensional array. The MR has been normalized for comparison purposes. The MR values ranged from 10-50% depending on the device. Figure 8(a) shows normalized MR field switching plots for a 70 nm x 100 nm nanomagnet for approximately 0, 10, and 20 nm spacing between nanomagnets. The coercivity of the contact nanomagnet increases as the spacing is increased. This is due to a decrease in the dipole field from neighboring elements.

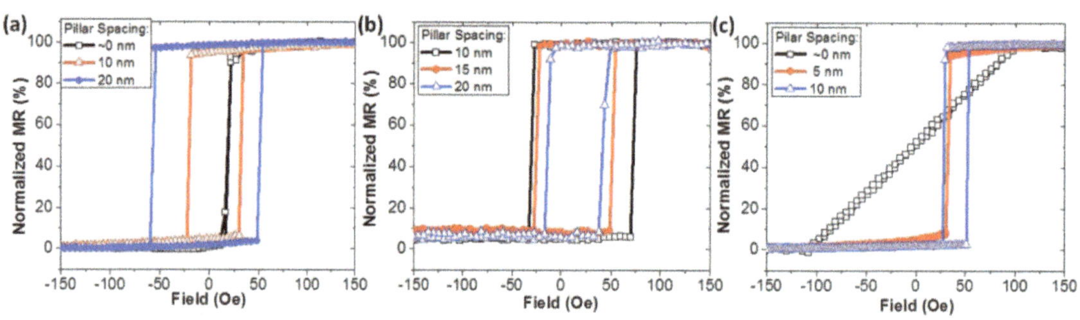

Figure. 8. Normalized MR loops for field switching for different pillar spacing in array with (a) 70 x 100 nm, (b) 100 x 100 nm, and (c) 50 x 80 nm nanomagnets.

Figure 8(b) shows the normalized MR plots for 100 x 100 nm nanomagnet for 10, 15, and 20 nm spacing. The opposite behavior is observed compared to the first case. In this case, the coercivity decreases as the pillar spacing is increased. This happens because the 100 x 100 nm have little or no shape anisotropy and so the dipole field sets the easy axis direction. As the dipole field strength decreases the nanomagnets become easier to switch and the coercivity decreases.

Figure 8(c) shows the normalized MR plots for the 50 x 80 nm nanomagnet arrays. Spacing between nanomagnets is approximately 0, 5, and 10 nm. The coercivity decreases as the spacing increases due to the shape anisotropy of the pillars. For pillar spacing near 0 nm, the easy axis resets along the array and a linear magnetic response is observed.

Figure 9 summarizes the coercivity as a function of spacing for the 70 x 100 nm and 100 x 100 nm nanomagnet arrays. Analytical calculations of the expected coercivity are compared to the experimental values, and reasonable agreement is observed. The calculations include the effect of neighboring elements, finite temperature, and shape anisotropy. If we make the simplifying assumption of an infinitely long periodic array of identical AF-coupled nanomagnets, then the shape anisotropy and dipole interactions compete such that the net uniaxial anisotropy is

$$K_u = M_s^2\left(\frac{\Delta N}{2} - \frac{3}{2}\sum_n \frac{V}{r_n^3}\right) \tag{1}$$

The first term represents shape anisotropy, and it favors an easy axis direction perpendicular to the chain of nanomagnets. This term was estimated by comparing demagnetizing energy values from OOMMF.[31] The second term represents the summation of dipole interactions, and it favors an easy axis direction parallel to the chain of nanomagnets. The center-to-center pitch of nanomagnets is set to r_1.

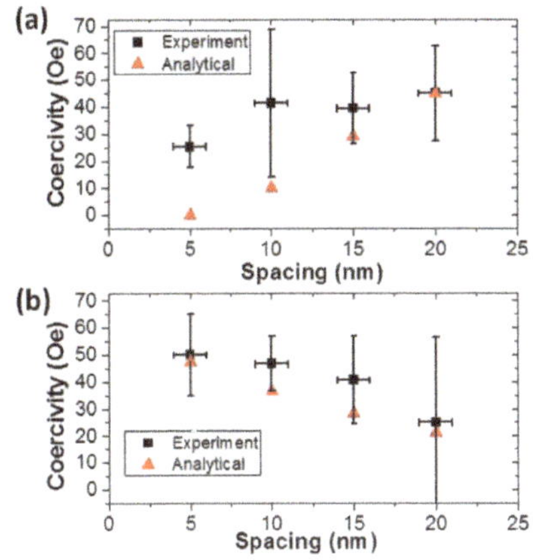

Figure. 9. Plot of the experimental and analytical coercivity as a function of spacing with (a) 70 x 100 nm and (b) 100 x 100 nm nanomagnets in array.

This simple analysis assumes a perfect right angle between the two competing energy terms. As such, the positive or negative sign on K_u acts as a placeholder for easy axis direction. As a point of interest, the direction of the nanomagnet array differs for the measurements used to construct plots in Figure 9(a) and 9(b). In the case of Figure 9(a), the nanomagnets are arrayed perpendicular to the fixed layer in anticipation of the positive value obtained for K_u. In the case of Figure 9(b), the direction of the array is parallel to the direction of the fixed layer to accommodate the negative value of K_u.

The sweep-rate (dH/dt) of measurements at finite temperature is included in the calculation with[32]

$$H_c = H_0\left[1 - \sqrt{\frac{kT}{M_s H_0 V}}Erf^{-1}(0,\Lambda)\right]$$

$$\Lambda = Erf\left(\sqrt{\frac{M_s H_0 V}{kT}}\right) - \left(\frac{2}{\sqrt{\pi}}\right)\left(\sqrt{\frac{M_s H_0 V}{kT}}\right)\left(\frac{\tau_0 ln(2)}{H_0}\right)\frac{dH}{dt}$$

$$Erf(z) = \left(\frac{2}{\sqrt{\pi}}\right)\int_0^z exp(-t^2)dt \qquad (2)$$

Here, we assume $kT = 4.14\times10^{-14}$ ergs, $\tau_0 = 10^{-9}$ sec, and the value of H_0 is set to $H_k = 2K_u/M_s$. The same analysis was used to predict the nanomagnet coercivity as a function of aspect ratio and spacing as shown in Figure 10. The major axis is fixed at 100 nm for the calculations. Also, the free layer thickness is set at 2 nm.

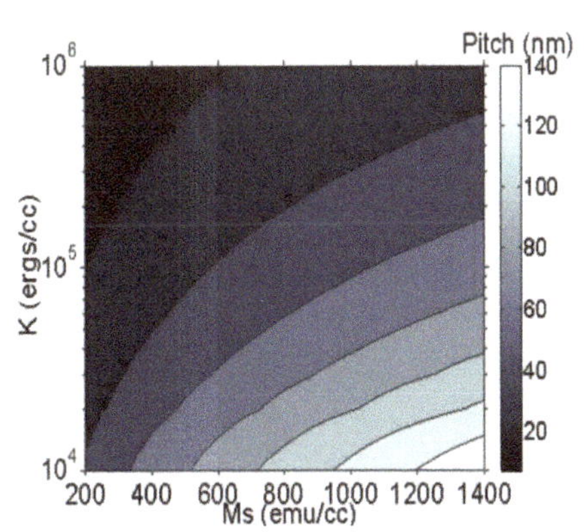

Figure. 10. Phase plot of the calculated pitch to reset the easy axis direction for different magnetic anisotropy and saturation magnetization for a nanomagnet with a volume of 6400 nm^3.

For sensing applications, a key parameter is the sensitivity which is normally compared in units of MR %/Oe. The devices that we fabricated had sensitivity between 1-5 %/Oe since our MR was in the range of 10-50%. This lower sensitivity is not a limitation of the design or concept but rather is due to a non-optimized film structure and fabrication process. MTJ elements easily reach 100-200% MR in production level devices.[18,19] So, once the self-biased nanomagnet design is optimized the sensitivity will increase by up to an order of magnitude.

A promising application of the nano-scale self-biased sensing scheme is two or three dimensional magnetic detection with a single die. An array of nanomagnetic elements can be tuned to be sensitive along a field direction for a desired working field range by controlling the aspect ratio, volume, spacing, and direction of the easy axis direction with respect to the fixed layer. Varying these parameters would allow two-dimensional detection as demonstrated with the 70 x 100 nm and 100 x 100 nm pillars. Pushing the idea it one step further three-dimensional detection is possible by reducing the free layer thickness. The CoFeB free layer magnetization direction becomes perpendicular to the film plane once the thickness is on the order of 1 nm. Thus the free layer may be tilted partially out-of-plane making it sensitive to perpendicular fields. All of this is would be on the nanoscale with a device footprint on the order of 100s of nm and not microns.

CONCLUSION

In conclusion, MTJ based nanomagnet arrays offer a highly tunable system with an electrical I/O for logic and sensing applications. The experimental demonstrations of spintronic I/O of individual nanomagnets, the relationship between the clocking field and STT, and coupling reliability of nanomagnets mark benchmark for logic applications. Demonstrating the ability for controlling the magnetic responses of the individual nanomagnet in the arrays provides an additional benchmark for logic and opens the door for other applications such as sensing including two or three-dimensional field sensing. Analytical calculations agree well with the experimental data of magnetic responses and provide a means to design the nanomagnet arrays in order to obtain a desired magnetic behavior to fit a variety of contexts. The demonstration of MTJ nanomagnetic arrays push MQCA forward as a post CMOS logic scheme and marks an appealing approach for sensing using the self-biased nature of the array.

ACKNOWLEDGMENT

This work was supported by the MRSEC Program of the National Science Foundation under Award Number DMR-0819885, DARPA Non-Volatile Logic program, NSF Minnesota Nano Fabrication Center and NSF NNIN program.

REFERENCE

1. P. Apte, B. Doering, P. Gargini, *et al.*, in *International Technology Roadmap for Semiconductor 2010 Uptdate*, edited by I. R. Committee (2010), http://www.itrs.net/links/2010itrs/2010Update/ToPost/2010_Update_Overview.pdf.
2. L. Qian and E. Winfree, *Science* 332(6034), 1196–1201 (2011).http://dx.doi.org/10.1126/science.1200520

3. J. H. Schon, H. Meng and Z. Bao, *Nature* 413(6857), 713–716 (2001).http://dx.doi.org/10.1038/35099520
4. S. Neusser and D. Grundler, *Advanced Materials* 21(28), 2927–2932 (2009).http://dx.doi.org/10.1002/adma.200900809
5. A. Khitun, M. Bao and K. L. Wang, *Journal of Physics D – Applied Physics* 43(26) (2010).
6. A. Ney, C. Pampuch, R. Koch and K. H. Ploog, *Nature* 425(6957), 485–487 (2003).http://dx.doi.org/10.1038/nature02014
7. R. P. Cowburn and M. E. Welland, *Science* 287(5457), 1466–1468 (2000).http://dx.doi.org/10.1126/science.287.5457.1466
8. A. Imre, G. Csaba, G. H. Bernstein, W. Porod and V. Metlushko, *Superlattices and Microstructures* 34(3-6), 513–518 (2004).http://dx.doi.org/10.1016/j.spmi.2004.03.051
9. D. B. Carlton, N. C. Emley, E. Tuchfeld and J. Bokor, *Nano Letters* 8(12), 4173–4178 (2008).http://dx.doi.org/10.1021/nl801607p
10. K. Walus, G. Schulhof, G. A. Jullien, R. Zhang and W. Wang, presented at the Signals, Systems and Computers, 2004. *Conference Record of the Thirty-Eighth Asilomar Conference on*, 2004 (unpublished).
11. M. Niemier, M. Alam, X. S. Hu, G. Bernstein, W. Porod, M. Putney and J. DeAngelis, in *Proceedings of the 2007 international symposium on Low power electronics and design* (ACM, Portland, OR, USA, 2007), pp. 26–31.
12. A. Imre, G. Csaba, L. Ji, A. Orlov, G. H. Bernstein and W. Porod, *Science* 311(5758), 205–208 (2006).http://dx.doi.org/10.1126/science.1120506
13. W. F. EgelhoffJr, P. W. T. Pong, J. Unguris, R. D. McMichael, E. R. Nowak, A. S. Edelstein, J. E. Burnette and G. A. Fischer, *Sensors and Actuators A: Physical* 155(2), 217–225 (2009).http://dx.doi.org/10.1016/j.sna.2009.08.016
14. W. F. Egelhoff, V. E. Höink, J. W. Lau, W. F. Shen, B. D. Schrag and G. Xiao, 107(9), 09C705 (2010).
15. P. W. T. Pong, B. Schrag, A. J. Shapiro, R. D. McMichael and J. W. F. Egelhoff, 105(7), 07E723 (2009).
16. J. M. Almeida, P. Wisniowski and P. P. Freitas, 103(7), 07E922 (2008).
17. L. R. Shah, N. Bhargava, S. Kim, R. Stearrett, X. Kou, X. Sun, S. Sun, J. Kolodzey, E. R. Nowak and J. Q. Xiao, 109(7), 07C731 (2011).
18. M. Durlam, D. Addie, J. Akerman, B. Butcher, P. Brown, J. Chan, M. DeHerrera, B. N. Engel, B. Feil, G. Grynkewich, J. Janesky, M. Johnson, K. Kyler, J. Molla, J. Martin, K. Nagel, J. Ren, N. D. Rizzo, T. Rodriguez, L. Savtchenko, J. Salter, J. M. Slaughter, K. Smith, J. J. Sun, M. Lien, K. Papworth, P. Shah, W. Qin, R. Williams, L. Wise and S. Tehrani, in *Electron Devices Meeting, 2003. IEDM '03 Technical Digest. IEEE International* (2003), pp. 34–3631–3436–33.
19. S. Parkin, J. Xin, C. Kaiser, A. Panchula, K. Roche and M. Samant, *Proceedings of the IEEE* 91(5), 661–680 (2003).http://dx.doi.org/10.1109/JPROC.2003.811807
20. S. A. Wolf, L. Jiwei, M. R. Stan, E. Chen and D. M. Treger, *Proceedings of the IEEE* 98(12), 2155–2168 (2010).http://dx.doi.org/10.1109/JPROC.2010.2064150
21. S. Moralejo, F. J. Castano, C. A. Ross, C. Redondo and F. Castano, *Journal of Physics D: Applied Physics* 41(19), 195003 (2008).http://dx.doi.org/10.1088/0022-3727/41/19/195003
22. M. Becherer, G. Csaba, R. Emling, P. Osswald, W. Porod, P. Lugli and D. Schmitt-Landsiedel, in *Nanoelectronics Conference, 2008. INEC 2008. 2nd IEEE International* (2008), pp. 1043–1046.
23. M. Becherer, J. Kiermaier, S. Breitkreutz, G. Csaba, X. Ju, J. Rezgani, T. Kießling, C. Yilmaz, P. Osswald, P. Lugli and D. Schmitt-Landsiedel, *Solid-State Electronics* 54(9), 1027–1032 (2010).http://dx.doi.org/10.1016/j.sse.2010.04.011
24. A. Fert, *Angewandte Chemie International Edition* 47(32), 5956–5967 (2008).http://dx.doi.org/10.1002/anie.200801093
25. P. A. Grunberg, *Reviews of Modern Physics* 80(4), 1531 (2008).http://dx.doi.org/10.1103/RevModPhys.80.1531
26. A. A. Tulapurkar, Y. Suzuki, A. Fukushima, H. Kubota, H. Maehara, K. Tsunekawa, D. D. Djayaprawira, N. Watanabe and S. Yuasa, *Nature* 438(7066), 339–342 (2005).http://dx.doi.org/10.1038/nature04207
27. M. Tsoi, A. G. M. Jansen, J. Bass, W. C. Chiang, M. Seck, V. Tsoi and P. Wyder, *Physical Review Letters* 80(19), 4281 (1998).http://dx.doi.org/10.1103/PhysRevLett.80.4281
28. E. B. Myers, D. C. Ralph, J. A. Katine, R. N. Louie and R. A. Buhrman, *Science* 285(5429), 867–870 (1999).http://dx.doi.org/10.1126/science.285.5429.867
29. A. Lyle, A. Klemm, J. Harms, Y. Zhang, H. Zhao and J.-P. Wang, *Applied Physics Letters* 98(9), 092502–092502 (2011).http://dx.doi.org/10.1063/1.3558915
30. See supplementary material at http://dx.doi.org/10.1063/1.3672177 for a full description of the fabrication process and STT/H_{Clock} phase measurements.
31. M. J. Donahue and D. G. Porter, in *Interagency Report* (National Institute of Standards and Technology, 1999), Vol. NISTIR 6376.
32. J. Moritz, C. Ó. Coileáin, G. Feng, K. Nakajima, S. van Dijken and J. M. D. Coey, *Journal of Magnetism and Magnetic Materials* 296(2), 118–123 (2006).http://dx.doi.org/10.1016/j.jmmm.2005.02.004

CHAPTER 16

SUPERCONDUCTING NANOWIRE SINGLE-PHOTON DETECTORS AT A WAVELENGTH OF 940 NM

W. J. Zhang, H. Li, L. X. You, Y. H. He, L. Zhang, X.Y. Liu, X. Y. Yang, J. J. Wu, Q. Guo, S. J. Chen, Z. Wang and X. M. Xie

The efficient generation and detection of single photons lies at the heart of quantum information (QI) processing.[1–4] Self-assembled InAs/GaAs quantum dots (QDs) have proven to be an efficient source of single photons with high purity (99.7%) and indistinguishability (99.5%).[5] They hold the promise of offering an excellent solid-state platform for investigations of multiphoton entanglement,[1] boson sampling,[6] and linear optical quantum computing.[7] However, a major experimental challenge[4] is the lack of high-performance detectors at the emission wavelength range of these QDs, which is typically around a wavelength of 940 nm. Silicon avalanche photodiodes (APDs) have low efficiencies (20%–30%) and long dead times (~45 ns) at a wavelength of 940 nm,[5,8] with inevitable afterpulsing counts. Thus, the development of faster, more efficient single-photon detectors at a wavelength of 940 nm is critical to the implementation of QI technologies in the future.

Superconducting nanowire single-photon detectors (SNSPDs) have proven to be promising detectors for near infrared photons because of their high detection efficiency (>90%), low dark count rate (<100 Hz), low timing jitter (<70 ps), and short recovery time (<40 ns). For instance, WSi-SNSPD demonstrated a record system detection efficiency (SDE) of 93% at a wavelength of 1550 nm,[9] though it needs a sub-Kelvin operating temperature. Nb(Ti)N-SNSPDs have demonstrated an SDE of over 75% at a moderate temperature over 2 K.[10,11] Waveguide-coupled SNSPDs have shown a fast-recovery time (<10 ns), but the SDE is still limited to <20%.[12,13] Indeed, SNSPDs may provide excellent performance not only at 1550 nm but also at shorter wavelengths, since photon energy scales as the inverse of wavelength. Recently, Liu et al demonstrated multimode fiber-coupled SNSPDs with SDEs of over 70% at a wavelength of 635 nm.[14] However, there has been little research on developing high-performance SNSPDs at wavelengths around 940 nm, though there were some reports on the characterization of QD single-photon sources using SNSPDs with relatively poor performance at these wavelengths[3,8,13]

In this letter, we report single-mode fiber-coupled SNSPDs at a wavelength of 940 nm with high SDE and fast recovery time. The detectors were fabricated on a Si substrate with a multilayer dielectric mirror to realize high optical absorptance with front-side illumination. By optimizing the design and the fabrication for photons with a wavelength of 940 nm, we demonstrate that SNSPDs can achieve an SDE of ~60% with a dark count rate (DCR) of 10 Hz, a recovery time <12 ns, and timing jitter of ~50 ps.

The optical absorption of the nanowire is one of the key parameters to consider for achieving a high SDE. Optical cavity structures on a Si substrate have been demonstrated to be an effective way to enhance the optical absorption of SNSPDs at a wavelength of 1550 nm.[10,11]However, the structure is not applicable for wavelengths below 1.2 μm since the photons are reflected and absorbed by the Si substrate. Consequently, another structure with a distributed Bragg reflector (DBR) on a Si[14,15] or GaAs[16] substrate was developed for SNSPD instead, with which high absorption can be guaranteed with front-side illumination. This structure also provided a high flexibility, which was independent of the substrate material, for various target wavelengths as long as the thicknesses of the DBR films were changed.

Citation: W. J. Zhang, H. Li, L. X. You, Y. H. He, L. Zhang, X.Y. Liu, X. Y. Yang, J. J. Wu, Q. Guo, S. J. Chen, Z. Wang and X. M. Xie, Superconducting nanowire single-photon detectors at a wavelength of 940 nm, doi: 10.1063/1.4922548.

Figure 1(a) illustrates the simulation results for the reflectance (R) of the DBR substrate without NbN thin film and for the absorption efficiency (A) of SNSPD of two polarization states as functions of wavelength. The full detector structure from top to bottom was NbN/DBR/Si, as shown in Fig. 1(b). The simulations were based on nanowires with a linewidth/spacing of 110/220 nm and a thickness of 6.5 nm by using the RF module of COMSOL multiphysics.[17] The simulated A showed a broadband enhancement around a wavelength of 940 nm. The absorptions in the parallel ($A_{//}$) and perpendicular ($A_{\perp}$) polarization states reached 94% and 49%, respectively, resulting in a PER (polarization extinction ratio, $A_{//}/A_{\perp}$) of 1.9.

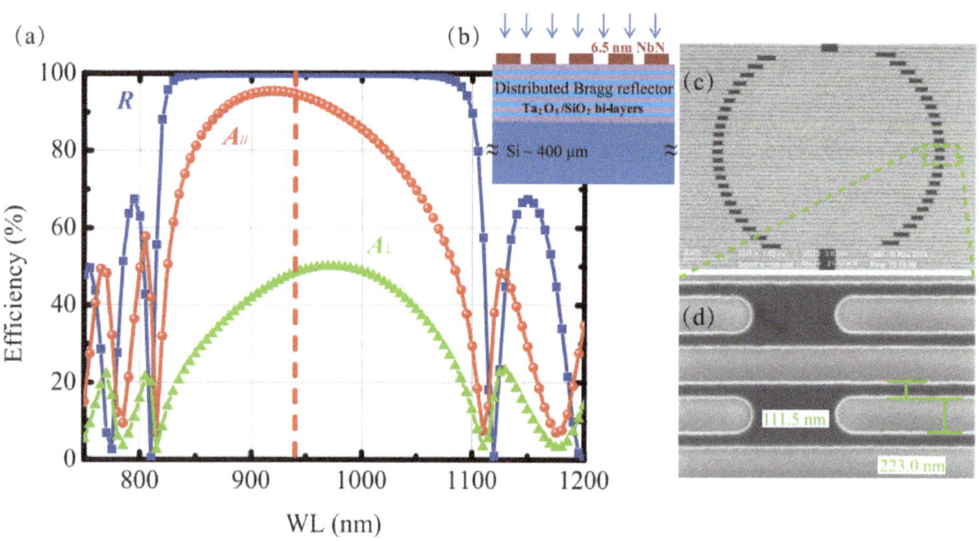

Figure. 1. (a) Simulated reflectance, R, of DBR/Si substrate (blue squares) and absorptance, A, of the SNSPD. $A_{\parallel}$ (red dots) and $A_{\perp}$ (green triangles) represent A for parallel and perpendicular polarizations, respectively; (b) Schematic of SNSPD on DBR/Si substrate; (c) Scanning electron micrograph of an NbN SNSPD with an active area diameter of 10 μm; (d) Magnified image of the nanowire around the corner showing a well-defined linewidth.

The DBR substrate was a 13 periodic Ta_2O_5/SiO_2 bi-layers deposited on a silicon wafer with 99% reflectance measured at a wavelength of 940 nm. 6.5-nm-thick NbN films were grown on DBR substrate by direct-current magneto-sputtering. The root-mean-square roughness of NbN thin films on DBR substrate were 0.18 nm, comparable to the ones on thermally oxidized silicon substrates. The meander nanowires were patterned into NbN films by electron-beam-lithography and reactive ion etching using CF_4 plasma. Aiming at QI applications at a wavelength of 940 nm, which need high SDE as well as fast recovery times, the nanowire was designed to have a low filling factor of 1/3 with a linewidth/spacing of 110/220 nm and a circular active area. SNSPDs of different active area diameters of $\Phi = 5$–18 μm were fabricated on the same wafer. In this way, we could study the SDE relations on the optical coupling and the size dependence of the recovery time. Figure1(c) shows the scanning electron micrograph of SNSPD with a diameter of 10 μm. The well-defined nanowire, shown in Fig. 1(d), guaranteed a high uniformity of SNSPD.

The devices were mounted on a fiber-coupled package optimized for front-side illumination. A 940-nm single-mode fiber (Thorlabs Inc., SM-800) with a core diameter of 5.6 μm was adopted for optical coupling. The packages were then installed in a cryostat based on a Gifford–McMahon cryo-cooler, which had a minimal temperature of 2.2 K. To optically probe the devices, a 940-nm continuous-wave fiber laser and variable attenuators were adopted. A fiber polarization controller was inserted after the attenuators to change the polarization of the photons. Since there are no commercially available variable attenuators for 940 nm light, we carefully calibrated the attenuators manually,[18] so that the light was attenuated to a typical photon flux of 0.1 M/s. The fiber was cut and spliced with the one attached to the detector in the cryostat. The photon response signal from SNSPD was amplified by a room temperature low-noise amplifier (RF Bay Inc., LNA650), and then counted using a pulse counter (SRS Inc., SR400).

The statistics of 50 devices indicated that SNSPDs had a critical temperature, $T_C \sim 7.6 \pm 0.4$ K. We measured the switching current (I_{sw}) of the devices with different diameters (10 devices for each size). The results indicated that the I_{sw}s ranged from 14.6 to 24.0 μA. Considering the wire width (w) ~ 110 nm, and thickness (t) ~ 6.5 nm, we obtained a switching current density $J_{sw} = I_{sw}/w/t = 2.0$–$3.4$ MA/cm,[2] i.e. 2.7 ± 0.7 MA/cm^2. Figure 2(a) shows the highest SDEs for various active areas (Φ = 5, 10, 15, and 18 μm) as functions of the bias current (I_b) for the parallel polarization. All the SDEs were saturated with the increase of the bias current, indicating high intrinsic detection efficiencies (IDE) of all the detectors. On the other hand, the SDEs increased with the increase of the active area and saturated for $\Phi \geq 15$ μm, indicating an excellent optical coupling efficiency (OCE) for large active area SNSPDs. The highest SDE reached 66.5% ± 2.5% for Φ = 18 μm. I_{sw} decreased as Φ increased, probably due to the increasing defects in nanowires. However, the SDEs were not influenced by the defects, because of the broad saturated plateau in the SDE–I_b curves. Moreover, the highest SDEs can be reached even at a low DCR of ~10 Hz indicated by the dashed line in Fig. 2(b). The SDEs included all losses in the system and the overall relative errors of the SDE values could be estimated to be 3.7%, resulting from the relative uncertainty of the power meter calibration (3.50%), the long-term stability of the light power (1.14%), and attenuator calibration (0.55%).

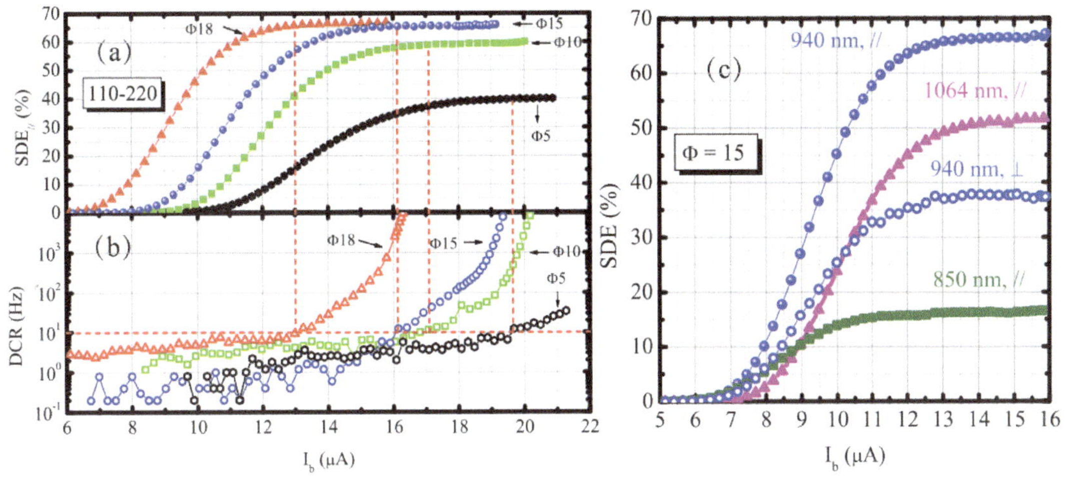

Figure. 2. (a) Parallel-polarized SDE ($SDE_{//}$) vs. the bias current (I_b) for SNSPDs with various sizes at 2.2 K under illumination at a wavelength of 940 nm; (b) DCR vs. I_b. The dashed line indicates the DCR at10 Hz; (c) $SDE_{//}$ at wavelengths of 850, 940, and 1064 nm and SDE of 940 nm vs.I_b for the device with Φ = 15 μm.

Figure 2(c) shows the wavelength dependence of SDE in parallel polarization ($SDE_{//}$) for the SNSPD with Φ = 15 μm. $SDE_{//}$ of 19.2% ± 0.7%, 65.4% ± 2.4%, and 51.1% ± 1.9% were obtained for wavelengths of 850, 940, and 1064 nm, respectively, which is qualitatively consistent with the simulation results in Figure 1(a). We observed saturated plateaus in SDE for all these wavelengths. As the wavelengths increased, the saturation plateau turned narrower and the transition range became wider. The $SDE_{\perp}$ for 940 nm with perpendicular polarization was also measured and shown in Figure 2(c), which gave the PER value of about 1.8, slightly smaller than the simulation result, probably due to the geometric parameter deviations of the practical devices with the design parameters.

Next, the counting rate (CR) and the recovery time of the SNSPDs were investigated. By increasing the photon intensity of the illumination, we obtained the CR dependence of the SDE at 0.90 I_{sw} (Figure 3(a)). When CR varied from 0.1 to 1 MHz, the SDEs were almost constant. As the CR increased, the detector could not respond to all the incoming photons limited by the slow current recovery process of the SNSPD, resulting in a decrease in the measured SDE. By defining CR_{exp} as the SDE with the 1/e suppression from its maximum value, the CR_{exp} values were found to be 22 ± 2, 34 ± 3, 68 ± 5, and 72 ± 6 MHz for Φ = 18, 15, 10, and 5, respectively. Figure 3(b) shows the averaged temporal response pulses recorded by a high-speed, real-time oscilloscope (Tek DSA 71254). The recovery times (τ_r), defined as the time required for the pulse to decay from 90% to 10% of the maximum of the pulse, were 44.4, 26.2, 11.9 and 4.8 ns, for Φ = 18, 15, 10, and 5 μm, respectively. Correspondingly, $CR_r = 1/\tau_r$ were 23, 38, 84, and 208 MHz.[19] In general, small active areas were evidently beneficial in obtaining a high CR. However, we noticed that CR_{exp} for SNSND with Φ ≤ 10 μm ($L_k \leq 345$ nH[20]) was not as high as the expectation CR_r, due to the increase of the latching possibility under bright light illumination. This similar latching effect was reported in low L_k devices on SiO_2/Si substrates.[10,19] Thus, a further increase in CR by reducing the active area or L_k could be stemmed by the latching effect.

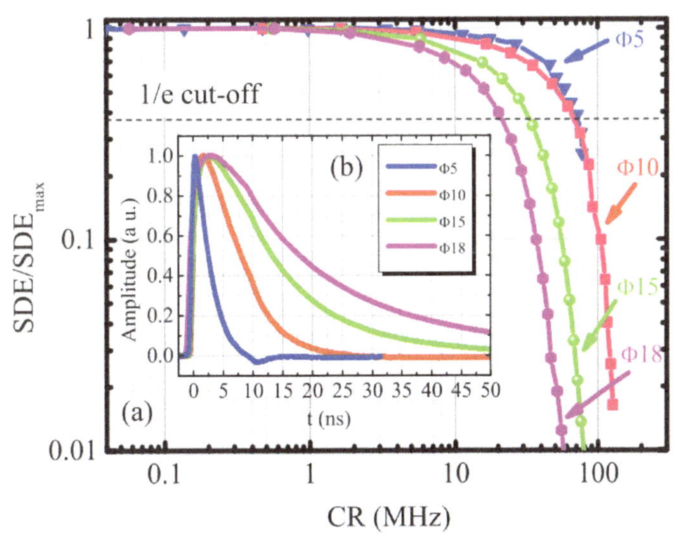

Figure. 3. (a) Counting rate (CR) dependence of the normalized SDE for devices with Φ = 5, 10, 15, and 18 μm. CR was read out by the conventional circuit with a 50 Ω shunting resistor;(b) Averaged response pulse waveform.

The timing jitter of the SNSPD was also evaluated using a time-correlated single-photon counting (TCSPC) module and 100-fs pulsed laser at a wavelength of 1550 nm and a repetition rate of 20 MHz.[21] The intensity of the laser was set to be 0.1 photon/pulse to assure the SNSPD was in single-photon-response regime. The timing jitter defined by the full width at half maximum of the histogram was 49.6 ps at the bias current of 18.0 μA (0.90 I_{sw}) for the SNSPD with $\Phi = 10$ μm. For all the detectors we measured, the timing jitter ranged in value from 40 to 60 ps, depending on the signal-to-noise ratio of the voltage pulse. It is noted here that the timing jitter measured at 1550 nm could be different from the value measured at 940 nm due to the different mechanisms (hot-spot or vortex-based mechanisms).[22]

SDE can be expressed as SDE = IDE × OCE × AE,[9] where AE is the absorption efficiency. The experimental results above indicated that IDE and OCE (for $\Phi \geq 15$ μm) are nearly 100%; thus, we could estimate that AE ~ SDE ~ 67%. The reasons for the deviation of AE with $A_{//}$ (94%) are complicated. For example, a non-ideal DBR structure may cause a lower $A_{//}$ compared with the calculated value. Geometric parameters of the practical nanowire also deviated from the designed values due to the nanofabrication process. Besides, the calculations were all based on the reflective indices of the materials obtained at room temperature. Different reflective indices at low temperature could also effectively change the final AE.[23,24] Indeed, both experiments and calculations have proven that SNSPD with a moderately high filling factor will give a high SDE.[25,26] The SDEs of ~60% (the maximal value is ~67% for $\Phi = 18$ μm) were demonstrated in our study with a low filling factor of 1/3 which is a trade-off between the high SDE and the short recovery time.[25] The maximal SDE is slightly lower than the values reported in other literatures for other wavelengths (>75% at 1550 nm[9,11] and >70% at 635 nm.[14]) where the SNSPDs with a high filling factor (~0.6) were studied. On the other hand, the SDE of 67% is comparable with the previous results (<70%) for the SNSPDs with the same filling factor of 1/3 at a 1550-nm wavelength.[25] Furthermore, a low filling factor nanowire is beneficial to reduce the current crowding effect,[26,27] giving a high yield and broaden the saturated plateau in the SDE–I_b curves (see Fig. 2). To further enhancing the CR, there are some useful methods, such as parallel nanowires,[28] a DC-coupled cryogenic circuit,[19] or increased load impedance, R_L.[2,29] However, the intrinsic solution should focus on decreasing the thermal relaxation time between SNSPD and the substrate to avoid latching.

In conclusion, we report the fabrication and characterization of SNSPDs at a wavelength of 940 nm. These SNSPDs showed high detection efficiencies (~60%) and fast-recovery times (~12 ns) beyond the available semiconducting APDs. The SNSPDs may contribute to the effective characterization of single-photon sources around a wavelength of 940 nm, which is promising for future QI processing. Further work to improve SDE and shorten the recovery time is ongoing.

ACKNOWLEDGMENTS

This work was funded by the NSFC (61401441 & 61401443), Strategic Priority Research Program (B) of the CAS (XDB04010200 & XDB04020100), 973 Program of China (2011CBA00202).

REFERENCES

1. J. W. Pan, Z. B. Chen, C. Y. Lu, H. Weinfurter, A. Zeilinger, and M. Zukowski, Rev. Mod. Phys. 84, 777 (2012).http://dx.doi.org/10.1103/RevModPhys.84.777
2. R. Valivarthi, I. Lucio-Martinez, A. Rubenok, P. Chan, F. Marsili, V. B. Verma, M. D. Shaw, J. A. Stern, J. A. Slater, D. Oblak, S. W. Nam, and W. Tittel, Opt. Exp. 22, 24497 (2014).http://dx.doi.org/10.1364/OE.22.024497
3. W. B. Gao, P. Fallahi, E. Togan, J. Miguel-Sanchez, and A. Imamoglu, Nature 491, 426 (2012).http://dx.doi.org/10.1038/nature11573
4. C. Y. Lu and J. W. Pan, Nat. Photon. 8, 174 (2014).http://dx.doi.org/10.1038/nphoton.2014.29
5. Y. J. Wei, Y. M. He, M. C. Chen, Y. N. Hu, Y. He, D. Wu, C. Schneider, M. Kamp, S. Höfling, C. Y. Lu, and J. W. Pan, Nano Lett. 14, 6515 (2014).http://dx.doi.org/10.1021/nl503081n
6. M. Tillmann, B. Dakic, R. Heilmann, S. Nolte, A. Szameit, and P. Walther, Nat. Photon. 7, 540 (2013). http://dx.doi.org/10.1038/nphoton.2013.102
7. E. Knill, R. Laflamme, and G. J. Milburn, Nature 409, 46 (2001).http://dx.doi.org/10.1038/35051009
8. R. H. Hadfield, M. J. Stevens, S. S. Gruber, A. J. Miller, R. E. Schwall, R. P. Mirin, and S. W. Nam, Opt. Exp. 13, 10846 (2005).http://dx.doi.org/10.1364/OPEX.13.010846
9. F. Marsili, V. B. Verma, J. A. Stern, S. Harrington, A. E. Lita, T. Gerrits, I. Vayshenker, B. Baek, M. D. Shaw, R. P. Mirin, and S. W. Nam, Nat. Photon. 7, 210 (2013). http://dx.doi.org/10.1038/nphoton.2013.13
10. D. Rosenberg, A. J. Kerman, R. J. Molnar, and E. A. Dauler, Opt. Exp. 21, 1440 (2013).http://dx.doi.org/10.1364/OE.21.001440
11. S. Miki, T. Yamashita, H. Terai, and Z. Wang, Opt. Exp. 21, 10208 (2013).http://dx.doi.org/10.1364/OE.21.010208
12. W. H. P. Pernice, C. Schuck, O. Minaeva, M. Li, G. N. Goltsman, A. V. Sergienko, and H. X. Tang, Nat. Commun. 3, 1325 (2012).http://dx.doi.org/10.1038/ncomms2307
13. G. Reithmaier, S. Lichtmannecker, T. Reichert, P. Hasch, K. Muller, M. Bichler, R. Gross, and J. J. Finley, Sci. Rep. 3(2013).http://dx.doi.org/10.1038/srep01901
14. D. K. Liu, S. Miki, T. Yamashita, L. X. You, Z. Wang, and H. Terai, Opt. Exp. 22, 21167 (2014).http://dx.doi.org/10.1364/OE.22.021167
15. B. Baek, J. A. Stern, and S. W. Nam, Appl. Phys. Lett. 95, 191110 (2009).http://dx.doi.org/10.1063/1.3263715
16. A. Gaggero, S. J. Nejad, F. Marsili, F. Mattioli, R. Leoni, D. Bitauld, D. Sahin, G. J. Hamhuis, R. Notzel, R. Sanjines, and A. Fiore, Appl. Phys. Lett. 97, 151108 (2010).http://dx.doi.org/10.1063/1.3496457
17. The refractive indices used in our simulation were $n\_Ta_2O_5 = 2.26$, $n\_SiO_2 = 1.47$, n_NbN = 4.81-4.77*i, n_Si = 3.46, measured by using an ellipsometry at 940 nm.
18. We fixed the optical input power and fed it into a variable attenuator (Agilent 81570A, operating at 1200 nm). We

measured the output power as function of attenuation with a calibrated high-sensitivity InGaAs power sensor (Agilent 81634B) at 940 nm. The output power showed a good linearity (slope ~ -1.08, standard error ~ 1.4×10^{-4}) with attenuation in dB.

19. A. J. Kerman, D. Rosenberg, R. J. Molnar, and E. A. Dauler, J. Appl. Phys. 113, 144511 (2013).http://dx.doi.org/10.1063/1.4799397
20. For Φ = 18, 15, 10, and 5 μm, L_k = 1288, 800, 345, 82 nH, respectively, measured by using a network analyzer.
21. L. X. You, X. Y. Yang, Y. H. He, W. X. Zhang, D. K. Liu, W. J. Zhang, L. Zhang, L. Zhang, X. Y. Liu, S. J. Chen, Z. Wang, and X. M. Xie, AIP Adv. 3 (2013).http://dx.doi.org/10.1063/1.4817581
22. M. Hofherr, D. Rall, K. Ilin, M. Siegel, A. Semenov, H. W. Hübers, and N. A. Gippius, J. Appl. Phys. 108, 014507 (2010).http://dx.doi.org/10.1063/1.3437043
23. X. Y. Yang, H. Li, W. J. Zhang, L. X. You, L. Zhang, X. Y. Liu, Z. Wang, W. Peng, X. M. Xie, and M. H. Jiang, Opt. Exp. 22, 16267 (2014).http://dx.doi.org/10.1364/OE.22.016267
24. X. Y. Yang, H. Li, L. X. You, W. J. Zhang, L. Zhang, Z. Wang, and X. M. Xie, Appl. Opt. 54, 96 (2015).http://dx.doi.org/10.1364/AO.54.000096
25. T. Yamashita, S. Miki, H. Terai, and Z. Wang, Opt. Exp. 21, 27177 (2013).http://dx.doi.org/10.1364/OE.21.027177
26. J. K. W. Yang, A. J. Kerman, E. A. Dauler, B. Cord, V. Anant, R. J. Molnar, and K. K. Berggren, IEEE Trans. on Appl. Supercond. 19(2009).http://dx.doi.org/10.1109/TASC.2009.2017953
27. H. L. Hortensius, E. F. C. Driessen, T. M. Klapwijk, K. K. Berggren, and J. R. Clem, Appl. Phys. Lett. 100, 182602 (2012).http://dx.doi.org/10.1063/1.4711217
28. M. Ejrnaes, A. Casaburi, O. Quaranta, S. Marchetti, A. Gaggero, F. Mattioli, R. Leoni, S. Pagano, and R. Cristiano, Supercond. Sci. Technol. 22, 055006 (2009). http://dx.doi.org/10.1088/0953-2048/22/5/055006
29. J. K. W. Yang, A. J. Kerman, E. A. Dauler, V. Anant, M. R. Kristine, and K. K. Berggren, IEEE Trans. on Appl. Supercond. 17, 5 (2007).

CHAPTER 17

MAGNETIC ORDERING OF NITROGEN-VACANCY CENTERS IN DIAMOND VIA RESONATOR-MEDIATED COUPLING

Bo-Bo Wei, Christian Burk, Jörg Wrachtrup and Ren-Bao Liu

INTRODUCTION

The negatively charged nitrogen-vacancy (NV) centers in high-purity diamond have been considered as a promising candidate for solid state quantum information processing due to their long coherence time [1–3] and high feasibility in initialization, control, and readout of their spin states [4]. Simulation of many-body physics using NV centers, in analogue to cold atom physics [5, 6], has been proposed [7–10]. To realize phase transitions in NV center qubit systems under realistic temperatures, however, sufficiently strong interactions between NV centers located sufficiently close (<30 nm) is till highly challenging [7, 8, 11]. A new opportunity is to use resonators as mediators [7, 12–15], which has potential of coupling NV centers at distance.

In this paper, we propose to realize long-range coupling between many separated NV centers via a mechanical resonator. A remarkable feature of the long-range interacting system is: the critical temperature for ferromagnetic phase transitions is proportional to the number of spins, so a ferromagnetic order could be formed at a temperature of tens of millikelvin (mK) for a sample with ~10^4 NV centers, even though the mediated coupling between two NV centers is less than 200 kHz (~10 μK).

MICROSCOPIC MODEL

The essential idea can be understood by considering a prototype system shown in Figure 1(a). A mechanical resonator with frequency ω_r is attached with a magnet. The NV sample is placed right under the magnet with a distance *d*. The oscillation of the mechanical resonator generates a time dependent magnetic field that causes Zeeman shift of the spins in the sample. In contrast to the schemes of direct dipolar coupling between NV centers, our scheme does not require short distance between NV centers. We use a diamond sample where the orientation axes of the NV color centers can lie along one of the four possible crystallographic axes in diamonds, [111] (Figure 1(b)) which can be grown by chemical vapor deposition method [16, 17]. The NV centers have a spin triplet ground state (S=1) with a large zero field splitting Δ=2.87 GHz. We apply a static magnetic field B_{NV} along the [111] direction of the sample so that two Zeeman states of the electron spin qubits with magnetic quantum number $|1\rangle$ and $|0\rangle$ are nearly degenerate, $\delta \equiv (\Delta - \gamma_e B_{NV}) \sim 0$ (Figure 1(c)).

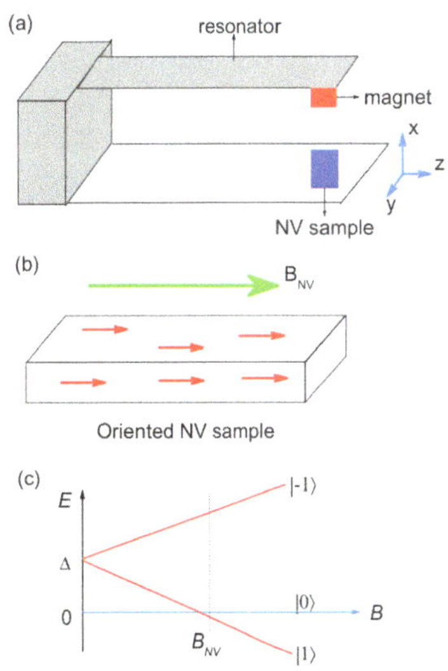

Figure 1 Schematic setup. (a) Schematic set up of a hybrid NV and mechanical resonator system. A magnet attached to the end of the me-

Citation: Bo-Bo Wei, Christian Burk, Jörg Wrachtrup and Ren-Bao Liu, Magnetic ordering of nitrogen-vacancy centers in diamond via resonator-mediated coupling, doi:10.1140/epjqt/s40507-015-0032-2.

chanical resonator is positioned at a distance 25 nm above the NV centers, coupled to the electronic spin of the defect centers. (b) The NV defects in the samples. The defects have one preferred orientations in the sample, [111]. A static magnetic field B_{NV} is applied along the [111] direction. (c) Schematic energy levels of NV centers as a function of the magnetic field. The B_{NV} is taken so that the states $|1\rangle$ and $|0\rangle$ of the NV centers are nearly degenerate.

We make use of the two Zeeman states of the electron spin $|1\rangle$ and $|0\rangle$ and define a pseudo-spin by $\sigma_z=|1\rangle\langle 1|-|0\rangle\langle 0|$ and the corresponding spin flip operators $\sigma^+=|1\rangle\langle 0|$ and $\sigma^-=|0\rangle\langle 1|$. In each NV center there is a ^{14}N nuclear spin which interacts with the on-site NV center spin. The Hamiltonian of the NV center near the degenerate point ($\delta\approx 0$) is $H_{NV} = \sum_j [\Delta_N (I_j^z)^2 - \gamma_N I_j^z B_{NV} + AI_j \cdot \sigma_j + \delta\sigma_j^z]$, where $\Delta_N=5.1$ MHz is the ^{14}N nuclear spin quadrupole splitting [18], γ_N is the gyromagnetic ratio of the nitrogen nuclear spin and $A\approx 2$ MHz is the hyperfine coupling between the electronic spin and the ^{14}N nuclear spin [18].

The motion of the mechanical resonator is described by the Hamiltonian $H_r = \omega_r b^\dagger b$, with ω_r as the frequency of the fundamental vibration mode of the resonator and b as the corresponding annihilation operator. For example a silicon nitride string resonator has dimensions ($325\times 0.35\times 0.1$) μm with $\omega_r=2\pi\times 1.0$ MHz and $Q=1.3\times 10^6$ at room temperature [19].

The magnetic field felt by the electronic spin of the NV centers can be approximated by a magnetic dipole [20], $\vec{B}(d_j - x(t)) = \vec{B}(d_j) - G_m\vec{x}(t) + o(x^2)$, where d_j is the distance between the equilibrium position of the resonator and jth NV center, G_m is the magnetic field gradient at the position of the NV centers and x is the amplitude of mechanical resonator oscillation ($x\sim 10^{-12}$ m at temperature of mK). This magnetic field will induce a Zeeman shift to the NV spins with Hamiltonian $H_z = \sum_j \eta_j \sigma_j^x (b + b^\dagger)$ and $\eta_j = \gamma_e |G_m| a_0$, where γ_e is the gyromagnetic ratio for electron and $a_0 = \sqrt{\hbar/2m\omega_r}$ is the amplitude of zero point fluctuations for a resonator of mass m.

Due to the distance distribution of the NV centers to the magnet, the coupling of the NV centers to the mechanical resonator have a distribution. A magnetic tip with size of 100 nm produces a magnetic gradient $G_m\sim 7.8\times 10^6$ T/m at a distance 25 nm away from the tip [21]. In such case $\eta/2\pi$ could reach 200 kHz. For a sample with typical distance of NV centers ~20 nm, the neighbor NV centers have a direct dipolar interaction with strength ~5 kHz (~0.25 μK), much less than the critical temperature to be discussed later. For the sake of simplicity, in the following we assume the coupling between the mechanical resonator and NV centers to be uniform and neglect the direct dipolar coupling between NV centers. Without these assumptions, however, the results in this paper would only be quantitatively affected. The coupling constant $\eta/2\pi\sim 200$ kHz considerably exceeds both the electronic spin decoherence rate (~kHz) of the NV centers and the intrinsic damping rate of the mechanical resonator, $\gamma=\omega_r/Q$ [19]. We neglect the interaction between the mechanical resonator and the ^{14}N nuclear spins since it is three orders of magnitude smaller than that between the NV center and the mechanical resonator.

Summarizing the three terms described above, the total microscopic Hamiltonian including the spin qubits in the NV centers, nitrogen nuclear spins, the mechanical resonator and the interaction among them is,

$$H = \omega_r b^\dagger b + \sum_{j=1}^{N} \left[\Delta_N (I_j^z)^2 - \gamma_N B_{NV} I_j^z + AI_j \cdot \sigma_j\right] + \sum_{j=1}^{N} \left[\eta(b + b^\dagger)\sigma_j^x + \delta\sigma_j^z\right]. \quad (1)$$

EFFECTIVE LONG-RANGE ISING MODEL

To obtain the effective interactions between NV centers, we calculate the partition function of the hybrid system $Z = \mathrm{Tr}[e^{-\beta H}]$. We shall neglect the Zeeman energy of the ^{14}N nuclear spin since its strength (~0.3 MHz) is much smaller than the zero field splitting ~5 MHz and the hyperfine interaction ~3 MHz. The hyperfine interaction between the electronic spin and the ^{14}N nuclear spin term $AI\cdot\sigma$ can be separated into the diagonal term $AI^x\sigma^x$ and the flip-flop term $A(I^y\sigma^y+I^z\sigma^z)$. Let us first neglect the flip-flop term and the qubit splitting ($\delta=0$). By partial trace of the phonon bath and the nuclear spins, the partition function can be factorised as

$$Z = \mathrm{Tr}\left[e^{-\beta H}\right] = Z_{\text{phonon}} \times Z_{^{14}N} \times \mathrm{Tr}\left[\exp(-\beta H_{\text{eff}})\right], \quad (2)$$

where Z_{phonon} is the partition function of an independent phonon bath, $Z14_N$ is the partition function of the nuclear spins. A unitary transformation of the form, $U = \exp(\sum_j \eta_j \sigma_j^x (b - b^\dagger)/\omega_r)$, can be applied to the Hamiltonian in equation (1), with the Zeeman splitting and flip-flop term neglected, to obtain an effective Hamiltonian for the NV electron spins up to the second order in η/ω_r,

$$H_{\text{eff}} = -\sum_{i<j} \frac{2\eta^2}{\omega_r}\sigma_i^x\sigma_j^x - h(\beta)\sum_j \sigma_j^x, \quad (3)$$

with $h(\beta) = \beta^{-1}\ln[(1 + 2\cosh(\beta A_x))/3]$ being an effective magnetic field resulting from the on-site interaction between the NV center and the ^{14}N nuclear spin. Note that in equation (3) all the NV center spins interact with each other, making an effectively N-dimensional model. The physics for the effective long-range interactions between the NV centers is that an NV center spin flips by virtually emitting a phonon, and then the phonon is virtually absorbed by another NV center spin far away with a spin flip. Therefore two NV spins at distance can flip-flop via virtual exchange of phonons, which leads to long-range $\sigma_i^x\sigma_j^x$ coupling. In the temperature range we are interested in (~10 mK), the effective field, $h(\beta)\sim 0.3$ μK is negligible. Therefore the distant NV centers interact through an effective long-range ferromagnetic Ising interaction with the Hamiltonian

$$H_{\text{electron}} = -\sum_{i<j} J\sigma_i^x\sigma_j^x, \quad (4)$$

where the ferromagnetic coupling strength $J = 2\eta^2/\omega_r$. If the flip-flop term of the hyperfine interaction is taken into account, it would contribute an effective transverse field to the electronic spins, which, however, is ~0.3 μK and hence negligible.

FERROMAGNETIC ORDERING

The long-range Ising model can be solved by mapping the ensemble onto to a single large spin. With the notation $S_x = \sum_{j=1}^{N}\sigma_j^x/2$, the Hamiltonian of the long-range Ising model can be written as

$$H_{\text{electron}} = -J\sum_{i<j}\sigma_i^x\sigma_j^x = -J\left(2S_x^2 - \frac{N}{2}\right). \tag{5}$$

The spin quantum number S takes values 0,1,2,…,N/2 for even N and 1/2,3/2,…,N/2 for odd N. Moreover, the spin degeneracy in each large spin subspace is $D[S] = C_N^{N/2-S} - C_N^{N/2-S-1}$ where C_N^j is the binomial coefficient. Then the electron spin partition function

$$Z = \mathrm{Tr}\left[e^{-\beta H_{\text{electron}}}\right] = e^{-N\beta J/2}\sum_{n=0}^{N} C_N^n e^{2\beta J N^2(n/N-1/2)^2}. \tag{6}$$

In the large N limit [22]

$$Z \approx e^{-\tilde{\beta}/2} N \int_{-1/2}^{1/2} dX e^{-N\varphi(\tilde{\beta},X)}, \tag{7}$$

where $\varphi = (1/2+X)\ln(1/2+X) + (1/2-X)\ln(1/2-X) - 2\tilde{\beta}X^2$ and $\tilde{\beta} = \beta NJ$. The integration in the partition function can be evaluated by the saddle point approximation method. For $\tilde{\beta} < 1$, the saddle point appears at X=0, and for $\tilde{\beta} > 1$, there are two symmetric saddle point located respectively within (−1/2,0) and (0,1/2). Therefore $\tilde{\beta} = 1$ is the critical point for the ferromagnetic phase transition in the long-range Ising model. So the critical temperature for the long-range ferromagnetic Ising model, k_BT_c=NJ, is proportional to the number of spins and the coupling strength [23]. The mean-field theory gives a good estimation of the phase transition temperature in the long-range Ising model, which becomes exact in the thermodynamic limit.

We present the scaled free energy of the long-range Ising model as a function of the spin magnetization $M_x \equiv \langle S_x \rangle / N$ for different temperatures in Figure 2. One can see that at high temperature (Figure 2(a)) the free energy minimum is situated at M_x=0 and a paramagnetic state is stable, while at low temperature (Figure 2(b)) the free energy minimum is doubly degenerated with $M_x \approx 1$ or $M_x \approx -1$, and a ferromagnetic state is preferred.

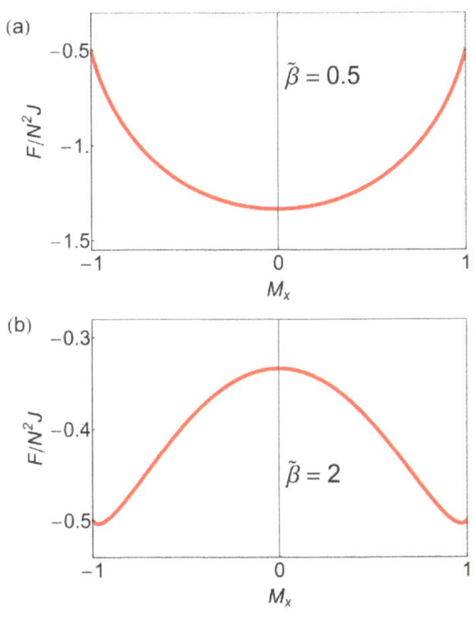

Figure 2 Free energy landscapes of the long-range Ising model. (a) The free energy as a function of the spin polarization along the x direction at high temperature $\tilde{\beta} \equiv NJ/k_BT = 0.5$. (b) The same as (a) but at a lower temperature $\tilde{\beta} \equiv NJ/k_BT = 2.0$.

For the hybrid system, the phase transition temperature can be controlled through the detuning from the degeneracy point between the NV spin states $|1\rangle$ and $|0\rangle$. If the NV qubits have a small splitting $\delta = \Delta - \gamma B_{\text{NV}}$, the resonator mediated interaction between the NV centers becomes

$$J = -\left(\frac{\eta^2}{\omega_r - \delta} + \frac{\eta^2}{\omega_r + \delta}\right). \tag{8}$$

The critical temperature becomes

$$k_BT_c = \frac{N\eta^2}{\omega_r - \delta} + \frac{N\eta^2}{\omega_r + \delta}. \tag{9}$$

Here the magnetic field is tuned so that the NV spin states $|1\rangle$ and $|0\rangle$ are nearly degenerate. Hence the splitting is much smaller than the frequency of the resonator. In Figure 3 we plot the phase diagrams of the NV centers as a function of splitting for N=10,000. In a diamond sample with NV concentration ~50 ppb, coupling between a resonator and such many NV centers is possible. A ferromagnetic order can be formed at the temperature of 50 mK. The magnetization of the large number of spins would cause a displacement of the resonator $\sim 2a_0\eta N\langle M_x\rangle/(\hbar\omega_r) \approx 2$ nm for N=10^4 NV centers, which can be detected by optical reflection as done in atomic force microscopy.

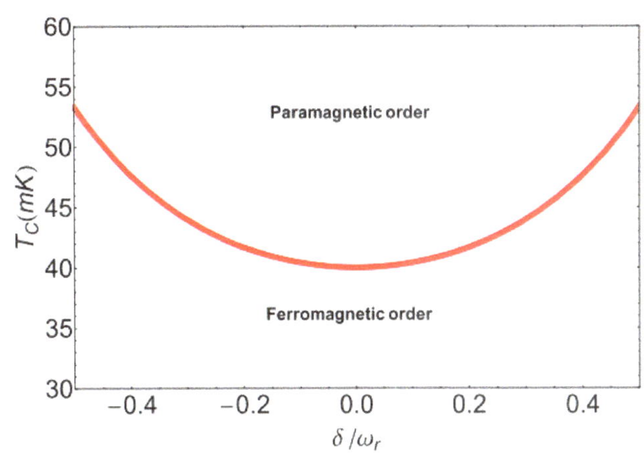

Figure 3 Phase diagram. The phase transition temperature of the NV-resonator hybrid system as a function of the scaled detuning δ/ω_r. We take the NV-mechanical resonator interaction strength η=200 kHz, eigen frequency of the mechanical resonator ω_r=1 MHz and the number of spin is N=10,000.

In Figure 4(a), we present the numerical and the mean field results of the order parameter as a function of reduced temperature, k_BT/NJ. One can see that the order parameter vanishes at and above the critical temperature k_BT/NJ=1 and also the finite size numerical results approach to that of the mean field predictions reasonably well as the number of spins increases. The specific heat shown in Figure 4(b) has a finite jump which signals a second order phase transitions occurs.

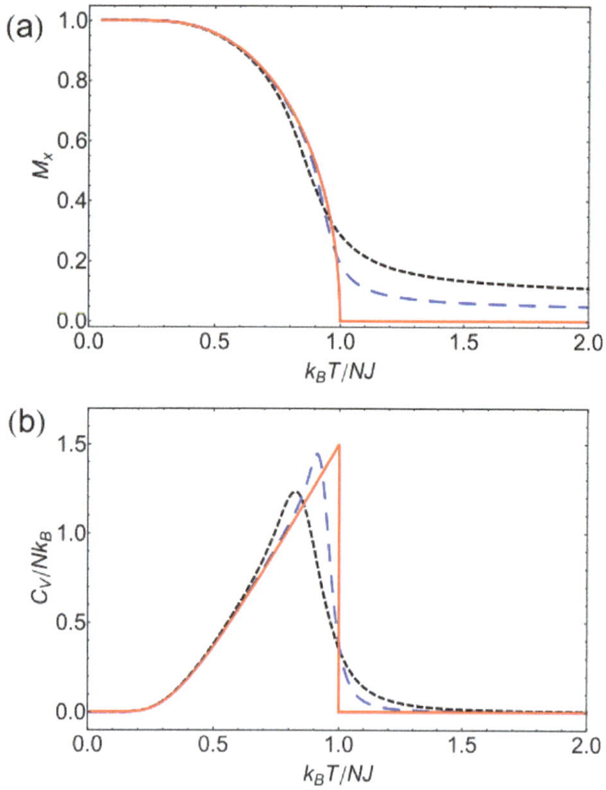

Figure 4 Comparison of the finite size numerical results and the mean field theory. (a) The order parameter, M_x, as a function of the reduced dimensionless temperature, k_BT/NJ for different sizes of the spin systems. The black-short-dashed line is for N=100, the blue-long-dashed line is for N=500 and the red-solid line is the mean field prediction. (b) The specific heat plot in the same style as in (a).

CONCLUSION

In summary, we propose a scheme to realize ferromagnetic ordering of distant nitrogen-vacancy centers by using a mechanical resonator to mediate long-range Ising-type interaction. The critical temperature for the ferromagnetic phase transition in the long-range Ising model is proportional to the number of spins, so the ferromagnetic order could be formed at the temperature of tens of millikelvin for a sample with ten thousand nitrogen-vacancy centers. In addition, it may also be possible to use a superconducting resonator [24, 25] as a medium to realize the long-range ferromagnetic coupling between the NV centers. Since the interactions between the NV centers mediated by a superconducting resonator is usually small compared to that by a mechanical resonator, high density NV samples are required to observe the magnetic ordering in the NV centers.

ACKNOWLEDGEMENTS

We are grateful to P Bertet for useful discussions. This work was supported by National Basic Research Program of China Grant 2014CB921402, Hong Kong RGC/GRF 401413, Hong Kong RGC/CRF CUHK4/CRF/12G, and CUHK VC's One-Off Discretionary Fund.

AUTHORS' CONTRIBUTIONS

RBL conceived the idea, RBL & BBW formulated the theory, RBL, BBW, CB & JW designed the physical system, BBW carried out the calculation, BBW & RBL wrote the paper, and all authors commented on the manuscript.

REFERENCES

1. Kennedy T-A, Colton J-S, Butler J-E, Linares R-C, Doering P-J. Long coherence times at 300 K for nitrogen-vacancy center spins in diamond grown by chemical vapor deposition. Appl Phys Lett. 2004;83:4190.
2. Gaebel T, Domhan M, Popa I, Wittmann C, Neumann P, Jelezko F, Rabeau J-R, Stavrias N, Greentree A-D, Prawer S, Meijer J, Twamley J, Hemmer P-R, Wrachtrup J. Room-temperature coherent coupling of single spins in diamond. Nat Phys. 2006;2:408.
3. Balasubramanian G, Neumann P, Twitchen D, Markham M, Kolesov R, Mizuochi M, Isoya J, Achard J, Beck J, Tissler J, Jacques V, Hemmer P-R, Jelezko F, Wrachtrup J. Ultralong spin coherence time in isotopically engineered diamond. Nat Mater. 2009;8:383.
4. Gruber A, Dräbenstedt A, Tietz C, Fleury L, Wrachtrup J, von Borczyskowski C. Scanning confocal optical microscopy and magnetic resonance on single defect centers. Science. 1997;276:2012.
5. Lewenstein M, Sanpera A, Ahufinger V, Damski B, Sen(De) A, Sen U. Ultracold atomic gases in optical lattices: mimicking condensed matter physics and beyond. Adv Phys. 2007;56:243.
6. Bloch I, Dalibard J, Zwerger W. Many-body physics with ultracold gases. Rev Mod Phys. 2008;80:885.
7. Rabl P, Kolkowitz S-J, Koppens F-H-L, Harris J-G-E, Zoller P, Lukin M-D. A quantum spin transducer based on nanoelectromechanical resonator arrays. Nat Phys. 2010;6:602.
8. Yao N-Y, Jiang L, Gorshkov A-V, Maurer P-C, Giedke G, Cirac J-I, Lukin M-D. Scalable architecture for a room temperature solid-state quantum information processor. Nat Commun. 2012;3:800.
9. Cai J-M, Retzker A, Jelezko F, Plenio M-B. A large-scale quantum simulator on a diamond surface at room temperature. Nat Phys. 2013;9:168.
10. Chen S-W, Jiang Z-F, Liu R-B. Quantum criticality at high temperature revealed by spin echo. New J Phys. 2013;15:043032.
11. Dolde F, Jakobi I, Naydenov B, Zhao N, Pezzagna S, Trautmann C, Meijer J, Neumann P, Jelezko F, Wrachtrup J. Room-temperature entanglement between single defect spins in diamond. Nat Phys. 2013;9:139.
12. Rabl P, Cappellaro P, Gurudev Dutt M-V, Jiang L, Maze J-R, Lukin M-D. Strong magnetic coupling between an electronic spin qubit and a mechanical resonator. Phys Rev B. 2009;79:041302.
13. Xu Z-Y, Hu Y-M, Yang W-L, Feng M, Du J-F. Deterministically entangling distant nitrogen-vacancy centers by a nanomechanical cantilever. Phys Rev A. 2009;80:022335.
14. Arcizet O, Jacques V, Siria A, Poncharal P, Vincent P, Seidelin S. A single nitrogen-vacancy defect coupled to a nanomechanical oscillator. Nat Phys. 2011;7:879.

15. Li P-B, Liu Y-C, Gao S-Y, Xiang Z-L, Rabl P, Li F-L, Xiao Y-F. arXiv:1503.02437 (2015).
16. Lesik M, Tetienne J-P, Tallaire A, Achard J, Mille V, Gicquel A, Roch J-F, Jacques V.arXiv:1401.2795 (2014).
17. Michl J, Teraji T, Zaiser S, Jakobi I, Waldherr G, Dolde F, Neumann P, Doherty M-W, Manson N-B, Isoya J, Wrachtrup J. arXiv:1401.4106 (2014).
18. He X-F, Manson N-B, Fisk P-T-H. Paramagnetic resonance of photoexcited N-V defects in diamond. II. Hyperfine interaction with the ^{14}N nucleus. Phys Rev B. 1993;47:8816.
19. Verbridge S-S, Craighead H-G, Parpia J-M. A megahertz nanomechanical resonator with room temperature quality factor over a million. Appl Phys Lett. 2008;92:013112.
20. Hunger D, Camerer S, Korppi M, Jöckel A, Hänsch T-W, Treutlein P. Coupling ultracold atoms to mechanical oscillators. C R Phys. 2011;12:871.
21. Mamin H-J, Poggio M, Degen C-L, Rugar D. Nuclear magnetic resonance imaging with 90-nm resolution. Nat Nanotechnol. 2007;2:301.
22. Wilms J, Vidal J, Verstraete F, Dusuel S. Finite-temperature mutual information in a simple phase transition. J Stat Mech. 2012;2012:P01023.
23. Huang K. Statistical mechanics. New York: Wiley; 1987. MATH
24. Zhu X-B, Saito S, Kemp A, Kakuyanagi K, Karimoto S-I, Nakano H, Munro W-J, Tokura Y, Everitt M-S, Nemoto K, Kasu M, Mizuochi N, Semba K. Coherent coupling of a superconducting flux qubit to an electron spin ensemble in diamond. Nature. 2011;478:221.
25. Kubo Y, Grezes C, Dewes A, Umeda T, Isoya J, Sumiya H, Morishita N, Abe H, Onoda S, Ohshima T, Jacques V, Dréau A, Roch J-F, Diniz I, Auffeves A, Vion D, Esteve D, Bertet P. Hybrid quantum circuit with a superconducting qubit coupled to a spin ensemble. Phys Rev Lett. 2011;107:220501.

CHAPTER 18

QUANTUM TRANSPORT IN GRAPHENE QUANTUM DOTS

Hai-Ou Li, Tao Tu, Gang Cao, Lin-Jun Wang, Guang-Can Guo and Guo-Ping Guo

INTRODUCTION

Graphene exhibits unique electrical properties and offers substantial potential as building blocks of nanodevices owing to its unique two-dimensional structure (Geim et al., 2007; Geim et al., 2009; Ihn et al., 2010). Besides being a promising candidate for high performance electronic devices, graphene may also be used in the field of quantum computation, which involves exploration of the extra degrees of freedom provided by electron spin, in addition to those due to electron charge. During the past few years, significant progress has been achieved in implementation of electron spin qubits in semiconductor quantum dots (Hanson et al., 2007; Hanson et al., 2008). To realize quantum computation, the effects of interactions between qubits and their environment must be minimized (Fischer et al., 2009). Because of the weak spin-orbit coupling and largely eliminated hyperfine interaction in graphene, it is highly desirable to coherently control the spin degree of freedom in graphene nanostructures for quantum computation (Trauzettel et al., 2007; Guo et al., 2009). However, the low energy quasiparticles in single layer grapheme behave as massless Dirac fermions (Geim et al., 2007; Geim et al., 2009), and the relativistic Klein tunneling effect leads to the fact that it is hard to confine electrons within a small region to form quantum dot in graphene using traditional electrostatical gates (Ihn et al., 2010; Trauzettel et al., 2007). It is now possible to etch a grapheme flake into nano-constrictions in size, which can obtain electron bound states and thus act as quantum dots. As a result, usually a diamond-like characteristic of suppressed conductance consisting of a number of sub-diamonds is clearly seen (Stampfer et al., 2009; Gallagher et al., 2010), indicating that charge transport in the single graphene quantum dot device may be described by the model of multiple graphene quantum dots in series along the nanoribbon. The formation of multiple quantum dot structures in the nanoribbons may be attributed to edge roughness or local potential. The rough edges also lift the valley degeneracy, which could suppress the exchange coupling between spins in the grapheme quantum dots (Trauzettel et al., 2007; Ponomarenko et al., 2008). Recently, there was a striking advance on experimental production of graphene single (Ponomarenko et al., 2008; Stampfer et al., 2008a;Stampfer et al., 2008 b; Schenz et al., 2009; Wang et al., 2010; Guttinger et al., 2011) or double quantum dots (Molitor et al., 2009; Molitor et al., 2010; Liu et al., 2010; Wang et al., 2011a; Volk et al., 2011; Wang et al., 2012;) which is an important first step towards such promise.

In this chapter, we introduce the design and fabrication of etched gate tunable single and double quantum dots in single-layer and bilayer graphene and present several important quantum transport measurements in these systems. A quantum dot with an integrated charge sensor is becoming a common architecture for a spin or charge based solid state qubit. To implement such a structure in graphene, we have fabricated a twin-dot structure in which the larger QD serves as a single electron transistor (SET) to read out the charge state of the nearby gate controlled small QD. A high SET sensitivity allowed us to probe Coulomb charging as well as excited state spectra of the QD, even in the regime where the current through the QD is too small to be measured by conventional transport means (Wang et al., 2010;Wang et al., 2011b). We also have measured quantum transport properties of gates controlled parallel-coupled double quantum dots (PDQD) and series-coupled double quantum dots (SDQD) device on both single layer and bilayer graphene (Wang et al., 2011a; Wang et al., 2012). The inter-dot coupling strength can be effectively

Citation: Hai-Ou Li, Tao Tu, Gang Cao, Lin-Jun Wang, Guang-Can Guo and Guo-Ping Guo (2013). Quantum Transport in Graphene Quantum Dots, New Progress on Graphene Research, Prof. Jian Ru Gong (Ed.), InTech, DOI: 10.5772/52873.

tuned from weak to strong by in-plane plunger gates. All the relevant energy scales and parameters can be extracted from the honeycomb charge stability diagrams. We precisely extract a large inter-dot tunnel coupling strength for the series-coupled quantum dots (SDQD) allowing for the observation of tunnel-coupled molecular states extending over the whole double dot. The present method of designing and fabricating graphene QD is demonstrated to be general and reliable and will enhance the realization of graphene nanodevice and desirable study of rich QD physical phenomena in grapheme. These results demonstrate that both single and double quantum dots in single-layer and bilayer graphene bode well for future quantum transport study and quantum computing applications. The clean, highly controllable systems serves as an essential building block for quantum devices in a nuclear-spin-free world.

A GRAPHENE QUANTUM DOT WITH A SINGLE ELECTRON TRANSISTOR AS AN INTEGRATED CHARGE SENSOR

The measurement of individual electrons or its spins in GaAs quantum dots (QDs) has been realized by so-called charge detection via a nearby quantum point contact (QPC) or single electron transistor (SET) (Lu et al., 2003; Elzerman et al., 2004a). In particular, the combination of high speed and high charge sensitivity has made SET useful in studying a wide range of physical phenomena such as discrete electron transport (Lu et al., 2003; Bylander et al., 2005; Gotz et al., 2008), qubit read out (Lehnert et al., 2003; Duty et al., 2004; Vijay et al., 2009) and nanomechanical oscillators (Knobel et al., 2003;Lahaye et al., 2004). So far, most SETs have been using Al/AlOx/AlAl/AlOx/Altunnel junctions. However, the graphene SET reported here is technologically simple and reliable, making it an attractive substitute for use in various charge detector applications.

In this section, we realize an all graphene nanocircuit integration with a SET as charge read out for a QD. In conventional semiconductor systems, the gate-defined structure limits the distance between the QD and the detector. However, in our device reported here, the QD and the SET in the same material are defined in a single etching step, and the distance between the graphene nanostructures is determined by the etched area, which enables optimized coupling and sensing ability. The SET is placed in close proximity to the QD giving rise to a strong capacitive coupling between the two systems. Once an additional electron occupies the QD, the potential in the neighboring SET is modified by capacitive interaction that gives rise to a measurable conductance change. Even if charge transport through the QD is too small to be measured by conventional transport means, the SET charge sensor also allows measurements. These devices demonstrated here provide robust building blocks in a practical quantum information processor.

The graphene flakes were produced by mechanical cleaving of graphite crystallites by Scotch tape and then were transferred to a highly doped Si substrate with a 100 nm SiO_2 top layer. Thin flakes were found by optical microscopy, and single layer graphene flakes were selected by the Raman spectroscopy measurement. We used the standard electron beam lithography and lift off technique to make the Ohmic contact (Ti/Au) on the present graphene devices. Next, a new layer of polymethyl methacrylate is exposed by electron beam to form a designed pattern. Then, the unprotected areas are removed by oxygen reactive ion etching. One of our defined sample structures with a quantum dot and proximity SET is shown in Fig. 1. The quantum dot is an isolated central island of diameter 90 nm, connected by 30nm wide tunneling barriers to source and drain contacts. Here, the Si wafer was used as the back gate and there is also a graphene side gate near the small dot. The SET has a similar pattern while the conducting island has a much larger diameter (180nm). Electronic transport through both the devices exhibits Coulomb blockade (CB) characteristics with back/side gate voltage. The distance between the CB peaks is determined by the sum of charging and quantum confinement energies, and the former contribution becomes dominant for our devices with diameter >100nm (Kouwenhoven et al., 1997). Accordingly, we refer to it as a SET rather than a QD. The device was first immersed into a liquid helium storage Dewar at 4.2K to test the functionality of the gates. The experiment was carried out in a top-loading dilution refrigerator equipped with filtered wiring and low-noise electronics at the base temperature of 10 mK. In the measurement, we employed the standard ac lock-in technique.

Fig. 2(a) shows the conductance through the dot Al/AlO_x/Al for applied side gate voltage V_{sg}. Clear CB peaks are observed related to charging of the tunable dot on the graphene. The dashed green lines in the range of 0.2-0.7V for side gate voltages show that the current through the dot becomes too small to be seen clearly. Fig. 2(b) shows the conductance through the SET versus side gate voltage V_{sg}. The SET is as close as possible to the QD and in this way charging signals of the dot were detected by tracking the change in the SET current. The addition of one electron to the QD leads to a pronounced change of the conductance of the charge detector by typically 30%. The slope of the SET conductance is the steepest at both sides of its CB resonances giving the best charge read-out signal. To offset the large current background, we used a lock-in detection method developed earlier for GaAs dot (Elzerman et al., 2004b). A square shaped pulse was superimposed on the dc bias on side gate voltage V_{sg}. A lock-in detector in sync with the pulse frequency measured the change of SET current due to the pulse modulation. Fig. 2(c) shows a typical trace of the lock-in signal of the transconductance through the SET dI_{SET}/dV_{sg}. These sharp spikes or dips originate from the change of the charge on the dot by one electron. It shows essentially the same features as Fig. 2(a), but is much richer, especially in the regime where the direct dot current is too small to be seen clearly. The vertical dashed lines in Fig. 2 illustrate the SET sensor signals correspond to the QD transport measurements perfectly and indicate the SET is reliable. We also notethat the individual charge events measurement has been demonstrated in a graphene QD with a QPC detector based on graphene nanoribbon (Güttinger et al., 2008).

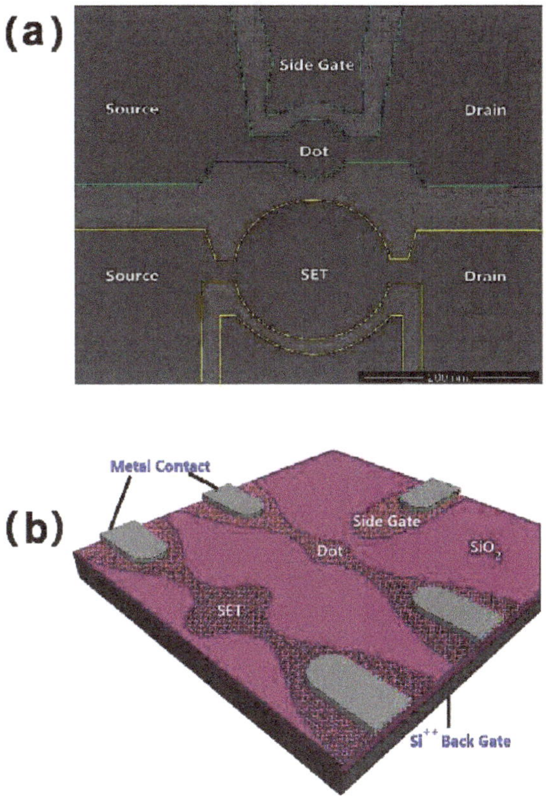

Figure 1. a) Scanning electron microscope image of the etched sample structure. The bar has a length of 200 nm. The upper small quantum dot as the main device has a diameter of 90 nm while the bottom single electron transistor as charge sensor has a diameter of 180 nm. The bright lines define barriers and the graphene side gate. (b) Schematic of a representative device.

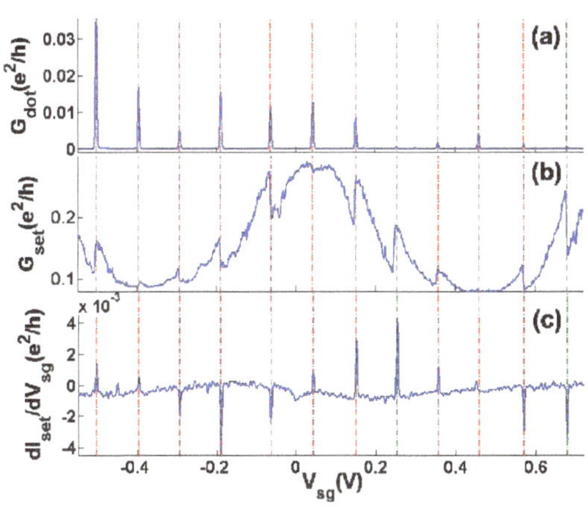

Figure 2. a) Conductance through the quantum dot vs the side gate voltage. (b) The example of conductance through the single electron transistor for the same parameter ranges as in panel (a). The steps in conductance have about 30% change of the total signal and are well aligned with the CB in panel (a). (c) Transconductance of the single electron transistor for the same parameters as in panel (a). The spikes and dips indicate the transitions in the charge states by addition of single electron in quantum dot. In particular, the dashed green lines show that the charge detection can allow measurement in the regime where the current through the dot is too small to be seen clearly by direct means. The vertical dashed red lines are a guide for the eyes to relate features in these graphs.

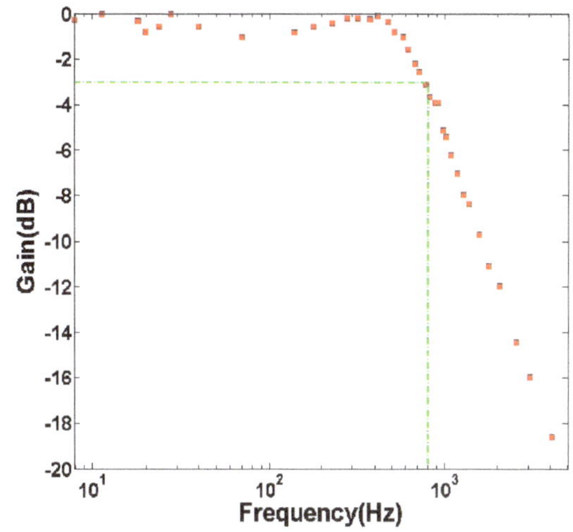

Figure 3. The magnitude of the SET signal *dISET /dVsg* as a function of the modulating pulse frequency. The dashed green line illustrates that the bandwidth of the SET device is about 800 Hz corresponding to a gain of 0.707(-3 dB). Due to the stray capacitances, the response decreases rapidly after 800 Hz.

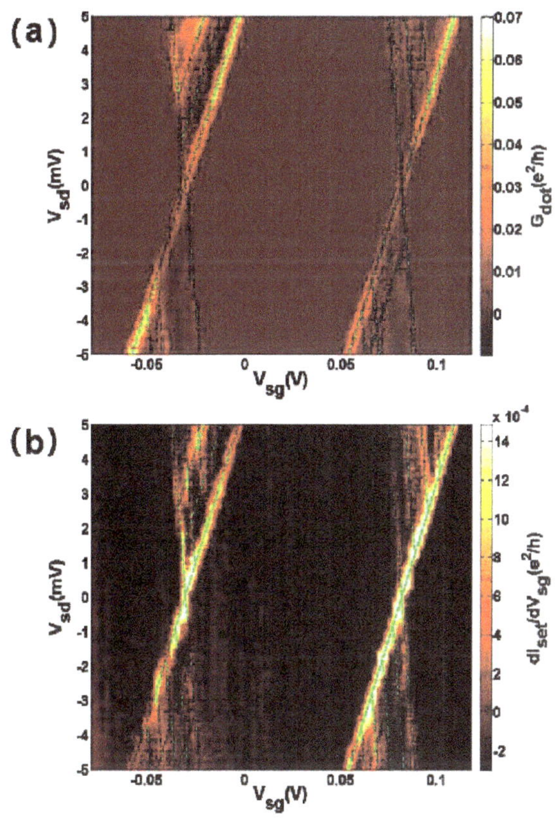

Figure 4. a) Plot of the differential conductance of the quantum dot as a function of the bias voltage and the side gate voltage applied on the dot. From the lines parallel to the edges of Coulomb diamonds, we can identify the excited states. (b) Transconductance of the single electron transistor with the same parameters as in panel (a). Perfect matching with panel (a) and resolving more excited states spectra indicate that the single electron transistor can be used as a highly sensitive charge detector. Data in panels (a) and (b) were recorded simultaneously during a single sweep. Dashed green lines are the guide for identifying the excited states.

More quantitative information on the system can be obtained from the measurement of the height response of the peak at 0.152 V in Fig. 2(c) as a function of the modulating pulse frequency on the side gate. The resulting diagram for the SET dI_{SET}/dV_{sg} gain magnitude is shown in Figure 3.The dashed green line indicates the gain of 0.707 (-3 dB), corresponding approximately to the bandwidth of 800 Hz of the SET device. By applying a signal of 5×10-25×10-2electrons on back gate of the SET and measuring the signal with a signal-to-noise ratio of 1, we achieved a charge sensitivity 10-3e/Hz---√10-3e/Hzof which is similar to that obtained previously in a GaAs QD and superconducting Al SET detector system (Berman et al., 1999). The system can be simply considered as a resistor-capacitor circuit (RC circuit), and the bandwidth is limited by the resistor and capacitance of the cable connecting the SET and the room temperature equipments. As a result, we would expect the bandwidth can be greatly improved by adding a cold amplifier (Vink et al., 2007). It is also expected that adding a side gate near the SET to independently set the SET operating point to about 25 kΩ can obviously enhance the bandwidth.

The information contained in the signal goes beyond simple charge counting. For instance, the stability diagram measurement can reveal excited states, which is crucial to get information of the spin state of electrons on a quantum dot (Hanson et al., 2003). Fig. 4(a) shows Coulomb diamonds for the conductance through the dot GQD versus bias voltage V_{sd} and side gate voltage V_{sg}. For comparison,Fig. 4(b) shows the transconductance of the SET dI_{SET}/dV_{sg} as a function of the same parameters. A perfect match between the QD transport measurements and the detector signal is observed. Moreover, the discrete energy spectra of the graphene quantum dot are revealed by the presence of additional lines parallel to the diamond edges. These lines indicate the quantum dot is in the high bias regime where the source-drain bias is high enough that the excited states can participate in electron tunneling (Hanson et al., 2007). The excited states become much more visible in the SET charge detector signal than the direct measurement. All of these features have been seen in GaAs QD with QPC (Hanson et al., 2007), but here we achieve the goal with an all graphene nanocircuit of QD with SET. In the previous reports, the QD and QPC detector are separated by typically 100 nm in width. In the present case, the SET detector is 50 nm from the edge of the QD. Therefore it is expected that the capacitance coupling between QD and SET is enhanced compared to the conventional case realized in semiconductor QD and QPC. This enhanced coupling leads to a larger signal-to-noise ratio of the SET detector signal that can be exploited for time resolved charge measurement or charge/spin qubit read out on the QD.

In summary, we have presented a simple fabrication process that produces a quantum dot and highly sensitive single electron transistor charge detector with the same material, graphene. Typically the addition of a single electron in QD would result in a change in the SET conductance of about 30%. The charging events measured by both the charge detector and direct transport through the dot perfectly match and more excited states information beyond the conventional transport means is also obtained. The devices demonstrated here represent a fascinating avenue towards realizing a more complex and highly controllable electronic nanostructure formed from molecular conductors such as graphene.

CONTROLLABLE TUNNEL COUPLING AND MOLECULAR STATES IN A GRAPHENE DOUBLE QUANTUM DOT

Previously, the charge stability diagram in coupled quantum-dot systems has been studied by the classical capacitance model (van der Wiel et al., 2003). However the quantum effect should also manifest itself (Yang et al., 2011). In particular, the tunnel coupling *t* between the two dots in a double dot is an important quantity, because it can affect the geometry of the overall charge stability diagram. Furthermore, several different spin qubit operations can be performed by controlling this tunnel rate as a function of time. For approaches based on single electron spin qubit, utilizing t enables the $\sqrt{\text{SWAP}}$ gate operations between two qubits (Petta et al., 2005). In an architecture in which each qubit is composed of two-electron single-triplet states, control of t in the presence of a non-uniform magnetic field enables universal single qubit rotations (Foletti et al., 2009).

In this Section, we report an experimental demonstration and electrical transport measurement in a tunable graphene double quantum dot device. Depending on the strength of the inter-dot coupling, the device can form atomic like states on the individual dots (weak tunnel coupling) or molecular like states of the two dots (strong tunnel coupling). We also extract the inter-dot tunnel coupling t by identifying and characterizing the molecule states with wave functions extending over the whole graphene double dot. The result implies that this artificial grapheme device may be useful for implementing two-electron spin manipulation.

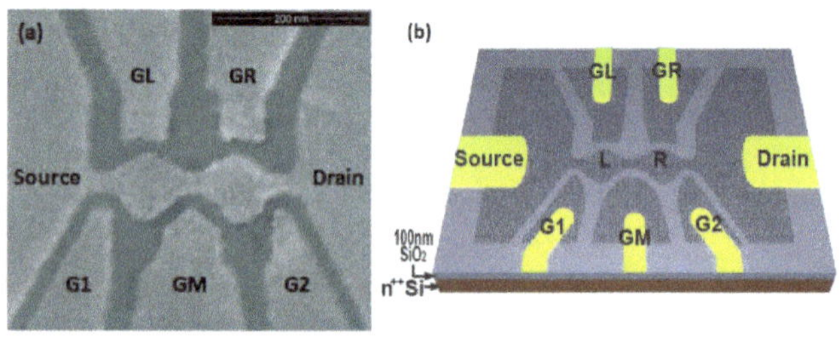

Figure 5. a) Scanning electron microscope image of the structure of the designed multiple gated sample studied in this work. The double quantum dot have two isolated central islands of diameter 100 nm in series, connected by 20×20 nm tunneling barriers to source and drain contacts (S and D) and 30×20 nm tunneling barrier with each other. These gates are labeled by G1, GL, GM, GR, G2, in which gate GM, G1 and G2, are used to control the coupling barriers between the dots as well as the leads. Gates GL and GR are used to control and adjust the energy level of each dot. (b) Schematic of a representative device.

A scanning electron microscope image of our defined sample structure with double quantum dot is shown in Fig. 5(a) and Fig. 5(b). The double quantum dot has two isolated central island of diameter 100 nm in series, connected by 20××20 nm narrow constriction to source and drain contacts (S and D electrodes) and 30××20 nm narrow constriction with each other. These constrictions are expected to act as tunnel barriers due to the quantum size effect. In addition, the highly P-doped Si substrate is used as a back gate and five lateral side gates, labeled the left gate G1, right gate G2, center gate GM and GL(R), which are expected for local control. All of side gates are effective, in which gates GL, GR and G2 have very good effect on two dots and middle barrier, while gates G1 and GM have weak effect on those. The device was first immersed into a liquid helium storage dewar at 4.2K to test the functionality of the gates. The experiment was carried out in a He3 cryostat equipped with filtered wiring and low-noise electronics at the base temperature of 300mK. In the measurement, we employed the standard AC lock-in technique with an excitation voltage 20μVμVat 11.3 Hz.

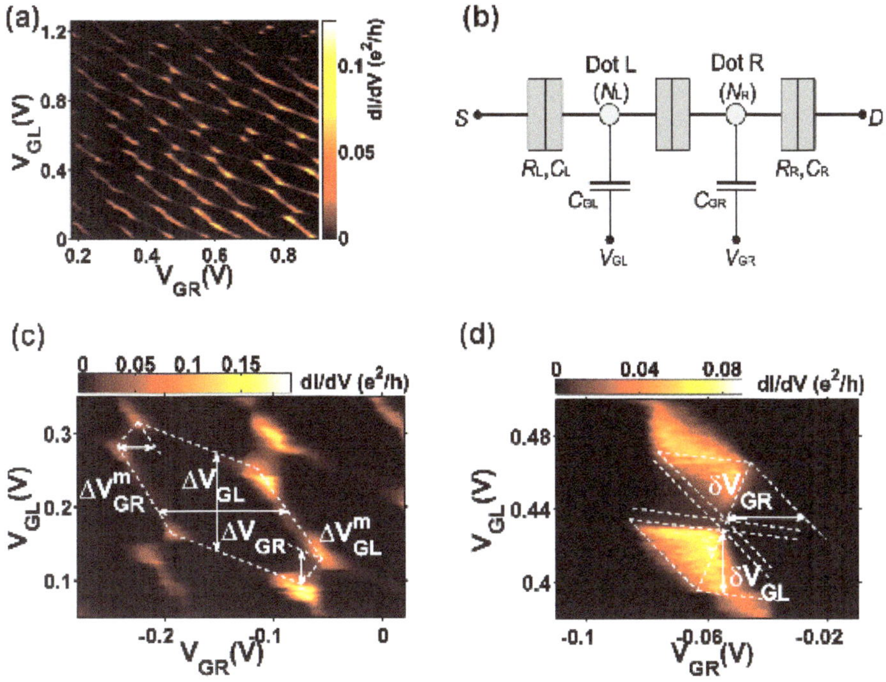

Figure 6. a) Color scale plot of the differential conductance versus voltage applied on gate GL (V_{GL}) and gate GR (V_{GR}) at V_{sd}= 20μVμV, V_{G1} = 0V, V_{GM} = 0V, V_{G2} = 0V and V_{bg} = 2.5 V. The honeycomb pattern we got stands for the typical charge stability diagram of coupled double quantum dots. (b) Pure capacitance model of a graphene double dot system. Zoom in of a honeycomb structure (c) and a vertex pair (d) at V_{sd} = 900μVμV.

Fig. 6(a) displays the differential conductance through the graphene double quantum dot circuit as a function of gate voltages V_{GL} and V_{GR}. Here the measurement was recorded at V_{sd}= 20μVμV, V_{G1} = 0V, V_{GM} = 0V, V_{G2} = 0V and V_{bg} = 2.5 V. The honeycomb pattern is clearly visible and uniforms over many times. Each cell of the honeycomb corresponds to a well-defined charge configuration (NL,NR) in the nearly independent dots, where NL and NR denote the number of electrons on the left and right dot, respectively. The conductance is large at the vertices, where the electrochemical potentials in both dots are aligned with each other and the Fermi energy in the leads and resonant sequential tunneling is available. These vertices are connected by faint lines of much smaller conductance along the edges of the honeycomb cells. At these lines, the energy level in one dot is aligned with the electrochemical potential in the corresponding lead and inelastic cotunneling processes occur. The observed honeycomb pattern resembles the charge stability diagram found for weakly coupled GaAs double quantum dot (van der Wiel et al., 2003). Such similarities indicate that graphene quantum dot devices will continue to share features with well-studied semiconductor quantum dot systems. The energy-level statistics of single graphene quantum dot was probed and shown to agree well with the theory of chaotic Dirac billiards (Ponomarenko et al., 2008). It is interesting and important to know whether these Dirac fermions' behaviors can be realized and observed in grapheme double quantum dot. Nevertheless, it will be studied in the future work.

More quantitative information such as double dot capacitances can be extracted using a electrostatic model as shown in Fig. 6(b) (van der Wiel et al., 2003). First, the capacitance of the dot to the side gate can be determined from measuring the size of the honeycomb in Fig. 6(c) as $C_{GL}=e/\Delta V_{GL}\approx 1.27$aF and $C_{GR}=e/\Delta V_{GR}\approx 1.49$aF. Next, the capacitance ratios can be determined from measuring the size of the vertices in Fig. 6(d) at finite bias V_{sd}=900μV as $\alpha_L=|V_{sd}|/\delta V_{GL}=0.029$ and $\alpha_R=|V_{sd}|/\delta V_{GR}=0.035$. Using the relation $C_{GL}/C_L=\alpha_L$ and $C_{GR}/C_R=\alpha_R$, we can obtain the typical values of dot capacitances as C_L=44.8aF and C_R=44.1aF, respectively. The amount of interdot coupling can be achieved by measuring the vertices splitting in Fig. 6(c). Assuming the capacitively coupling is dominant in the weakly coupled dots regime (van der Wiel et al., 2003; Mason et al., 2004), the mutual capacitance between dots is calculated as

$$C_M = \Delta V^m_{GL}\ C_{GL}\ C_R / e = \Delta V^m_{GR} C_{GR} C_L\ / e = 9.2\text{aF}$$

It has been expected that opening the interdot constriction by gate voltage will cause the tunnel coupling to increase exponentially faster than the capacitive coupling (Kouwenhoven et al., 1997). Fig. 7(a)-(c) represent a selection of such measurements by holding the same V_{GR} and V_{bg} and scanning different ranges of V_{GL} between -0.5 V to 0.35 V. An evolution of conductance pattern indicates that the stability diagram changes from weak to strong tunneling regimes (van der Wiel et al., 2003; Mason et al., 2004). The conductance near the vertices depends on the relative contributions of the capacitive coupling and tunnel coupling. For the former, the vertices become a sharpened point, while for the latter, the vertices become blurred along the edges of the honeycomb cell

(Graber et al., 2007). In Fig. 7(b), the vertices is not obvious as those in Fig. 7(a), which indicates a stronger tunnel coupling. The results suggest that two graphene dots are interacting with each other through the large quantum mechanical tunnel coupling, which is analogous to covalent bonding. We will analyze it in details below. An increase in inter-dot coupling also leads to much larger separation of vertices in Fig. 7(b) (Mason et al., 2004), and finally, to a smearing of honeycomb features in Fig. 7(c). In this case, the double dots behave like a single dot, as illustrated in Fig. 7(g). We note that a similar evolution is observed for four different values of V_{bg} from 2.5 V to 2.0 V at the same V_{GL} and V_{GR} regimes as shown in Fig. 7(d)-(f). Thus the inter-dot tunnel coupling could also be changed by V_{GL} or V_{bg}. This can be explained by the fact that the side gates and back gate may influence the central barrier through the existing capacitances between the gates and the central barrier.

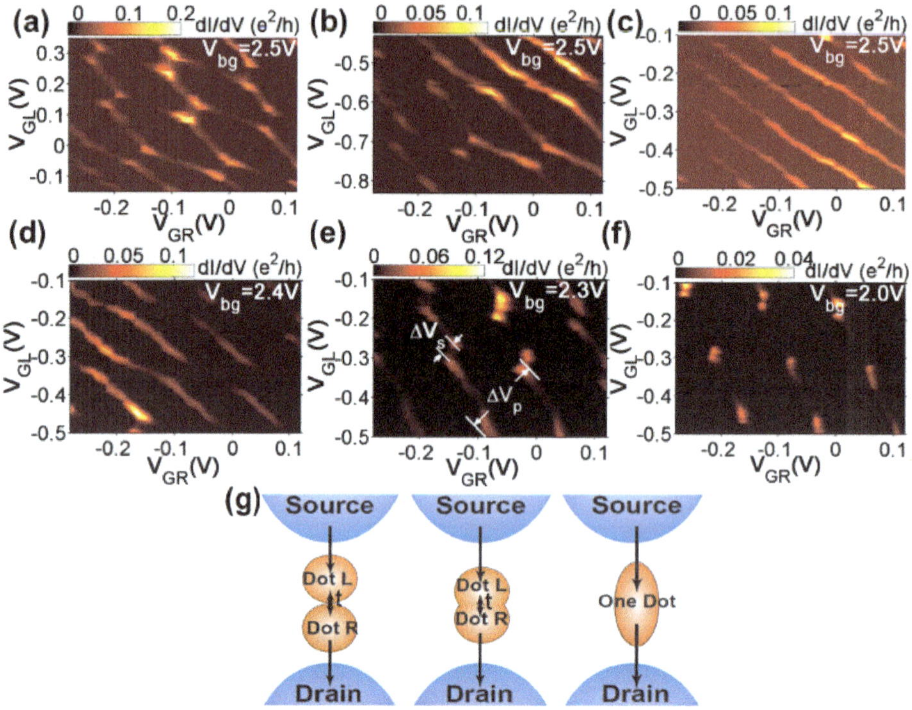

Figure 7. a)-(c) Colorscale plot of the differential conductance versus voltage applied on gate GL (V_{GL}) and gate GR (V_{GR}) at V_{bg} = 2.5 V for different V_{GL} regimes. (d)-(f) Color scale plot of the differential conductance versus voltage applied on gate L (V_{GL}) and gate R(V_{GR}) for different back gate voltage V_{bg}. The trend of interdot tunnel coupling changing from weak to strong can be seen clearly. (g) Sketches of the characteristic electronic configurations with interdot tunnel coupling *t*.

Similar to the definitions in Ref. (Livermore et al., 1996), we define $f = 2\Delta V_S / \Delta V_P$ with ΔV_S representing the splitting between vertices in the diagonal direction and ΔV_P the vertex pairs distance (Fig. 7(e)). Thus, the case f=1 stands for strong coupling limit where the double dots behave like a single dot, while the case f=0 represents weak coupling limit where the double dots behave like two isolated dots. This way, f should have a certain relationship with tunnel couplings which offers us a method to measure the contribution of the interdot tunneling to the splitting of the vertex. In our double-dot sample, a clear evolution of ffis obtained through scanning different regimes of V_{GL} with fixed V_{GR} (Fig. 7(a)-(c)). Through extracting ΔV_S and ΔV_P, we get f=0.5 for (a) and f=0.65 for (b) and f=1 for (c) respectively. These values indicate that control of tunnel coupling as a function of such a gate voltage is conceivable.

Having understood the qualitative behavior of the graphene device in the strong coupling regime, we extract the quantitative properties based on a quantum model of graphene artificial molecule states (Yang et al., 2003; Graber et al., 2007; Hatano et al., 2005). Here we only take into account the topmost occupied state in each dot and treat the other electrons as an inert core (van der Wiel et al., 2003;Golovach et al., 2004). In the case of neglected tunnel coupling, the nonzero conductance can only occur right at the vertices which are energy degenerate points as $E(N_L + 1, N_R) = E(N_L, N_R + 1)$. When an electron can tunnel coherently between the two dots, the eigenstates of the double dot system become the superposed states of two well-separated dot states with the form

$$|\psi_B\rangle = -\sin\frac{\theta}{2}e^{-\frac{i\varphi}{2}}|N_L+1, N_R\rangle + \cos\frac{\theta}{2}e^{\frac{i\varphi}{2}}|N_L, N_R+1\rangle \tag{1}$$

$$|\psi_A\rangle = \cos\frac{\theta}{2}e^{-\frac{i\varphi}{2}}|N_L+1, N_R\rangle + \sin\frac{\theta}{2}e^{\frac{i\varphi}{2}}|N_L, N_R+1\rangle \tag{2}$$

Where $\theta = \arctan(\frac{2t}{\varepsilon})$, $\varepsilon = E_L - E_R$, E_L and E_R are the energies of state $|N_L+1, N_R\rangle$ and $|N_L, N_R+1\rangle$, respectively. Thus $|\psi_B\rangle$ and $|\psi_A\rangle$ are the bonding and anti-bonding state in terms of the uncoupled dot, and the energy difference between these two states can be expressed by

$$E_\Delta = U' + \sqrt{\varepsilon^2 + (2t^2)} \tag{3}$$

Here $U' = \frac{2e^2 C_m}{C_L C_R - C_m^2}$ is the contribution from electrostatic coupling between dots (Ziegler et al., 2000).

Provided that the graphene double-dot molecule eigenstate $|\psi\rangle$ participates in the transport process, sequential tunneling is also possible along the honeycomb edges. In Fig. 8(a) and Fig. 8(b), a colorscale plot of the differential conductance is shown at $V_{sd} = 20$ μVin the vicinity of a vertex. As expected the visible conductance is observed at both the position of the vertex and the honeycomb edges extending from the vertex. Fig. 8(c) shows a fit of the energy difference E_Δ from the measured mount of splitting of the positions of the differential conductance resonance peak in the ε-direction. Here we use $\varepsilon = E_L - E_R = e\alpha_L V_{GL} - e\alpha_R V_{GR}$ to translate the gate voltage detuning $V_{GL}-V_{GR}$ with the conversion factors α_L and α_R determined above. The fitting with Eq. (3) yields the values of tunnel coupling strength t≈727μeV and U'≈209μeV. Similar measurements have been performed in a carbon nanotube double dots with t≈358μeV and U'≈16μeV (Graber et al., 2007) and semiconductor double dots with t≈80μeV and U'≈175μeV (Hatano et al., 2005). The fact that the tunnel coupling t is dominant than capacitive coupling U' implies the inter-dot tunnel barrier in the etched grapheme double dot is much more transparent than those gated carbon nanotube or semiconductor double dot.

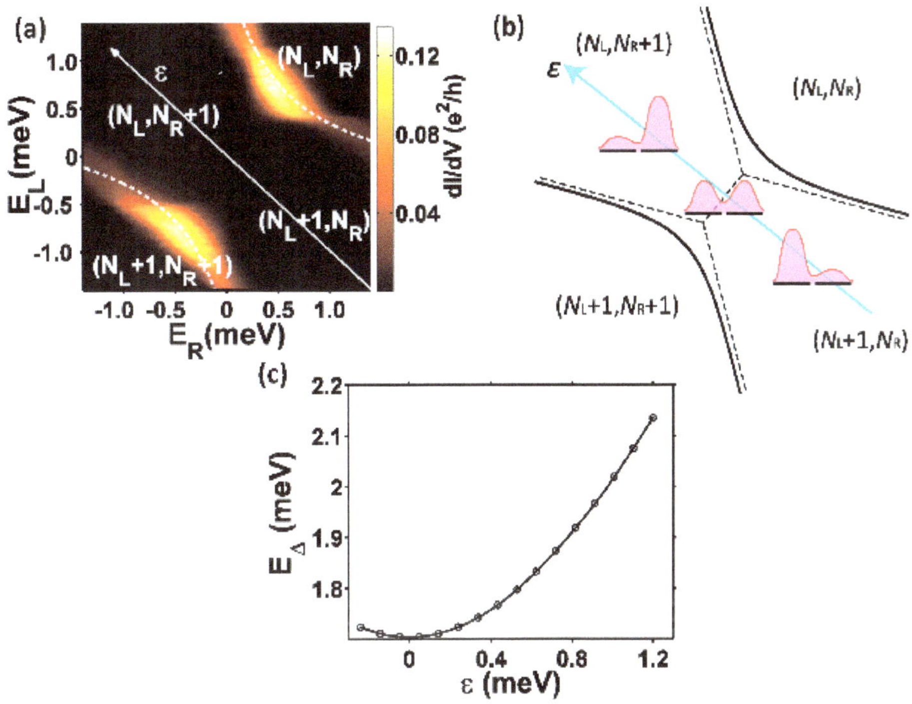

Figure 8. a) Colorscale plot of the differential conductance versus the energies of each dotE_L and E_R at $V_{sd} = 20$ μV near the selected two vertices with dashed lines as guides to the eye. (b) Schematic of a single anticrossing and the evolution from the state localized in each dot to a molecule state extending across both dots (Hatano et al., 2005). (c) E_Δ dependence of the detuning $\varepsilon = E_L - E_R$ E_Δ(circles) is measured from the separation of the two high conductance wings in Fig. 4(a). The line illustrates a fit of the data to Eq. (3).

Finally, we discuss the relevance of graphene double dot device for implementing a quantum gate and quantum entanglement of coupled electron spins. A $\sqrt{SWAP}$ operation has already been demonstrated in a semiconductor double dot system using the fast control of exchange coupling J(Petta et al., 2005). The operation time τ is about 180 ps for J≈0.4meV corresponding to t≈0.16meV. In the present graphene device, we have obtained much larger t≈0.72meV and the estimated τ≈50ps is much shorter than the predicted decoherence time (μsμs) (Fischer et al., 2009). The results indicate the ability to carry out two-electron spin operations in nanosecond timescales on a graphene device, four times faster than perviously shown for semiconductor double dot.

In summary, we have measured a graphene double quantum dot with multiple lectrostatic gates and observed the transport pattern evolution in different gate configurations. This way offers us a method to identify the molecular states as a quantum-mechanical superposition of double dot and measure the contribution of the interdot tunneling to the splitting of the differential conductance vertex. The precisely extracted values of inter-dot tunnel coupling t for this system is much larger than those in previously reported semiconductor device. These short operation times due to large tunneling strength together with the predicted very long coherence times suggest that the requirements for implementing quantum information processing in graphene nanodevice are within reach.

GATES CONTROLLED PARALLEL-COUPLED DOUBLE QUANTUM DOT ON BOTH SINGLE LAYER AND BILAYER GRAPHEME

In contrast to DQD in series, where the applied current passes through the double dot in serial, the parallel-coupled double quantum dot (PDQD) requires two sets of entrances and exits, one for each dot. In addition, the source and drain must maintain coherence of the electron waves through both dots, in a manner analogous to a Young's double slit. Thus PDQD is an ideal artificial system for investigating the interaction and interference. Rich physical phenomena, such as Aharonov-Bohm (AB) effect, Kondo regimes and Fano effect, have been predicted to be observed in parallel DQD (Holleitner et al., 2001;Lo´pez et al., 2002; Ladro´n de Guevara et al., 2003; Orellana et al., 2004; Chen et al., 2004). Particularly excitement is the prospect of accessing theoretically predicted quantum critical points in quantum phase transitions (Dias da Silva et al., 2008). The grapheme PDQD is an attractive system for investigating the quantum phase transitions due to its intrinsically large energy separation between on-dot quantum levels, thus offering a significant advantage over conventional systems as GaAs or silicon based quantum dots.

In this section, we present the design, fabrication, and quantum transport measurement of double dot structure coupled in parallel, on both bilayer and single layer grapheme flakes, which may open a door to study the rich PDQD physical phenomena in this material the parallel graphene structure can be tuned from a strong-coupling resulted artificial molecule state to a weak-coupling resulted two-dot state by adjusting in plane plunger gates. The tuning is found to be very reliable and reproducible, with good long-term stability on the order of days.

Graphene flakes are produced by mechanical cleaving of bulk graphite crystallites by Scotch tape (Novoselov et al., 2004). For this kind of exfoliated graphene flakes on SiO_2 substrate, the mobility is normally about 15000 $cm^2/(Vs)$ (Geim et al., 2007). By using heavily doped Si substrate with 100 nm thick SiO_2 on top, we can identify monolayer, bilayer, and few layer graphenes through optical microscope. Monolayer and bilayer graphenes were further checked by Raman spectrum. Firstly, graphene flakes are transferred to the substrate with gold markers. Then, a layer of 50nm thick polymethyl methacrylate (PMMA) is spun on the substrate for electron beam lithography (EBL) to form a designed pattern. After that, O_2/Ar (50:50) plasma is used to remove unprotected parts of graphene. Next, an area of over exposed PMMA is used to separate a bridge plunger gate from the drain part of graphene (Chen et al., 2004; Huard et al., 2007). The final step is to make the metal contacts, which are defined by the standardized EBL process, followed by the E-beam evaporation of Ti/Au (2 nm/50 nm).

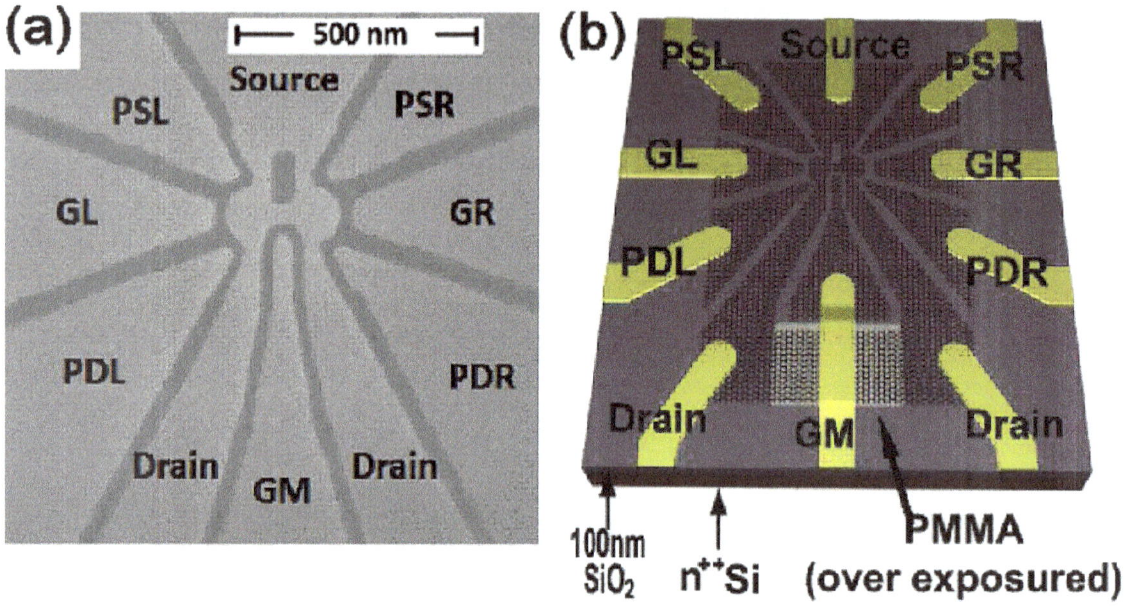

Figure 9. Color online) (a) Scanning electron microscope image of the etched parallel coupled graphene double dot sample structure. The bar has a length of 500 nm. The diameters of the two dots are both 100 nm, constriction between the two dots is 35 nm in width and length. The four narrow parts connecting the dot to source and drain parts have a width of 30 nm. Seven in-plane plunger gates labeled as GL, GR, GM, PSL, PDL, PSR, and PDR are integrated around the dot for fine tuning. (b) Schematic picture of the device. N-type heavily doped silicon substrate is used as a global back gate. A layer of overexposed PMMA is used as a bridge to make gate GM separated from the drain part of graphene.

Fig. 9(a) shows a scanning electron microscope (SEM) image of one sample with the same structure as the bilayer device we measured. Two central islands with diameter of 100 nm connect through 30 nm wide narrow constrictions to the source and the drain regions. Another narrow constriction (35nm in both width and length) connects the two central islands. Seven in-plane plunger gates labeled as GL, GR, GM, PSL, PDL, PSR, and PDR are integrated in close proximity to the dots. GL, GR, and GM are, respectively, designed to adjust the energy level of left dot, right dot, and inter-dot coupling strength. And PSL, PDL (PSR, PDR) are used for the tuning of the coupling of the left (right) dot to source and drain. The n-type heavily doped silicon substrate is used as a global back gate. The bridge plunger gate GM is separated from the drain part of graphene by a layer of over exposed PMMA. All the devices were primarily tested to check the functionality of all the gates in a liquid helium storage dewar at 4.2 K. Then the samples were mounted on a dilution refrigerator equipped with filtering wirings and low-noise electronics at the base temperature of 10 mK. To maintain consistency, we will use the data from one sample only in the following.

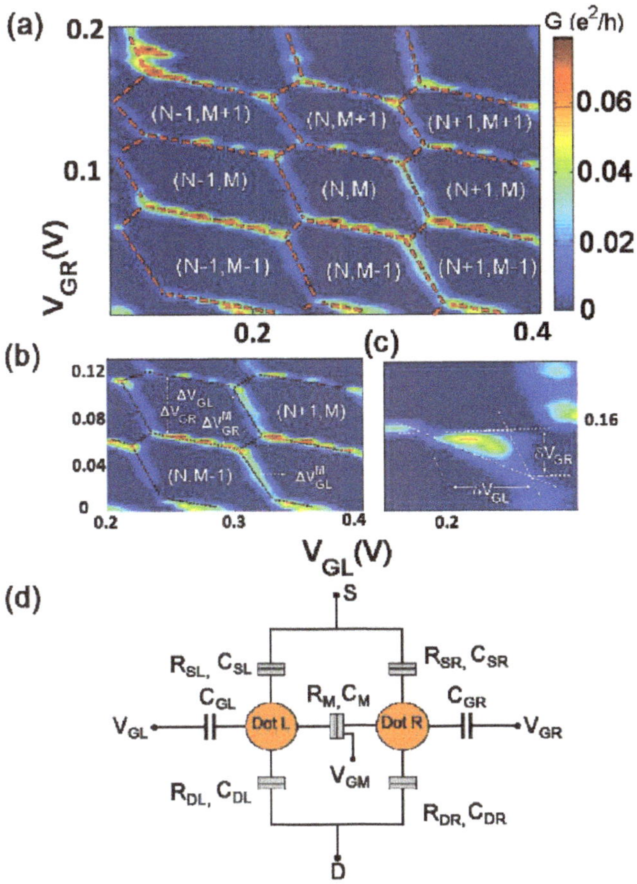

Figure 10. Color online) (a), (b), and (c) PDQD differential conductance as a function of plunger gate voltage V_{GL} and V_{GR}. The red dash lines are guides to the eyes showing the honeycomb pattern. (N, M) represents the carriers in the left and right dot, respectively. (b) Zoom-in of the area (N,M) of the honeycomb pattern. (c) Zoom-in of a vertex pair with white dash lines. (d) Capacitance model for the analysis of the double dot system. Graphene nanoconstrictions behave as tunneling barriers, which are presented, for example, as RSL, CSL (a capacitance and a resistance coupled in parallel). Gate GL and GR are capacitively coupled to the dots; C_{GL} and C_{GR} represent the capacitance.

Fig. 10(a) shows color scale plot of the measured differential conductance of the double dot as a function of V_{GL} and V_{GR} detected in standard ac lock-in technique with anexcitation ac voltage 20μμV at frequency of 11.3 Hz. A dc bias of 0.3 mV is applied, the back gate voltage V_{bg} is fixed at 5 V and the middle plunger gate V_{GM} is -0.45 V. The hexagon pattern characteristic for double dot coupled in parallel is clearly visible. Figure 2(b) Zoom-in of the area (N, M) of the honeycomb pattern, Figure 2(c) Zoom-in of a vertex pair with white dashed lines. From the model of purely capacitively coupled dots as illuminated by Figure 2(d), the energy scales of the system can be extracted (van der Wiel et al., 2003;Molitor et al., 2010; Moriyama et al., 2009). The capacitance of the dot to the side gate can be determined from measuring the size of the honeycomb as shown in Fig. 10(a) and 10(b), ΔV_{GL}=0.087V, ΔV_{GR}=0.053V, $\Delta V_{GL}^{m}=0.0261\text{V}$ $\Delta V_{GR}^{m}=0.$ therefore, $C_{GL}=e/\Delta V_{GL}=1.84\text{aF}, C_{GR}=e/\Delta V_{GR}=3.0\text{aF}$. With a large DC bias of 0.3 mV, we can get δV_{GL}=0.013V and δV_{GR}=0.01V as shown in Figure 2(c). The lever arm between the left (right) gate V_{GL} and the left (right) dot can be calculated as $\alpha_{GL}=V_{bias}/\delta V_{GL}=0.023(\alpha_{GR}=V_{bias}/\delta V_{GR}=0.03)$. The total capacitances of the dots can then be calculated as $C_L=C_{GL}/\alpha_L$=79.8aF and $C_R=C_{GR}/\alpha_R$=100.4aF, the corresponding charging energy $E_{CL}=\alpha_{GL}.\Delta V_{GL}$=2.0meV and $E_{CR}=\alpha_{GR}.\Delta V_{GR}$=1.6meV, the coupling energy between the two dots $E_{CM}=\alpha_{GL}.\Delta V_{GL}^{m}=0.3\text{meV}$. It is also noted that the lever arms between the left gate and the right dot and vice versa can be determined from the slope of the co-tunneling lines delimiting the hexagons. These crossing couplings only modify the results slightly and are neglected usually (Molitor et al., 2010; Moriyama et al., 2009; Liu et al., 2010). Here, by calculating dots area and carrier density (related to V_{BG}), or from the Coulomb charging period, we estimate each dot contains more than 20 electrons when V_{BG}=5V.

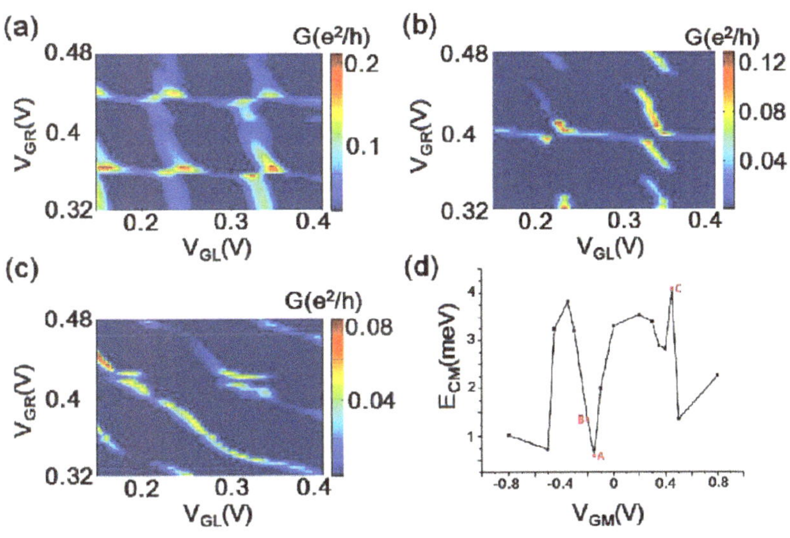

Figure 11. Color online) Interdot coupling vs middle gate voltage V_{GM}. Conductance as a function of gate voltage V_{GL} and V_{GR} at V_{BG}=3 V, V_{bias}=- 1 mV, the scan regions of GL and GR are the same. (a), (b), and (c) represent three different coupling regimes of the two dots. (a) weak coupling regime, V_{GM}= -0.15 V, (b) medium coupling regime, V_{GM}= -0.2 V (c) strong coupling regime, V_{GM}= 0.45 V. (d) shows coupling energy E_{CM} (V) as a non-monotonic function of the middle gate voltage V_{GM}. A, B, C point here represent the corresponding coupling energy in (a), (b), and (c).

By applying voltage to the middle plunger gate GM, the interdot coupling can be tuned efficiently Fig.11(a), 11(b),and 11(c) show the charge stability diagrams of the PDQD in three different coupling regimes. [(a) weak, (b) medium, and (c) strong]. In these measurements, back gate voltage V_{BG} =3V, Source-Drain DC bias V_{bias} is set to-1.0mV, the scan regions of GL and GR are the same. Only the voltage applied to the gate GM is adjusted as (a) V_{GM}= -0.15 V, (b) V_{GM}=0.2 V, and (c) V_{GM}=0.45 V. By using the same model as in Figure 2, we can calculate the corresponding coupling energy between the dots: (a) E_{CM}=0.58 meV, (b) E_{CM}=1.34 meV, and (c) E_{CM}=4.07 meV. The honeycomb diagrams of the parallel and serial DQD look similar except for the weak coupling regime, as shown in Figure3(a). In this case, the lines delimiting the hexagons are more visible in comparison with serial DQD, because the leads have two parallel accesses to the dots in parallel DQD, which also enables correlated tunneling of two valence electrons simultaneously (Holleitner et al., 2002). Fig. 11(d) indicates the coupling energy changes with the gate voltage VGM. As in the previous reports of graphene DQD in series (Molitor et al., 2010; Liu et al., 2010), the inter-dot coupling is non-monotonically depended on the applied gate voltage. Although the detailed reasons for this non-monotony are undetermined, we assumed that one key factor will be the disorders in graphene introduced by either fabrication steps or substrate (Todd et al., 2009). Many more efforts are still needed to address this issue for the realization of practical graphene based nanodevices.

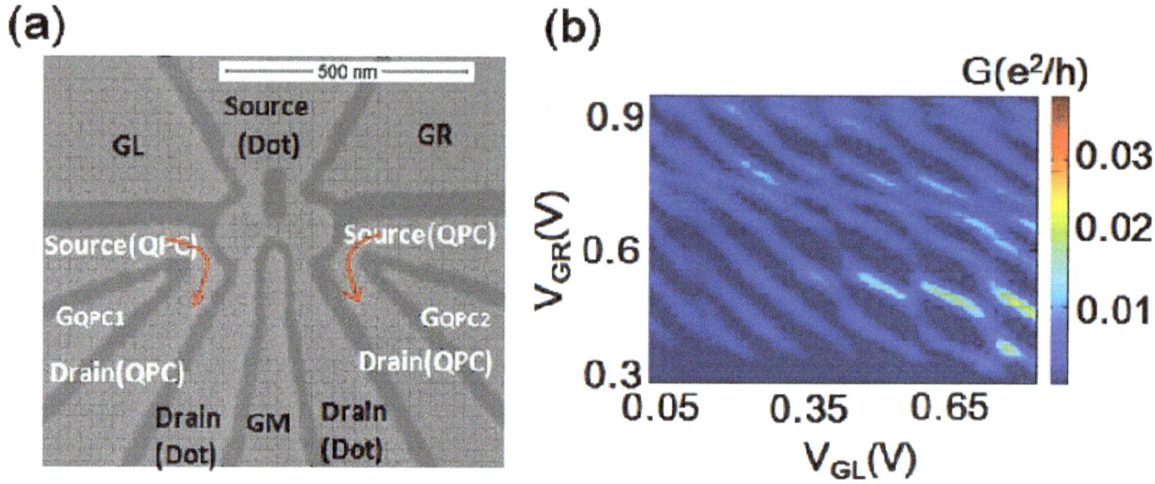

Figure 12. Color online) (a) SEM image of single layer graphene PDQD integrated with two QPCs. The bar has a length of 500 nm. (b) Characteristic honeycomb structure of the conductance through the PDQD as a function of two in-plane plunger gates voltage V_{GL} and V_{GR}, revealed by direct transport measurement of the PDQD at 4.2 K.

We have designed and fabricated an alternative structure of a PDQD integrated with two quantum point contact sensors (QPCs) in single layer graphene, as shown in Figure 4(a). The integrated QPCs can be used as a non-invasive charge detector which may have various applications (van der Wiel et al., 2003;Hanson et al., 2007; Guo et al., 2001; Zhang et al., 2007). As primary tests of the present structure, we can get similar charge stability diagram of the PDQD as in Fig.12 (b) by the direct quantum transport tests at 4.2 K. Although the non-invasive measurements by QPC are still under processing, no remarkable difference is founded between PDQD in bilayer and monolayer graphenes from direct transport measurement. Making tunable coupling double dot is the first step towards the quantum dot based quantum computation bits, the architectonics with integrated charge detector around double quantum dot demonstrated here offers the chance to achieve the charge or spin reading out, which is essential for the quantum computation device. Therefore, a lot of extended and follow-up works can be done on this basis in the future. Both bilayer and single layer graphenes can be exploited in this application.

In summary, we have discussed low temperature quantum transport measurement of gate-controlled parallel coupled double quantum dot on both bilayer and single layer graphenes. The inter-dot coupling strength can be largely tuned by graphene in-plane gates. With the quantum transport honeycomb charge stability diagrams, a common model of purely capacitively coupled double dot is used to extract all the relevant energy scales and parameters of grapheme PDQD. Although many more effects are still needed to further upgrade and exploit the present designed grapheme quantum dot system, the results have intensively demonstrated the promise of the realization of graphene nanodevice and desirable study of rich PDQD physical phenomena in graphene.

CONCLUSION

To conclude, we have discussed the design and fabrication of etched gate tunable single and double quantum dots in single-layer and bilayer graphene and present several important quantum transport measurements in these systems. A quantum dot with an integrated charge sensor is becoming a common architecture for a spin or charge based solid state qubit. To implement such a structure in graphene, we have fabricated a twin-dot structure in which the larger QD serves as a single electron transistor (SET) to read out the charge state of the nearby gate controlled small QD. A high SET sensitivity of $10^{-3}e/\sqrt{Hz}$ allowed us to probe Coulomb charging as well as excited state spectra of the QD, even in the regime where the current through the QD is too small to be measured by conventional transport means. We also have measured quantum transport properties of gates controlled parallel-coupled double quantum dots (PDQD) and series-coupled double quantum dots (SDQD) device on both single layer and bilayer graphene with multiple electrostatic gates that are used to enhance control to investigate it. At low temperatures, the transport measurements reveal honeycomb charge stability diagrams which can be tuned from weak to strong inter-dot tunnel coupling regimes. We precisely extract a large inter-dot tunnel coupling strength for this system allowing for the observation of tunnel-coupled molecular states extending over the whole series-coupled double dot. The inter-dot coupling strength also can be effectively tuned from weak to strong by in-plane plunger gates for parallel-coupled double quantum dots. All the relevant energy scales and parameters can be extracted from the honeycomb charge stability diagrams. The present method of designing and fabricating graphene DQD is demonstrated to be general and reliable and will enhance the realization of graphene nanodevice and desirable study of rich DQD physical phenomena in graphene, and highly controllable system serves as an essential building block for quantum devices in a nuclear-spin-free world.

ACKNOWLEDGMENTS

This work was supported by the National Basic Research Program of China (Grants No. 2011CBA00200 and 2011CB921200), and the National Natural Science Foundation of China (Grants No. 10934006, 11074243, 11174267, 91121014, and 60921091)

REFERENCES

1. Berman D, Zhitenev N. B, Ashoori R. C, and Shayegan M, Phys. Rev. Lett. 82, 161 (1999).
2. Bylander J, Duty T, and Delsing P, Nature London 434, 361 (2005).
3. Chen J. C, Chang A. M, and Melloch M. R., Phys. Rev. Lett. 92, 176801 (2004).
4. Dias da Silva L. G. G. V, Ingersent K, Sandler N, and Ulloa S. E, Phy. Rev. B 78, 153304 (2008).
5. Duty T, Gunnarsson D, Bladh K, and Delsing P, Phys. Rev. B 69, 40503(R) (2004).
6. Elzerman J. M, Hanson R, L. van Beveren H. W, L. Vandersypen M. K, and Kouwenhoven L. P, Appl. Phys. Lett. 84, 4617 (2004b).
7. Elzerman J. M, Hanson R, van Beveren L. H. Witkamp W, B, Vandersypen L. M. K, and L. P. Kouwenhoven, Nature London 430, 431 (2004a).
8. Fischer J, Trauzettel B, and Loss D, Phys. Rev. B 80, 155401 (2009).
9. Fischer J. and Loss D, Science 324, 1277 (2009).
10. Foletti S, Bluhm H, Mahalu D, Umansky V, and Yacoby A, Nat. Phys. 5, 903 (2009).
11. Gallagher P., Todd K., and Goldhaber-Gordon D., Phys. Rev. B 81, 115409 (2010).
12. Geim A. K and Novoselov K. S, Nature Mater. 6, 183 (2007).
13. Geim A. K., Science 324, 1530 (2009).
14. Golovach V. N, and Loss D, Phys. Rev. B 69, 245327 (2004).
15. Gotz G, Steele G. A, Vos W. J, and Kouwenhoven L. P, Nano Lett. 8, 4039 (2008).
16. Graber M. R, Coish W. A, Hoffmann C, Weiss M, Furer J, Oberholzer S, Loss D, and Scho°nenberger C, Phys. Rev. B 74, 075427 (2007).
17. Guo G. P, Li C. F, and Guo G. C, Phys. Lett. A 286, 401 (2001).
18. Guo G. P, Lin Z. R, Tu T, Cao G, Li X. P, and Guo G. C, New J. Phys.11, 123005 (2009).
19. Güttinger J, Seif J, Stampfer C, Capelli A, Ensslin K, and Ihn T, Phys. Rev. B 83, 165445 (2011).
20. Güttinger J, Stampfer C, Hellmüller S, Molitor F, Ihn T, and Ensslin K, Appl. Phys. Lett. 93, 212102 (2008).
21. Hanson R, Kouwenhoven L. P, Petta J. R, Tarucha S, and Vandersypen L. M. K, Rev. Mod. Phys. 79, 1217 (2007).
22. Hanson R, Witkamp B, Vandersypen L. M. K, van Beveren L. H. W, Elzerman J. M, and Kouwenhoven L. P, Phys. Rev. Lett. 91, 196802 (2003).
23. Hanson R. and Awschalom D, Nature 453, 1043 (2008)
24. Hatano T, Stopa M, and Tarucha S, Science 309, 268 (2005).
25. Holleitner A. W, Blick R. H, Hu° ttel A. K, Eberl K, and Kotthaus J. P, Science 297, 70 (2002).
26. Holleitner A. W, Decker C. R, Qin H, Eberl K, and Blick R. H, Phys. Rev. Lett. 87, 256802 (2001).
27. Huard B, Sulpizio J. Stander A, Todd N. K, Yang B, and Goldhaber Gordon D, Phys. Rev. Lett. 98, 236803 (2007).
28. Ihn T., Guttinger J, Molitor F, Schnez S. Schurtenberger E, Jacobsen A, Hellmu°ller S, Frey T, Droscher S, Stampfer C. et al., Mater. Today 13, 44 (2010).
29. Knobel G. and Cleland A. N., Nature London 424, 291 (2003).
30. Kouwenhoven L. P, Marcus C, McEuen P. L, Tarucha S, Westervelt R. M, and N. S. Wingreen, in Mesoscopic Electron Transport, Series E:Applied Sciences Vol. 345, edited by Sohn L. L, Kouwenhoven L. P, and Schon G, Dordrecht Kluwer, 1997, pp. 105–214.
31. L. Ponomarenko, F. Schedin, Katsnelson M, Yang R, Hill E, Novoselov K, and Geim A, Science 320, 356 (2008).
32. Ladro´n de Guevara M. L, Claro F, and Orellanal P. A, Phys. Rev. B 67, 195335 (2003).
33. LaHaye D, Buu O, Camarota B, and Schwab K. C, Science 304, 74 (2004).
34. Lehnert K. W, Bladh K, Spietz L. F, Gunnarsson D, Schuster D. I, Delsing P, and Schoelkopf R. J, Phys. Rev. Lett. 90, 027002 (2003).
35. Liu X. L, Hug D, and L. Vandersypen M. K. Nano Lett. 10, 1623 (2010).
36. Livermore C, Crouch C. H, Westervelt R. M, Campman K. L, and A. Gossard C, Science 274, 1332 (1996).
37. Lo´pez R, Aguado R, and Platero G, Phys. Rev. Lett. 89, 136802 (2002).
38. Lu W, Ji Z. Q, Pfeiffer L, K. W. West, and A. J. Rimberg, Nature London 423, 422 (2003).
39. Mason N, Biercuk M. J, and Marcus C. M, Science 303, 655 (2004).

40. Molitor F, Droscher S, Guttinger J, Jacobson A, Stampfer C, Ihn T, and Ensslin K, Appl. Phys. Lett. 94, 222107 (2009).

41. Molitor F, Knowles H, Droscher S, Gasser U, Choi T, Roulleau P, Guttinger J, Jacobsen A, Stampfer C, Ensslin K. et al., Europhys. Lett. 89, 67005 (2010).

42. Moriyama S, Tsuya D, Watanabe E, Uji S, Shimizu M, T. Mori, T. Yamaguchi, and Ishibashi K, Nano Lett. 9, 2891 (2009).

43. Novoselov K. S, Geim A. K, Morozov S. V, Jiang D, Zhang Y, Dubonos S. V, Grigorieva I. V, and Firsov A. A, Science 306, 666 (2004).

44. Orellana P. A, Ladron de Guevara M. L, and Claro F, Phys. Rev. B 70, 233315 (2004).

45. Petta J. R, Johnson A. C, Taylor J. M, Laird E. A, Yacoby A, Lukin M. D , Marcus C. M, Hanson M. P, and Gossard A. C, Science 309, 2180 (2005).

46. Ponomarenko L, Schedin F, Katsnelson M, Yang R. Hill E, Novoselov K, and Geim A. K, Science 320, 356 (2008).

47. Schnez S, Molitor F, Stampfer C, Guttinger J, Shorubalko I, Ihn T, and Ensslin K, Appl. Phys. Lett. 94, 012107 (2009).

48. Stampfer C, Guttinger J, F. Molitor, D. Graf, T. Ihn, and K. Ensslin, Appl. Phys. Lett. 92, 012102 (2008a).

49. Stampfer C, Schurtenberger E, Molitor F, Guettinger J, Ihn T, and Ensslin K, Nano Lett. 8, 2378 (2008b).

50. Stampfer C., Güttinger J., Hellmüller S., Molitor F., Ensslin K., and Ihn T., Phys. Rev. Lett. 102, 056403 (2009).

51. Todd K, Chou H. Amasha T, S, and Goldhaber-Gordon D, Nano Lett. 9, 416 (2009).

52. Trauzettel B, Bulaev D. V, Loss D, and Burkard G, Nat. Phys. 3, 192 (2007).

53. van der Wiel W. de Francheschi G, S, Elzermann J. M, Fujisawa T, Tarucha S, and Kouwenhoven L P, Rev. Mod. Phys. 75, 1 (2003).

54. Vijay R, M. H. Devoret, and I. Siddiqi, Rev. Sci. Instrum. 80, 111101 (2009).

55. Vink I. T, Nooitgedagt T, Schouten R. N, and Vandersypen L. M. K, Appl. Phys. Lett. 91, 123512 (2007).

56. Volk C, Fringes S, Terres B, Dauber J, Engels S, Trellenkamp S, and Stampfer C,.Nano Lett. 11, 3581 (2011).

57. Wang L. J, Cao G, Li H. O, Tu T, Zhou C, Hao X. J, Guo G. C, and Guo G. P, Chinese .Physics. Letters. 28, 067301 (2011b).

58. Wang L. J, Cao G, Tu T, Li H. O, Zhou C, Hao X. J, Su Z, Guo G. C, Jiang H. W, and Guo G. P, Appl. Phys. Lett. 97, 262113 (2010).

59. Wang L. J, Guo G. P, Wei D, Cao G, Tu T, Xiao M, Guo G. C, and Chang A.M, Appl. Phys. Lett. 99, 112117 (2011a).

60. Wang L. J, Li H. O, Tu T, Cao G, Zhou C, Hao X. J, Su Z, Xiao M, Guo G. C, Chang A.M, and Guo G. P, Appl. Phys. Lett. 100, 022106(2012).

61. Yang S, Wang X, and Das Sarma S, Phys. Rev. B 83, 161301(R) (2011).

62. Zhang H, Guo G. P, Tu T, and Guo G. C, Phys. Rev. A 76, 012335 (2007).

63. Ziegler R, Bruder C, and Schoeller H, Phys. Rev. B 62, 1961 (2000).

CHAPTER 19

SILICON CARBIDE FOR NOVEL QUANTUM TECHNOLOGY DEVICES

Stefania Castelletto, Lorenzo Rosa and Brett C. Johnson

Keywords: Silicon carbide deep defects, Paramagnetic properties, Optical-detected magnetic resonance, Single-photon sources

INTRODUCTION

The most common and technologically advanced SiC polytypes are 4H-SiC and 6H-SiC with hexagonal structures and 3C-SiC with a zinc-blende crystal structure (cubic). High-quality bulk single-crystal 3C-SiC can be grown epitaxially on different substrates, most notably on silicon. This has led to many exciting devices fabricated with this particular polytype[1]. SiC has a wide bandgap (2.4-3.2 eV depending on the polytype), a high thermal conductivity, the ability to sustain high electric fields before breakdown and the highest maximum current density, making it ideal for high-power electronics [2]. More recently, it has become a notable material in the field of quantum computing and spintronics as several of its intrinsic defects are associated with an electron spin that can be used as quantum bit [3, 4]. To enhance solid-state quantum systems scalability, a fully integrated device with quantum control should be built; thus, quantum systems should be part of the material used to fabricate the final device. Other solid-state quantum systems fully integrated into a functional device[5–7] operate at cryogenic temperatures (4 K or below), limiting their engineering and scalability. A room temperature solid-state "qubit" is the nitrogen vacancy (NV) centre[8] in diamond; yet diamond is not mature for standard device fabrication protocols. SiC, on the other hand, is widely used in LEDs (commonly as a substrate for GaN films), power electronics and microelectromechanical and nano-electromechanical systems (MEMs, NEMs) [9, 10] and has well-developed device processing protocols which are compatible with industry standards. In addition, nanostructures can be formed in SiC such as nanoparticles, quantum dots, nanowires and nanopillars. The ability to grow SiC on silicon provides an unprecedented advantage which facilitates the fabrication of nanophotonic cavities[11, 12]. As a compound semiconductor, SiC harbours a rich assortment of optically active intrinsic and extrinsic defects that can be used as quantum systems. In this chapter, we will summarise key findings and the main properties of deep optical and paramagnetic defects in bulk SiC that very recently have prompted novel quantum effects such as single-photon source (SPS) and quantum coherence control of their spin[13–17]. We will review defects with optical emission used in quantum spintronics and quantum optics, optically detected resonance methods applied to these defects, the achievement of coherent control of ensemble and single spins and their present spin coherence times. We will review single-photon emission from nanostructures of silicon carbide[18, 19]. We will review recent nanophotonics advances[20–30] in this material to achieve defect integration and enhancement. Other applications of the engineered defects in quantum technologies are also summarised. Some challenges still remain before this material can fully realise its potential in advancing quantum technology. In the summary, we critically analyse present challenges based on the reported results and provide an outlook of likely future investigations that can lead to successful application of SiC for quantum devices.

DEEP DEFECTS IN SiC With Quantum Properties

Many deep-energy-level defects within the bandgap give rise to radiative recombination in SiC with photoluminescence (PL) from the UV to the infrared. The dual compound nature and the existence of nonequivalent lattice sites in different polytypes (quasi-cubic, k, and hexagonal, h) give rise to a large variety of possible PL. Here, we will discuss only recently

Citation: Stefania Castelletto, Lorenzo Rosa and Brett C. Johnson, Silicon Carbide for Novel Quantum Technology Devices, DOI: 10.5772/61166.

investigated defects that provided so far experimental evidence of quantum effects [14–16]. These defects are intrinsic defects, known as silicon vacancies (V_{Si})[31–33], divacancies ($V_{Si}VC$) [34] and carbon antisite-vacancy pair ($C_{Si}VC$)[35, 36]. In Figures 1-3, we show the atomistic defect structure and the presently understood energy levels corresponding to so far identified quantum systems in 4H, 6H and 3C polytypes. These defects coincide with the most commonly previously observed intrinsic defects in the material. They can be created in ensemble by neutron or electron irradiation, followed by annealing, as well as they can be created during growth. Recent experimental challenges rely in their creation as isolated systems or individual defects with their correlation to the ensemble level for their identification in bulk SiC. For this to be reliable, due to the large variability properties of the bulk material, which also can present many defects at the same time, the purity of the pristine wafer resulted to be essential. Therefore, the isolation of single defects as quantum systems was performed in high purity 4H intrinsic SiC. These defects act as radiative recombination centres within the bandgap with a PL achieved by out-of-resonance excitation with a laser. The PL is typically characterised by a sharp zero-phonon line (ZPL) and phonon side bands (PSB) at longer emission wavelengths. Recombination centres are modelled as a quantum system with ground, excited and metastable states (a two- or three-level system). These defects can have different ZPLs in the first instance according to the polytype, which provides nonequivalent crystallographic sites in the matrix. Additionally, the material doping can influence the emission wavelength. They are often characterised by a charge state which also can influence the spectral location of the ZPL peak. The charge state is also associated to the spin state (mostly of the ground state) of the defect, which can be observed from PL measurements as a reduction/enhancement (depending on the mechanism of the population of the energy levels) of the PL, when probed with the application of an additional magnetic field or a microwave excitation in resonance with the spin sublevels. Other defects also yielding quantum properties are not yet presently completely identified. The main effect on the variation of the defects ZPLs is related to the polytype, whether the defect occupies an axial position in the lattice (orientation along the c-axis, hh or kk lattice location) or an off-axis position (hk, kh orientation basal); thus, at least 2^n ZPLs can be present for n-nonequivalent crystallographic site positions. In addition, the symmetry of the defects can give rise to other ZPLs with an excited state split by Jahn-Teller distortion, reducing the symmetry and introducing additional ZPLs. VSi consists of a missing silicon atom in the matrix. In 4H-SiC, V_{Si} shows two ZPLs (two nonequivalent sites) stable up to 500-600°C; in addition, two possible charge states are known. The ZPL, known as V_1 , emits at 862.2 nm, while V_2 emits at 917.0 nm. These lines have been associated to a negative charge state, while the neutral charge state has ZPLs at 859.0 nm and 861.7 nm, respectively. Both defects are quite common in 4H- and 6H-SiC, as well as in 3C. The VSi negative charge in 4H and 6H has a high spin (S=3/2) ground state, shown in Figure 1(c)[32]. VSi ZPL emissions in 6H-SiC are known as V_1 ,V_2 and V_3 lines, owning to the three nonequivalent sites (h,k_1 ,k_2; see Figure 1) at 865 nm, 887 nm and 906 nm, respectively.

The closest pair of V_{Si} and carbon vacancy VC, the divacancy $V_{Si}VC$, can form during the migration of the isolated vacancies. PL arising from $V_{Si}VC$ is attributed to its neutral charge state ($V_{Si}V$ 0 C) and has several ZPLs in 4H- and 6H-SiC. In 4H-SiC, the $V_{Si}VC$ defects possess four ZPLs (stable up to 1,500°C) associated to specific PL at 997.5 nm, 1013.6 nm, 1050.7 nm and 1053.9 nm. These ZPLs are related the axial (hh, kk) and basal (hk, kh) location in the lattice (see Figure 2). Other ZPLs in a similar wavelength range have been recently found [38] denominated PL5 (1043 nm) and PL6 (1039 nm) due to their unknown origin. In 6H-SiC, the $V_{Si}VC$-defect-known ZPLs are at 998.7nm, 1010.9 nm, 1030 nm, 1048.7 nm and 1074.6nm. ZPLs obtained in irradiated 6H samples at 1,139 nm; 1,135 nm; 1,124 nm; 1,108 nm; and 1,093 nm were found to have quantum characteristics similar to the ($V_{Si}V$ 0 C)[39] and were associated with UD2 defect, previously known in 4H- and 6H-SiC[40, 41].

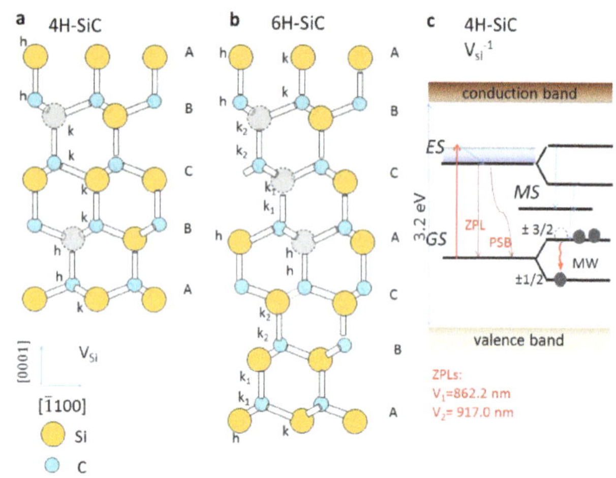

Figure 1. Representation of the atomic structure of the V_{Si} in the SiC lattice in a 4H and b 6H polytypes, located in different nonequivalent sites (k,h in 4H) and (h, k_1,k_2 in 6H). c Visualisation of the energy levels in the negatively charged state of the V_{Si} for 4H, where the ground state (GS), excited state (ES) and metastable state (MS) structures are visualised, with the ZPL, PSB and spin level of the high spin S=3/2 ground state. A microwave transition in the GS is responsible for the PL enhancement of the ZPL in the presence of a microwave excitation (MW). Both 4H and 6H polytypes containing these defects have been used for quantum control of the ground state spin [16, 37]

In a dual element semiconductor, the antisite-vacancy complex is an intrinsic defect. The defect possesses C_{3v} symmetry if C_{Si} is along the c-axis (on-axis configurations, $C_{Si}VC$(kk) and $C_{Si}VC$(hh)), while it has C_{1h} symmetry if C_{Si} is out of the c-axis (off-axis configurations, $C_{Si}VC$(kh) and $C_{Si}VC$(hk)) in 4H-SiC (see Figure 3a). Its PL was found to be in the 648-678 nm range and initially attributed to its neutral charge state [36]. It has been known as AB lines found in both 4H- and 6H-SiC with also different doping. According to recent modelling, the neutral charged state of the $C_{Si}VC$ defect associated to the AB lines should not have any visible PL; therefore, the visible PL is attributed to the positive charge state of this defect with S=1/2. Eight ZPLs were associated with the $C_{Si}VC$ in 4H-SiC, labelled as A_1 = 648.7 nm, A_2 =651.8 nm, A_3 = 665.1 nm, A_4=668.5 nm, B_1 = 671.7 nm, B_2 =673.0 nm, B_3 =675.2 nm and B_4 = 676.5 nm. The eight AB PL lines in 4H-SiC can naturally account for the four ground-state configurations (hh, kk, hk, kh) where each possesses two ZPLs due to the splitting of the excited state.

SINGLE-PHOTON SOURCE IN BULK MATERIAL

Over the past decade, single-photon generation has been observed in many physical systems such as single molecules [42], quantum dots[43, 44], diamond colour centres[45] and others[46, 47]. The generation and detection of single photons play a central role in the foundation of quantum mechanics. Additionally, an efficient and high-quality single-photon source operating at room temperature is needed to ensure secure communication channels via the implementation of protocols of quantum key distribution. SPSs are needed to establish quantum photonics communication networks [48]. At the present, SiC has provided in bulk materials SPS from all the defects discussed in Section 1, isolated in ultrapure 4H-SiC. The first observation of single-photon statistics from SiC was associated with the PL from a single C antisite-vacancy pair. This defect is the brightest single-photon emitter in a bulk solid state system operating at room temperature, showing up to 106 counts/s at saturation, indicating a high quantum efficiency and occurring in the region 650-700 nm. The single-photon emission corresponds to a 3-level atomic system, due to the presence of a metastable state. The brightness is associated to the polarisation selection rules of these defects, most of the time parallel to the excitation electric field, as well as to the dipole strength (short radiative lifetime of 1.5 ns). The defect has been isolated using confocal microscopy in electron-irradiated high-purity, semi-insulating (HPSI) SiC along the main axis (1,000) from a 3-inch wafer, containing VC ~ $10^{14}/cm^3$, B~ $10^{14}/cm^3$, N and other defects< $10^{14}/cm^3$. Electron irradiation with energy of 2Mev was used to create defects deeper in the material, and fluences of 10^{13}-$10^{14}cm^{-3}$ were used for the single-defect creation. In Figure 4, the confocal maps, experimental set-up, typical PL and photon correlation measured using a Hanbury Brown and Twiss interferometer indicate SPS. PL at room temperature in single emitters was compared to ensemble generated with electron fluences of $10^{17}cm^{-3}$, showing high similarity when excited at 532 nm.

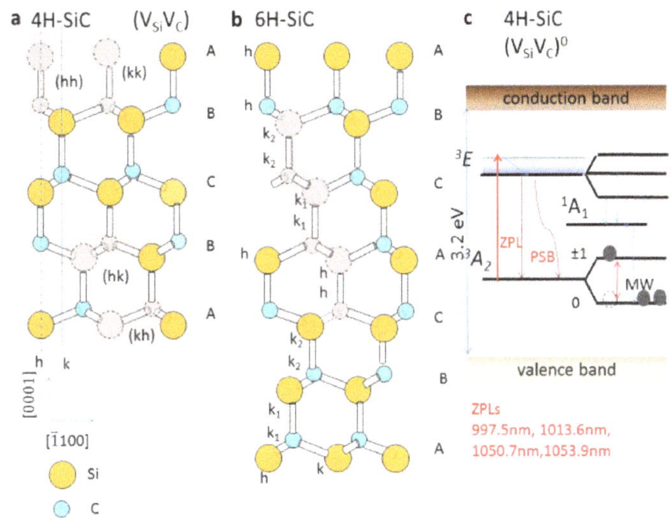

Figure 2. Representation of the atomic structure of the V_{Si}VC in the SiC lattice in a 4H and b 6H polytypes, located in different nonequivalent sites (k, h in 4H) and (h, k_1,k_2 in 6H).c Visualisation of the energy levels in the neutral state of the (V_{Si}V 0 C) for 4H, where the ground state ($3A_2$), excited state (3E) and metastable state ($1A_1$) structures are visualised, with the ZPL, PBS and spin level of the high spin S=1 ground state. A microwave transition in the GS is responsible for the PL enhancement of the ZPL in presence of a microwave excitation (MW). Both 4H and 6H polytypes containing these defects have been used for quantum control of the ground state spin [15, 38, 39].

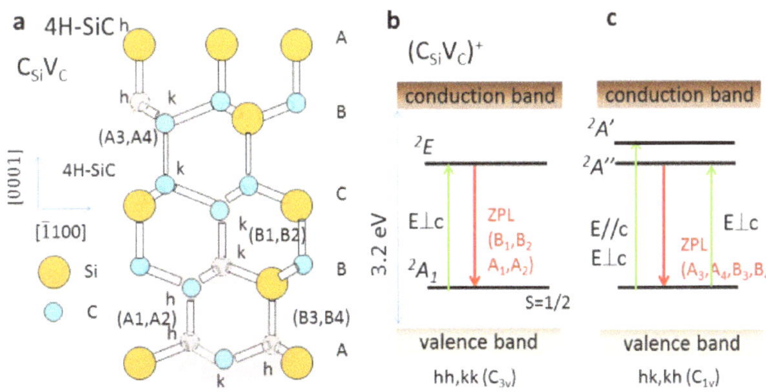

Figure 3. Representation of the atomic structure of the C_{Si}VC in the SiC lattice in a 4H, located in different nonequivalent sites (k, h in 4H). Visualisation of the energy levels in the positive charge state of the (C_{Si}V + C) for 4H, where the ground state ($2A_1$) and excited state (2E) are visualised, with the ZPLs in the axial b and basal c defects position [14].

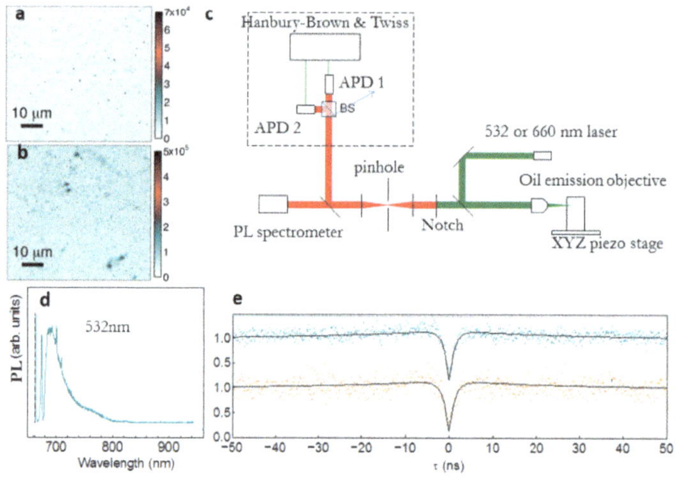

Figure 4. Confocal maps of a untreated sample that occasionally shows native defects whose origin is yet unclear and b confocal map showing irradiated generated single defects with a low temperature annealing of 300°C. The colour code indicates much brighter emitters in the irradiated samples. c Experimental set-up.d Typical room temperature PL of the SPS. e Photon correlation of two SPSs in the irradiated sample excited at 532 and 660 nm. The reduction of photon coincidence at zero delay time below 0.5 indicates the presence of single-photon statistics.

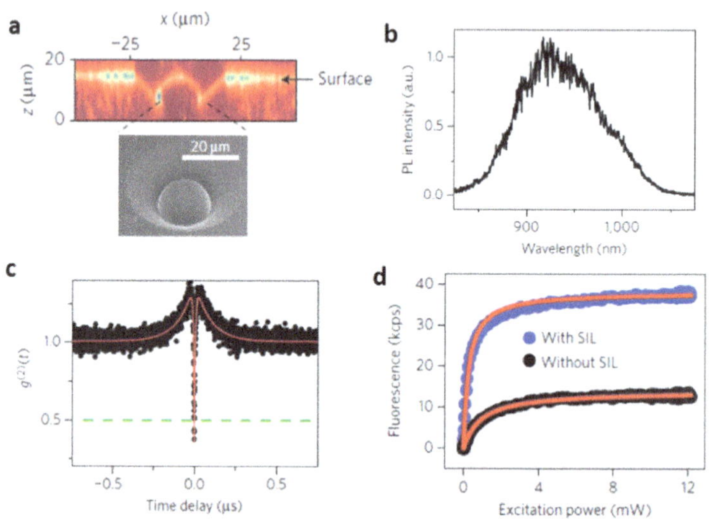

Figure 5. Confocal map of a a solid immersion lens (SIL) containing a single V_{Si} defect and the SIL SEM image. b PL at room temperature of the V_{Si}.c Photon correlation of defect PL in the irradiated sample. The reduction of photon coincidence at zero delay time below 0.5 indicates the presence of single-photon statistic. d Photon count rate from a single defect versus excitation power with and without the SIL. Images reproduced with permission from Macmillan Publishers Ltd: Nature Materials [16], copyright (2015).

V_{Si} denoted as V_2 lines with ZPL at 917 nm in bulk 4H-SiC has also been isolated [16], using confocal microscopy exciting the substrate by a 730 nm laser. The defect at the single level was created by electron irradiation and low-temperature annealing in vacuum. Due to the geometry of the defect location in the material cut along (1,000) axis, this defect could not be excited with the laser polarisation along to the defect axis, yielding a very low count rate at the single-photon level (15,000counts/s) at saturation. A solid immersion lens of 20 µm diameter was milled using a Ga focussed ion beam in the material to increase the collection efficiency of the single photons, proving an enhancement of a 2.5 factor. Similar results are shown of isolation of this defect in 4H by [17]; in this case, the defect was created using neutron irradiation, and a clear increase of the number of defects with radiation fluences was observed. Similar count rate and PL broad emission in the 900 nm region was observed for the single emitter. A lifetime of 6.1 ns was determined and perfect photo-stability was observed. In Figure 5, we show the main properties of this defect as a SPS.

Divacancies were also isolated from 120 µm 4H-SiC single-crystal epitaxial film grown on an n-type 4H-SiC substrate. The substrate was grown by hotwall chemical vapour deposition, a technique used to create commercial quality, multilayer electronic structures at the wafer scale. The epilayer is optimised to have no basal plane dislocations or polytype inclusions and a very low (5 $10^{13}cm^{-3}$) unintentional dopant density. The epilayer was mechanically separated from the substrate, polished and diced and then irradiated with 2 MeV electrons at a range of fluences 5 $\times 10^{12}$ cm^{-2} to $10^{15}cm^{-2}$. Annealing was used to achieve the migration of the vacancy to form the defects. Superconducting nanowire-based single-photon detectors able to detect the infrared emission were integrated into a home-built confocal microscope.

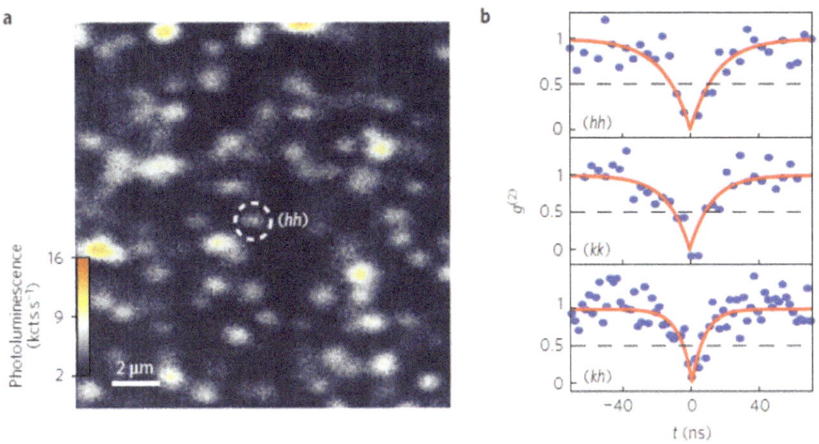

Figure 6. Confocal map of a a 4H-SiC membrane irradiated at 10^{13} cm^{-3} electron fluence. PL is collected by a confocal at a depth of 20 μm at a temperature of 20 K. PL spots were mostly identified as divacancies, though not all bright spots correspond to isolated single defects. b Photo-correlation $g^{(2)}(\tau)$ measurements for single divacancy defects located corresponding to (hh), (kk) and (kh) sites. Fits to a simple two-level model are shown. Images reproduced with permission from Macmillan Publishers Ltd: Nature Materials [15], Copyright (2015)

Excitation was performed with 975 nm continuous-wave laser. The sample was cooled to 20 K to observe distinct bright spots in a scanning PL image, as shown in Figure 6. The optical lifetimes of the neutral divacancies is 14±3 ns. The measured count rates were very low in the range of 3000-5000 counts/s, due to detector losses (~ 28% detection efficiency) in the spectral region of their ZPLs and the low collection efficiency from low NA objective in the cryostat. In addition, because of a high background due to polishing artifacts it was required to focus deep (20 μm) below the SiC surface to observe isolated single emitters, which added another 20-30% optical losses.

SINGLE-PHOTON SOURCES IN NANOMATERIALS

SiC nanomaterial scientific interest is motivated by the material larger potentials at the nanoscale in optoelectronics as an efficient ultraviolet emitter. SiC nanomaterial has also generated interest for its biocompatibility in nanomedicine or as emitting nanoprobes for imaging[49–51]. PL spectra of SiC nanostructures and nanopowder have been largely studied. The most common PL origin due to subgap emission is radiative recombination of defects and surface states[52], while above-gap emission is attributed to quantum confinement phenomenon in small nanocrystals such as quantum dots (QDs). At present the peak emission wavelengths of the SiC QDs are in the UV-blue-green region, typically between 380 nm and 550 nm[53–56]. This spectral emission region, however, is not ideal for biomedical applications due to the cell autofluorescence in the same spectral band. Additionally, infrared would be desirable for in-depth biological tissue imaging.

Recently the first demonstration of a deep defect within the bandgap has been shown in SiC nanocrystals obtained from 3C polytype[18], showing an emission in the red spectral region (650 nm). This red emission has been demonstrated to emit single photons. Two sources of SiC nanoparticles were explored: commercial 3C SiC nanomaterial with an average size of 45 nm suspended in ethanol and large (size: 200-500nm) in-house synthesised SiC nanocrystals suspended in MilliQ, dried on glass cover slip[57]. A room temperature custom-built confocal microscope combined with an atomic force microscope (Figure 7c) was used to correlate the size of the nanocrystals with PL, while photon correlation was used to determine the quantum properties of the PL. Low-temperature Raman spectroscopy and cathodoluminescence were used to aid the study of the origin of the SPS PL in 3C nanocrystals, indicating that the PL was not due to surface defects. A distinctive ZPL was observed at 648 nm at 80 K from ensemble measurements. A CW laser pump at 532 nm was used to excite the samples, while a pulsed 532 nm laser was used to measure the fluorescence lifetime, which results in the range of (2 ÷ 5.3)±0.2 ns, depending on the treatment. Some nanocrystals provided SPSs with partial polarisation in absorption and emission. We tentatively attributed the defect responsible for SPS in 3C to the carbon antisite-vacancy pair by comparing to ensemble low-temperature PL in bulk 3C heavily irradiated sample. This defect has C_{3v} symmetry in 3C-SiC lattice (see Figure 7a). Because of the relatively small bandgap of 3C-SiC, the double-positive charge state was considered as possible candidates for the PL. The calculated (2+7→+) charge transition level is at 1.95 eV which is very close to the ZPL energy of the bulk material defect ZPL.

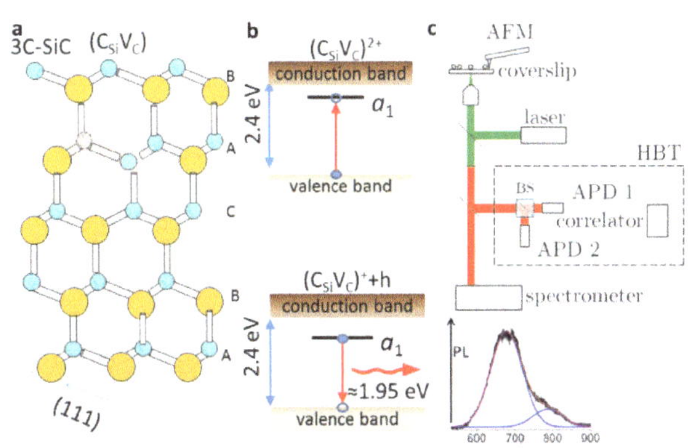

Figure 7. Representation of the atomic structure of the $C_{Si}VC$ in the 3C-SiC lattice in a. b Visualisation of the energy levels in the double-positive charge state of the $(C_{Si}V_C^{2+})$ for 3C, where the ground state and excited state are visualised. The

defect ZPL is around 648 nm from ensemble measurements (not clearly observed at the single level at room temperature).c Experimental set-up used to study the 3C-SiC nanoparticle defect SPS and room temperature PL measured from SPS [18].

SPS has also been observed in SiC nanotetrapods[19]. The tetrapod crystal phase structure consists of 3C core and 4H arms. The coexistence of 3C and 4H polytypes in these nanostructures led first to the observation of PL in the spectral region between 650 and 780 nm, not observed before in SiC nanostructures. This has been understood as a new quantum confined system accessible at room temperature. In this case, the quantum confinement and the nonclassical emission originated from the homogeneous heterostructure of 3C core and 4H legs, forming a quantum well in SiC nanotetrapod. The schematic of the SiC tetrapods is shown in Figure 8, accompanied by a high-resolution scanning electron microscope (SEM) image of the tetrapods. The nanotetrapods were measured to have an average leg length of 100 nm and leg diameter smaller than 50 nm, while the core was estimated as small as several nanometres. The possible band structure in SiC tetrapods is analysed by means of quantum mechanical simulations on a simplified model. The exciton is confined along the quasi one-dimensional potential curve created by 4H(leg)- 3C(core)- 4H(leg) structure. The emission wavelength of the individual tetrapods depends on their geometry, particularly, on their global symmetry, and much less on the diameter of 3C core (Figure 8 c). If the length of 4H legs was the same for all tetrapods (global Td symmetry), then a classical rectangular quantum well forms for the electrons in the conduction band (0.92 eV) and a minor potential barrier for the holes in the valence band (0.05 eV), so that the potential curve does not show any steepness ($\Delta V = 0$) (Figure 8 a (top)). The symmetric tetrapods should show no polarisation of light. However, if the length of the 4H legs was different (i.e. not all the legs have the same length) in a SiC tetrapod, then this induces different polarisations of surface charges at the end of 4H legs in these tetrapods, so a steep potential curve ($\Delta V > 0$) both for the electrons and holes in the conduction and valence band edges (Figure 8 a(bottom)). This effect creates a triangular potential well for the electron in the conduction band and starts to push the hole away from 3C region.

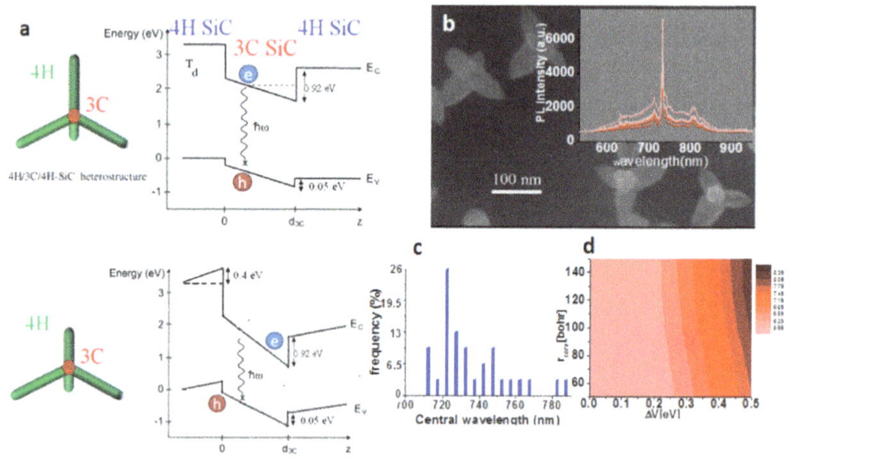

Figure 8. a Schematics of the tetrapod structure where the diameter of 3C core is d_{3C}. Top: Band line-up for 4H/3C/4H-SiC heterostructures for the ideal symmetry case T_d. Bottom: Quantum confinement model of SiC tetrapods when an asymmetric structure is considered and a polarisation potential change on both legs is ΔV=0.4 eV. b Scanning electron microscope of grown tetrapods. Every tetrapod has an average leg length of ~ 100 nm. Inset: exemplary PL form a single tetrapod corresponding to an SPS.c Measured ZPL central emission for SPS in single nanotetrapods. d Calculated emission wavelengths from this model where ΔV is the potential difference along the arms and rcore is the radius of the 3C core, respectively. The steep potential curve with a height of ΔV arises from the different lengths of the 4H-SiC legs of a tetrapod [19].

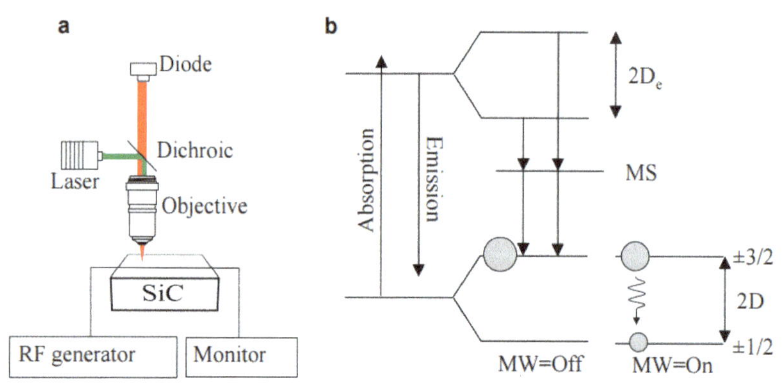

Figure 9. a Schematic of the optical set-up used to measure the ODMR signal. b Energy-level diagram of a spin system with a quartet ground state based on the V_{Si} [16, 37]. D and De denote the zero-field splitting of the ground singlet state and the excited state, respectively. The population of the spin substate levels is indicated by the grey circles. Further details in the text.

OPTICAL CONTROL OF SPIN STATE

Spin-dependent luminescence can be used to study magnetic and hyperfine parameters for excited states of defects in SiC. A wealth of knowledge about the atomic origin of defects and their properties has been extracted from such optical measurements. More recently, with the emergence of quantum information processing technologies, optical methods have been developed to coherently control and read out the spin of a defect centre. Figure 9a shows a schematic of a basic set-up often used to measure spin-dependent luminescence in SiC by the optically detected magnetic resonance (ODMR) technique. The main components are the excitation laser, a dichroic mirror and a diode to collect the filtered light in the spectral region of interest. A microwave carrying wire lays in close proximity to the probed defects, and an external magnetic field may also be applied. The electron paramagnetic resonance (EPR) spectrum is recorded by monitoring the PL intensity, while the frequency of the applied microwave excitation is varied. Figure 9 b shows an example of an energy-level scheme for a defect (based on the V_{Si} [16, 37]). Firstly, the defect is optically excited. In this system, intersystem crossing (MS) causes the higher-lying spin sublevel of the ground state ($m_s = \pm 3/2$) to be preferentially populated. When resonance with an allowed EPR transition occurs on the application of a microwave field, such as $|m_s = \pm 3/2i \rightarrow |m_s = \pm 1/2i$, the steady-state population of the sublevels is redistributed. This leads to an enhanced singlet ground-state absorption and an increase in the PL intensity. Decreases in PL can also be observed depending on the relative values of the populating and decay rates. ODMR has only been observed for a limited number of systems, most notably for single NV centres in diamond [58] and more recently single V_{Si} [16] and V_{Si}VC [15] centres in SiC which will be discussed further below. Optically induced alignment (polarisation) of the ground-state spin sublevels of the V_{Si} in 4H- and 6H-SiC was observed recently for the first time at room temperature [59, 60]. The alignment schemes vary depending on the crystal polytype and crystallographic position of V_{Si} in the crystal lattice, as well as their zero-field splitting parameters. For 6H-SiC, the zero-field splitting parameter of V_{Si} is 9×10^{-4} cm^{-1} (26.9 MHz for the k-site) and 42.8×10^{-4} cm^{-1} (128.3 MHz for the h-site), while for 4H, it was found to be 22×10^{-4} cm^{-1} (65.9 MHz for the h-site). The spin states of the UD-2 lines have recently been optically addressed and coherently controlled [38, 39]. In this work, with the four UD-2 lines labelled as PL1-PL4, the spin-1 ground state of each line was investigated as well as a pair of defect spin states of unidentified origin (PL5 and PL6). Room temperature Hahn echo measurements of these latter two defects was demonstrated with T_2^* times of 214 ns and 1248 ns, respectively, similar to the V_{Si} defect. The spin coherence times were found to be comparable to those of the NV in diamond[61]. Such measurements have been performed in 3C, 4H and 6H polytypes with defects existing in the as-grown material or those formed by ion implantation [39]. The UD-2 family of PL lines in 6H-SiC also show ODMR signals and are labelled QL1-QL6. The QL1 line is associated with a spin system with a zero-field splitting of 1.299 GHz. In contrast to conventional EPR where the transverse magnetisation of a large spin ensemble is detected, ODMR can have single spin sensitivity due to linking weakly allowed magnetic dipole transitions to highly allowed electric dipole emission processes. Optically addressing and coherently controlling the spin state of single defect is a potential basis for quantum information and nanoscale sensing applications. This has recently been demonstrated [15, 16] by using a range of pulsed ODMR techniques. Many of the detection and pulse sequences developed for EPR can be applied to ODMR experiments with minor modifications. Such measurements have allowed unprecedented insight into the spin of defects at the ensemble concentration level and now to the single-defect level. Figure 10 shows two ODMR modes of operation as applied to a single V_{Si} defect in SiC [16]. For this centre, spin transitions between ground state sublevels are induced by the microwave excitation giving rise to an enhancement of the PL intensity (Figure 10a and b). The line width is 6 MHz and is limited by the inhomogeneous spin coherence time T_2^* with laser and microwave power broadening effects superimposed. Pulsed methods can be used to circumvent this. Figure 10c shows a simple sequence of pulses that can be used to measure the Rabi oscillations, the characteristic signature of a coherent system interacting with an electromagnetic field. Oscillations reflect the behaviour of a population of an excited state as a function of the input pulse area or qubit rotations. The microwave frequency is set to resonance and the oscillation period is proportional to the microwave field strength. The largest Rabi frequency measured was 8.4 MHz ($B_1 = 0.16$ mT), corresponding to a $\pi/2$ rotation in about 30 ns. Interactions of the spin system with its environment cause a relaxation of polarisation and coherence and are characterised by two time constants, the longitudinal time T1 (spin-lattice relaxation time) and the transverse relaxation time T2 (spin-spin relaxation time). Long spin coherence is essential for quantum information and sensing applications. Using a Hahn-echo pulse sequence ($\pi/2 - \tau - \pi - \tau'$–projection), Widmann et al. placed a lower bound to the spin coherence time of 160 μs and an upper bound of 1 ms, limited by the spin relaxation time. The accuracy was limited by modulations in the signals possibly due to hyperfine coupling to a ^{29}Si nuclear spin close by. These long coherence times collected at room temperature are an indication that SiC is a promising platform for quantum technologies. The main issues are the weak PL intensities (40 kcts/s when integrated with a solid immersion lens) and the poor ODMR contrast (< 1%). Christle et al.[15] report the coherent control of the electronic spin of individual neutral divacancies with emission in the NIR compatible with telecommunication wavelengths. At a temperature of 20 K, a coherence time of 1.2 ms was obtained. The inhomogeneous spin dephasing time, T_2^*, measured by a Ramsey pulse sequence, $\pi/2 - \tau - \pi/2$, was found to be up to 4.4 μs. These were longer than that observed earlier for ensembles [38]. A spin coherence time of 1.2 ms was also deduced for a spin ensemble rather than on a single-defect level owing to limitations in their photon collection efficiency. This duration is remarkably long and much greater than that found for the NV centre in isotropically pure diamond [61]. In addition, still greater spin coherence times might be achieved if the SiC is also isotropically purified [62]. This result may be counterintuitive since the natural abundance of ^{29}Si is greater than 13C. Theoretical calculations attribute this phenomenon to three factors: the longer bond length in SiC, the smaller gyromagnetic ratio of ^{29}Si and the suppression of heteronuclear spin pair flip-flop processes in the strong magnetic field regime [63]. Long-lived spin coherence is essential to implement quantum information and sensing technologies; in

SiC, it seems to be a common feature of the spins in all three polytypes. Single-spin coherent control has been achieved for both the V_{Si} and $V_{Si}VC$ centres. Although the NV centre in diamond dominates this field, SiC shows some advantages being compatible with standard device processing protocols, available in high-quality 3-inch wafers and available in a wide range of polytypes. SiC-based devices in which single spins act as the active device element may be employed in emerging quantum technologies.

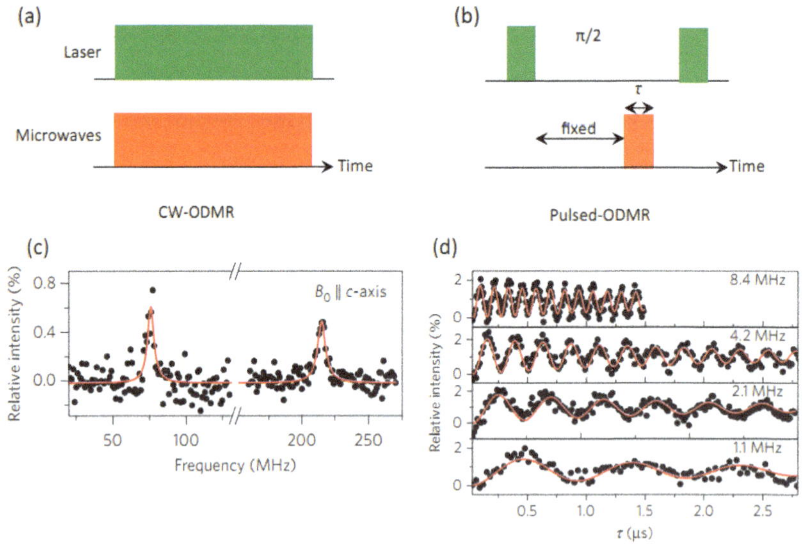

Figure 10. a Schematic of the continuous wave ODMR method with the microwave frequency varied while the PL is measured. b Pulse sequence used to detect spin Rabi oscillations. b ODMR spectrum of the V_{Si}, in a magnetic field of 50 G parallel to the c-axis. d Spin Rabi oscillations of the $|m_s = +3/2\rangle \rightarrow |m_s = +1/2\rangle$ transition of a single V_{Si}; for various field strengths B1. c and d reprinted by permission from Macmillan Publishers Ltd: Nature Materials 14, 164, copyright (2015)[16].

PHOTONICS NANOCAVITIES AND MICRO-CAVITIES

Spontaneous emission from an SPS can be enhanced if the defect or an equivalent 2-level system is placed inside a cavity. The magnitude of the enhancement depends on the coupling (g) between the cavity and the atom's emission modes. The enhancement of the spontaneous emission of a dipole in a cavity is the Purcell factor, given by $F = 3/4\pi^2 Q/V(\lambda/n)^3$, where the quality factor $Q = 2\pi\nu_c/k$ is related to the cavity resonance frequency ν_c and the cavity field decay rate k, while V is the mode volume of the cavity. Nanophotonics is recently an expanding area for quantum technologies as it can provide a strong confinement of photons in a tiny space and potential applications such as ultrasmall and integrated photonic chips, slowing and stopping lights, quantum information processing and environmental sensing. Quantum emitters in bulk materials have inherently poor directional emission, prohibiting high collection and coupling efficiencies to fibre networks for long-range communication. One approach is to utilise cavities to enhance coupling to a specific optical mode. Using a photonic-crystal waveguide yields extraordinarily high coupling efficiencies. Similarly, optical microresonators offer the possibility to increase the coupling of quantum emission with the cavity mode. SiC-based nanophotonics is recently an expanding area as it could further improve present technologies based on Si or GaAs. One reason is due to the suppression of two-photon absorption at high input powers, because of its larger bandgap. Additionally, SiC can provide broadband operation, even in the visible range. The ability to construct nanocavities and optical microdiscs in SiC will possibly allow relevant technological advance in room temperature SPSs based on SiC defects and the coupling of multi-emitters to reach strong coupling regime. The main nanofabrication advantage available in SiC is based on its routinely achievable p- and n-doped in different polytypes and in its possibility to be grown epitaxially on a sacrificial silicon substrate; these features allow dopant-selective photoelectrochemical etching[64, 65] and more amenable application of reactive ion etching procedures to nano-fabricate photonics or micro-mechanical structures[21].

SiC Photonics Cavities

SiC 2D-photonic nanocavity crystals were initially fabricated on thin SiC-on-insulator wafers, which were specially prepared using the smart-cut technique[66]. The main challenge for 4H and 6H is related to the fact that SiC is much harder than silicon, and therefore, the important challenges here are the preparation of SiC membrane structure and the etching of SiC material as the conventional plasma etching technique for Si cannot be used. The wafers consisted of a SiC (6H crystalline structure) surface layer with a thickness of 180 nm above an SiO_2 layer (680 nm), on top of a Si substrate (300 nm) that was used for handling. Quality factors up to 1,300 were achieved with such crystals, and their resonance tuning from the visible to the infrared can allow the selective enhancement of the SiC SPSs[12]. The PCh cavities were L3 nanocavities with three missing air holes. Designs and realisation based on schematics of air hole-shifted and heterostructured nanocavities[67] showed higher Q factors of 6×10^3 and up to 5×10^5 , demonstrating that the design concepts previously used in Si can be applied to the SiC hexagonal polytypes. Regardless these were the first attempts, these PCh nanocavities showed so far the highest Q (see Figure 11). More recently 3C-SiC polytype is used for better integration with Si photonics, by heteroepitaxial growth on sacrificial Si substrate, allowing for direct patterning without ion implantation damage. With this method, an L3 PhC cavity has been fabricated in a thin-film (200 nm) 3C-SiC slab, obtaining Q-factor of 1,000 and modal volume of $0.75(\lambda/n)^3$ over the band between 1250 and 1600 nm in the telecom IR range[22]. The possibility to incorporate colour centres with optically addressable spins similar to NV centres in diamond has been shown by fabricating H1 and L3 PhC cavities in 300-nm-thick 3C-SiC, as shown in Figure 12, with experimental Q of 1,000 for the strongly coupled cavity and (λ/n) 3 mode volume, tuned to the zero phonon line of the Ky5 colour centre (emission band 1,100 to 1,300 nm at 20 K), enhancing PL collection up to 10 times[23]. PhC structures made by a square or hexagonal lattice of SiC rods in air have also been proposed for telecom band applications[28], resulting in the confinement of the cavity mode in air, with the aim of reducing modal dispersion and temperature sensitivity. Theoretical modelling of a defect A1 cavity in SiC hexagonal lattice (photonic bandgap, PBG, from 1,380 to 1,850 nm) showed a thermal shift of the cavity TE mode seven times smaller as compared to a Si-based lattice with similar PBG, for a temperature range between 25 and 200 ◦C, with a maximum Q of 224. The highest Q reported to date of 7.69 × 104 has been achieved using a 45-μm-long 1D PhC nanobeam cavity at 1.5 μm wavelength [29]. It was fabricated on a 280-nm-thick amorphous SiC (a-SiC) film, deposited on an Si substrate by plasma-enhanced chemical vapour deposition (PECVD) in a gas mixture of silane and methane and patterned by e-beam lithography and reactive-ion etching with $CF4/O_2$ plasma. The nanobeam was released in KOH with subsequent critical point drying.

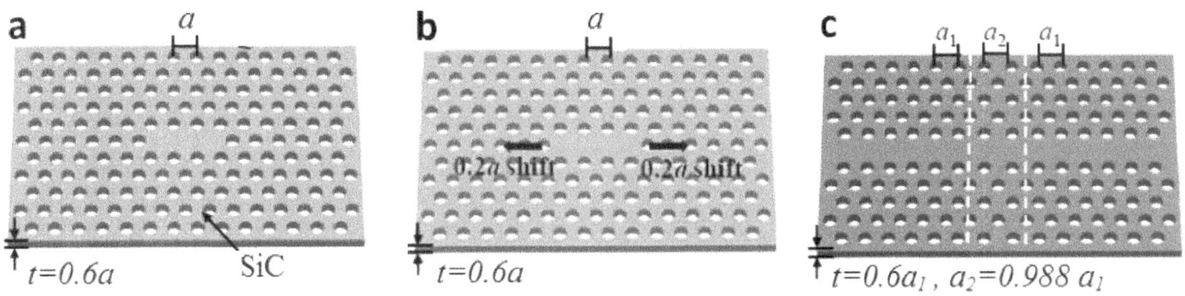

Figure 11. Schematics of a SiC-based (L3) nanocavity with three missing air holes and a Q ≈1200; b air hole-shifted Q ≈6000 and ˛Ac hetero-structured PCh nanocavity with higher Q obtained in 6H polytype, Q ≈500000. a is the cavity lattice constant related to the holes radius that for NIR range of 1380÷1590 nm is a = 550 nm[11]. Lattice constant as small as a=150 nm was achieved for resonance frequencies at 550-600 nm. Image adapted from ref[11].

SiC Microresonators

High-Q optical microresonators can also dramatically enhance optical fields inside a small volume and are considered relevant in applications to enhance nonlinear optics effects and their implication to quantum optics. In addition, they are important in fields such as cavity optomechanics[68, 69]. In general for microresonators, the materials play a crucial role based on its properties such as linear/nonlinear optical susceptibilities and mechanical property. High-Q microresonators generally exhibit large thermal and mechanical sensitivities, making them desired for practical application. Microresonators are commonly fabricated in disc shapes resting on a central pedestal; however, recently suspended ring resonators have been proposed in order to obtain better waveguide dispersion engineering properties for nonlinear processes[20]. Some materials, such as Si and GaAs, have small bandgaps that limit the operating spectral range. Others, such as lithium niobate and chalcogenide glass, exhibit photorefractive effect or two-photon absorption and strong Raman scattering, making it challenging to handle high optical powers. For microphotonic/ nanophotonic applications, it is crucial to search for a material platform with combined superior optical, mechanical and thermal properties. SiC seems to match some of these requirements for an optimal microresonator performance. Since strong coupling between a point defect optical source and the resonant cavity is challenging, due to having to place the defect at the field maximum of the cavity mode, a disc resonator in this respect has the good feature of supporting multiple modes, which cover almost the whole volume of the disc, making overlapping and tuning to the zero phonon line of the defect easier[26]. The single-crystalline 3C-SiC in Lu et al.[21] was epitaxially grown on a (100) silicon substrate by the two-step atmospheric pressure chemical vapour deposition. A disc geometry was patterned by electron-beam lithography and etched by reactive ion etching with CF_4/Ar plasma, optimised to form the well-defined device structure. The silicon substrate was undercut by XeF_2 to form the supporting pedestal. The intrinsic Q factor is shown to be 6.19×10^3. The damage to the crystal incurred during ion etching can impair the optical efficiency and the spin coherence of the defect sources, so an alternative is sought by low-damage selective chemical etch processes, such as photoelectrochemical etching, exploiting the higher etch rate of n-doped to p-doped material under direct UV illumination in HF or KOH solutions[24]. Epitaxial p-doped SiC on n +-SiC was patterned by reactive-ion etching (RIE) in SF_6/O_2 plasma using 3 μm alumina microspheres as masking agents; then the p-layer was undercut by KOH etching under UV illumination. Well-undercut microdiscs reached a measured Q-factor of 9,200 at 617.4 nm resonance, with a theoretical maximum of 10^5 , limited by light leakage into the supporting pillar. The issue of light loss in the supporting Si substrate was addressed by exploiting the high stiffness of an 860-nm-thick 3C-SiC layer, building a 20 μm radius ring microresonator suspended in air from a central pedestal by 200-nm-wide spokes and by e-beam lithography followed by CHF_3/O_2 RIE[20]. An intrinsic Q factor of 14,100 was measured at 1,543 nm wavelength. PL from 3C-SiC in the visible range was first demonstrated for multi-emitter cavity quantum electrodynamics at room temperature in disc microresonators of 210 nm thickness and diameter below 2 μm, fabricated by two-step chemical vapour deposition (CVD) followed by e-beam patterning, HBr/C12 plasma etching and XeF_2 undercut to form the pedestal[30]. Characterisation by laser scanning confocal microscopy between 650 and 850 nm wavelength evidenced whispering gallery modes with Q-factor up to 2,300 and mode volume around $2 \times (\lambda/n)^3$. The highest Q-factor reported to date for a disc microresonator is 5.12×10^4 , obtained at 1,551 nm wavelength with e-beam-patterned heteroepitaxial 3C-SiC on Si, ion etched in CF4/Ar plasma with subsequent KOH undercut[26]. The resulting disc had a thickness of 700 nm and a radius of 6.25 μm, and simulations showed a strong coupling regime with a cooperativity C=20 and a maximum theoretical radiation-limited Q-factor above 108 .

Nonlinear Effects

SiC is a non-centrosymmetric material, which means that it has nonzero second-order susceptibility, permitting the observation of nonlinear effects such as second-harmonic generation (SHG), and Pockels effect, which are not available in the more common centrosymmetric materials (SiO_2, SiNx and others whose molecule's symmetry allows only third-order susceptibility). Differently from other non-centrosymmetric semiconductors and crystals such as $LiNbO_3$, SiC has the ability to house optically active defects for quantum processing. Nonlinear effects in this material combined with colour centres have the potential to increase the applications in other fields such as nonlinear photonic, including all-optical switching, wavelength conversion and harmonic generation. These effects are important for the development of quantum devices based on nonlinear effects[70]. SiC second order, $\chi^{(2)}_{i,j,k}$, and electro-optical $r_{i,j,k}$ coefficients were predicted from ab initio calculations in 3C, 4H and 6H [71] and measured [72–75] in 4H and 6H polytypes. Second-order nonlinear optical coefficients are at the core of the physics of nonlinear optical processes; their knowledge and high values are essential to increase the efficiency of frequency conversion. Hexagonal SiC is characterised by three independent components of the second-order nonlinear optical tensor, recently measured accurately as $d_{31} = d_{32} = 6.7(6.5)$pm/V, $d_{15} = d_{24} = 6.5(6.7)$pm/V and $d_{33} = -12.5(-11.7)$pm/V in 6H (4H). The refractive indices and the birefringence are used to determine if phase matching can be realised; they have been remeasured from the visible to the mid-infrared region, and Sellmeier equations for ordinary and extraordinary refractive index in 4H and 6H have been provided [76]. This has led to mid-infrared region laser output tunable from 3.90 to 5.60 μm by phase-matched difference frequency generation in 4H-SiC. In this case, the material can withstand high power and for laser higher outputs in this spectral region. For applications, it is important to enhance SHG in photonics cavities. The first demonstration of SHG in SiC-based nanophotonic structures used a 6H-SiC PhC cavity tuned to 1560 nm with a Q of 10,000, taking advantage of the high refractive index of SiC (2.5) to increase confinement and of the large electronic bandgap (3.0 eV) to extend the operating bandwidth from IR to visible and to suppress two-photon absorption. A conversion efficiency of 2.59×10^{-5} was obtained for 0.17 mW average input power[25]. A high-Q amorphous SiC disc microresonator has been demonstrated for nonlinear applications, with an optical Q of 1.3×10^5 in the 1,550 nm telecom band[27]. With PECVD deposition followed by e-beam patterning and plasma etching, a smooth-sidewall disc with 570 nm thickness and 6 μm radius was fabricated, and its Kerr nonlinearity was characterised with a pump-probe self-/cross-phase modulation scheme using two high-Q whispering-gallery modes. The resonator was pumped at 1545 nm with sinusoidally modulated light, transferring the modulation on the refractive index due to optical Kerr effect, which was then measured by a weak probe signal coupled at 1,498 nm. This gave a nonlinear coefficient $n_2 = 5.9 \times 10^{-15}$ cm²/W, higher than any other material in use to generate frequency combs. Recently, SHG microscopy was used as a non-invasive method for imaging and identification of structural defects such as polytype inclusions and stacking faults in SiC epilayers, grown on hexagonal SiC by the vapour-liquid-solid techniques. This technique is competitive with destructive sample preparation and severely limited area of analysis to tens of microns associated instead with high-resolution TEM. By combining the SHG-based imaging with X-ray diffraction and SHG rotational anisotropy, the growth of 3C polytype on the 4H-SiC substrate was confirmed, and the polytype of the imaged defects was identified [77].

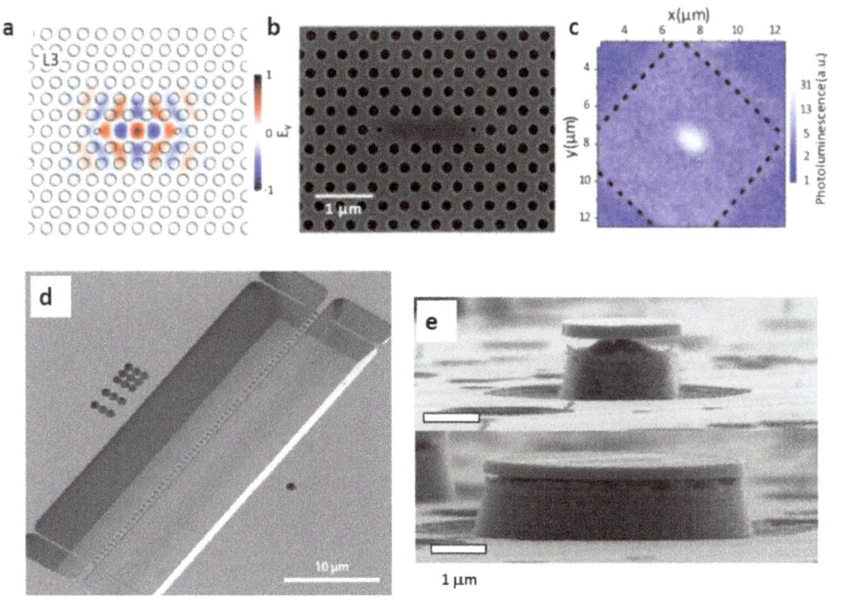

Figure 12. a Simulated E_y field distribution, b SEM image and c PL at 20K from an L3 photonic-crystal cavity in 3C-SiC. The cavity resonance is tuned by size reduction and shift of the neighbouring crystal holes, for 1,060 nm excitation of the embedded Ky5 colour centre[23]. d SEM image of a fabricated a-SiC 45-µm-long PhC nanobeam cavity, achieving the highest Purcell factor reported of 10^4 at 1.5 µm wavelength[29]. e SEM micrographs of high-Q disc resonators obtained by low-damage photoelectrochemical selective etching of p-SiC over n^+-SiC. Notice the different undercuts achieved by wet etching and the surface porosity of the supporting pillar, which does not affect the etched microdisc[24]. Images reproduced from [23, 24, 29] with Copyright permission 2014-2015, AIP Publishing LLC.

OTHER APPLICATIONS

Paramagnetic defects in SiC have been integrated into some devices as a proof of principle demonstration of their electrical control or of their ability to sense changes in strain, magnetic field and temperature. The discovery of single-photon emission and its engineering and isolating single defects with proper transition energy on demand can open a route for an efficient electrical single-photon source, particularly in the IR region, due to the described subgap defects. To fabricate devices, standard semiconductor manufacturing technology in combination with high-energy electron irradiation of the material can allow, for instance, the construction of LEDs with radiative recombination occurring at irradiation-induced intrinsic defects. Such LEDs have been demonstrated with two strong PL emission bands in the visible and near infrared (NIR), associated with two different intrinsic defects, one being the Si vacancy[78]. Selective manipulation of individual defect spin can be achieved. The spin separation required to achieve strong dipolar coupling between spins is of the order of tens of nanometres. Electric fields can be confined on similar length scales; therefore, electrically driven spin resonance methods can be used to manipulate the defect spin state [79]. Divacancy defects in 4H- and 6H-SiC possess a spin-dependent optical cycle, which allows non-resonant laser illumination to polarise first and then read out its ground-state spin. Since the defect PL depends on whether its spin state is $m_s = 0$ or $m_s = \pm 1$, it is possible to control the defect spin dynamics by the measuring differential PL (ΔIPL) between an initial state and one that has been evolved by applying a magnetic or electric field pulses. These ΔIPL measurements, in addition to enabling conventional (magnetically driven) ODMR, can also be used for electrically driven ODMR (EODMR). In Figure 13, we show the electrical control of the QL1 (unknown yet) defect in 6H-SiC with similar spin properties of the neutral divacancies, with ZPL at 1.088 eV and ODMR at 1.3GHz [39]. Ac electric fields are used to drive the Rabi oscillations across a magnetic-dipole forbidden spin transition ($m_s = \pm 2$) of the optically addressable electronic spin. A flow cryostat cools the device illustrated in Figure 13b to 20 K, and a permanent magnet is used to provide a static magnetic field parallel to the defect magnetic dipole. QL1 colour centres were produced in 6H-SiC substrates via a carbon implantation and annealing process introducing defects immediately below the surface. The spins were localised within a 400-nm-thick layer immediately beneath the 6H-SiC surface and were optically pumped with a 976 nm laser addressing 10^4 defects in the optical excitation volume. We show in Figures 13 c and d the effect observed from the application of an electric field via the electrode to the transition corresponding to the prohibited optical transition. Another PL resonance appears in the presence of the $m_s = \pm 2$ transition indicating a population transfer between the two-spin states.

Falk et al.[80] state that the spin states of neutral divacancies in 4H-SiC are highly sensitive to electrical and mechanical perturbations of material. The experiments were done in high-purity semi-insulating 4H-SiC wafers, purchased from Cree Inc., where neutral divacancies were incorporated during crystal growth. A 50 µm membrane from the sample was cut and mounted on top of a piezo actuator. The piezo actuation applies tensile strain to the SiC membrane perpendicular to the c-axis. The device was cooled to 20 K. The neutral divacancy's ZPLs were excited optically at 976 nm and monitored. Microwave excitation for electron spin resonance was supplied by waveguide antennae on the chip. A variation of the ODMR signal was observed in the presence of strain. The strain sensitivity is inferred to be $10^{-7}/\sqrt{HzN}$, where N is the number of spins. Also, the electric field response exhibited significant spin-dependent PL, 2-7 times stronger than with NV centres in diamond. Electric field pulses were applied across the SiC membrane, using patterned electrodes that transmit light, and the ODMR signal was monitored. If these techniques were extended to single defects, these properties could be applicable to the nanometre scale, and they could be applied for sensing intracellular electric fields, integrating nanoscale sensing into SiC bioelectronics [81], or coupling spins to SiC nano-mechanical resonators [9]. In another work, the spin resonances of the VSi defect at 128 MHz and 28 MHz associated with the V2 and V3 ZPLs in 6H SiC were used to monitor magnetic field and temperature variation [82].

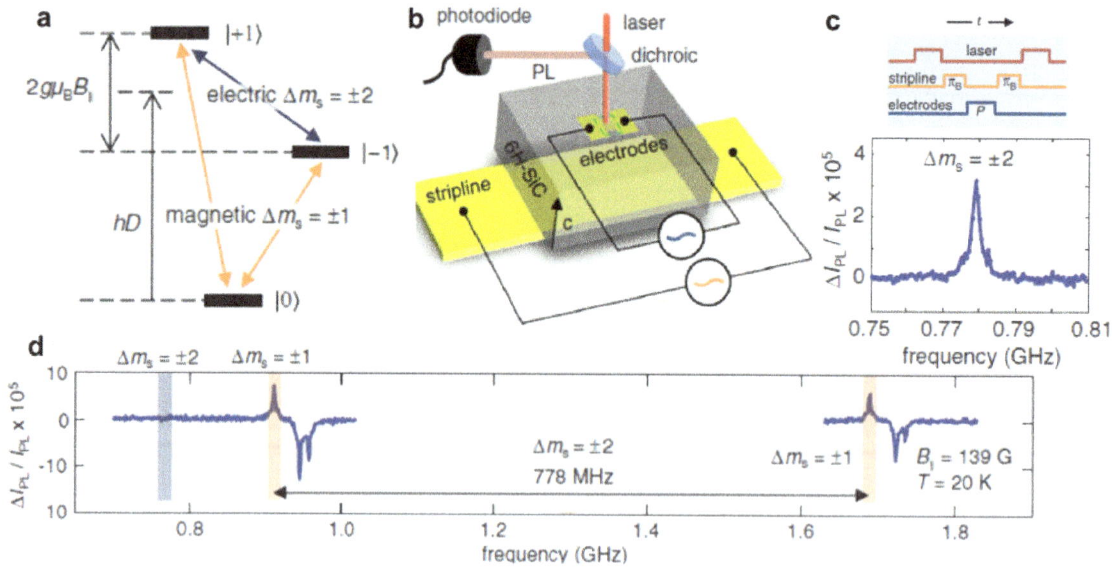

Figure 13. a The ground state spin structure of QL1 defect in 6H-SiC, where the spin levels $m_s = 0, \pm 1$ are shown in the presence of a magnetic field Bk . The allowed optical transition $m_s = \pm 1$ are in orange, while the $m_s = \pm 2$ in blue are not allowed. b Experimental set-up used for PL collection via a dichroic mirror and measured with a photodiode. The spins are driven electrically by the electrodes and magnetically by the stripline directly on the bulk material.c A clear EODMR feature is observed at the frequency difference (778 MHz) of the $m_s = | + 1i$ and $m_s = | - 1i$, indicating population transfer across the two levels. On the top of the panel, the time sequence of the microwave pulse and the electric field are applied to achieve EODMR. The length of P is fixed and its frequency is swept. The electrode power was 0.09 W and the magnetic Bk=139 G.d ODMR signal when the stripline is driven at Bk=139 G. The $m_s = \pm 1$ resonances are shaded orange, and the $m_s = \pm 2$ resonance (at 778 MHz, shaded blue) is not seen in ODMR only. Images reproduced from [79] with Copyright permission 2014 American Physical Society.

An important milestone is to achieve the integration of these defects in nanoparticles. The first demonstration of optical spin control of silicon vacancy in 600-nm-size material was achieved in [83]. High-quality, defect-free 6H-SiC bulk material using the well-established sublimation technique in argon atmosphere at a high temperature (2500÷2600°C) was grown from a specially prepared source pure SiC powder synthesised from silicon and carbon mixture of spectral purity. Macroscopic crystal fragments in the mm range were synthesised and then placed in the central irradiation tube of a TRIGA Mark II nuclear reactor, where vacancies were formed by neutron irradiation followed by annealing, achieving an initial concentration of V_{Si} defects of the order of 10^{15} cm^{-3} . High-energy milling process was used to achieve size from 60 to 600 nm. PL from V_{Si} ($V_{1,2,3}$) was observed in 60 nm and 600 nm crystals at room and cryogenic temperature for comparison to identify the ZPLs. No single-photon emission was observed while V_3 and V_2 from VSi spin resonances were observed at 27 MHz and 127 MHz with positive and negative ODMR contrast, respectively.

SUMMARY AND OUTLOOK

We have summarised SiC paramagnetic defects with quantum properties observed following single-defect creation and isolation. These defects are present in various polytypes, and their variety in different crystal sites increases their space of applications. We also reviewed the state of the art of quantum effects observed in SiC nanostructures. SiC as a host for quantum systems is just at its inception, though the beginning indicates it could be one of the most prominent and rich material to investigate or to employ in the future. A large variety of deep defects were investigated from the point of view of paramagnetic and optical properties in the past, and now other quantum properties have been revealed, providing additional information not available at the ensemble level and novel insights relative to previously known defects. The challenge of engineering a desired defect in a specific location in the material is still an open problem, though it has been successfully demonstrated so far with electron, neutron and ion irradiation. SiC single defects in bulk and in the red part of the spectrum is one of the best SPS operating at room temperature, surpassing in terms of brightness other materials. The properties of these defects are also important for their spin coherent control and their use as spin qubits. The spin coherence time in ensemble and at the single level of intrinsic defects in SiC is very promising and comparable to NV in diamond. Isolation of intrinsic single spin has been achieved. However, for the most interesting spectral region in the infrared, it appears that the emission is not very bright and needs further enhancement via integration of the defects in photonics crystal cavities or microresonator. The spin coherence control still requires cryogenic temperature for the IR emitters. Since both Si and C have spin-free isotopes ($^{28}Si^{12}C$), it is expected that the measured coherence times could be enhanced through the sublimation crystal growth to reduce the abundance of ^{29}Si and ^{13}C isotopes having nonzero nuclear spins. While the challenge of the nano-fabrication of this material seems to indicate its feasibility and viability, for this to be successful, a major challenge needs to be addressed, specifically, the creation of the defects in a specific location close to the material surface within the photonic devices. Another option could be to integrate SiC QDs in hybrid photonics systems. This will require the successful incorporation of the desired defects with high yield in QDs and NPs. Additionally, many other potential defects in the material should be explored, particularly extrinsic defects that could facilitate their on-demand integration in devices. Alternatively for intrinsic defects originated from damage, more precise damage techniques could aid to the accurate location of the defects in the lattice. Nanostructures are ready to be integrated with deep optical and paramagnetic defects, though better design of cavities and choice of higher quality material would allow to take full advantage of functionalities related to single-photon emission and spin sensing to further advance their current applications. If all these challenges will be overcome in the next years, we envision the construction of SiC-defect-based LEDs and photonic crystal structures with in-built active quantum systems, for next-generation quantum technology devices.

REFERENCES

1. C. A. Zorman, A. J. Fleischman, A. S. Dewa, M. Mehregany, C. Jacob, S. Nishino, and P. Pirouz, "Epitaxial growth of 3C-SiC films on 4 in. diam (100) silicon wafers by atmospheric pressure chemical vapor deposition," J. Appl. Phys. 78, 5136–5138 (1995).
2. R. Madar, "Materials science: Silicon carbide in contention," Nature 430, 974–975 (2004).
3. A. Dzurak, "Quantum computing: Diamond and silicon converge," Nature 479, 47–48 (2011).
4. D. DiVincenzo, "Quantum bits: Better than excellent," Nat. Mater. 9, 468–469 (2010).
5. J. R. Petta, A. C. Johnson, J. M. Taylor, E. A. Laird, A. Yacoby, M. D. Lukin, C. M. Marcus, M. P. Hanson, and A. C. Gossard, "Coherent manipulation of coupled electron spins in semiconductor quantum dots," Science 309, 2180–2184 (2005).
6. M. Hofheinz, H. Wang, M. Ansmann, R. C. Bialczak, E. Lucero, M. Neeley, A. D. O'Connell, D. Sank, J. Wenner, J. M. Martinis, and A. N. Cleland, "Synthesizing arbitrary quantum states in a superconducting resonator," Nature 459, 546–549 (2009).
7. J. J. Pla, K. Y. Tan, J. P. Dehollain, W. H. Lim, J. J. L. Morton, D. N. Jamieson, A. S. Dzurak, and A. Morello, "A single-atom electron spin qubit in silicon," Nature 489, 541–545 (2012).
8. G. Davies and M. F. Hamer, "Optical studies of the 1.945 ev vibronic band in diamond," Proc. R. Soc. Lond., Ser. A 348, 285–298 (1976).
9. Y. T. Yang, K. L. Ekinci, X. M. H. Huang, L. M. Schiavone, M. L. Roukes, C. A. Zorman, and M. Mehregany, "Monocrystalline silicon carbide nanoelectromechanical systems," Appl. Phys. Lett. 78, 162–164 (2001).
10. C. A. Zorman and R. J. Parro, "Micro- and nanomechanical structures for silicon carbide mems and nems," Physica Status Solidi B 245, 1404–1424 (2008).
11. B.-S. Song, S. Yamada, T. Asano, and S. Noda, "Demonstration of two-dimensional photonic crystals based on silicon carbide," Opt. Express 19, 11084–11089 (2011).
12. S. Yamada, B.-S. Song, T. Asano, and S. Noda, "Silicon carbide-based photonic crystal nanocavities for ultra-broadband operation from infrared to visible wavelengths," Appl. Phys. Lett. 99, 201102 (2011).
13. S. Castelletto, B. C. Johnson, and A. Boretti, "Quantum effects in silicon carbide hold promise for novel integrated devices and sensors," Adv. Opt. Mater. 1, 609–625 (2013).
14. S. Castelletto, B. Johnson, V. Ivády, N. Stavrias, T. Umeda, A. Gali, and T. Ohshima, "A silicon carbide room-temperature single-photon source," Nat. Mater. 13, 151–156 (2014).
15. D. J. Christle, A. L. Falk, P. Andrich, P. V. Klimov, J. U. Hassan, N. T. Son, E. Janzén, T. Ohshima, and D. D. Awschalom, "Isolated electron spins in silicon carbide with millisecond coherence times," Nat. Mater. 14, 160–163 (2015).
16. M. Widmann, S.-Y. Lee, T. Rendler, N. T. Son, H. Fedder, S. Paik, L.-P. Yang, N. Zhao, S. Yang, I. Booker, A. Denisenko, M. Jamali, S. A. Momenzadeh, I. Gerhardt, T. Ohshima, A. Gali, E. Janzén, and J. Wrachtrup, "Coherent control of single spins in silicon carbide at room temperature," Nat. Mater. 14, 164–168 (2015).
17. F. Fuchs, B. Stender, M. Trupke, J. Pflaum, V. Dyakonov, and G. V. Astakhov, "Engineering near infrared single photon emitters in ultrapure silicon carbide," Nat. Comm. (2015). doi:10.1038/ncommos8578 .
18. S. Castelletto, B. C. Johnson, C. Zachreson, D. Beke, I. Balogh, T. Ohshima, I. Aharonovich, and A. Gali, "Room temperature quantum emission from cubic silicon carbide nanoparticles," ACS Nano 8, 7938–7947 (2014).
19. S. Castelletto, Z. Bodrog, A. P. Magyar, A. Gentle, A. Gali, and I. Aharonovich, "Quantum-confined single photon emission at room temperature from SiC tetrapods," Nanoscale 6, 10027–10032 (2014).
20. J. Cardenas, M. Zhang, C. T. Phare, S. Y. Shah, C. B. Poitras, B. Guha, and M. Lipson, "High Q SiC microresonators," Opt. Express 21, 16882–16887 (2013).
21. X. Lu, J. Y. Lee, P. X.-L. Feng, and Q. Lin, "Silicon carbide microdisk resonator," Opt. Lett. 38, 1304–1306 (2013).
22. M. Radulaski, T. M. Babinec, S. Buckley, A. Rundquist, J. Provine, K. Alassaad, G. Ferro, and J. Vuˇckovi´c, "Photonic crystal cavities in cubic (3c) polytype silicon carbide films," Opt. Express 21, 32623–32629 (2013).
23. G. Calusine, A. Politi, and D. D. Awschalom, "Silicon carbide photonic crystal cavities with integrated color centers," Appl. Phys. Lett. 105, 011123 (2014).
24. A. P. Magyar, D. Bracher, J. C. Lee, I. Aharonovich,

and E. L. Hu, "High quality sic microdisk resonators fabricated from monolithic epilayer wafers," Appl. Phys. Lett. 104, 051109 (2014).

25. S. Yamada, B.-S. Song, S. Jeon, J. Upham, Y. Tanaka, T. Asano, and S. Noda, "Second-harmonic generation in a silicon-carbide-based photonic crystal nanocavity," Opt. Lett. 39, 1768–1771 (2014).

26. X. Lu, J. Y. Lee, P. X.-L. Feng, and Q. Lin, "High q silicon carbide microdisk resonator," Appl. Phys. Lett. 104, 181103 (2014).

27. X. Lu, J. Y. Lee, S. Rogers, and Q. Lin, "Optical kerr nonlinearity in a high-q silicon carbide microresonator," Opt. Express 22, 30826–30832 (2014).

28. J. Boruah, Y. Kalra, and R. Sinha, "Demonstration of temperature resilient properties of 2D silicon carbide photonic crystal structures and cavity modes," Optik - Int. J. Light Electron Opt. 125, 1663–1666 (2014).

29. J. Y. Lee, X. Lu, and Q. Lin, "High-Q silicon carbide photonic-crystal cavities," Appl. Phys. Lett. 106, 041106 (2015).

30. M. Radulaski, T. M. Babinec, K. Müller, K. G. Lagoudakis, J. L. Zhang, S. Buckley, Y. A. Kelaita, K. Alassaad, G. Ferro, and J. Vuˇckovi´c, "Visible photoluminescence from cubic (3c) silicon carbide microdisks coupled to high quality whispering gallery modes," ACS Photonics 2, 14–19 (2015).

31. E. Sörman, N. T. Son, W. M. Chen, O. Kordina, C. Hallin, and E. Janzén, "Silicon vacancy related defect in 4h and 6h sic," Phys. Rev. B 61, 2613–2620 (2000).

32. N. Mizuochi, S. Yamasaki, H. Takizawa, N. Morishita, T. Ohshima, H. Itoh, and J. Isoya, "Continuous-wave and pulsed epr study of the negatively charged silicon vacancy with s = 3 2 and C3v symmetry in n-type 4h − SiC," Phys. Rev. B 66, 235202 (2002).

33. J. Steeds, G. Evans, L. Danks, S. Furkert, W. Voegeli, M. Ismail, and F. Carosella, "Transmission electron microscope radiation damage of 4h and 6h sic studied by photoluminescence spectroscopy," Diam. Relat. Mater. 11, 1923–1945 (2002).

34. N. T. Son, P. Carlsson, J. ul Hassan, E. Janzén, T. Umeda, J. Isoya, A. Gali, M. Bockstedte, N. Morishita, T. Ohshima, and H. Itoh, "Divacancy in 4H-SiC," Phys. Rev. Lett. 96, 055501 (2006).

35. T. Umeda, N. T. Son, J. Isoya, E. Janzén, T. Ohshima, N. Morishita, H. Itoh, A. Gali, and M. Bockstedte, "Identification of the carbon antisite-vacancy pair in 4h-sic," Phys. Rev. Lett. 96, 145501 (2006).

36. J. W. Steeds, "Photoluminescence study of the carbon antisite-vacancy pair in 4h- and 6h-sic," Phys. Rev. B 80, 245202 (2009).

37. H. Kraus, V. Soltamov, D. Riedel, S. Väth, F. Fuchs, A. Sperlich, P. Baranov, V. Dyakonov, and G. Astakhov, "Room-temperature quantum microwave emitters based on spin defects in silicon carbide," Nat. Phys. 10, 157–162 (2014).

38. W. F. Koehl, B. B. Buckley, F. J. Heremans, G. Calusine, and D. D. Awschalow, "Room temperature coherent control of defect spin qubits in silicon carbide," Nature 479, 84 (2011).

39. A. L. Falk, B. B. Buckley, G. Calusine, W. F. Koehl, V. V. Dobrovitski, A. Politi, C. A. Zorman, P. X.-L. Feng, and D. D. Awschalom, "Polytype control of spin qubits in silicon carbide," Nat. Commun. 4, 1819 (2013).

40. B. Magnusson and E. Janzén, "Optical characterization of deep level defects in SiC," Mater. Sci. Forum 340342, 341–346 (2005).

41. N. T. Son, P. N. Hai, M. Wagner, W. M. Chen, A. Ellison, C. Hallin, B. Monemar, and E. Janzén, "Optically detected magnetic resonance studies of intrinsic defects in 6H-SiC," Semicond. Sci. Technol. 14, 1141 (1999).

42. T. Basché, W. E. Moerner, M. Orrit, and H. Talon, "Photon antibunching in the fluorescence of a single dye molecule trapped in a solid," Phys. Rev. Lett. 69, 1516–1519 (1992).

43. P. Michler, A. Imamoglu, M. D. Mason, P. J. Carson, G. F. Strouse, and S. K. Buratto, "Quantum correlation among photons from a single quantum dot at room temperature," Nature 406, 968–970 (2000).

44. C. Santori, D. Fattal, J. Vuckovic, G. S. Solomon, and Y. Yamamoto, "Indistinguishable photons from a single-photon device," Nature 419, 594–597 (2002).

45. C. Kurtsiefer, S. Mayer, P. Zarda, and H. Weinfurter, "Stable solid-state source of single photons," Phys. Rev. Lett. 85, 290–293 (2000).

46. A. Tribu, G. Sallen, T. Aichele, R. Andre, J.-P. Poizat, C. Bougerol, S. Tatarenko, and K. Kheng, "A high-temperature single-photon source from nanowire quantum dots," Nano Lett. 8, 4326–4329 (2008).

47. R. Kolesov, K. Xia, R. Reuter, R. Stöhr, A. Zappe, J. Meijer, P. Hemmer, and J. Wrachtrup, "Optical detection of a single rare-earth ion in a crystal," Nat. Commun. 3, 1029 (2012).

48. C. Santori, F. D., and Y. Yamamoto, Single Photon Device and Applications (Wiley, 2010).

49. J. Fan, H. Li, J. Jiang, L. K. Y. So, Y. W. Lam, and P. K. Chu, "3c-sic nanocrystals as fluorescent biological labels," Small 4, 1058–1062 (2008).

50. B. Mognetti, A. Barberis, S. Marino, F. Di Carlo, V. Lysenko, O. Marty, and A. Géloën, "Preferential killing of cancer cells using silicon carbide quantum dots," J. Nanosci. Nanotechnol. 10, 7971–7975 (2010).

51. T. Serdiuk, V. Lysenko, B. Mognetti, V. Skryshevsky, and A. Géloën, "Impact of cell division on intracellular uptake and nuclear targeting with fluorescent sic-based nanoparticles," J. Biophotonics 6, 291–297 (2013).

52. D. Dai, X. Guo, and J. Fan, "Identification of luminescent surface defect in SiC quantum dots," Appl. Phys. Lett. 106, 053115 (2015).

53. X. L. Wu, J. Y. Fan, T. Qiu, X. Yang, G. G. Siu, and P. K. Chu, "Experimental evidence for the quantum confinement effect in 3c-sic nanocrystallites," Phys. Rev. Lett. 94, 026102 (2005).

54. J. Botsoa, V. Lysenko, A. Geloen, O. Marty, J. M. Bluet, and G. Guillot, "Application of 3C-SiC quantum dots for living cell imaging," Appl. Phys. Lett. 92, 173902 (2008).

55. D. Beke, Z. Szekrényes, I. Balogh, M. Veres, É. Fazakas, L. K. Varga, K. Kamarás, Z. Czigány, and A. Gali, "Characterization of luminescent silicon carbide nanocrystals prepared by reactive bonding and subsequent wet chemical etching," Appl. Phys. Lett. 99, 213108 (2011).

56. S. Yang, B. Kiraly, W. Y. Wang, S. Shang, B. Cao, H. Zeng, Y. Zhao, W. Li, Z.-K. Liu, W. Cai, and T. J. Huang, "Fabrication and characterization of beaded sic quantum rings with anomalous red spectral shift," Adv. Mater. 24, 5598–5603 (2012).

57. D. Beke, Z. Szekrényes, D. Pálfi, G. Róna, I. Balogh, P. A. Maák, G. Katona, Z. Czigány, K. Kamarás, B. Rózsa, L. Buday, B. Vértessy, and A. Gali, "Silicon carbide quantum dots for bioimaging," J. Mater. Res. FirstView, 1 (2012).

58. A. Gruber, A. Dräbenstedt, C. Tietz, L. Fleury, J. Wrachtrup, and C. von Borczyskowski, "Scanning confocal optical microscopy and magnetic resonance on single defect centers," Science 276, 2012 (1997).

59. V. A. Soltamov, A. A. Soltamova, P. G. Baranov, and I. I. Proskuryakov, "Room temperature coherent spin alignment of silicon vacancies in 4h- and 6h-sic," Phys. Rev. Lett. 108, 226402 (2012).

60. D. Riedel, F. Fuchs, H. Kraus, S. Väth, A. Sperlich, V. Dyakonov, A. A. Soltamova, P. G. Baranov, V. A. Ilyin, and G. V. Astakhov, "Resonant addressing and manipulation of silicon vacancy qubits in silicon carbide," Phys. Rev. Lett. 109, 226402 (2012).

61. P. L. Stanwix, L. M. Pham, J. R. Maze, D. Le Sage, T. K. Yeung, P. Cappellaro, P. R. Hemmer, A. Yacoby, M. D. Lukin, and R. L. Walsworth, "Coherence of nitrogen-vacancy electronic spin ensembles in diamond," Phys. Rev. B 82, 201201 (2010).

62. I. G. Ivanov, M. Yazdanfar, B. Lundqvist, J.-T. Chen, J. Hassan, P. Stenberg, R. Liljedahl, N. T. Son, J. W. Ager, III, O. Kordina, and E. Janzén, "High-Resolution Raman and Luminescence Spectroscopy of Isotope-Pure(SiC)-Si-28-C-12, Natural and C-13 - Enriched 4H-SIC," (TRANS TECH PUBLICATIONS LTD, 2014) pp. 471–474.

63. L.-P. Yang, C. Burk, M. Widmann, S.-Y. Lee, J. Wrachtrup, and N. Zhao, "Electron spin decoherence in silicon carbide nuclear spin bath," Phys. Rev. B 90, 241203 (2014).

64. F. Zhao, M. M. Islam, and C.-F. Huang, "Photoelectrochemical etching to fabricate single-crystal sic {MEMS} for harsh environments," Mater. Lett. 65, 409–412 (2011).

65. M. M. Islam, C.-F. Huang, and F. Zhao, "Single-crystal sic resonators by photoelectrochemical etching," Mater. Sci. Forum 717-720, 529–532 (2012).

66. L. Di Cioccio, Y. Le Tiec, F. Letertre, C. Jaussaud, and M. Bruel, "Silicon carbide on insulator formation using the smart cut process," Electron. Lett. 32, 1144–1145 (1996), .

67. B.-S. Song, S. Noda, T. Asano, and Y. Akahane, "Ultra-high-q photonic double-heterostructure nanocavity," Nat. Mater. 4, 207–210 (2005).

68. K. J. Vahala, "Optical microcavities," Nature 424, 839–846 (2003).

69. T. J. Kippenberg and K. J. Vahala, "Cavity optomechanics: Back-action at the mesoscale," Science 321, 1172–1176 (2008).

70. J. L. O'Brien, A. Furusawa, and J. Vuˇckovi´c, "Photonic quantum technologies," Nat. Photonics 3, 687 (2009).

71. I. J. Wu and G. Y. Guo, "Second-harmonic generation and linear electro-optical coefficients of sic polytypes and nanotubes," Phys. Rev. B 78, 035447 (2008).

72. S. Singh, J. R. Potopowicz, L. G. V. Uitert, and S. H. Wemple, "Nonlinear optical properties of hexagonal silicon carbide," Appl. Phys. Lett. 19, 53 (1971).

73. P. M. Lundquist, W. P. Lin, G. K. Wong, M. Razeghi, and J. B. Ketterson, "Second harmonic generation in hexagonal silicon carbide," Appl. Phys. Lett. 66, 1883 (1995).

74. S. Niedermeier, H. Schillinger, R. Sauerbrey, B. Adolph, and F. Bechstedt, "Second-harmonic generation in silicon carbide polytypes," Appl. Phys. Lett. 75, 618 (1999).

75. H. Sato, M. Abe, I. Shoji, J. Suda, and T. Kondo, "Accurate measurements of second-order nonlinear optical coefficients of 6h and 4h silicon carbide," J. Opt. Soc. Am. B 26, 1892–1896 (2009).

76. S. Wang, M. Zhan, G. Wang, H. Xuan, W. Zhang, C. Liu, C. Xu, Y. Liu, Z. Wei, and X. Chen, "4h-sic: a new nonlinear material for midinfrared lasers," Laser Photonics Rev. 7, 831–838 (2013).

77. R. Hristu, S. Stanciu, D. Tranca, A. Matei, and G. Stanciu, "Nonlinear optical imaging of defects in cubic silicon carbide epilayers," Sci. Rep. 4 (2014), 10.1038/srep05258.

78. F. Fuchs, V. A. Soltamov, S. Väth, P. G. Baranov, E. N. Mokhov, G. V. Astakhov, and V. Dyakonov, "Silicon carbide light-emitting diode as a prospective room temperature source for single photons," Sci. Rep. 3 (2013).

79. P. Klimov, A. Falk, B. Buckley, and D. D. Awschalom, "Electrically driven spin resonance in silicon carbide color centers," Phys. Rev. Lett. 112, 087601 (2014).

80. A. L. Falk, P. V. Klimov, B. B. Buckley, V. Ivády, I. A. Abrikosov, G. Calusine, W. F. Koehl, A. Gali, and D. D. Awschalom, "Electrically and mechanically tunable electron spins in silicon carbide color centers," Phys. Rev. Lett. 112, 187601 (2014).

81. S. E. Saddow, in Silicon Carbide Biotechnology, edited by S. E. Saddow (Elsevier, Oxford, 2012).

82. H. Kraus, V. Soltamov, F. Fuchs, D. Simin, A. Sperlich, P. Baranov, G. Astakhov, and V. Dyakonov, "Magnetic field and temperature sensing with atomic-scale spin defects in silicon carbide," Sci. Rep. 4 (2014).

83. A. Muzha, F. Fuchs, N. V. Tarakina, D. Simin, M. Trupke, V. A. Soltamov, E. N. Mokhov, P. G. Baranov, V. Dyakonov, A. Krueger, and G. V. Astakhov, "Room-temperature near-infrared silicon carbide nanocrystalline emitters based on optically aligned spin defects," Appl. Phys. Lett. 105, 243112 (2014).

INDEX

L

M

N

O

P

Q

R

S